Springer Monographs in Mathematics

Siegfried Carl
Vy Khoi Le
Dumitru Motreanu

Nonsmooth Variational Problems and Their Inequalities

Comparison Principles and Applications

Springer

Siegfried Carl
Institut für Mathematik
Martin-Luther-Universität
  Halle-Wittenberg
D-06099 Halle
Germany
siegfried.carl@mathematik.uni-halle.de

Vy Khoi Le
Department of Mathematics and
  Statistics
University of Missouri-Rolla
Rolla, MO 65409
U.S.A
vy@umr.edu

Dumitru Motreanu
Département de Mathématiques
Université de Perpignan
66860 Perpignan
France
motreanu@univ-perp.fr

Mathematics Subject Classifications (2000): (Primary) 35B05, 35J20, 35J85, 35K85, 35R70, 47J20, 47J35, 49J52, 49J53; (Secondary) 35J60, 35K55, 35R05, 35R45, 49J40, 58E35

ISBN-13: 978-1-4419-4033-9   e-ISBN-13: 978-0-387-46252-3

Printed on acid-free paper.

springer.com   (TXQ/SB)

Preface

Nonsmooth variational problems have their origin in the study of nondifferentiable energy functionals, and they arise as necessary conditions of critical points of such functionals. In this way, *variational inequalities* are related with convex energy or potential functionals, whereas the new class of *hemivariational inequalities* arise in the study of nonconvex potential functionals that are, in general, merely locally Lipschitz. The foundation of variational inequalities is from Fichera, Lions, and Stampacchia, and it dates back to the 1960s. Hemivariational inequalities were first introduced by Panagiotopoulos about two decades ago and are closely related with the development of the new concept of Clarke's generalized gradient. By using this new type of inequalities, Panagiotopoulos was able to solve various open questions in mechanics and engineering.

This book focuses on nonsmooth variational problems not necessarily related with some potential or energy functional, which arise, e.g., in the study of boundary value problems with nonsmooth data and/or nonsmooth constraints such as multivalued elliptic problems with multifunctions of Clarke's subgradient type, variational inequalities, hemivariational inequalities, and their corresponding evolutionary counterparts. The main purpose is to provide a systematic and unified exposition of comparison principles based on a suitably extended sub-supersolution method. This method manifests as an effective and flexible technique to obtain existence and comparison results of solutions. Moreover, it can be employed for the investigation of various qualitative properties such as location, multiplicity, and extremality of solutions. In the treatment of the problems under consideration, a wide range of methods and techniques from nonlinear and nonsmooth analysis are applied; a brief outline of which has been provided in a preliminary chapter to make the book self-contained. The book is an outgrowth of the authors' research on the subject during the past 10 years. A great deal of the material presented here has been obtained only in recent years and appears for the first time in book form.

The materials presented in our book are accessible to graduate students in mathematical and physical sciences, researchers in pure and applied mathematics, physics, mechanics, and engineering.

It is our pleasure to acknowledge a debt of gratitude to Dr. Viorica Motreanu for her competent and dedicated help during the preparation of this book at its various stages. Finally, the authors are grateful to the very professional editorial staff of Springer, particularly to Ana Bozicevic and Vaishali Damle for their effective and productive collaboration.

Halle   Siegfried Carl
Rolla   Vy K. Le
Perpignan   Dumitru Motreanu
September 2005

The original version of this book was revised: The first name and last name of author Vy Khoi Le was unfortunately published with an error. The First Name should be Vy Khoi and Le should be the Last Name. The initially published version has now been corrected. A correction to this book can be found at DOI 10.1007/978-0-387-46252-3_8.

Contents

Preface v

1 Introduction 1

2 Mathematical Preliminaries 11
  2.1 Basic Functional Analysis 11
      2.1.1 Operators in Normed Linear Spaces 11
      2.1.2 Duality in Banach Spaces 15
      2.1.3 Convex Analysis and Calculus in Banach Spaces 20
      2.1.4 Partially Ordered Sets 27
  2.2 Sobolev Spaces 28
      2.2.1 Spaces of Lebesgue Integrable Functions 28
      2.2.2 Definition of Sobolev Spaces 30
      2.2.3 Chain Rule and Lattice Structure 34
      2.2.4 Some Inequalities 36
  2.3 Operators of Monotone Type 39
      2.3.1 Main Theorem on Pseudomonotone Operators 39
      2.3.2 Leray–Lions Operators 41
      2.3.3 Multivalued Pseudomonotone Operators 45
  2.4 First-Order Evolution Equations 49
      2.4.1 Motivation 50
      2.4.2 Vector-Valued Functions 53
      2.4.3 Evolution Triple and Generalized Derivative 55
      2.4.4 Existence Results for Evolution Equations 59
      2.4.5 Multivalued Evolution Equations 62
  2.5 Nonsmooth Analysis 63
      2.5.1 Clarke's Generalized Gradient 63
      2.5.2 Some Calculus 68
      2.5.3 Critical Point Theory 73
      2.5.4 Linking Theorem 77

3 Variational Equations 81
　3.1 Semilinear Elliptic Equations 81
　　　3.1.1 Comparison Principle 82
　　　3.1.2 Directed and Compact Solution Set 84
　　　3.1.3 Extremal Solutions 91
　3.2 Quasilinear Elliptic Equations 93
　　　3.2.1 Comparison Principle 94
　　　3.2.2 Directed and Compact Solution Set 97
　　　3.2.3 Extremal Solutions 103
　3.3 Quasilinear Parabolic Equations 105
　　　3.3.1 Parabolic Equation with p-Laplacian 110
　　　3.3.2 Comparison Principle for Quasilinear Equations ... 112
　　　3.3.3 Directed and Compact Solution Set 116
　　　3.3.4 Extremal Solutions 122
　3.4 Sign-Changing Solutions via Fučik Spectrum 123
　　　3.4.1 Introduction 124
　　　3.4.2 Preliminaries 125
　　　3.4.3 Main Result 130
　3.5 Quasilinear Elliptic Problems of Periodic Type 134
　　　3.5.1 Problem Setting 134
　　　3.5.2 Sub-Supersolutions 136
　　　3.5.3 Existence Result 138
　3.6 Notes and Comments 141

4 Multivalued Variational Equations 143
　4.1 Motivation and Introductory Examples 143
　　　4.1.1 Motivation 144
　　　4.1.2 Comparison Principle: Subdifferential Case 146
　　　4.1.3 Comparison Principle: Clarke's Gradient Case 149
　4.2 Inclusions with Global Growth on Clarke's Gradient 155
　　　4.2.1 Preliminaries 157
　　　4.2.2 Comparison and Compactness Results 160
　4.3 Inclusions with Local Growth on Clarke's Gradient 167
　　　4.3.1 Comparison Principle 167
　　　4.3.2 Compactness and Extremality Results 176
　4.4 Application: Difference of Multifunctions 180
　　　4.4.1 Hypotheses and Main Result 181
　　　4.4.2 A Priori Bounds 182
　　　4.4.3 Proof of Theorem 4.36 186
　4.5 Parabolic Inclusions with Local Growth 190
　　　4.5.1 Comparison Principle 191
　　　4.5.2 Extremality and Compactness Results 201
　4.6 An Alternative Concept of Sub-Supersolutions 208
　4.7 Notes and Comments 209

5   Variational Inequalities 211
   5.1   Variational Inequalities on Closed Convex Sets 213
      5.1.1   Solutions and Extremal Solutions above Subsolutions . . 214
      5.1.2   Comparison Principle and Extremal Solutions 226
   5.2   Variational Inequalities with Convex Functionals 234
      5.2.1   General Settings—Sub- and Supersolutions 235
      5.2.2   Existence and Comparison Results................... 238
      5.2.3   Some Examples 242
   5.3   Evolutionary Variational Inequalities....................... 246
      5.3.1   General Settings 247
      5.3.2   Comparison Principle 249
      5.3.3   Obstacle Problem 255
   5.4   Sub-Supersolutions and Monotone Penalty Approximations . . . 257
      5.4.1   Hypotheses and Preliminary Results 258
      5.4.2   Obstacle Problem 260
      5.4.3   Generalized Obstacle Problem 262
   5.5   Systems of Variational Inequalities......................... 267
      5.5.1   Notations and Assumptions......................... 268
      5.5.2   Preliminaries 269
      5.5.3   Comparison Principle for Systems 272
      5.5.4   Generalization, Minimal and Maximal Solutions 274
      5.5.5   Weakly Coupled Systems and Extremal Solutions 275
   5.6   Notes and Comments 277

6   Hemivariational Inequalities 279
   6.1   Notion of Sub-Supersolution 281
   6.2   Quasilinear Elliptic Hemivariational Inequalities 285
      6.2.1   Comparison Principle 286
      6.2.2   Extremal Solutions and Compactness Results 290
      6.2.3   Application 293
   6.3   Evolutionary Hemivariational Inequalities 299
      6.3.1   Sub-Supersolutions and Equivalence of Problems 301
      6.3.2   Existence and Comparison Results................... 303
      6.3.3   Compactness and Extremality Results 310
   6.4   Notes and Comments 316

7   Variational–Hemivariational Inequalities 319
   7.1   Elliptic Variational–Hemivariational Inequalities 319
      7.1.1   Comparison Principle 320
      7.1.2   Compactness and Extremality 328
   7.2   Evolution Variational–Hemivariational Inequalities........... 336
      7.2.1   Definitions and Hypotheses 338
      7.2.2   Preliminary Results................................ 340
      7.2.3   Existence and Comparison Result 343
      7.2.4   Compactness and Extremality 351

7.3   Nonsmooth Critical Point Theory 355
7.4   A Constraint Hemivariational Inequality.................... 362
7.5   Eigenvalue Problem for a Variational–Hemivariational
       Inequality 368
7.6   Notes and Comments 375

Correction to: Nonsmooth Variational Problems and **C1**
       Their Inequalities

List of Symbols 379

References 381

Index 393

1

Introduction

A powerful and fruitful tool for proving existence and comparison results for a wide range of nonlinear elliptic and parabolic boundary value problems is the method of sub- and supersolutions.

In one of its simplest forms, this method is a consequence of the classic maximum principle for sub- and superharmonic functions that can be seen in the following classic example. Consider the homogeneous Dirichlet boundary value problem

$$-\Delta u = f \quad \text{in} \quad \Omega, \quad u = 0 \quad \text{on} \quad \partial\Omega, \tag{1.1}$$

where $\Omega \subset \mathbb{R}^N$ is a bounded domain with smooth boundary $\partial\Omega$, $f : \overline{\Omega} \to \mathbb{R}$ is some given smooth function, and assume the existence of a classic subsolution $\underline{u}$ and supersolution $\bar{u}$ of (1.1), i.e., $\underline{u}, \bar{u} \in C^2(\Omega) \cap C(\overline{\Omega})$ satisfying

$$-\Delta\underline{u} \leq f \quad \text{in} \quad \Omega, \quad \underline{u} \leq 0 \quad \text{on} \quad \partial\Omega, \tag{1.2}$$

$$-\Delta\bar{u} \geq f \quad \text{in} \quad \Omega, \quad \bar{u} \geq 0 \quad \text{on} \quad \partial\Omega. \tag{1.3}$$

Then $w = \underline{u} - \bar{u}$ is readily seen as a subharmonic function in Ω with nonpositive boundary values, i.e.,

$$-\Delta w \leq 0 \quad \text{in} \quad \Omega, \quad w \leq 0 \quad \text{on} \quad \partial\Omega, \tag{1.4}$$

and thus, by the classic maximum principle (see [187]), it follows that $w \leq 0$ in Ω, i.e., $\underline{u} \leq \bar{u}$ in $\overline{\Omega}$. Moreover, because any solution u of (1.1) satisfies both (1.2) and (1.3), it must be at the same time a subsolution and a supersolution of (1.1), which implies the unique solvability of the Dirichlet problem (1.1). Thus, in view of the maximum principle, any pair of sub-supersolutions of (1.1) must be ordered, and the solution u of (1.1) must be unique and must be contained in the ordered interval $[\underline{u}, \bar{u}]$. In this way, the maximum principle enables us to obtain a priori bounds for the solution of problem (1.1). Also, an immediate consequence of the maximum principle is the order-preserving property of solutions of (1.1), which means that if u_1 and u_2 are the solutions of (1.1) corresponding to right-hand sides f_1 and f_2, respectively, satisfying $f_1 \leq f_2$, then $u_1 \leq u_2$ in Ω.

Unfortunately, maximum principles do not hold in many nonlinear elliptic problems written in the abstract form

$$Au = f \quad \text{in} \ \ \Omega, \quad Bu = 0 \quad \text{on} \ \ \partial\Omega. \tag{1.5}$$

However, if $\underline{u}$ and $\bar{u}$ are appropriate (weak) sub- and supersolutions of (1.5) satisfying, in addition, $\underline{u} \leq \bar{u}$, then (weak) solutions of (1.5) (not necessarily unique) exist within the interval $[\underline{u}, \bar{u}]$ formed by the ordered pair of sub- and supersolutions. It is basically this property that we will refer to as a *comparison principle* for the problems under consideration. For example, consider the following prototype of (1.5):

$$-\Delta_p u + g(u) = f \quad \text{in} \ \ \Omega, \quad u = 0 \quad \text{on} \ \ \partial\Omega, \tag{1.6}$$

where $\Delta_p u = \text{div}\,(|\nabla u|^{p-2}\nabla u)$ is the p-Laplacian, $1 < p < \infty$, $f \in L^q(\Omega)$ with q being the Hölder conjugate to p satisfying $1/p + 1/q = 1$, and $g : \mathbb{R} \to \mathbb{R}$ is a continuous function with some growth condition. As is well known, in general, problem (1.6) does not admit classic solutions, and therefore, it has to be treated within the framework of weak solutions. Let $V = W^{1,p}(\Omega)$ and $V_0 = W_0^{1,p}(\Omega)$ denote the usual Sobolev spaces with their dual spaces V^* and V_0^*, respectively, then a weak solution of the Dirichlet problem (1.6) is defined as follows:

$$u \in V_0 : \quad -\Delta_p u + g(u) = f \quad \text{in} \ \ V_0^*, \tag{1.7}$$

where due to the continuous embedding $L^q(\Omega) \subset V_0^*$, f has to be interpreted as a dual element of V_0^*. As $Au = -\Delta_p u + g(u)$ defines a bounded and continuous mapping from V_0 into V_0^*, (1.7) provides an appropriate functional analytic framework for the boundary value problem (1.6), which is equivalent with the following *variational equation:*

$$u \in V_0 : \quad \langle -\Delta_p u + g(u), \varphi \rangle = \langle f, \varphi \rangle \quad \text{for all} \ \varphi \in V_0, \tag{1.8}$$

where $\langle \cdot, \cdot \rangle$ denotes the duality pairing. It follows from standard integration by parts that the variational equation (1.8) is equivalent to

$$u \in V_0 : \quad \int_\Omega |\nabla u|^{p-2}\nabla u \nabla \varphi \, dx + \int_\Omega g(u)\,\varphi\,dx = \langle f, \varphi \rangle \quad \text{for all} \ \varphi \in V_0. \tag{1.9}$$

A natural extension of the classic notion of sub- and supersolution to the weak formulation (1.7) of the boundary value problem (1.6) is defined as follows. The function $\bar{u} \in V$ is a weak supersolution of (1.7) if

$$\bar{u} \geq 0 \ \text{on} \ \partial\Omega \quad \text{and} \quad -\Delta_p \bar{u} + g(\bar{u}) \geq f \quad \text{in} \ \ V_0^*, \tag{1.10}$$

where the inequality in V_0^* has to be taken with respect to the dual-order cone $V_{0,+}^*$ of V_0^*, defined by

$$V_{0,+}^* = \{u^* \in V_0^* : \langle u^*, \varphi \rangle \geq 0 \text{ for all } \varphi \in V_0 \cap L_+^p(\Omega)\},$$

where $L_+^p(\Omega)$ is the positive cone of all nonnegative elements of $L^p(\Omega)$ by which the natural partial ordering of functions in $L^p(\Omega)$ is defined. Due to (1.10), we obtain the following well-known equivalent definition of a weak supersolution of (1.7). The function $\bar{u} \in V$ is a weak supersolution if $\bar{u} \geq 0$ on $\partial \Omega$ and

$$\int_\Omega |\nabla \bar{u}|^{p-2} \nabla \bar{u} \nabla \varphi \, dx + \int_\Omega g(\bar{u}) \, \varphi \, dx \geq \langle f, \varphi \rangle \quad \text{for all } \varphi \in V_0 \cap L_+^p(\Omega).$$

(1.11)

Similarly, $\underline{u} \in V$ is a weak subsolution of (1.7) if $\underline{u} \leq 0$ on $\partial \Omega$ and

$$\int_\Omega |\nabla \underline{u}|^{p-2} \nabla \underline{u} \nabla \varphi \, dx + \int_\Omega g(\underline{u}) \, \varphi \, dx \leq \langle f, \varphi \rangle \quad \text{for all } \varphi \in V_0 \cap L_+^p(\Omega).$$

(1.12)

Comparison principles for solutions of nonlinear elliptic and parabolic variational equations including the special case (1.7) are well known and can be found, e.g., in the monographs [43, 66, 83]. Thus, we have, e.g., if $\underline{u}$ and $\bar{u}$ are sub- and supersolutions of (1.7), respectively, and if $\underline{u} \leq \bar{u}$, then solutions exist within the ordered interval $[\underline{u}, \bar{u}]$. Moreover, the solution set S enclosed by an ordered pair of sub- and supersolutions can be shown to be compact and to possess greatest and smallest elements with respect to the natural partial ordering of functions induced by the order cone $L_+^p(\Omega)$. A review and detailed proofs of these results will be given in Chap. 3.

The existence and comparison results along with the topological and order related characterization of the solution set S obtained for nonlinear elliptic and parabolic variational equations generalize the following elementary result on the real line $\mathbb{R}$. Consider the real equation

$$F(u) = 0, \quad u \in \mathbb{R}, \tag{1.13}$$

and assume that:

(i) The function $F : \mathbb{R} \to \mathbb{R}$ is continuous.
(ii) $\underline{s}, \bar{s} \in \mathbb{R}$ satisfying $\underline{s} \leq \bar{s}$ exist such that $F(\underline{s}) \leq 0$ and $F(\bar{s}) \geq 0$.

Then solutions of (1.13) exist within the real interval $[\underline{s}, \bar{s}]$, and the set of all solutions of (1.13) is closed and bounded and, thus, compact. Moreover, the solution set has a greatest and smallest element s^* and s_*, respectively (see Fig. 1.1).

This classic existence and enclosure result follows from the intermediate value theorem for continuous functions, whereas the existence of greatest and

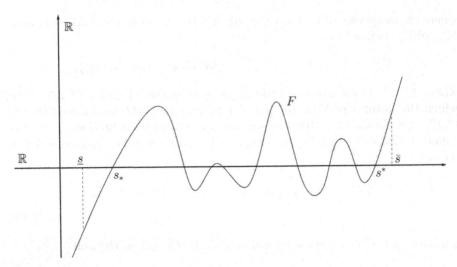

Fig. 1.1. Sub-supersolution

smallest solutions is an immediate consequence of the order property of the real line $\mathbb{R}$, which, speaking in abstract terms, is a completely ordered Banach space.

Now the results above concerning (weak) solutions of the nonlinear problem (1.6) nicely fit into this elementary picture. Let $F : V_0 \to V_0^*$ be defined by

$$F(u) = -\Delta_p u + g(u) - f.$$

Then the equivalent elliptic variational equation (1.7) can be rewritten as

$$u \in V_0 : \quad F(u) = 0 \quad \text{in } V_0^*.$$

Assume that:

(i*) The function $g : \mathbb{R} \to \mathbb{R}$ is continuous and satisfies a certain growth condition.

(ii*) $\underline{u}, \bar{u} \in V$ satisfying $\underline{u} \leq \bar{u}$ exist with

$$\underline{u} \leq 0 \quad \text{on } \partial\Omega, \quad \bar{u} \geq 0 \quad \text{on } \partial\Omega \quad \text{such that } F(\underline{u}) \leq 0 \text{ and } F(\bar{u}) \geq 0.$$

Then the existence and comparison result as well as the characterization of the solution set for (1.6) given above hold. Note that in view of (i*), the operator $F : V_0 \to V_0^*$ is continuous, bounded, and pseudomonotone, but not necessarily coercive. As will be seen in Chap. 3, the existence of sub- and supersolutions supposed in (ii*) will be used to compensate this drawback.

In this monograph, we focus primarily on *nonsmooth variational problems*. Just as "nonlinear" in mathematics stands for "not necessarily linear," we use "nonsmooth" to refer to certain situations in which smoothness is not necessarily assumed. The relaxed smoothness requirements have often been motivated by the needs of disciplines other than mathematics, such as mechanics and engineering.

Our main goal is to extend the idea of sub-supersolutions and to provide a systematic and unified approach for obtaining comparison principles for both nonsmooth stationary and evolutionary variational problems. We shall demonstrate that much of the idea of the method of sub-supersolutions that has been known for elliptic and parabolic variational equations can be developed in a general nonsmooth setting. To give an idea of what we mean by nonsmooth variational problems, let us consider a few examples.

A nonsmooth variational problem arises, e.g., when the nonlinearity g in (1.9) is no longer continuous. If $g : \mathbb{R} \to \mathbb{R}$ satisfies some growth condition but is only supposed to be Borel-measurable, then problem (1.9) becomes a discontinuous variational equation. Even though the operator A of the equivalent operator equation (1.7) given by $Au = -\Delta_p u + g(u)$ is still well defined and bounded from V_0 into its dual space V_0^*; it is, however, no longer continuous. In this case, the sub-supersolution method, in general, fails as shown by the following simple example.

Let us consider (1.7) with $p = 2$, $f(x) \equiv 1$, and g the Heaviside step function given by $g(s) = 0$ for $s \le 0$, and $g(s) = 1$ for $s > 0$; i.e., we consider

$$u \in V_0 = W_0^{1,2}(\Omega) : \quad -\Delta u + g(u) = 1 \quad \text{in } V_0^*. \tag{1.14}$$

One readily verifies that the constant functions $\underline{u} = -c$ and $\bar{u} = c$ with c any positive constant provide an ordered pair of sub-supersolutions of (1.14). However, problem (1.14) has no solutions within the order interval $[-c, c]$. Furthermore, (1.14) does not possess solutions at all. In fact, if u was a solution, then it satisfies the variational equation

$$\int_\Omega \nabla u \nabla \varphi \, dx = \int_\Omega (1 - g(u)) \, \varphi \, dx \quad \text{for all } \varphi \in V_0.$$

Taking as a special test function the solution u, we obtain in view of the definition of g the following inequality:

$$\int_\Omega |\nabla u|^2 \, dx = \int_\Omega (1 - g(u)) \, u \, dx \le 0,$$

and hence it follows that $u = 0$. This result is a contradiction, because $u = 0$ is apparently not a solution of (1.14).

Problem (1.14) with g being the Heaviside function is embedded into a relaxed multivalued setting replacing the discontinuous function g by an associated multivalued function $s \mapsto [\underline{g}(s), \bar{g}(s)]$, where $\underline{g}(s)$ and $\bar{g}(s)$ denote the left-sided and right-sided limits of g at $s \in \mathbb{R}$. It turns out that this multifunction that, roughly speaking, arises from g by filling in the gap at the point of discontinuity, coincides with the multifunction $s \mapsto \partial j(s)$, where $\partial j(s)$ denotes

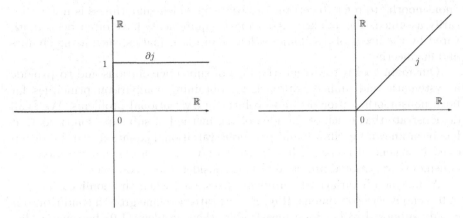

Fig. 1.2. Subdifferential of j

Fig. 1.3. Primitive of Heaviside function

the subdifferential of the primitive $j : \mathbb{R} \to \mathbb{R}$ of g given by $j(u) = \int_0^u g(s)\,ds$, which is a convex and Lipschitz continuous function, (see Fig. 1.2 and Fig. 1.3).

Thus, the relaxed multivalued problem (1.14) reads as follows:

$$u \in V_0 : \quad -\Delta_p u + \partial j(u) \ni 1 \quad \text{in } V_0^*, \tag{1.15}$$

where $j : \mathbb{R} \to \mathbb{R}$ is the above primitive of the Heaviside function. As j is convex and even Lipschitz continuous, one can easily show that (1.15) is equivalent to

$$u \in V_0 : \quad \partial \hat{E}(u) \ni 0,$$

where $\partial \hat{E}(u)$ is the subdifferential at u of the nonsmooth, convex, continuous, and coercive functional $\hat{E} : V_0 \to \mathbb{R}$ defined by

$$\hat{E}(u) = \frac{1}{p} \int_\Omega |\nabla u|^p \, dx + \int_\Omega j(u) \, dx - \langle 1, u \rangle.$$

As $\hat{E}$ in our example is even strictly convex, a unique solution of the optimization problem exists

$$u \in V_0 : \quad \hat{E}(u) = \inf_{v \in V_0} \hat{E}(v),$$

which in turn is equivalent to $\partial \hat{E}(u) \ni 0$. Thus, problem (1.15) has only one solution, which is the minimum point of the nonsmooth functional $\hat{E}$.

To motivate other types of nonsmooth variational problems, consider the functional E:

$$E(u) = \frac{1}{p} \int_\Omega |\nabla u|^p \, dx + \int_\Omega j(u) \, dx - \langle f, u \rangle, \quad u \in V_0, \tag{1.16}$$

where $f \in V_0^*$ and $j : \mathbb{R} \to \mathbb{R}$ is the primitive of a continuous function g that satisfies some growth condition. Then $E : V_0 \to \mathbb{R}$ is a C^1-functional whose critical points are the solutions of the variational problem (1.9). In this sense, (1.9) may be considered as a smooth variational problem in case g is continuous.

A nonsmooth variational problem already occurs if we are looking for critical points of the C^1-functional E of (1.16) under some constraint, which is represented, for example, by a closed convex subset $K \subset V_0$. This leads to the following well-known *variational inequality* for the operator $Au = -\Delta_p u + g(u)$:

$$u \in K : \quad \langle Au - f, \varphi - u \rangle \geq 0, \quad \text{for all} \quad \varphi \in K. \tag{1.17}$$

Introducing the indicator function I_K of the set K, we see that (1.17) is equivalent to the variational inequality

$$u \in K : \quad \langle Au - f, \varphi - u \rangle + I_K(\varphi) - I_K(u) \geq 0, \quad \text{for all} \quad \varphi \in V_0, \tag{1.18}$$

which in turn is equivalent to the *differential inclusion*

$$u \in K : \quad -Au + f \in \partial I_K(u),$$

where ∂I_K is the subdifferential of the indicator function $I_K : V_0 \to [0, +\infty]$, which is proper if $K \neq \emptyset$, convex, and lower semicontinuous.

Another type of nonsmooth variational problems arises if we consider critical points of the functional E above when j is the primitive of a not necessarily continuous function g satisfying only some growth and measurability conditions. Under these assumptions, $E : V_0 \to \mathbb{R}$ is, in general, no longer convex, but only locally Lipschitz, and u is called a critical point of E if

$$0 \in \partial E(u), \tag{1.19}$$

where $\partial E(u) \subset V_0^*$ denotes Clarke's generalized gradient. For example, if u is a minimum point of E over V_0, then u is a critical point, and it satisfies (1.19). Applying basic facts from nonsmooth analysis, we see that (1.19) is equivalent to

$$u \in V_0 : \quad \langle -\Delta_p u - f, \varphi \rangle + J^\circ(u; \varphi) \geq 0, \quad \text{for all} \quad \varphi \in V_0, \tag{1.20}$$

where $J^\circ(u; v)$ denotes the generalized directional derivative at u in direction v of the locally Lipschitz functional $J : V_0 \to \mathbb{R}$ given by

$$J(u) = \int_\Omega j(u) \, dx.$$

Problem (1.20) is called a *hemivariational inequality,* which is equivalent to the inclusion

$$\Delta_p u + f \in \partial J(u),$$

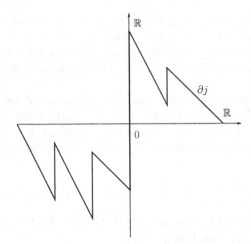

Fig. 1.4. Zig-zag friction law

where $\partial J(u) \subset V_0^*$ is Clarke's generalized gradient of the integral functional J at u. Closely related but, in general, not equivalent to (1.20) is the following differential inclusion:

$$u \in V_0: \quad -\Delta_p u + \partial j(u) \ni f \quad \text{in} \quad V_0^*, \tag{1.21}$$

where $\partial j : \mathbb{R} \to 2^{\mathbb{R}} \setminus \{\emptyset\}$ is Clarke's generalized gradient of the locally Lipschitz integrand $j : \mathbb{R} \to \mathbb{R}$ of J. An example of multifunctions ∂j that appear in applications is shown in Fig. 1.4.

Finally, if we try to find solutions of the hemivariational inequality under constraints, we arrive at the nonsmooth variational problem

$$u \in K: \quad \langle -\Delta_p u - f, \varphi - u \rangle + I_K(\varphi) - I_K(u) + J^o(u; \varphi - u) \geq 0,$$
$$\text{for all} \quad \varphi \in V_0, \tag{1.22}$$

which is called a *variational-hemivariational inequality*. The field of hemivariational inequalities, initiated with the pioneering work of Panagiotopoulos (cf. [179, 180]), has attracted increasing attention over the last decade mainly due to its many applications in mechanics and engineering. This new type of variational inequalities arises, e.g., in mechanical problems governed by nonconvex, possibly nonsmooth energy functionals (so-called superpotentials), which appear if nonmonotone, multivalued constitutive laws are taken into account.

However, note that the multivalued problems (1.15) and (1.21), the variational inequality (1.17), the hemivariational inequality (1.20), and the variational-hemivariational inequality (1.22) only serve as prototypes of nonsmooth variational problems of elliptic type that will be treated in this book. Comparison principles will be obtained for more general nonsmooth variational problems that are not necessarily related to some potential functional, and for their evolutionary counterparts. It should be noted also that the treatment of evolutionary nonsmooth variational problems is by no means a straightforward extension of nonsmooth (stationary) elliptic variational problems, and it requires different tools. Moreover, not only scalar but also systems of nonsmooth variational problems will be treated.

As shown, the notion of sub- and supersolution for nonlinear elliptic variational equations is an almost direct extension of the classic notion of sub- and supersolution for the Laplace equation. A similar statement can be made for parabolic variational equations, for which the notion of sub- and supersolution is a natural extension of the one for the heat equation.

The situation is, however, different for variational and hemivariational inequalities. Because of the intrinsic asymmetry of these problems (where the problems are stated as inequalities rather than as equalities), it is much more difficult to define sub- and supersolutions for variational and hemivariational inequalities. As an indispensable requirement, this notion should be an extension of the well-known notion of sub- and supersolution for variational equations. It seems to be the main reason that this powerful method and the comparison principles related with it have not been employed so far to investigate nonsmooth variational problems.

The rapid development of the theory of variational and hemivariational inequalities and the prolific growth of its numerous applications (see [124, 177]) made evident to us the need for a detailed and systematic exposition of the sub-supersolution method for nonsmooth variational problems that covers the one for variational equations in a natural way. We have made efforts to define a notion of sub- and supersolution in such a way that will allow us to establish comparison principles for nonsmooth variational problems similar to the corresponding concepts for variational equations. The comparison principles based on the new notion of sub- and supersolution will be seen to preserve many characteristic features of the elementary example on the real line considered above; i.e., we will be able to prove not only existence and enclosure of solutions for nonsmooth variational problems but also qualitative properties of the solution set, such as compactness and existence of smallest and greatest solutions. In addition, these new comparison principles will be shown to provide effective tools to study noncoercive nonsmooth variational problems and permit more flexible requirements on the growth rates of certain nonlinear data involved.

This book is basically an outgrowth of the authors' research on the subject during the past 10 years. It consists of seven chapters, including the introductory chapter. Each chapter begins with a short overview, and notes and remarks are added at the end. Chapter 2 provides needed mathematical prerequisites to make the book self-contained. Chapter 3 deals with the sub-supersolution method for weak solutions of nonlinear elliptic and parabolic variational equations, and it may be considered in some sense as a preparatory chapter to get to know some methods and techniques used also in later chapters. Chapter 4 to Chapter 7 form the core of the book dealing with

nonsmooth variational problems. Chapter 4 deals with multivalued elliptic
and parabolic problems that involve multifunctions of Clarke's subgradient
type. The key notion of sub-supersolution for variational inequalities is developed in Chapter 5. In Chapter 6, we deal with comparison principles for
hemivariational inequalities and reveal their connection with the multivalued
problems considered in Chapter 4. Finally, in Chapter 7, we treat variational–
hemivariational inequalities and related problems such as eigenvalue problems
for this kind of variational problems.

Some important features of the monograph are as follows:

- Presenting a systematic and unified exposition of the sub-supersolution
  method for nonsmooth stationary and evolutionary variational problems,
  including variational and hemivariational inequalities.
- Proving existence and comparison results, and characterizing the solution
  set topologically and order theoretically.
- Inclusion of numerous new results, some of which have never been published.
- Efforts have been made to make the presentation self-contained by providing the necessary mathematical background and theories in an extra
  chapter.
- Attempts to draw a broad audience by writing the first section of each
  chapter in a manner that emphasizes simple cases and ideas more than
  complicated refinements.
- Being accessible to graduate students in mathematics and engineering.
- The power of the developed methodology is demonstrated through various
  examples and applications.

2

Mathematical Preliminaries

In this chapter, we provide the mathematical background as it will be used in later chapters.

2.1 Basic Functional Analysis

The purpose of this section is to provide a survey of basic results from functional analysis that will be used in the sequel. However, we will assume that the reader is familiar with some elementary notions such as metric spaces, Banach spaces, and Hilbert spaces, as well as notions related with the topological structure of these spaces. Unless otherwise indicated, all linear spaces considered in this book are assumed to be defined over the real number field $\mathbb{R}$. The proofs of the results presented in this section can be found in standard textbooks, e.g., [5, 13, 24, 129, 200, 222].

2.1.1 Operators in Normed Linear Spaces

Let $(X, \| \cdot \|_X)$ and $(Y, \| \cdot \|_Y)$ be normed linear spaces, and let

$$A : D(A) \subset X \to Y$$

be an operator with domain $D(A)$ and range denoted by range(A). When $D(A) = X$, we write

$$A : X \to Y.$$

Note that usually we drop the subscripts X and Y in the notation of the norms $\| \cdot \|_X$ and $\| \cdot \|_Y$, respectively, if no ambiguity exists.

Definition 2.1. Let $A : D(A) \subset X \to Y$.

(i) A is continuous at the point $u \in D(A)$ iff for each sequence (u_n) in $D(A)$,

$$u_n \to u \quad implies \quad Au_n \to Au.$$

The operator $A : D(A) \subset X \to Y$ is called continuous iff it is continuous at each point $u \in D(A)$.

(ii) *A is called compact iff A is continuous, and A maps bounded sets into relatively compact sets.*

Note that one sometimes uses the notion *completely continuous* for compact. For compact operators, the following fixed-point theorem from Schauder holds.

Theorem 2.2 (Schauder's Fixed-Point Theorem). *Let X be a Banach space, and let*

$$A : M \to M$$

be a compact operator that maps a nonempty subset M of X into itself. Then A has a fixed point provided M is bounded, closed, and convex.

In finite-dimensional normed linear spaces, Theorem 2.2 reduces to Brouwer's fixed-point theorem.

Corollary 2.3 (Brouwer's Fixed-Point Theorem). *If the operator*

$$A : M \to M$$

is continuous, then A has a fixed point provided M is a compact, convex, nonempty subset in a finite-dimensional normed linear space.

Let

$$A : D(A) \subset X \to Y$$

be a *linear* operator, which means that the domain $D(A)$ of the operator A is a linear subspace of X and A satisfies

$$A(\alpha u + \beta v) = \alpha A u + \beta A v \quad \text{for all } u, v \in D(A), \ \alpha, \beta \in \mathbb{R}.$$

Proposition 2.4. *Let $A : X \to Y$ be a linear operator. Then the following two conditions are equivalent:*

(i) *A is continuous.*

(ii) *A is bounded; i.e., there is a constant $c > 0$ such that*

$$\|Au\| \leq c\|u\| \quad \text{for all } u \in X.$$

For a linear continuous operator $A : X \to Y$, the operator norm $\|A\|$ is defined by

$$\|A\| = \sup_{\|u\| \leq 1} \|Au\|,$$

which can easily be shown to be equal to

$$\|A\| = \sup_{\|u\| = 1} \|Au\|.$$

Proposition 2.5. *Let $L(X, Y)$ denote the space of linear continuous operators $A : X \to Y$, where X is a normed linear space and Y is a Banach space. Then $L(X, Y)$ is a Banach space with respect to the operator norm.*

Definition 2.6. *Let*

$$A : D(A) \subset X \to Y$$

be a linear operator. The graph of A denoted by $\mathrm{Gr}(A)$ is defined by the subset

$$\mathrm{Gr}(A) = \{(u, Au) : u \in D(A)\}$$

of the product space $X \times Y$. The operator A is called closed (or graph-closed) iff $\mathrm{Gr}(A)$ is closed in $X \times Y$, which means that for each sequence (u_n) in $D(A)$, it follows from

$$u_n \to u \quad \text{in } X \quad \text{and} \quad Au_n \to v \quad \text{in } Y$$

that $u \in D(A)$ and $v = Au$. Finally, on $D(A)$, the so-called graph norm $\|\cdot\|_A$ is defined by

$$\|u\|_A = \|u\| + \|Au\| \quad \text{for } u \in D(A).$$

Corollary 2.7. *If X and Y are Banach spaces and $A : D(A) \subset X \to Y$ is closed, then $D(A)$ equipped with the graph norm, i.e., $(D(A), \|\cdot\|_A)$, is a Banach space.*

Theorem 2.8 (Banach's Closed Graph Theorem). *Let X and Y be Banach spaces. Then each closed linear operator $A : X \to Y$ is continuous.*

For completeness, we shall recall the Uniform Boundedness Theorem and the Open Mapping Theorem, which together with Banach's Closed Graph Theorem are all consequences of Baire's Theorem.

Theorem 2.9 (Uniform Boundedness Theorem). *Let $\mathcal{F}$ be a nonempty set of continuous maps*

$$F : X \to Y,$$

where X is a Banach space and Y is a normed linear space. Assume that

$$\sup_{F \in \mathcal{F}} \|Fu\| < \infty \quad \text{for each } u \in X.$$

Then a closed ball $\overline{B}$ in X of positive radius exists such that

$$\sup_{u \in \overline{B}} \left(\sup_{F \in \mathcal{F}} \|Fu\| \right) < \infty.$$

Corollary 2.10 (Banach–Steinhaus Theorem). *Let $\mathcal{L} \subset L(X,Y)$ be a nonempty set of linear continuous operators*

$$A : X \to Y,$$

where X is a Banach space and Y is a normed linear space. Assume that

$$\sup_{A \in \mathcal{L}} \|Au\| < \infty \quad \text{for each} \quad u \in X.$$

Then $\sup_{A \in \mathcal{L}} \|A\| < \infty$.

Theorem 2.11 (Banach's Open Mapping Theorem). *Let X and Y be Banach spaces and $A : X \to Y$ be a linear continuous operator. Then the following two conditions are equivalent:*

(i) *A is surjective.*
(ii) *A is open, which means that A maps open sets onto open sets.*

Corollary 2.12 (Banach's Continuous Inverse Theorem). *Let X and Y be Banach spaces and $A : X \to Y$ be a linear continuous operator. If the inverse operator*

$$A^{-1} : Y \to X$$

exists, then A^{-1} is continuous.

Definition 2.13 (Embedding Operator). *Let X and Y be normed linear spaces with*

$$X \subset Y.$$

The embedding operator $i : X \to Y$ is defined by $i(u) = u$; i.e., i is the identity operator from X into Y.

(i) *The embedding $X \subset Y$ is called continuous iff the embedding operator $i : X \to Y$ is continuous; i.e., a constant $c > 0$ exists such that*

$$\|u\|_Y \leq c \|u\|_X \quad \text{for all} \quad u \in X,$$

which is equivalent with

$$u_n \to u \text{ in } X \quad \text{implies} \quad u_n \to u \text{ in } Y.$$

(ii) *The embedding $X \subset Y$ is called compact iff the embedding operator $i : X \to Y$ is compact; i.e., i is continuous and each bounded sequence (u_n) in X has a subsequence that converges in Y.*

Remark 2.14. More generally, one can define a continuous embedding of a normed linear space X into a normed linear space Y, whenever a linear, continuous, and injective operator $i : X \to Y$ exists. Similarly, X is compactly embedded into Y iff a linear, compact, and injective operator $i : X \to Y$ exists.

2.1.2 Duality in Banach Spaces

Definition 2.15. *Let X be a normed linear space. A linear continuous functional on X is a linear continuous operator*

$$f : X \to \mathbb{R}.$$

The set of all linear continuous functionals on X is called the dual space X^* of X; i.e., $X^* = L(X, \mathbb{R})$. For the image $f(u)$ of the functional f at $u \in X$, we write

$$\langle f, u \rangle = f(u) \quad u \in X, \quad f \in X^*,$$

and $\langle \cdot, \cdot \rangle$ is called the duality pairing.

According to the operator norm defined in Sect. 2.1.1, the norm of f is given through

$$\|f\| = \sup_{\|u\| \leq 1} |\langle f, u \rangle|.$$

As a consequence of Proposition 2.5, we get the following result.

Corollary 2.16. *Let X be a normed linear space. Then the dual space X^* is a Banach space with respect to the norm $\|f\|$ for $f \in X^*$.*

The most important theorem about the structure of linear functionals on normed linear spaces is the Hahn–Banach Theorem. For real linear spaces, the Hahn–Banach Theorem reads as follows (see [24]).

Theorem 2.17 (Hahn–Banach Theorem). *Let $p : E \to \mathbb{R}$ be a function on a real linear space E satisfying*

$$p(\lambda x) = \lambda p(x), \quad \forall \, x \in E, \, \forall \, \lambda \geq 0,$$
$$p(x + y) \leq p(x) + p(y), \quad \forall \, x, y \in E.$$

Let G be a linear subspace of E, and let $g : G \to \mathbb{R}$ be a linear functional such that

$$g(x) \leq p(x), \quad \forall \, x \in G.$$

Then a linear functional $f : E \to \mathbb{R}$ exists with the properties

$$f(x) = g(x), \quad \forall \, x \in G$$

and

$$f(x) \leq p(x), \quad \forall \, x \in E.$$

As an immediate consequence from Theorem 2.17, we obtain the following theorem, which is the Hahn–Banach Theorem for normed linear spaces.

Theorem 2.18. *Let X be a normed linear space. Assume M is a linear subspace of X, and $F : M \to \mathbb{R}$ is a linear functional such that*

$$|F(u)| \le c\,\|u\| \quad for\ all\ \ u \in M,$$

where c is some positive constant. Then F can be extended to a linear continuous functional $f : X \to \mathbb{R}$ that satisfies

$$|\langle f, u\rangle| \le c\,\|u\| \quad for\ all\ \ u \in X.$$

First consequences from the Hahn–Banach Theorem are given in the following corollary.

Corollary 2.19. *Let X be a normed linear space.*

(i) *For each given $u_0 \in X$ with $u_0 \ne 0$, a functional $f \in X^*$ exists such that*

$$\langle f, u_0\rangle = \|u_0\| \quad and \quad \|f\| = 1.$$

(ii) *For all $u \in X$, one has*

$$\|u\| = \sup_{f \in X^*,\ \|f\| \le 1} |\langle f, u\rangle|.$$

(iii) *If for $u \in X$ the condition*

$$\langle f, u\rangle = 0 \quad for\ all\ \ f \in X^*$$

holds, then $u = 0$.

We set

$$X^{**} = (X^*)^*,$$

which is called the bidual space and which consists of all linear continuous functionals $F : X^* \to \mathbb{R}$.

Proposition 2.20. *Let X be a normed linear space. The operator $j : X \to X^{**}$ defined by*

$$j(u)(f) = \langle f, u\rangle \quad for\ all\ \ u \in X,\ f \in X^*$$

has the following properties:

(i) *j is linear and*
$$\|j(u)\| = \|u\| \quad for\ all\ \ u \in X.$$

(ii) *$j(X)$ is a closed subspace of X^{**} if and only if X is a Banach space.*

The operator $j : X \to X^{**}$ is called the canonical embedding of X into X^{**}.

Definition 2.21. *A normed linear space X is called reflexive if the canonical embedding $j : X \to X^{**}$ is surjective; i.e., $j(X) = X^{**}$.*

We readily observe that every reflexive normed linear space X is in fact a Banach space, which is isometrically isomorphic to X^{**}, and thus, we may write $X = X^{**}$.

Corollary 2.22. (i) *Each Hilbert space is reflexive.*

(ii) *Every closed linear subspace of a reflexive Banach space X is again reflexive.*

(iii) *The product of a finite number of reflexive Banach spaces is a reflexive Banach space.*

(iv) *Let X and Y be two isomorphic normed linear spaces. If X is a reflexive Banach space, then Y is also a reflexive Banach space.*

(v) *Let X be a Banach space. Then X is reflexive if and only if X^* is reflexive.*

(vi) *If X is a separable and reflexive Banach space, then X^* is separable.*

Next we define the *dual* or *adjoint operator* of a linear operator $A : D(A) \subset X \to Y$, where X and Y are two Banach spaces.

Definition 2.23. *Assume $D(A)$ is dense in X. Then the dual operator*

$$A^* : D(A^*) \subset Y^* \to X^*$$

is defined by the following relation:

$$\langle A^*v, u \rangle = \langle v, Au \rangle \quad \text{for all} \quad v \in D(A^*), \quad u \in D(A),$$

where $v \in Y^*$ belongs to $D(A^*)$ if and only if a $w \in X^*$ exists such that

$$\langle w, u \rangle = \langle v, Au \rangle \quad \text{for all} \quad u \in D(A).$$

To verify that A^* is well defined, we note first that according to Definition 2.23, an element $v \in Y^*$ belongs to $D(A^*)$ if and only if a $w \in X^*$ exists such that

$$\langle w, u \rangle = \langle v, Au \rangle \quad \text{for all} \quad u \in D(A).$$

We set $A^*v = w$. As $D(A)$ is dense in X, the element w is uniquely determined by v, and thus, the operator A^* is well defined. Moreover, one readily observes that A^* is linear and graph-closed. In the special case that $D(A) = X$, we have the following results.

Proposition 2.24. *Let X and Y be two Banach spaces, and let $A : X \to Y$ be a linear and continuous operator. Then the dual operator*

$$A^* : Y^* \to X^*$$

is also linear and continuous, and we have

$$\|A^*\| = \|A\|.$$

Moreover, if the linear operator $A : X \to Y$ is compact, then so is the dual operator $A^* : Y^* \to X^*$.

The following facts about the duality of embeddings are important, e.g., for the understanding of the concept of evolution triple, which will be introduced in Sect. 2.4.3.

Proposition 2.25. *Let X and Y be Banach spaces with $X \subset Y$ such that X is dense in Y, and the embedding*

$$i : X \to Y$$

is continuous. Then the following is true:

(i)   *The embedding $Y^* \subset X^*$ is continuous, and the embedding operator $\hat{i} :$ $Y^* \to X^*$ is identical with the dual operator of i; i.e., $\hat{i} = i^*$.*
(ii) *If X is, in addition, reflexive, then Y^* is dense in X^*.*
(iii) *If the embedding $X \subset Y$ is compact, then so is the embedding $Y^* \subset X^*$.*

Proof: As for (i), density arguments show that each element of Y^* can be uniquely identified with an element of X^*, and the continuity of the embedding $Y^* \subset X^*$ follows from the continuity of i. The proof of (ii) makes use of the Hahn–Banach Theorem in connection with the reflexivity of X. (see [222, Chap. 18, 21]), and (iii) follows from Proposition 2.24.   □

In finite-dimensional Banach spaces, closed and bounded sets are compact. This result is no longer true for infinite-dimensional Banach spaces because of the following famous theorem due to Riesz.

Theorem 2.26 (Riesz' Lemma). *Let X be a normed linear space. Then, the closed unit ball in X is compact if and only if X is finite-dimensional.*

According to Theorem 2.26, in infinite-dimensional Banach spaces, there are bounded sequences that have no convergent subsequence. This lack of compactness in infinite-dimensional spaces is one of the main reasons for many difficulties in the functional analytical treatment of variational problems. To overcome these difficulties, new concepts of convergence (or new topologies) have been introduced with respect to which the unit ball is compact (respectively, sequentially compact).

Definition 2.27. *Let X be a Banach space. A sequence $(u_n) \subset X$ is called weakly convergent in X to an element $u \in X$ iff*

$$\langle f, u_n \rangle \to \langle f, u \rangle \quad \text{for all } f \in X^*.$$

The weak convergence is denoted by

$$u_n \rightharpoonup u \quad as \quad n \to \infty \quad or \quad \text{w--}\lim_{n \to \infty} u_n = u.$$

Note, in contrast to the weak convergence, we call the usual convergence with respect to the norm $(u_n \to u)$ sometimes the strong convergence. The following theorem provides a compactness result with respect to the topology introduced by the weak convergence.

Theorem 2.28 (Eberlein–Smulian Theorem). *Let X be a reflexive Banach space. Then, each bounded sequence $(u_n) \subset X$ has a weakly convergent subsequence.*

A few properties of weak convergence are summarized in the next proposition.

Proposition 2.29. *Let X be Banach spaces, and $(u_n) \subset X$.*

(i) $u_n \to u$ *implies* $u_n \rightharpoonup u$.

(ii) *If X is finite-dimensional, then strong and weak convergence are equivalent.*

(iii) *If $u_n \rightharpoonup u$, then (u_n) is bounded and*

$$\|u\| \le \liminf_{n \to \infty} \|u_n\|.$$

(iv) *If $u_n \rightharpoonup u$ in X and $f_n \to f$ in X^*, then it follows that*

$$\langle f_n, u_n \rangle \to \langle f, u \rangle.$$

(v) *If $u_n \to u$ in X and $f_n \rightharpoonup f$ in X^*, then it follows that*

$$\langle f_n, u_n \rangle \to \langle f, u \rangle.$$

The reverse of the Eberlein–Smulian Theorem is also true; i.e, a Banach space is reflexive if and only if every bounded sequence has a weakly convergent subsequence. Thus, the compactness result given by Theorem 2.28 is only valid in reflexive Banach spaces. To deal with nonreflexive Banach spaces, the following so-called weak* convergence has been introduced.

Definition 2.30. *Let X be a Banach space. A sequence $(f_n) \subset X^*$ is called weakly* convergent to an element $f \in X^*$ iff*

$$\langle f_n, u \rangle \to \langle f, u \rangle \quad \text{for all} \ \ u \in X.$$

The weak convergence is denoted by*

$$f_n \rightharpoonup^* f \ \ as \ \ n \to \infty, \quad or \ \ w^* - \lim_{n \to \infty} f_n = f.$$

Proposition 2.31. *Let X be a Banach space, and let (f_n) be a sequence in the dual space X^*.*

(i) $f_n \to f$ *in X^* implies* $f_n \rightharpoonup^* f$.

(ii) *If $f_n \rightharpoonup^* f$, then (f_n) is bounded in X^* and*

$$\|f\| \le \liminf_{n \to \infty} \|f_n\|.$$

(iii) *If $u_n \to u$ in X and $f_n \rightharpoonup^* f$ in X^*, then it follows that*

$$\langle f_n, u_n \rangle \to \langle f, u \rangle.$$

(iv) $f_n \rightharpoonup f$ in X^* implies $f_n \rightharpoonup^* f$.

(v) If X is reflexive, then $f_n \rightharpoonup^* f$ is equivalent to $f_n \rightharpoonup f$.

Definition 2.32. Let $A : X \to Y$ be a linear operator, where X and Y are Banach spaces. A is called *weakly sequentially continuous* iff

$$u_n \rightharpoonup u \quad implies \quad Au_n \rightharpoonup Au.$$

A is called *strongly continuous* iff

$$u_n \rightharpoonup u \quad implies \quad Au_n \to Au.$$

A few simple consequences are provided in the next proposition.

Proposition 2.33. Let $A : X \to Y$ be a linear operator, where X and Y are Banach spaces.

(i)   If A is continuous, then A is weakly sequentially continuous.
(ii) If A is compact, then A is strongly continuous.
(iii) If A is strongly continuous and X is reflexive, then A is compact.

2.1.3 Convex Analysis and Calculus in Banach Spaces

Let X be a normed linear space. A subset K of X is convex iff

$$u, v \in K \quad implies \quad tu + (1-t)v \in K \quad for \ all \ \ 0 \le t \le 1.$$

Theorem 2.34. Let H be a Hilbert space with inner product $(\cdot, \cdot)$, and let K be a nonempty, closed, and convex subset of H. Then to each $u \in H$, a uniquely defined $v \in K$ closest to u exists, that is,

$$v \in K : \quad \|u - v\| = \inf_{w \in K} \|u - w\|.$$

Equivalently, $v \in K$ is the uniquely defined solution of the variational inequality

$$v \in K : \quad (u - v, w - v) \le 0 \quad for \ all \ \ w \in K.$$

Consequences of Theorem 2.34 are the well-known Orthogonal Projection Theorem and the Riesz Representation Theorem of linear continuous functionals on Hilbert spaces. The latter implies that a Hilbert space H is isometrically isomorphic with its dual space H^*. A generalization of the Riesz Representation Theorem is the Lax–Milgram Theorem (see Sect. 2.3).

Important consequences of the Hahn–Banach Theorem are various separation theorems, such as the following one.

Theorem 2.35 (Separation Theorem). Let X be a normed linear space, and let $K \subset X$ be a closed and convex subset. If $u_0 \in X \setminus K$, then a linear continuous functional $f \in X^*$ and an $\alpha \in \mathbb{R}$ exists such that

$$\langle f, u \rangle \le \alpha \quad for \ all \ u \in K, \quad and \quad \langle f, u_0 \rangle > \alpha.$$

Definition 2.36. *A subset M of a normed linear space X is called weakly sequentially closed if the limit of every weakly convergent sequence $(u_n) \subset M$ belongs to M; i.e.,*

$$(u_n) \subset M \text{ and } u_n \rightharpoonup u \text{ implies } u \in M.$$

Simple examples show that, in general, closed sets of a normed linear space need not be weakly sequentially closed. However, by means of Theorem 2.35, one gets the following equivalence.

Proposition 2.37. *Let M be a convex subset of a normed linear space X. Then, M is closed if and only if M is weakly sequentially closed.*

Next we present some convexity and smoothness properties of the norm in Banach spaces that are important for proving existence results for abstract operator equations involving operators of monotone type (see Theorem 2.156 in Sect. 2.4.4).

Definition 2.38. *A Banach space X is called strictly convex if and only if*

$$\|tu + (1-t)v\| < 1 \quad \text{provided that } \|u\| = \|v\| = 1, \ u \neq v, \ \text{and } 0 < t < 1.$$

A Banach space X is called locally uniformly convex if and only if for each $\varepsilon \in (0, 2]$, and for each $u \in X$ with $\|u\| = 1$, a $\delta(\varepsilon, u) > 0$ exists such that for all v with $\|v\| = 1$ and $\|u - v\| \geq \varepsilon$, the following holds:

$$\frac{1}{2}\|u + v\| \leq 1 - \delta(\varepsilon, u).$$

A Banach space X is called uniformly convex if and only if X is locally uniformly convex and δ can be chosen to be independent of u.

Obviously we have the following implications:

uniformly convex $\implies$ locally uniformly convex $\implies$ strictly convex.

Example 2.39. Each Hilbert space is uniformly convex. This readily follows from the parallelogram identity

$$\left\|\frac{1}{2}(u-v)\right\|^2 + \left\|\frac{1}{2}(u+v)\right\|^2 = \frac{1}{2}(\|u\|^2 + \|v\|^2).$$

Example 2.40. Let $1 < p < \infty$ and $\Omega \subset \mathbb{R}^N$ be a domain; then from Clarkson's inequality (see Sect. 2.2.4), it follows that $L^p(\Omega)$ is uniformly convex. By using this result, one readily sees that the Sobolev spaces $W^{m,p}(\Omega)$ are uniformly convex too, for $1 < p < \infty$ and $m = 0, 1, \ldots$.

Furthermore, the following theorems hold.

Theorem 2.41 (Milman–Pettis Theorem). *Each uniformly convex Banach space is reflexive.*

Convexity properties of the norm are closely related with smoothness properties of the norm, i.e., the smoothness of the function $u \mapsto \|u\|$.

Theorem 2.42. *Let X be a reflexive Banach space. Then the following holds:*

(i) *If X^* is strictly convex, then the function $u \mapsto \|u\|$ is Gâteaux-differentiable on $X \setminus \{0\}$.*

(ii) *If X^* is locally uniformly convex, then the function $u \mapsto \|u\|$ is Fréchet-differentiable on $X \setminus \{0\}$.*

(iii) *(Troyanski) In every reflexive Banach space X, an equivalent norm can be introduced so that both X and X^* are locally uniformly convex.*

The notions of Gâteaux and Fréchet derivatives that occur in Theorem 2.42 are natural generalizations of the directional and total derivative of functions $f : \mathbb{R}^n \to \mathbb{R}^m$, respectively, to mappings between Banach spaces. In particular, in the calculus of variations, these notions allow us to generalize the classic criteria in the study of extrema for real-valued functions in $\mathbb{R}^n$ to real-valued functionals $F : D(F) \subset X \to \mathbb{R}$ defined on a subset of a Banach space X.

Definition 2.43 (Gâteaux Derivative). *Let X and Y be Banach spaces, and let $f : U \subset X \to Y$ be a map whose domain $D(f) = U$ is an open subset of X. The directional derivative of f at $u \in U$ in the direction $h \in X$ is given by*

$$\delta f(u; h) = \lim_{t \to 0} \frac{f(u + th) - f(u)}{t}$$

provided this limit exists. If $\delta f(u; h)$ exists for every $h \in X$, and if the mapping $D_G f(u) : X \to Y$ defined by

$$D_G f(u) h = \delta f(u; h)$$

is linear and continuous, then we say that f is Gâteaux-differentiable at u, and we call $D_G f(u)$ the Gâteaux derivative of f at u.

Definition 2.44 (Fréchet Derivative). *Let X and Y be Banach spaces, and let $f : U \subset X \to Y$, where the domain $D(f) = U$ is an open subset of X. Then f is called Fréchet-differentiable at $u \in U$ if and only if a linear and continuous mapping $A : X \to Y$ exists such that*

$$\lim_{\|h\| \to 0} \frac{\|f(u + h) - f(u) - Ah\|}{\|h\|} = 0$$

or equivalently

$$f(u + h) - f(u) = Ah + o(\|h\|), \quad (h \to 0).$$

If such a mapping A exists, then we call $D_F f(u) = A$ (or simply $f'(u) = A$) the Fréchet derivative of f at u.

Corollary 2.45. *Let X and Y be Banach spaces, and let $f : U \subset X \to Y$. Then the following relations between Gâteaux and Fréchet derivative hold:*

(i) *If f is Fréchet-differentiable at $u \in U$, then f is Gâteaux-differentiable at u.*

(ii) *If f is Gâteaux-differentiable in a neighborhood of u_0 and $D_G f$ is continuous at u_0, then f is Fréchet-differentiable at u_0 and $f'(u_0) = D_G f(u_0)$.*

Remark 2.46. If $f : U \subset X \to Y$ is Fréchet-differentiable in U and $f' : U \to L(X, Y)$ is continuous, then we write $f \in C^1(U; Y)$ or simply $f \in C^1(U)$ if $Y = \mathbb{R}$. In a similar way as for mappings from $\mathbb{R}^n$ into $\mathbb{R}^m$, one can prove chain rules for both the Fréchet and the Gâteaux derivative.

Example 2.47. Let $X = L^p(\Omega)$, where $1 < p < \infty$. We will compute the Gâteaux derivative of the pth power L^p-norm, i.e., of the function $f : X \to \mathbb{R}$ defined by

$$f(u) = \|u\|_{L^p(\Omega)}^p.$$

After elementary calculations, we get

$$D_G f(u)h = \delta f(u; h) = p \int_\Omega |u|^{p-2} u h \, dx$$

if we consider real-valued functions $u : \Omega \to \mathbb{R}$. In case the functions are complex-valued, we get

$$\delta f(u; h) = \frac{p}{2} \int_\Omega |u|^{p-2} (\bar{u} h + u \bar{h}) \, dx.$$

We introduce next the notions of convex and semicontinuous functions (or functionals).

Definition 2.48 (Semicontinuous, Convex Functionals). *Let X be a Banach space and $\phi : M \subset X \to [-\infty, \infty]$ with $M = D(\phi)$.*

(i) *The functional ϕ is called sequentially lower semicontinuous at $u \in M$ if and only if*

$$\phi(u) \leq \liminf_{n \to \infty} \phi(u_n) \tag{2.1}$$

holds for each sequence $(u_n) \subset M$ such that $u_n \to u$ as $n \to \infty$.

(ii) *The functional ϕ is called lower semicontinuous if and only if the set M_r is closed relative to M for all $r \in \mathbb{R}$, where*

$$M_r = \{u \in M : \phi(u) \leq r\}.$$

(iii) *The functional ϕ is called weak sequentially lower semicontinuous at $u \in M$ if and only if (2.1) holds for each weakly convergent sequence (u_n) to u, i.e., $u_n \rightharpoonup u$.*

(iv) *The functional ϕ is called sequentially upper semicontinuous (respectively, weak sequentially upper semicontinuous, upper semicontinuous) if and only if $-\phi$ is sequentially lower semicontinuous (respectively, weak sequentially lower semicontinuous, lower semicontinuous).*

(v) *The functional ϕ is called convex if and only if M is convex and*

$$\phi(tu + (1-t)v) \le t\phi(u) + (1-t)\phi(v), \quad 0 \le t \le 1, \tag{2.2}$$

for all $u, v \in M$ for which the right-hand side of (2.2) is meaningful; ϕ is called strictly convex if and only if for all t with $0 < t < 1$ and for all $u, v \in M$ with $u \ne v$ inequality (2.2) holds strictly; i.e., (2.2) holds with $\le$ replaced by $<$.

The following proposition provides the connection between the above notions.

Proposition 2.49. *Let X be a Banach space and $\phi : M \subset X \to [-\infty, \infty]$ with $M = D(\phi)$.*

(i) *ϕ is sequentially lower semicontinuous on M if and only if ϕ is lower semicontinuous on M.*

(ii) *Assume $u \in M$ with $\phi(u) \ne \pm\infty$. Then ϕ is sequentially lower semicontinuous at u if and only if, for each $\varepsilon > 0$, a $\delta(\varepsilon) > 0$ exists such that for all $v \in M$ with*

$$\|v - u\| < \delta(\varepsilon) \quad implies \quad \phi(u) < \phi(v) + \varepsilon.$$

(iii) *ϕ is continuous if and only if ϕ is both lower and upper semicontinuous.*

(iv) *If, in addition, M is closed and convex, and ϕ is convex, then lower semicontinuous, sequentially lower semicontinuous and weak sequentially lower semicontinuous are mutually equivalent.*

Let X be a Banach space. In what follows we consider only convex functionals $\phi : X \to \mathbb{R} \cup \{+\infty\}$; i.e., we do not allow "$-\infty$" as a value for the convex functional ϕ. The reason is that if $\phi(u_0) = -\infty$ at some point u_0 and if, in addition, ϕ is lower semicontinuous, then ϕ would be nowhere finite. This can readily be seen by the following arguments. Assume there is some $u \in X$ with $\phi(u) \in \mathbb{R}$. Then from the convexity we get for all $t \in (0, 1)$, $\phi(tu_0 + (1-t)u) = -\infty$. Taking the limit $t \to 0$, the lower semicontinuity yields $\phi(u) = -\infty$, a contradiction.

Definition 2.50. *Let X be a Banach space and $\phi : X \to \mathbb{R} \cup \{+\infty\}$ be a convex functional.*

(i) *The effective domain of ϕ is the set $\mathrm{dom}(\phi)$ defined by*

$$\mathrm{dom}(\phi) = \{u \in X : \phi(u) < +\infty\}.$$

(ii) *ϕ is said to be proper if $\mathrm{dom}(\phi) \ne \emptyset$.*

(iii) *The epigraph of ϕ, denoted by* epi(ϕ), *is given by*

$$\mathrm{epi}(\phi) = \{(u, \lambda) \in X \times \mathbb{R} : \phi(u) \leq \lambda\}.$$

We summarize some elementary properties of convex functionals as follows.

Corollary 2.51. *Let X be a Banach space, and let $\phi, \phi_i : X \to \mathbb{R} \cup \{+\infty\}$, $i = 1, 2$, be convex functionals. Then the following holds:*

(i)   *dom(ϕ) is convex.*
(ii) *If $\lambda \geq 0$, then $\lambda\phi$ is convex.*
(iii) *If ϕ_1 and ϕ_2 are convex, then $\phi_1 + \phi_2$ is convex.*
(iv) *ϕ is convex, proper, and lower semicontinuous if and only if epi(ϕ) is, respectively, convex, nonempty, and closed in $X \times \mathbb{R}$.*

Proposition 2.52. *Let X be a Banach space, and let $\phi : X \to \mathbb{R} \cup \{+\infty\}$ be a convex, proper, and lower semicontinuous functional. Then ϕ is locally Lipschitz on the interior of* dom(ϕ).

Theorem 2.53 (Weierstrass' Theorem). *Let X be a reflexive Banach space. If $\phi : X \to \mathbb{R} \cup \{+\infty\}$ is a convex, proper, and lower semicontinuous functional satisfying*

$$\lim_{\|u\| \to \infty} \phi(u) = +\infty,$$

then the problem

$$u \in X : \quad \phi(u) = \inf_{v \in X} \phi(v)$$

admits at least one solution.

The following notion of subgradient generalizes the classic concept of a derivative.

Definition 2.54 (Subdifferential). *Let X be a Banach space, and let $\phi : X \to \mathbb{R} \cup \{+\infty\}$ be a convex and proper functional. An element $u^* \in X^*$ is called a subgradient of ϕ at $u \in$ dom(ϕ) if and only if the following inequality holds:*

$$\phi(v) \geq \phi(u) + \langle u^*, v - u \rangle \quad \text{for all } v \in X. \tag{2.3}$$

The set of all $u^* \in X^*$ satisfying (2.3) is called the subdifferential of ϕ at $u \in$ dom(ϕ), and is denoted by $\partial\phi(u)$.

First properties of the subdifferential are given in the following proposition.

Proposition 2.55. *Let X be a Banach space, and let $\phi : X \to \mathbb{R} \cup \{+\infty\}$ be a convex and proper functional. Then we have the following properties of $\partial\phi$:*

(i) $\partial\phi(u)$ *is convex and weak*-closed.*
(ii) *If ϕ is continuous at $u \in \mathrm{dom}(\phi)$, then $\partial\phi(u)$ is nonempty, convex, bounded, and weak*-compact.*

Note, in (i) of Proposition 2.55 $\partial\phi(u) = \emptyset$ is possible.

Proposition 2.56. *Let X be a Banach space, and let $\phi : X \to \mathbb{R} \cup \{+\infty\}$ be a convex and proper functional. If ϕ is Gâteaux-differentiable at $u \in \mathrm{int}(\mathrm{dom}(\phi))$, then $\partial\phi(u) = \{D_G\phi(u)\}$. If ϕ is continuous at u and $\partial\phi(u)$ is a singleton, then ϕ is Gâteaux-differentiable at u.*

The following sum rule for the subdifferential is due to Moreau and Rockafellar.

Proposition 2.57 (Sum Rule). *Let X be a Banach space, and let ϕ_1, $\phi_2 : X \to \mathbb{R} \cup \{+\infty\}$ be convex functionals. If there is a point $u_0 \in \mathrm{dom}(\phi_1) \cap \mathrm{dom}(\phi_2)$ at which ϕ_1 is continuous, then the following holds:*

$$\partial(\phi_1 + \phi_2)(u) = \partial\phi_1(u) + \partial\phi_2(u) \quad \textit{for all } u \in X.$$

Example 2.58. Let $f : \mathbb{R} \to \mathbb{R}$ be a nondecreasing function with its one-sided limits $\underline{f}$ and $\bar{f}$. Define $\phi : \mathbb{R} \to \mathbb{R}$ by

$$\phi(x) = \int_{x_0}^{x} \underline{f}(s)\,ds = \int_{x_0}^{x} \bar{f}(s)\,ds.$$

Note that ϕ is convex and finite on $\mathbb{R}$, i.e., $\mathrm{dom}(\phi) = \mathbb{R}$, and thus ϕ is even locally Lipschitz. Elementary calculations show that the subdifferential is given by

$$\partial\phi(x) = [\underline{f}(x), \bar{f}(x)].$$

Example 2.59. Let $\phi : \mathbb{R} \to \mathbb{R} \cup \{+\infty\}$ be a convex, proper, lower semicontinuous function, and $\Omega \subset \mathbb{R}^N$ a Lebesgue-measurable set such that either $0 = \phi(0) = \min_{s \in \mathbb{R}} \phi(s)$ or the measurable set Ω has finite measure. Define $\Phi : L^p(\Omega) \to \mathbb{R} \cup \{+\infty\}$, $1 < p < \infty$, by

$$\Phi(u) = \int_{\Omega} \phi(u(x))\,dx \quad \text{if } \phi(u) \in L^1(\Omega), \quad +\infty \text{ otherwise.}$$

Then $\Phi : L^p(\Omega) \to \mathbb{R} \cup \{+\infty\}$ is convex, proper, lower semicontinuous, and $u^* \in \partial\Phi(u)$ if and only if

$$u^* \in L^q(\Omega), \quad \text{and} \quad u^*(x) \in \partial\phi(u(x)), \quad \text{for a.e. } x \in \Omega,$$

where q is the Hölder conjugate; i.e., $1/p + 1/q = 1$.

2.1.4 Partially Ordered Sets

Definition 2.60 (Partially Ordered Set). *Let P be a nonempty set. We say that a relation $x \leq y$ between certain pairs of elements of P is a partial ordering in P, and that $(P, \leq)$ is a partially ordered set, if "$\leq$" has the following properties:*

(i) *$x \leq x$ for all $x \in P$ (reflexivity).*
(ii) *If $x \leq y$ and $y \leq x$, then $x = y$ (antisymmetry).*
(iii) *If $x \leq y$ and $y \leq z$, then $x \leq z$ (transitivity).*

Note that $x < y$ stands for $x \leq y$ and $x \neq y$. Next we define several notions based on the partial ordering introduced above.

Definition 2.61. *Let $(P, \leq)$ be a partially ordered set.*

(i) *An element b of P is called an upper bound of a subset A of P if $x \leq b$ for each $x \in A$. If $b \in A$, we say that b is the greatest element of A. A lower bound of A and the smallest element of A are defined similarly, replacing $x \leq b$ above by $b \leq x$.*

(ii) *If the set of all upper bounds of A has the minimum, we call it a least upper bound of A and denote it by $\sup A$. The greatest lower bound, $\inf A$, of A is defined similarly.*

(iii) *An element $x \in A$ is called a maximal element of $A \subset P$, if there is no $y \neq x$ in A for which $x \leq y$. Similarly, a minimal element of A is defined. Obviously, every greatest element of A is a maximal element of A.*

(iv) *We say that a partially ordered set P is a lattice if $\inf\{x, y\}$ and $\sup\{x, y\}$ exist for all $x, y \in P$.*

(v) *A subset C of P is said to be upward directed if for each pair $x, y \in C$ there is a $z \in C$ such that $x \leq z$ and $y \leq z$, and C is downward directed if for each pair $x, y \in C$ there is a $w \in C$ such that $w \leq x$ and $w \leq y$. If C is both upward and downward directed, it is called directed.*

(vi) *A subset C of a partially ordered set P is called a chain if $x \leq y$ or $y \leq x$ for all $x, y \in C$.*

(vii) *We say that C is well ordered if each nonempty subset of C has a minimum, and inversely well ordered if each nonempty subset of C has a maximum. Obviously, each (inversely) well-ordered set is a chain and each chain is directed.*

Theorem 2.62 (Zorn's Lemma). *If in a partially ordered set P, every chain has an upper bound, then P possesses a maximal element.*

2.2 Sobolev Spaces

In this section, we summarize the main properties of Sobolev spaces. These properties include, e.g., the approximation of Sobolev functions by smooth functions (density theorems), continuity properties and compactness conditions (embedding theorems), the definition of the boundary values of Sobolev functions (trace theorem), and calculus for Sobolev functions (chain rule).

2.2.1 Spaces of Lebesgue Integrable Functions

Let $\mathbb{R}^N$, $N \geq 1$, be equipped with the Lebesgue measure, and let $\Omega \subset \mathbb{R}^N$ be a domain; i.e., Ω is an open and connected subset of $\mathbb{R}^N$. For $1 \leq p < \infty$, we denote by $L^p(\Omega)$ the Banach space of measurable functions $u : \Omega \to \mathbb{R}$ with respect to the norm

$$\|u\|_{L^p(\Omega)} = \left(\int_\Omega |u|^p dx \right)^{1/p} < \infty.$$

For a measurable function u, we put

$$\|u\|_{L^\infty(\Omega)} = \inf\{\alpha \in \mathbb{R} : \operatorname{meas}(\{x \in \Omega : |u(x)| > \alpha\}) = 0\}.$$

We denote by $L^\infty(\Omega)$ the Banach space of all measurable functions f satisfying $\|u\|_{L^\infty(\Omega)} < \infty$.

We also introduce the *local L^p-spaces*, denoted by $L^p_{\text{loc}}(\Omega)$. A function u belongs to $L^p_{\text{loc}}(\Omega)$ if it is measurable and

$$\int_K |u|^p dx < \infty$$

for every compact subset K of Ω.

The following main theorems can be found in standard textbooks on real analysis and measure theory (see [201, 114]).

Theorem 2.63 (Lebesgue's Dominated Convergence Theorem). *Suppose (u_n) is a sequence in $L^1(\Omega)$ such that*

$$u(x) = \lim_{n \to \infty} u_n(x)$$

exists almost everywhere (a.e.) on Ω. If there is a function $g \in L^1(\Omega)$ such that, for a.e. $x \in \Omega$, and for all $n = 1, 2, \ldots$,

$$|u_n(x)| \leq g(x)$$

then $u \in L^1(\Omega)$ and

$$\lim_{n \to \infty} \int_\Omega |u_n - u| dx = 0.$$

In some sense the following reverse statement of Theorem 2.63 holds.

Theorem 2.64. *Let u_n, $u \in L^1(\Omega)$, $n \in \mathbb{N}$, be such that*

$$\lim_{n \to \infty} \int_\Omega |u_n - u| \, dx = 0.$$

Then a subsequence (u_{n_k}) of (u_n) exists with

$$u_{n_k}(x) \to u(x) \quad \text{for a.e. } x \in \Omega.$$

Theorem 2.65 (Fatou's Lemma). *Let (u_n) be a sequence of measurable functions, and let $g \in L^1(\Omega)$. If*

$$u_n \geq g \quad \text{a.e. on } \Omega,$$

then we have

$$\int_\Omega \liminf_{n \to \infty} u_n \, dx \leq \liminf_{n \to \infty} \int_\Omega u_n \, dx.$$

If $\Omega \subset \mathbb{R}^N$ is a measurable subset, we denote its Lebesgue measure by

$$\text{meas}(\Omega) = |\Omega|.$$

Theorem 2.66 (Egorov's Theorem). *Let (u_n), u be measurable functions, and*

$$u_n \to u \quad \text{a.e. on } \Omega,$$

where $\Omega \subset \mathbb{R}^N$ is measurable with $|\Omega| < \infty$. Then for each $\varepsilon > 0$, a measurable subset $E \subset \Omega$ exists such that

(i) $|\Omega \setminus E| < \varepsilon$.
(ii) $u_n \to u$ *uniformly on E.*

A characterization of the dual spaces of $L^p(\Omega)$ is given in the next theorem.

Theorem 2.67 (Dual Space). *Let $\Omega \subset \mathbb{R}^N$ be a domain, and let Φ be a linear continuous functional on $L^p(\Omega)$, $1 < p < \infty$. Then a uniquely defined function $g \in L^q(\Omega)$ exists with q satisfying $1/p + 1/q = 1$ such that*

$$\langle \Phi, u \rangle = \int_\Omega g \, u \, dx \quad \text{for all } u \in L^p(\Omega)$$

and

$$\|\Phi\|_{(L^p(\Omega))^*} = \|g\|_{L^q(\Omega)}.$$

If Φ is a linear continuous functional on $L^1(\Omega)$, then a uniquely defined function $g \in L^\infty(\Omega)$ exists such that

$$\langle \Phi, u \rangle = \int_\Omega g \, u \, dx \quad \text{for all } u \in L^1(\Omega)$$

and

$$\|\Phi\|_{(L^1(\Omega))^*} = \|g\|_{L^\infty(\Omega)}.$$

In view of Theorem 2.67, the dual space of $L^p(\Omega)$ is isometrically isomorphic to $L^q(\Omega)$ for $1 \leq p < \infty$ with $q = \infty$ if $p = 1$.

We summarize some important properties of L^p-spaces in the following theorem.

Theorem 2.68. *Let $\Omega \subset \mathbb{R}^N$ be a domain.*

(i)   *For $1 \leq p < \infty$, the spaces $L^p(\Omega)$ are separable.*
(ii) *$L^\infty(\Omega)$ is not separable.*
(iii) *For $1 < p < \infty$, the spaces $L^p(\Omega)$ are reflexive.*
(iv) *$L^1(\Omega)$ and $L^\infty(\Omega)$ are not reflexive.*
(v)   *For $1 < p < \infty$, the spaces $L^p(\Omega)$ are uniformly convex.*

2.2.2 Definition of Sobolev Spaces

Let $\alpha = (\alpha_1, \ldots, \alpha_N)$ with nonnegative integers $\alpha_1, \ldots, \alpha_N$ be a multi-index, and denote its order by $|\alpha| = \alpha_1 + \cdots + \alpha_N$. Set $D_i = \partial/\partial x_i$, $i = 1, \ldots, N$, and $D^\alpha u = D_1^{\alpha_1} \cdots D_N^{\alpha_N} u$, with $D^0 u = u$. Let Ω be a domain in $\mathbb{R}^N$ with $N \geq 1$. Then $w \in L^1_{\text{loc}}(\Omega)$ is called the αth *weak* or *generalized derivative* of $u \in L^1_{\text{loc}}(\Omega)$ if and only if

$$\int_\Omega u D^\alpha \varphi \, dx = (-1)^{|\alpha|} \int_\Omega w\varphi \, dx, \quad \text{for all } \varphi \in C_0^\infty(\Omega),$$

holds, where $C_0^\infty(\Omega)$ denotes the space of infinitely differentiable functions with compact support in Ω. The generalized derivative w denoted by $w = D^\alpha u$ is unique up to a change of the values of w on a set of Lebesgue measure zero.

Definition 2.69. *Let $1 \leq p \leq \infty$ and $m = 0, 1, 2, \ldots$. The Sobolev space $W^{m,p}(\Omega)$ is the space of all functions $u \in L^p(\Omega)$, which have generalized derivatives up to order m such that $D^\alpha u \in L^p(\Omega)$ for all α: $|\alpha| \leq m$. For $m = 0$, we set $W^{0,p}(\Omega) = L^p(\Omega)$.*

With the corresponding norms given by

$$\|u\|_{W^{m,p}(\Omega)} = \left(\sum_{|\alpha| \leq m} \|D^\alpha u\|_{L^p(\Omega)}^p \right)^{1/p}, \quad 1 \leq p < \infty,$$

$$\|u\|_{W^{m,\infty}(\Omega)} = \max_{|\alpha| \leq m} \|D^\alpha u\|_{L^\infty(\Omega)},$$

$W^{m,p}(\Omega)$ becomes a Banach space.

Definition 2.70. *$W_0^{m,p}(\Omega)$ is the closure of $C_0^\infty(\Omega)$ in $W^{m,p}(\Omega)$.*

$W_0^{m,p}(\Omega)$ is a Banach space with the norm $\| \cdot \|_{W^{m,p}(\Omega)}$.

Before we summarize some basic properties of Sobolev spaces, we need to classify the regularity of boundaries.

Definition 2.71. *Let* $\Omega \subset \mathbb{R}^N$ *be a bounded domain, with boundary* $\partial\Omega$. *We say that the boundary* $\partial\Omega$ *is of class* $C^{k,\lambda}$, $k \in \mathbb{N}_0$, $\lambda \in (0,1]$, *if there are* $m \in \mathbb{N}$ *Cartesian coordinate systems* C_j, $j = 1, \ldots, m$,

$$C_j = (x_{j,1}, \ldots, x_{j,N-1}, x_{j,N}) = (x'_j, x_{j,N})$$

and real numbers α, $\beta > 0$, *as well as* m *functions* a_j *with*

$$a_j \in C^{k,\lambda}([-\alpha,\alpha]^{N-1}), \quad j = 1, \ldots, m,$$

such that the sets defined by

$$\Lambda^j = \{(x'_j, x_{j,N}) \in \mathbb{R}^N : |x'_j| \leq \alpha, \ x_{j,N} = a_j(x'_j)\},$$
$$V_+^j = \{(x'_j, x_{j,N}) \in \mathbb{R}^N : |x'_j| \leq \alpha, \ a_j(x'_j) < x_{j,N} < a_j(x'_j) + \beta\},$$
$$V_-^j = \{(x'_j, x_{j,N}) \in \mathbb{R}^N : |x'_j| \leq \alpha, \ a_j(x'_j) - \beta < x_{j,N} < a_j(x'_j)\},$$

possess the following properties:

$$\Lambda^j \subset \partial\Omega, \quad V_+^j \subset \Omega, \quad V_-^j \subset \mathbb{R}^N \setminus \Omega, \quad j = 1, \ldots, m,$$

and

$$\bigcup_{j=1}^m \Lambda^j = \partial\Omega.$$

Remark 2.72. If $\partial\Omega \in C^{0,1}$, then we call $\partial\Omega$ a *Lipschitz boundary*, which means that $\partial\Omega$ is locally the graph of a Lipschitz continuous function. In this case, the $(N-1)$-dimensional surface measure is well defined, on the basis of which $L^p(\partial\Omega)$-spaces can be introduced (see [66]). As Lipschitz continuous functions admit a.e. a gradient, the outer unit normal on $\partial\Omega$ exists for a.a. $x \in \partial\Omega$ (see [94]), which allows us to extend the integration by parts formula to Sobolev functions on Lipschitz domains.

Theorem 2.73. *Let* $\Omega \subset \mathbb{R}^N$ *be a bounded domain,* $N \geq 1$. *Then we have the following:*

(i) $W^{m,p}(\Omega)$ *is separable for* $1 \leq p < \infty$.
(ii) $W^{m,p}(\Omega)$ *is reflexive for* $1 < p < \infty$.
(iii) *Let* $1 \leq p < \infty$. *Then* $C^\infty(\Omega) \cap W^{m,p}(\Omega)$ *is dense in* $W^{m,p}(\Omega)$, *and if* $\partial\Omega$ *is a Lipschitz boundary, then* $C^\infty(\overline{\Omega})$ *is dense in* $W^{m,p}(\Omega)$, *where* $C^\infty(\Omega)$ *and* $C^\infty(\overline{\Omega})$ *are the spaces of infinitely differentiable functions in* Ω *and* $\overline{\Omega}$, *respectively (cf. [99]).*

As for the proofs of these properties we refer to [99].

Now we state some Sobolev embedding theorems. Let X, Y be two normed linear spaces with $X \subset Y$. We recall the operator $i : X \to Y$ defined by $i(u) = u$ for all $u \in X$ is called the embedding operator of X into Y. We say X is continuously (compactly) embedded in Y if $X \subset Y$ and the embedding operator $i : X \to Y$ is continuous (compact).

Theorem 2.74 (Sobolev Embedding Theorem). *Let $\Omega \subset \mathbb{R}^N$, $N \geq 1$, be a bounded domain with Lipschitz boundary $\partial\Omega$. Then the following holds:*

(i) *If $mp < N$, then the space $W^{m,p}(\Omega)$ is continuously embedded in $L^{p^*}(\Omega)$, $p^* = Np/(N - mp)$, and compactly embedded in $L^q(\Omega)$ for any q with $1 \leq q < p^*$.*

(ii) *If $0 \leq k < m - \frac{N}{p} < k + 1$, then the space $W^{m,p}(\Omega)$ is continuously embedded in $C^{k,\lambda}(\overline{\Omega})$, $\lambda = m - \frac{N}{p} - k$, and compactly embedded in $C^{k,\lambda'}(\overline{\Omega})$ for any $\lambda' < \lambda$.*

(iii) *Let $1 \leq p < \infty$, then the embeddings*

$$L^p(\Omega) \supset W^{1,p}(\Omega) \supset W^{2,p}(\Omega) \supset \cdots$$

are compact.

Here $C^{k,\lambda}(\overline{\Omega})$ denotes the *Hölder space*; cf. [99]. As for the proofs we refer to, e.g., [99, 222].

The proper definition of boundary values for Sobolev functions is based on the following theorem.

Theorem 2.75 (Trace Theorem). *Let $\Omega \subset \mathbb{R}^N$ be a bounded domain with Lipschitz ($C^{0,1}$) boundary $\partial\Omega$, $N \geq 1$, and $1 \leq p < \infty$. Then exactly one continuous linear operator exists*

$$\gamma : W^{1,p}(\Omega) \to L^p(\partial\Omega)$$

such that:

(i) *$\gamma(u) = u|_{\partial\Omega}$ if $u \in C^1(\overline{\Omega})$.*
(ii) *$\|\gamma(u)\|_{L^p(\partial\Omega)} \leq C \, \|u\|_{W^{1,p}(\Omega)}$ with C depending only on p and Ω.*
(iii) *If $u \in W^{1,p}(\Omega)$, then $\gamma(u) = 0$ in $L^p(\partial\Omega)$ if and only if $u \in W_0^{1,p}(\Omega)$.*

Definition 2.76 (Trace). *We call $\gamma(u)$ the trace (or generalized boundary function) of u on $\partial\Omega$.*

Remark 2.77. We note that the trace operator

$$\gamma : W^{1,p}(\Omega) \to L^p(\partial\Omega)$$

in Theorem 2.75 is not surjective; i.e., there are functions $\varphi \in L^p(\partial\Omega)$ that are not the traces of functions u from $W^{1,p}(\Omega)$. To describe precisely the range of the trace operator, Sobolev spaces of fractional order, usually referred to as Sobolev–Slobodeckij spaces, have to be taken into account (see [90, 132, 213, 219]). From [132, Theorem 6.8.13, Theorem 6.9.2], we obtain the following result.

Theorem 2.78. *Let $\Omega \subset \mathbb{R}^N$ be a bounded domain with Lipschitz boundary $\partial\Omega$, $N \geq 1$, and $1 < p < \infty$. Then*

$$\gamma(W^{1,p}(\Omega)) = W^{1-\frac{1}{p},p}(\partial\Omega).$$

The following compactness result of the trace operator holds (see [132]).

Theorem 2.79. *Let $\Omega \subset \mathbb{R}^N$ be a bounded domain with Lipschitz boundary $\partial\Omega$, $N \geq 1$.*

(i) *If $1 < p < N$, then*
$$\gamma : W^{1,p}(\Omega) \to L^q(\partial\Omega)$$

is completely continuous for any q with $1 \leq q < (Np - p)/(N - p)$.
(ii) *If $p \geq N$, then for any $q \geq 1$,*
$$\gamma : W^{1,p}(\Omega) \to L^q(\partial\Omega)$$

is completely continuous.

Sobolev–Slobodeckij spaces form a scale of continuous and even compact embeddings with respect to their fractional order of regularity. More precisely, we can deduce the following compact embedding result for the spaces $W^{l,2}(\Omega)$ with $l \in \mathbb{R}_+$ from [219, Theorem 7.9, Theorem 7.10].

Theorem 2.80. *Let $\Omega \subset \mathbb{R}^N$ be a bounded domain with Lipschitz boundary $\partial\Omega$, $N \geq 1$, and let $l_2 < l_1 \leq 1$, where $l_1, l_2 \in \mathbb{R}_+$. Then the embedding*
$$W^{l_1,2}(\Omega) \subset W^{l_2,2}(\Omega)$$

is compact.

If M is a $C^{k,\kappa}$-manifold ($C^{0,1}$ stands for Lipschitz-manifold) and $l_2 < l_1 < k + \kappa$ with $l_1, l_2 \in \mathbb{R}_+$ (for l_1 integer, $l_1 = k + \kappa$ is admissible), then the embedding
$$W^{l_1,2}(M) \subset W^{l_2,2}(M)$$

is compact.

In a similar way as for Sobolev spaces we have the following trace theorem, which can be deduced from [219, Theorem 8.7].

Theorem 2.81 (Trace Theorem). *Let $\Omega \subset \mathbb{R}^N$ be a bounded domain with Lipschitz boundary $\partial\Omega$, $N \geq 1$, and let $1/2 < l \leq 1$ with $l \in \mathbb{R}_+$. Then a uniquely defined continuous linear operator exists*
$$\gamma : W^{l,2}(\Omega) \to W^{l-1/2,2}(\partial\Omega)$$

such that
$$\gamma(u) = u|_{\partial\Omega} \quad if \ u \in C^1(\overline{\Omega}).$$

Theorem 2.80 and Theorem 2.81 hold likewise in the general case of the spaces $W^{l,p}(\Omega)$ with $l \in \mathbb{R}_+$, $1 < p < \infty$, and can be found, e.g., in [90, 132, 212, 213, 219].

The following extension result is useful in the study of unbounded domain problems.

Lemma 2.82. *Let $\Omega_0 \subset\subset \Omega$, that is, Ω_0 is compactly contained in Ω. Assume $g \in W^{1,p}(\Omega)$, $u \in W^{1,p}(\Omega_0)$, and $u - g \in W_0^{1,p}(\Omega_0)$, $1 \le p < \infty$. Then the function w defined by*

$$w(x) = \begin{cases} u(x) & \text{if} \quad x \in \Omega_0 \,, \\ g(x) & \text{if} \quad x \in \Omega \setminus \Omega_0 \end{cases}$$

is in $W^{1,p}(\Omega)$, and its generalized derivative $D_i w = \partial w / \partial x_i$, $i = 1, \dots, N$, is given by

$$D_i w(x) = \begin{cases} D_i u(x) & \text{if} \quad x \in \Omega_0 \,, \\ D_i g(x) & \text{if} \quad x \in \Omega \setminus \Omega_0 \,. \end{cases}$$

For the proof of Lemma 2.82, see [120, Lemma 20.14]. Its proof is based on the density property (iii) of Theorem 2.73 and the characterization of the traces of $W_0^{1,p}(\Omega)$ function.

2.2.3 Chain Rule and Lattice Structure

In this section, we assume that $\Omega \subset \mathbb{R}^N$ is a bounded domain with Lipschitz boundary $\partial \Omega$.

Lemma 2.83 (Chain Rule). *Let $f \in C^1(\mathbb{R})$ and $\sup_{s \in \mathbb{R}} |f'(s)| < \infty$. Let $1 \le p < \infty$ and $u \in W^{1,p}(\Omega)$. Then the composite function $f \circ u \in W^{1,p}(\Omega)$, and its generalized derivatives are given by*

$$D_i(f \circ u) = (f' \circ u) D_i u, \quad i = 1, \dots, N.$$

Lemma 2.84 (Generalized Chain Rule). *Let $f : \mathbb{R} \to \mathbb{R}$ be continuous and piecewise continuously differentiable with $\sup_{s \in \mathbb{R}} |f'(s)| < \infty$, and $u \in W^{1,p}(\Omega)$, $1 \le p < \infty$. Then $f \circ u \in W^{1,p}(\Omega)$, and its generalized derivative is given by*

$$D_i(f \circ u)(x) = \begin{cases} f'(u(x)) D_i u(x) & \text{if } f \text{ is differentiable at } u(x) \,, \\ 0 & \text{otherwise.} \end{cases}$$

The chain rule may further be extended to Lipschitz continuous f; see [99, 222].

Lemma 2.85 (Generalized Chain Rule). *Let $f : \mathbb{R} \to \mathbb{R}$ be a Lipschitz continuous function and $u \in W^{1,p}(\Omega)$, $1 \le p < \infty$. Then $f \circ u \in W^{1,p}(\Omega)$, and its generalized derivative is given by*

$$D_i(f \circ u)(x) = f_B(u(x)) D_i u(x) \quad \text{for a.e.} \quad x \in \Omega,$$

where $f_B : \mathbb{R} \to \mathbb{R}$ is a Borel-measurable function such that $f_B = f'$ a.e. in $\mathbb{R}$.

The generalized derivative of the following special functions are frequently used in later chapters.

Example 2.86. Let $1 \leq p < \infty$ and $u \in W^{1,p}(\Omega)$. Then $u^+ = \max\{u, 0\}$, $u^- = \max\{-u, 0\}$, and $|u|$ are in $W^{1,p}(\Omega)$, and their generalized derivatives are given by

$$(D_i u^+)(x) = \begin{cases} D_i u(x) & \text{if} \quad u(x) > 0, \\ 0 & \text{if} \quad u(x) \leq 0, \end{cases}$$

$$(D_i u^-)(x) = \begin{cases} 0 & \text{if} \quad u(x) \geq 0, \\ -D_i u(x) & \text{if} \quad u(x) < 0, \end{cases}$$

$$(D_i |u|)(x) = \begin{cases} D_i u(x) & \text{if} \quad u(x) > 0, \\ 0 & \text{if} \quad u(x) = 0, \\ -D_i u(x) & \text{if} \quad u(x) < 0. \end{cases}$$

As for the traces of u^+ and u^-, we have (cf. [66])

$$\gamma(u^+) = (\gamma(u))^+, \quad \gamma(u^-) = (\gamma(u))^-.$$

Lemma 2.87 (Lattice Structure). *Let u, $v \in W^{1,p}(\Omega)$, $1 \leq p < \infty$. Then $\max\{u, v\}$ and $\min\{u, v\}$ are in $W^{1,p}(\Omega)$ with generalized derivatives*

$$D_i \max\{u, v\}(x) = \begin{cases} D_i u(x) & \text{if} \quad u(x) > v(x), \\ D_i v(x) & \text{if} \quad v(x) \geq u(x), \end{cases}$$

$$D_i \min\{u, v\}(x) = \begin{cases} D_i u(x) & \text{if} \quad u(x) < v(x), \\ D_i v(x) & \text{if} \quad v(x) \leq u(x). \end{cases}$$

Proof: The assertion follows easily from the above examples and the generalized chain rule by using $\max\{u, v\} = (u-v)^+ + v$ and $\min\{u, v\} = u - (u-v)^+$; see [112, Theorem 1.20]. □

Lemma 2.88. *If (u_j), $(v_j) \subset W^{1,p}(\Omega)$ $(1 \leq p < \infty)$ are such that $u_j \to u$ and $v_j \to v$ in $W^{1,p}(\Omega)$, then $\min\{u_j, v_j\} \to \min\{u, v\}$ and $\max\{u_j, v_j\} \to \max\{u, v\}$ in $W^{1,p}(\Omega)$ as $j \to \infty$.*

For the proof, see [112, Lemma 1.22]. By means of Lemma 2.88, we readily obtain the following result.

Lemma 2.89. *Let $\underline{u}$, $\bar{u} \in W^{1,p}(\Omega)$ satisfy $\underline{u} \leq \bar{u}$, and let T be the truncation operator defined by*

$$Tu(x) = \begin{cases} \bar{u}(x) & \text{if} \quad u(x) > \bar{u}(x), \\ u(x) & \text{if} \quad \underline{u}(x) \leq u(x) \leq \bar{u}(x), \\ \underline{u}(x) & \text{if} \quad u(x) < \underline{u}(x). \end{cases}$$

Then T is a bounded continuous mapping from $W^{1,p}(\Omega)$ [respectively, $L^p(\Omega)$] into itself.

Proof: The truncation operator T can be represented in the form

$$Tu = \max\{u, \underline{u}\} + \min\{u, \bar{u}\} - u.$$

Thus, the assertion easily follows from Lemma 2.88.   □

Lemma 2.90 (Lattice Structure). *If* u, $v \in W_0^{1,p}(\Omega)$, *then* $\max\{u, v\}$ *and* $\min\{u, v\}$ *are in* $W_0^{1,p}(\Omega)$.

Lemma 2.90 implies that $W_0^{1,p}(\Omega)$ has a lattice structure as well; see [112].
   A partial ordering of traces on $\partial\Omega$ is given as follows.

Definition 2.91. *Let* $u \in W^{1,p}(\Omega)$, $1 \le p < \infty$. *Then* $u \le 0$ *on* $\partial\Omega$ *if* $u^+ \in W_0^{1,p}(\Omega)$.

2.2.4 Some Inequalities

In this section, we recall some well-known inequalities that are frequently used and that can be found in standard textbooks; see [93, 132, 222].

Young's Inequality

Let $1 < p, q < \infty$, and $1/p + 1/q = 1$. Then

$$ab \le \frac{a^p}{p} + \frac{b^q}{q} \quad (a, b \ge 0).$$

Proof: For a, $b \in \mathbb{R}_+$ satisfying $ab = 0$, the inequality is trivially satisfied. Let a, $b > 0$. As the function $x \mapsto e^x$ is convex, it follows that

$$ab = e^{\log a + \log b} = e^{\frac{1}{p}\log a^p + \frac{1}{q}\log b^q} \le \frac{1}{p}e^{\log a^p} + \frac{1}{q}e^{\log b^q} = \frac{a^p}{p} + \frac{b^q}{q}$$

□

Young's Inequality with Epsilon

Let $1 < p, q < \infty$, and $1/p + 1/q = 1$. Then

$$ab \le \varepsilon a^p + C(\varepsilon)b^q \quad (a, b \ge 0, \ \varepsilon > 0)$$

with $C(\varepsilon) = (\varepsilon p)^{-q/p}\frac{1}{q}$.

Proof: Again we only need to consider the case where a, $b > 0$. In this case, we set $ab = ((\varepsilon p)^{1/p}a)(\frac{b}{(\varepsilon p)^{1/p}})$ and apply Young's inequality.   □

Equivalent Norms

Let $1 \leq s < \infty$, and $\xi_i \in \mathbb{R}$, $\xi_i \geq 0$, $i = 1, \ldots, N$, then we have the following inequality:

$$a \left(\sum_{i=1}^{N} \xi_i^s \right)^{1/s} \leq \sum_{i=1}^{N} \xi_i \leq b \left(\sum_{i=1}^{N} \xi_i^s \right)^{1/s},$$

where a and b are some positive constants depending only on N and s.

Proof: The inequality is an immediate consequence of the fact that all norms in $\mathbb{R}^N$ are equivalent to each other. $\qquad \square$

Monotonicity Inequality

Let $1 < p < \infty$. Consider the vector-valued function $a : \mathbb{R}^N \to \mathbb{R}^N$ defined by

$$a(\xi) = |\xi|^{p-2} \xi \quad \text{for} \quad \xi \neq 0, \quad a(0) = 0.$$

If $1 < p < 2$, then we have

$$(a(\xi) - a(\xi')) \cdot (\xi - \xi') > 0 \quad \text{for all} \quad \xi, \xi' \in \mathbb{R}^N, \ \xi \neq \xi'.$$

If $2 \leq p < \infty$, then a constant $c > 0$ exists such that

$$(a(\xi) - a(\xi')) \cdot (\xi - \xi') \geq c |\xi - \xi'|^p \quad \text{for all} \quad \xi \in \mathbb{R}^N.$$

Hölder's Inequality

Let $1 \leq p, q \leq \infty$, $\frac{1}{p} + \frac{1}{q} = 1$. If $u \in L^p(\Omega)$, $v \in L^q(\Omega)$, then one has

$$\int_\Omega |uv| \, dx \leq \|u\|_{L^p(\Omega)} \|v\|_{L^q(\Omega)}.$$

Minkowski's Inequality

Let $1 \leq p \leq \infty$ and $u, v \in L^p(\Omega)$; then

$$\|u + v\|_{L^p(\Omega)} \leq \|u\|_{L^p(\Omega)} + \|v\|_{L^p(\Omega)}.$$

Clarkson's Inequalities

Let u, $v \in L^p(\Omega)$. If $2 \leq p < \infty$, then

$$\|u + v\|_{L^p(\Omega)}^p + \|u - v\|_{L^p(\Omega)}^p \leq 2^{p-1}\left(\|u\|_{L^p(\Omega)}^p + \|v\|_{L^p(\Omega)}^p\right).$$

If $1 < p < 2$, then

$$\|u + v\|_{L^p(\Omega)}^p + \|u - v\|_{L^p(\Omega)}^p \leq 2\left(\|u\|_{L^p(\Omega)}^p + \|v\|_{L^p(\Omega)}^p\right).$$

Proof: Use the function $\varphi : [0,1] \to \mathbb{R}$ defined by

$$\varphi(t) = \frac{(1+t)^p + (1-t)^p}{1+t^p}, \quad t \in [0,1].$$

$\square$

Remark 2.92. It follows immediately from Clarkson's inequalities that the spaces $L^p(\Omega)$ and the Sobolev spaces $W^{m,p}(\Omega)$ are uniformly convex for $1 < p < \infty$, and $m = 0, 1, \ldots,$.

Poincaré–Friedrichs Inequality

Let $\Omega \subset \mathbb{R}^N$ be a bounded domain, $1 \leq p < \infty$, and $u \in W_0^{1,p}(\Omega)$. Then we have the estimate

$$\|u\|_{L^p(\Omega)} \leq C \|\nabla u\|_{L^p(\Omega)},$$

where the constant C only depends on p, N, and Ω.

Remark 2.93. The Poincaré–Friedrichs inequality implies that

$$\|u\|_{W_0^{1,p}(\Omega)} = \|\nabla u\|_{L^p(\Omega)}$$

defines an equivalent norm on $W_0^{1,p}(\Omega)$. Equivalent norms on $W^{1,p}(\Omega)$ play an important role in the treatment of boundary value problems. The following general result provides a tool to identify equivalent norms on $W^{1,p}(\Omega)$.

Proposition 2.94. *Let $\Omega \subset \mathbb{R}^N$, $N \geq 1$, be a bounded domain with Lipschitz boundary $\partial\Omega$. Assume $\varphi : W^{1,p}(\Omega) \to \mathbb{R}_+$, $1 \leq p < \infty$, is a seminorm that satisfies the following conditions:*

(i) *A positive constant d exists such that*

$$\varphi(u) \leq d \|u\|_{W^{1,p}(\Omega)} \quad \text{for all} \ \ u \in W^{1,p}(\Omega).$$

(ii) *If $u = $ constant, then $\varphi(u) = 0$ implies $u = 0$.*

Then $\| \cdot \|_\sim$ *defined by*

$$\|u\|_\sim = \left(\|\nabla u\|_{L^p(\Omega)}^p + \varphi(u)^p \right)^{\frac{1}{p}}$$

defines an equivalent norm in $W^{1,p}(\Omega)$.

As an application of Proposition 2.94, we obtain, e.g., an equivalent norm on the closed subspace V_Γ of $W^{1,p}(\Omega)$ defined by

$$V_\Gamma = \{u \in W^{1,p}(\Omega) : \gamma(u) = 0 \ \text{on} \ \Gamma\},$$

where $\Gamma \subset \partial\Omega$ is some part of the boundary $\partial\Omega$ with strictly positive surface measure $|\Gamma| > 0$. To this end, define φ by

$$\varphi(u) = \left(\int_\Gamma |\gamma(u)|^p \, d\Gamma \right)^{\frac{1}{p}} \quad \text{for all} \ \ u \in W^{1,p}(\Omega),$$

where γ is the trace operator. We observe that (i) and (ii) of Proposition 2.94 are satisfied, and thus $\| \cdot \|_\sim$ defined above gives an equivalent norm on $W^{1,p}(\Omega)$. As $\varphi(u) = 0$ for $u \in V_\Gamma$, we see that

$$\|u\|_\sim = \|\nabla u\|_{L^p(\Omega)} \quad \text{for all} \ \ u \in V_\Gamma$$

is an equivalent norm on the subspace V_Γ.

2.3 Operators of Monotone Type

In this section, we provide the basic results on pseudomonotone operators from a Banach space X into its dual space X^*.

2.3.1 Main Theorem on Pseudomonotone Operators

Let X be a real, reflexive Banach space with norm $\| \cdot \|$, X^* its dual space, and denote by $\langle \cdot, \cdot \rangle$ the duality pairing between them. The norm convergence in X and X^* is denoted by "$\rightarrow$" and the weak convergence by "$\rightharpoonup$".

Definition 2.95. *Let* $A : X \rightarrow X^*$; *then* A *is called*

(i)   *continuous (respectively, weakly continuous) iff* $u_n \rightarrow u$ *implies* $Au_n \rightarrow Au$ *(respectively,* $u_n \rightharpoonup u$ *implies* $Au_n \rightharpoonup Au$)
(ii) *demicontinuous iff* $u_n \rightarrow u$ *implies* $Au_n \rightharpoonup Au$
(iii) *hemicontinuous iff the real function* $t \rightarrow \langle A(u+tv), w \rangle$ *is continuous on* $[0,1]$ *for all* $u, v, w \in X$
(iv) *strongly continuous or completely continuous iff* $u_n \rightharpoonup u$ *implies* $Au_n \rightarrow Au$

(v)   *bounded iff A maps bounded sets into bounded sets*
(vi) *coercive iff* $\lim_{\|u\|\to\infty} \frac{\langle Au,u\rangle}{\|u\|} = +\infty$

Definition 2.96 (Operators of Monotone Type). *Let $A : X \to X^*$; then A is called*

(i)   *monotone (respectively, strictly monotone) iff $\langle Au - Av, u - v\rangle \geq$ (respectively, $>$)0 for all $u, v \in X$ with $u \neq v$*
(ii) *strongly monotone iff there is a constant $c > 0$ such that $\langle Au - Av, u - v\rangle \geq c\|u - v\|^2$ for all $u, v \in X$*
(iii) *uniformly monotone iff $\langle Au - Av, u - v\rangle \geq a(\|u - v\|)\|u - v\|$ for all $u, v \in X$ where $a : [0, \infty) \to [0, \infty)$ is strictly increasing with $a(0) = 0$ and $a(s) \to +\infty$ as $s \to \infty$*
(iv) *pseudomonotone iff $u_n \rightharpoonup u$ and $\limsup_{n\to\infty}\langle Au_n, u_n - u\rangle \leq 0$ implies $\langle Au, u - w\rangle \leq \liminf_{n\to\infty}\langle Au_n, u_n - w\rangle$ for all $w \in X$*
(v)   *to satisfy (S_+)-condition iff $u_n \rightharpoonup u$ and $\limsup_{n\to\infty}\langle Au_n, u_n - u\rangle \leq 0$ imply $u_n \to u$*

We can show (cf. [18]) that the pseudomonotonicity according to (iv) of Definition 2.96 is equivalent to the following definition.

Definition 2.97. *The operator $A : X \to X^*$ is pseudomonotone iff $u_n \rightharpoonup u$ and $\limsup_{n\to\infty}\langle Au_n, u_n - u\rangle \leq 0$ implies $Au_n \rightharpoonup Au$ and $\langle Au_n, u_n\rangle \to \langle Au, u\rangle$.*

For the following result, see [222, Proposition 27.6].

Lemma 2.98. *Let $A, B : X \to X^*$ be operators on the real reflexive Banach space X. Then the following implications hold:*

(i)   *If A is monotone and hemicontinuous, then A is pseudomonotone.*
(ii) *If A is strongly continuous, then A is pseudomonotone.*
(iii) *If A and B are pseudomonotone, then $A + B$ is pseudomonotone.*

The main theorem on pseudomonotone operators due to Brézis is given by the next theorem (see [222, Theorem 27.A]).

Theorem 2.99 (Main Theorem on Pseudomonotone Operators). *Let X be a real, reflexive Banach space, and let $A : X \to X^*$ be a pseudomonotone, bounded, and coercive operator, and $b \in X^*$. Then a solution of the equation $Au = b$ exists.*

Remark 2.100. Theorem 2.99 contains several important surjectivity results as special cases, such as Lax–Milgram's theorem and the Main Theorem on Monotone Operators, which will be formulated in the following corollaries.

Corollary 2.101 (Main Theorem on Monotone Operators). *Let X be a real, reflexive Banach space, and let $A : X \to X^*$ be a monotone, hemicontinuous, bounded, and coercive operator, and $b \in X^*$. Then a solution of the equation $Au = b$ exists.*

For the proof of Corollary 2.101, we have only to mention that in view of Lemma 2.98, a monotone and hemicontinuous operator is pseudomonotone.

Corollary 2.102 (Lax–Milgram's Theorem). *Let X be a real Hilbert space, and let $a : X \times X \to \mathbb{R}$ be a bilinear form. Assume that*

(i) *a is bounded; i.e., there is a $C > 0$ such that*

$$|a(x, y)| \le C\|x\|\|y\| \quad \text{for } x, y \in X.$$

(ii) *a is coercive, i.e., there is a $C_0 > 0$ such that*

$$a(x, x) \ge C_0 \|x\|^2 \quad \text{for } x \in H.$$

Then, for each f in X^*, there is a unique element u in X such that

$$a(u, v) = \langle f, v \rangle \quad \text{for } v \in X.$$

The mapping $f \mapsto u$ is one-to-one, continuous, and linear from X^* onto X.

As for the proof, note that the bilinear form a of Corollary 2.102 defines a linear, bounded, and strongly monotone operator $A : X \to X^*$ acccording to

$$\langle Au, v \rangle = a(u, v) \quad \text{for all } u, v \in X,$$

and thus the equation $a(u, v) = \langle f, v \rangle$ of Corollary 2.102 is equivalent with the operator equation $Au = f$ in X^*. The existence result for the latter follows immediately from Corollary 2.101, because A is strongly monotone and continuous and therefore, in particular, also coercive. The uniqueness is a consequence of the strong monotonicity of A.

2.3.2 Leray–Lions Operators

An important class of operators of monotone type is the so-called Leray–Lions operators (see [215, 152]). These kinds of operators occur in the functional analytical treatment of nonlinear elliptic and parabolic problems.

Definition 2.103 (Leray–Lions Operator). *Let X be a real, reflexive Banach space. We say that $A : X \to X^*$ is a Leray–Lions operator if it is bounded and satisfies*

$$Au = \mathcal{A}(u, u), \quad \text{for } u \in X,$$

where $\mathcal{A} : X \times X \to X^*$ has the following properties:

(i)   *For any $u \in X$, the mapping $v \mapsto A(u, v)$ is bounded and hemicontinuous from X to its dual X^*, with*

$$\langle A(u, u) - A(u, v), u - v \rangle \geq 0 \quad \text{for } v \in X.$$

(ii) *For any $v \in X$, the mapping $u \mapsto A(u, v)$ is bounded and hemicontinuous from X to its dual X^*.*

(iii) *For any $v \in X$, $A(u_n, v)$ converges weakly to $A(u, v)$ in X^* if $(u_n) \subset X$ is such that $u_n \rightharpoonup u$ in X and*

$$\langle A(u_n, u_n) - A(u_n, u), u_n - u \rangle \to 0.$$

(iv) *For any $v \in X$, $\langle A(u_n, v), u_n \rangle$ converges to $\langle F, u \rangle$ if $(u_n) \subset V$ is such that $u_n \rightharpoonup u$ in X, and $A(u_n, v) \rightharpoonup F$ in X^*.*

As for the proof of the next theorem, see [215].

Theorem 2.104. *Every Leray–Lions operator $A : X \to X^*$ is pseudomonotone.*

Next we will see that quasilinear elliptic operators satisfying certain structure and growth conditions represent Leray–Lions operators. To this end, we need to study first the mapping properties of superposition operators, which are also called Nemytskij operators.

Definition 2.105 (Nemytskij Operator). *Let $\Omega \subset \mathbb{R}^N$, $N \geq 1$, be a nonempty measurable set, and let $f : \Omega \times \mathbb{R}^m \to \mathbb{R}$, $m \geq 1$, and $u : \Omega \to \mathbb{R}^m$ be a given function. Then the superposition or Nemytskij operator F assigns $u \mapsto f \circ u$; i.e., F is given by*

$$Fu(x) = (f \circ u)(x) = f(x, u(x)) \quad \text{for } x \in \Omega.$$

Definition 2.106 (Carathéodory Function). *Let $\Omega \subset \mathbb{R}^N$, $N \geq 1$, be a nonempty measurable set, and let $f : \Omega \times \mathbb{R}^m \to \mathbb{R}$, $m \geq 1$. The function f is called a Carathéodory function if the following two conditions are satisfied:*

(i)   *$x \mapsto f(x, s)$ is measurable in Ω for all $s \in \mathbb{R}^m$.*
(ii) *$s \mapsto f(x, s)$ is continuous on $\mathbb{R}^m$ for a.e. $x \in \Omega$.*

Lemma 2.107. *Let $f : \Omega \times \mathbb{R}^m \to \mathbb{R}$, $m \geq 1$, be a Carathéodory function that satisfies a growth condition of the form*

$$|f(x, s)| \leq k(x) + c \sum_{i=1}^{m} |s_i|^{p_i/q}, \quad \forall\, s = (s_1, \ldots, s_m) \in \mathbb{R}^m, \ a.e. \ x \in \Omega,$$

for some positive constant c and some $k \in L^q(\Omega)$, and $1 \leq q, p_i < \infty$ for all $i = 1, \ldots, m$. Then the Nemytskij operator F defined by

$$Fu(x) = f(x, u_1(x), \ldots, u_m(x))$$

is continuous and bounded from $L^{p_1}(\Omega) \times \cdots \times L^{p_m}(\Omega)$ into $L^q(\Omega)$. Here u denotes the vector function $u = (u_1, \ldots, u_m)$. Furthermore,

$$\|Fu\|_{L^q(\Omega)} \leq c \left(\|k\|_{L^q(\Omega)} + \sum_{i=1}^{m} \|u_i\|_{L^{p_i}(\Omega)}^{p_i/q} \right).$$

Definition 2.108. Let $\Omega \subset \mathbb{R}^N$, $N \geq 1$, be a nonempty measurable set. A function $f : \Omega \times \mathbb{R}^m \to \mathbb{R}$, $m \geq 1$, is called superpositionally measurable (or sup-measurable) if the function $x \mapsto Fu(x)$ is measurable in Ω whenever the component functions $u_i : \Omega \to \mathbb{R}$ of $u = (u_1, \ldots, u_m)$ are measurable.

Now let $\Omega \subset \mathbb{R}^N$ be a bounded domain with Lipschitz boundary $\partial\Omega$, let A_1 be the second-order quasilinear differential operator in divergence form given by

$$A_1 u(x) = -\sum_{i=1}^{N} \frac{\partial}{\partial x_i} a_i(x, u(x), \nabla u(x)),$$

and let A_0 denote the operator

$$A_0 u(x) = a_0(x, u(x), \nabla u(x)).$$

Let $1 < p < \infty$, $1/p + 1/q = 1$, and assume for the coefficients $a_i : \Omega \times \mathbb{R} \times \mathbb{R}^N \to \mathbb{R}$, $i = 0, 1, \ldots, N$ the following conditions.

(H1) Carathéodory and Growth Condition: Each $a_i(x, s, \xi)$ satisfies Carathéodory conditions, i.e., is measurable in $x \in \Omega$ for all $(s, \xi) \in \mathbb{R} \times \mathbb{R}^N$ and continuous in (s, ξ) for a.e. $x \in \Omega$. A constant $c_0 > 0$ and a function $k_0 \in L^q(\Omega)$ exist so that

$$|a_i(x, s, \xi)| \leq k_0(x) + c_0(|s|^{p-1} + |\xi|^{p-1})$$

for a.e. $x \in \Omega$ and for all $(s, \xi) \in \mathbb{R} \times \mathbb{R}^N$, with $|\xi|$ denoting the Euclidian norm of the vector ξ.

(H2) Monotonicity Type Condition: The coefficients a_i satisfy a monotonicity condition with respect to ξ in the form

$$\sum_{i=1}^{N} (a_i(x, s, \xi) - a_i(x, s, \xi'))(\xi_i - \xi_i') > 0$$

for a.e. $x \in \Omega$, for all $s \in \mathbb{R}$, and for all $\xi, \xi' \in \mathbb{R}^N$ with $\xi \neq \xi'$.

(H3) Coercivity Type Condition:

$$\sum_{i=1}^{N} a_i(x, s, \xi)\xi_i \geq \nu|\xi|^p - k(x)$$

for a.e. $x \in \Omega$, for all $s \in \mathbb{R}$, and for all $\xi \in \mathbb{R}^N$ with some constant $\nu > 0$ and some function $k \in L^1(\Omega)$.

Let V be a closed subspace of $W^{1,p}(\Omega)$ such that $W_0^{1,p}(\Omega) \subset V \subset W^{1,p}(\Omega)$, then under condition (H1) the differential operators A_1 and A_0 generate mappings from V into its dual space (again denoted by A_1 and A_0, respectively) defined by

$$\langle A_1 u, \varphi \rangle = \sum_{i=1}^N \int_\Omega a_i(x, u, \nabla u) \frac{\partial \varphi}{\partial x_i} \, dx \,, \quad \langle A_0 u, \varphi \rangle = \int_\Omega a_0(x, u, \nabla u) \, \varphi \, dx \,.$$

Theorem 2.109. *Set $A = A_1 + A_0$. Then the operators A, A_0, and A_1 have the following properties:*

(i)   *If (H1) is satisfied, then the mappings $A, A_1, A_0 : V \to V^*$ are continuous and bounded.*
(ii) *If (H1) and (H2) are satisfied, then $A : V \to V^*$ is pseudomonotone.*
(iii) *If (H1), (H2), and (H3) are satisfied, then A has the (S_+)-property.*

Conditions (H1) and (H2) are the so-called Leray–Lions conditions that guarantee that A is pseudomonotone. In their original paper, Leray and Lions [149] showed the pseudomonotonicity under conditions (H1), (H2), and the following additional condition.

(H4) $\limsup_{|\xi| \to \infty, \, s \in B} \sum_{i=1}^N \frac{a_i(x,s,\xi)\xi_i}{|\xi| + |\xi|^{p-1}} = +\infty$, for a.e. $x \in \Omega$ and all bounded sets B.

However, Landes and Mustonen have shown in [136] that condition (H4) is redundant for the pseudomonotonicity of A. As for the proof of the results stated in Theorem 2.109 as well as on existence theorems involving pseudomonotone operators, we refer to [17, 18] and [23, 27, 105, 152, 208, 222].

Example 2.110. Let $\Omega \subset \mathbb{R}^N$ be a bounded domain. A prototype of a monotone elliptic operator in Ω is the negative of the p-Laplacian Δ_p, $1 < p < \infty$, defined by

$$\Delta_p u = \text{div}(|\nabla u|^{p-2}\nabla u) \quad \text{where} \quad \nabla u = (\partial u/\partial x_1, \ldots, \partial u/\partial x_N).$$

This operator coincides with the Laplacian Δ if $p = 2$, and is of the form A_1 with the coeffients a_i, $i = 1, \ldots, N$, given by

$$a_i(x, s, \xi) = |\xi|^{p-2}\xi_i.$$

Thus, hypothesis (H1) is satisfied with $k_0 = 0$, $c_0 = 1$, and $a_0 = 0$. Hypothesis (H2) follows from the inequalities satisfied by the vector-valued function $\xi \mapsto |\xi|^{p-2}\xi$, (see Sect. 2.2.4) and (H3) is obviously true with $\nu = 1$ and $k = 0$ due to

$$\sum_{i=1}^N a_i(x, s, \xi)\xi_i = \sum_{i=1}^N |\xi|^{p-2}\xi_i\,\xi_i = |\xi|^p.$$

Therefore, hypotheses (H1)–(H3) are satisfied by the negative p-Laplacian, and in view of Theorem 2.109, we see that $-\Delta_p : V \to V^*$ is continuous, bounded, pseudomonotone, and has the (S_+)-property. Moreover, from the inequality

$$\langle -\Delta_p u - (-\Delta_p v), u - v \rangle = \int_\Omega (|\nabla u|^{p-2}\nabla u - |\nabla v|^{p-2}\nabla v)(\nabla u - \nabla v)\, dx \geq 0,$$

for all $u, v \in V$, we infer that $-\Delta_p : V \to V^*$ is, in particular, also a monotone operator. Depending on the domain of definition of $-\Delta_p$, we can say even more. For example, let $V = W_0^{1,p}(\Omega)$. According to Sect. 2.2.4,

$$\|u\|_V = \left(\int_\Omega |\nabla u|^p \, dx \right)^{1/p}$$

defines an equivalent norm in V. From the inequalities for the function $\xi \mapsto |\xi|^{p-2}\xi$, we see that the operator $-\Delta_p : W_0^{1,p}(\Omega) \to (W_0^{1,p}(\Omega))^*$ has the mapping properties given in the following lemma.

Lemma 2.111. *Let V be a closed subspace of $W^{1,p}(\Omega)$ such that $W_0^{1,p}(\Omega) \subset V \subset W^{1,p}(\Omega)$. Then one has:*

(i) $-\Delta_p : V \to V^*$ *is continuous, bounded, pseudomonotone, and has the (S_+)-property.*
(ii) $-\Delta_p : W_0^{1,p}(\Omega) \to (W_0^{1,p}(\Omega))^*$ *is*
    (a) *strictly monotone if $1 < p < \infty$.*
    (b) *strongly monotone if $p = 2$ (Laplacian).*
    (c) *uniformly monotone if $2 < p < \infty$.*

2.3.3 Multivalued Pseudomonotone Operators

In this section, we briefly recall the main results of the theory of pseudomonotone multivalued operators developed by Browder and Hess to the extent it will be needed in the study of variational and hemivariational inequalities. For the proofs and a more detailed presentation, we refer to the monographs [222, 177].

First we present basic results about the continuity of multivalued functions (multifunctions) and provide useful equivalent descriptions of these notions. Even though these notions can be defined in a much more general context, we confine ourselves to mappings between Banach spaces, which is sufficient for our purpose.

Definition 2.112 (Semicontinuous Multifunctions). *Let X, Y be Banach spaces and $A : X \to 2^Y$ be a multifunction.*

(i) *A is called upper semicontinuous at x_0, if for every open subset $V \subset Y$ with $A(x_0) \subset V$, a neighborhood $U(x_0)$ exists such that $A(U(x_0)) \subset V$. If A is upper semicontinuous at every $x_0 \in X$, we call A upper semicontinuous in X.*

(ii) *A is called lower semicontinuous at x_0 if for every neighborhood $V(y)$ of every $y \in A(x_0)$, a neighborhood $U(x_0)$ exists such that*

$$A(u) \cap V(y) \neq \emptyset \quad \text{for all } u \in U(x_0).$$

If A is lower semicontinuous at every $x_0 \in X$, we call A lower semicontinuous in X.

(iii) *A is called continuous at x_0 if A is both upper and lower semicontinuous at x_0. If A is continuous at every $x_0 \in X$, we call A continuous in X.*

Alternative equivalent continuity criteria are given in the following propositions. To this end, we introduce the *preimage* of a multifunction.

Definition 2.113 (Preimage). *Let $M \subset Y$ and $A : X \to 2^Y$ be a multifunction. The preimage $A^{-1}(M)$ is defined by*

$$A^{-1}(M) = \{x \in X : A(x) \cap M \neq \emptyset\}.$$

Proposition 2.114. *Let X, Y be Banach spaces and $A : X \to 2^Y$ be a multifunction. Then the following statements are equivalent:*

(i) *A is upper semicontinuous.*
(ii) *For all closed sets $C \subset Y$, the preimage $A^{-1}(C)$ is closed.*
(iii) *If $x \in X$, (x_n) is a sequence in X with $x_n \to x$ as $n \to \infty$, and V is an open set in Y such that $A(x) \subset V$, then $n_0 \in \mathbb{N}$ exists depending on V such that for all $n \geq n_0$, we have $A(x_n) \subset V$.*

Proposition 2.115. *Let X, Y be Banach spaces and $A : X \to 2^Y$ be a multifunction. Then the following statements are equivalent:*

(i) *A is lower semicontinuous.*
(ii) *For all open sets $O \subset Y$, the preimage $A^{-1}(O)$ is open.*
(iii) *If $x \in X$, (x_n) is a sequence in X with $x_n \to x$ as $n \to \infty$, and $y \in A(x)$, then for every $n \in \mathbb{N}$, we can find a $y_n \in A(x_n)$, such that $y_n \to y$, as $n \to \infty$.*

Remark 2.116. For a single-valued operator $A : X \to Y$, upper semicontinuous and lower semicontinuous in the multivalued setting is identical with continuous. For $A : M \to 2^N$ having the same corresponding properties, where M and N are subsets of the Banach spaces X and Y, respectively, then M and N have to be equipped with the induced topology.

Next we introduce the notion of multivalued monotone and pseudomonotone operators from a real, reflexive Banach space X into its dual space and formulate the main surjectivity result for these kinds of operators.

Definition 2.117 (Graph). *Let X be a real Banach space, and let $A : X \to 2^{X^*}$ be a multivalued mapping; i.e., to each $u \in X$, there is assigned a subset*

$A(u)$ of X^*, which may be empty if $u \notin D(A)$, where $D(A)$ is the domain of A given by

$$D(A) = \{u \in X : A(u) \neq \emptyset\}.$$

The graph of A denoted by $\mathrm{Gr}(A)$ is given by

$$\mathrm{Gr}(A) = \{(u, u^*) \in X \times X^* : u^* \in A(u)\}.$$

Definition 2.118 (Monotone Operator). *The mapping* $A : X \to 2^{X^*}$ *is called*

(i) *monotone iff*

$$\langle u^* - v^*, u - v \rangle \geq 0 \quad \text{for all } (u, u^*), (v, v^*) \in \mathrm{Gr}(A)$$

(ii) *strictly monotone iff*

$$\langle u^* - v^*, u - v \rangle > 0 \quad \text{for all } (u, u^*), (v, v^*) \in \mathrm{Gr}(A), \ u \neq v$$

(iii) *maximal monotone iff A is monotone and there is no monotone mapping $\tilde{A} : X \to 2^{X^*}$ such that $\mathrm{Gr}(A)$ is a proper subset of $\mathrm{Gr}(\tilde{A})$, which is equivalent to the following implication:*

$$(u, u^*) \in X \times X^* : \quad \langle u^* - v^*, u - v \rangle \geq 0 \ \text{ for all } (v, v^*) \in \mathrm{Gr}(A)$$

implies $(u, u^*) \in \mathrm{Gr}(A)$

The notions of strongly and uniformly monotone multivalued operators are defined in a similar way as for single-valued operators.

Example 2.119. If $X = \mathbb{R}$, then a maximal monotone mapping $\beta : \mathbb{R} \to 2^{\mathbb{R}}$ is called *maximal monotone graph* in $\mathbb{R}^2$. For example, an increasing function $f : \mathbb{R} \to \mathbb{R}$ generates a maximal monotone graph β in $\mathbb{R}^2$ given by

$$\beta(s) := [f(s - 0), f(s + 0)],$$

where $f(s \pm 0)$ are the one-sided limits of f in s.

A single-valued operator

$$A : D(A) \subset X \to X^*$$

is to be understood as a multivalued operator $A : X \to X^*$ by setting $Au = \{Au\}$ if $u \in D(A)$ and $Au = \emptyset$ otherwise. Thus, A is monotone iff

$$\langle Au - Av, u - v \rangle \geq 0 \quad \text{for all } u, v \in D(A),$$

and $A : D(A) \subset X \to X^*$ is maximal monotone iff A is monotone and the condition

$$(u, u^*) \in X \times X^* : \quad \langle u^* - Av, u - v \rangle \geq 0 \ \text{ for all } v \in D(A)$$

implies $u \in D(A)$ and $u^* = Au$.

Definition 2.120 (Pseudomonotone Operator). *Let X be a real reflexive Banach space. The operator $A : X \to 2^{X^*}$ is called pseudomonotone if the following conditions hold:*

(i)   *The set $A(u)$ is nonempty, bounded, closed, and convex for all $u \in X$.*
(ii) *A is upper semicontinuous from each finite-dimensional subspace of X to the weak topology on X^*.*
(iii) *If $(u_n) \subset X$ with $u_n \rightharpoonup u$, and if $u_n^* \in A(u_n)$ is such that*

$$\limsup \langle u_n^*, u_n - u \rangle \leq 0,$$

then to each element $v \in X$, $u^*(v) \in A(u)$ exists with

$$\liminf \langle u_n^*, u_n - v \rangle \geq \langle u^*(v), u - v \rangle.$$

Definition 2.121 (Generalized Pseudomonotone Operator). *Let X be a real reflexive Banach space. The operator $A : X \to 2^{X^*}$ is called generalized pseudomonotone if the following holds:*
Let $(u_n) \subset X$ and $(u_n^*) \subset X^*$ with $u_n^* \in A(u_n)$. If $u_n \rightharpoonup u$ in X and $u_n^* \rightharpoonup u^*$ in X^* and if $\limsup \langle u_n^*, u_n - u \rangle \leq 0$, then the element u^* lies in $A(u)$ and

$$\langle u_n^*, u_n \rangle \to \langle u^*, u \rangle.$$

The next two propositions provide the relation between pseudomonotone and generalized pseudomontone operators.

Proposition 2.122. *Let X be a real reflexive Banach space. If the operator $A : X \to 2^{X^*}$ is pseudomonotone, then A is generalized pseudomonotone.*

Under the additional assumption of boundedness, the following converse of Proposition 2.122 is true.

Proposition 2.123. *Let X be a real reflexive Banach space, and assume that $A : X \to 2^{X^*}$ satisfies the following conditions:*

(i)   *For each $u \in X$, we have that $A(u)$ is a nonempty, closed, and convex subset of X^*.*
(ii) *$A : X \to 2^{X^*}$ is bounded.*
(iii) *If $u_n \rightharpoonup u$ in X and $u_n^* \rightharpoonup u^*$ in X^* with $u_n^* \in A(u_n)$ and if $\limsup \langle u_n^*, u_n - u \rangle \leq 0$, then $u^* \in A(u)$ and $\langle u_n^*, u_n \rangle \to \langle u^*, u \rangle$.*

Then the operator $A : X \to 2^{X^*}$ is pseudomonotone.

As for the proof of Proposition 2.123 we refer to [177, Chap. 2]. Note that the notion of boundedness of a multivalued operator is exactly the same as for single-valued operators; i.e., the image of a bounded set is again bounded.

The relation between maximal monotone and pseudomonotone operators as well as the invariance of pseudomonotonicity under addition is given in the following theorem.

Theorem 2.124. *Let X be a real reflexive Banach space, and let A, $A_i : X \to 2^{X^*}$, $i = 1, 2$.*

(i) *If A is maximal monotone with $D(A) = X$, then A is pseudomonotone.*
(ii) *If A_1 and A_2 are two pseudomonotone operators, then the sum $A_1 + A_2 : X \to 2^{X^*}$ is pseudomonotone.*

The main theorem on pseudomonotone multivalued operators is formulated in the next theorem.

Theorem 2.125. *Let X be a real reflexive Banach space, and let $A : X \to 2^{X^*}$ be a pseudomonotone and a bounded operator, which is coercive in the sense that a real-valued function $c : \mathbb{R}_+ \to \mathbb{R}$ exists with*

$$c(r) \to +\infty, \quad as \quad r \to +\infty$$

such that for all $(u, u^*) \in Gr(A)$, we have

$$\langle u^*, u - u_0 \rangle \geq c(\|u\|_X)\|u\|_X$$

for some $u_0 \in X$. Then A is surjective; i.e., $\mathrm{range}(A) = X$.

Remark 2.126. We remark that the boundedness condition supposed in Theorem 2.125 can be dropped (see [177, Theorem 2.6]). This is because by definition of a multivalued pseudomonotone operator A according to Definition 2.120 the operator A has to be upper semicontinuous from each finite-dimensional subspace X_n of X to the weak topology on X^*. This latter condition along with the coercivity and the properties of the images allows us to get a surjectivity result on finite-dimensional subspaces X_n.

Theorem 2.127. *Let X be a real reflexive Banach space, $\Phi : X \to 2^{X^*}$ a maximal monotone operator, and $u_0 \in D(\Phi)$. Let $A : X \to 2^{X^*}$ be a pseudomonotone operator, and assume that either A_{u_0} is quasi-bounded or Φ_{u_0} is strongly quasi-bounded. Assume further that $A : X \to 2^{X^*}$ is u_0-coercive; i.e., a real-valued function $c : \mathbb{R}_+ \to \mathbb{R}$ exists with $c(r) \to +\infty$ as $r \to +\infty$ such that for all $(u, u^*) \in \mathrm{Gr}\,(A)$, we have $\langle u^*, u - u_0 \rangle \geq c(\|u\|_X)\|u\|_X$. Then $A + \Phi$ is surjective; i.e., $\mathrm{range}(A + \Phi) = X^*$.*

The operators A_{u_0} and Φ_{u_0} that appear in Theorem 2.127 are defined by $A_{u_0}(v) := A(u_0 + v)$ and similarly for Φ_{u_0}. As for the notion of *quasi-bounded* and *strongly quasi-bounded*, we refer to [177, p. 51]. In particular, one has that any bounded operator is quasi-bounded and strongly quasi-bounded.

2.4 First-Order Evolution Equations

In this section we present the basic functional analytic tools needed in the study of first-order single- and multivalued evolution equations in the form

$$u \in X, \ u' \in X^*: \quad u' + Au \ni f \ \text{ in } \ X^*, \ u(0) = u_0, \qquad (2.4)$$

where $X = L^p(0, \tau; V)$, $1 < p < \infty$, with $\tau > 0$ is the L^p-space of vector-valued functions $u : (0, \tau) \to V$ defined on the interval $(0, \tau)$ with values in some Banach space V, and u' is the generalized or distributional derivative of the function $t \mapsto u(t)$ with respect to $t \in (0, \tau)$. The right-hand side $f \in X^*$ is given, and $A : X \to 2^{X^*}$ is some (in general) multivalued operator. The initial values u_0 are taken from some Hilbert space H such that the embedding $V \subset H$ is continuous and dense. Problem (2.4) provides an abstract framework for the functional analytic treatment of initial-boundary value problems for parabolic differential equations and inclusions.

2.4.1 Motivation

To give a motivation for the study of the abstract problem (2.4), let us consider the classic initial-boundary value problem for the heat equation.

Let $\Omega \subset \mathbb{R}^N$ be a bounded domain with smooth boundary $\partial\Omega$, and denote $Q = \Omega \times (0, \tau)$ and $\Gamma = \partial\Omega \times (0, \tau)$ for some $\tau > 0$. We are looking for a function $(x, t) \mapsto u(x, t)$ defined in $\overline{\Omega} \times [0, \tau)$ such that

$$
\begin{aligned}
u_t - \Delta u &= f &&\text{in } Q, \\
u &= 0 &&\text{on } \Gamma, \\
u(\cdot, 0) &= u_0(\cdot) &&\text{in } \Omega,
\end{aligned}
\qquad (2.5)
$$

where the right-hand side $f : Q \to \mathbb{R}$ and the initial values $u_0 : \Omega \to \mathbb{R}$ are given functions. A classic (or strong) solution of (2.5) is a function that satisfies all equations of (2.5) pointwise in the usual sense. This, however, requires sufficient smoothness assumptions on the data f and u_0 as well as on the domain Ω. To be able to deal with (2.5) under relaxed regularity assumptions on the data, one tries instead to consider an appropriate generalized problem corresponding to (2.5), which in turn leads to the notion of weak solutions. To make plausible the definition of weak solutions of (2.5), we temporarily suppose that u is in fact a smooth solution of (2.5). In a similar way as in the explanation of weak solutions of the Dirichlet problem for elliptic equations (see Chap. 1), we formally multiply the heat equation by $v \in C_0^\infty(\Omega)$ and subsequently integrate by parts, which yields

$$\frac{d}{dt} \int_\Omega u(x, t) v(x) \, dx + \int_\Omega \nabla u(x, t) \nabla v(x) \, dx = \int_\Omega f(x, t) v(x) \, dx, \qquad (2.6)$$

for all $v \in C_0^\infty(\Omega)$. As $V = W_0^{1,2}(\Omega)$ is the closure of $C_0^\infty(\Omega)$ in $W_0^{1,2}(\Omega)$, we see that (2.6) makes perfect sense for $v \in V$ and $u(\cdot, t) \in V$ with $f \in L^2(Q)$. Now we change our viewpoint concerning the function u in that we deal with the space variable x and the time variable t in different ways. We associate with $u = u(x, t)$ a mapping (again denoted by u) $u : [0, \tau) \to V$ defined by

$$(u(t))(x) = u(x, t), \quad x \in \Omega, \ t \in [0, \tau),$$

which means that we are going to consider u not as a function of x and t together, but as a mapping $u : [0, \tau) \to V$. Let $H = L^2(\Omega)$, and denote by $(\cdot, \cdot)$ the inner product in H. Similarly as for u, we interpret the right-hand side function f as a mapping $f : [0, \tau) \to H$ according to

$$(f(t))(x) = f(x, t), \quad x \in \Omega, \ t \in [0, \tau).$$

By means of the bilinear form $a : V \times V \to \mathbb{R}$ defined by

$$a(w, v) = \int_\Omega \nabla w \nabla v \, dx,$$

we can rewrite (2.6) in the form

$$\frac{d}{dt}(u(t), v) + a(u(t), v) = (f(t), v) \quad \text{for all } v \in V, \tag{2.7}$$

which together with $u(0) = u_0 \in H$ represents a weak formulation of the initial-boundary value problem (2.5) [note that the homogeneous boundary values are taken into account by $u(t) \in V$]. For fixed t, let us consider the mapping $v \mapsto (f(t), v)$. Apparently this mapping is linear, and in view of the continuous embedding $V \subset H$, we have

$$|(f(t), v)| \leq \|f(t)\|_H \|v\|_H \leq c\|f(t)\|_H \|v\|_V,$$

which shows that the mapping is bounded. Thus, the mapping $v \mapsto (f(t), v)$ belongs to V^*; i.e., there is a $b \in V^*$ such that

$$\langle b, v \rangle = (f(t), v), \quad \text{for all } v \in V,$$

where $\langle \cdot, \cdot \rangle$ denotes the duality pairing between V and V^*. Next we will see that the functional b is defined in a unique way. Assume there is another $h \in H$ that generates the same functional b. It yields

$$(f(t) - h, v) = 0 \quad \text{for all } v \in V,$$

and thus $f(t) = h$, because $V \subset H$ is densely embedded. It allows us to identify b with $f(t)$. In this way, the element $f(t) \in H$ has to be considered as an element of V^*, and thus, we have

$$\langle f(t), v \rangle = (f(t), v) \quad \text{for all } v \in V. \tag{2.8}$$

The bilinear form a defined above, which can easily be seen to be bounded, generates a linear and bounded operator $A : V \to V^*$ through

$$\langle Aw, v \rangle = a(w, v) \quad \text{for all } w, v \in V. \tag{2.9}$$

Thus, by (2.8) and (2.9), we can rewrite (2.7) in the form

$$\frac{d}{dt}(u(t), v) + \langle Au(t), v \rangle = \langle f(t), v \rangle \quad \text{for all} \;\; v \in V. \tag{2.10}$$

We will later introduce the generalized or distributional derivative of a vector-valued function $t \mapsto u(t)$, whose derivative $u'(t)$ turns out to have the property

$$\langle u'(t), v \rangle = \frac{d}{dt}(u(t), v) \quad \text{for all} \;\; v \in V, \tag{2.11}$$

where d/dt is the generalized derivative of the real-valued function $t \mapsto (u(t), v)$ on $(0, \tau)$. Equations (2.10) and (2.11) result in the operator equation

$$u'(t) + Au(t) = f(t) \quad \text{in} \;\; V^*, \tag{2.12}$$

which is only required to be satisfied for a.e. $t \in (0, \tau)$.

Let $X = L^2(0, \tau; V)$, and denote by X^* its dual space, which is given by $X^* = L^2(0, \tau; V^*)$ (see Sect. 2.4.2). Furthermore, by means of the operator $A : V \to V^*$, we define an operator $\hat{A} : X \to X^*$ by

$$(\hat{A}u)(t) = Au(t), \quad t \in (0, \tau).$$

Thus, in view of (2.12) and the definition of $\hat{A}$, a generalized formulation of the initial-boundary value problem (2.5) reads as follows: For given $u_0 \in H$ and $f \in X^*$, we seek a function $u \in X$ such that $u' \in X^*$ and

$$u' + \hat{A}u = f \quad \text{in} \;\; X^*, \quad u(0) = u_0, \tag{2.13}$$

which is of the abstract form of the (single-valued) evolution equation (2.4).

We observe a few particularities that are typical in the functional analytic setting of parabolic problems.

(i)   The space and time variables x and t are treated differently, and the function $(x, t) \mapsto u(x, t)$ is considered as a vector-valued function.

(ii) The formulation of the given initial-boundary value problem as an abstract operator equation of the form (2.13) requires the use of two spaces H and V with the need that $V \subset H$ is densely and continuously embedded. It leads to the concept of *evolution triple*: $V \subset H \subset V^*$.

(iii) The solution space for the operator equation (2.13) is given by

$$W = \{u \in X : u' \in X^*\},$$

where u' has to be understood as the distributional derivative of the vector-valued function u.

In the following subsections, we will give the basic notions and existence results for the abstract evolution equation (2.4).

2.4.2 Vector-Valued Functions

Let B be a Banach space with norm $\| \cdot \|$, B^* its dual space, and $0 < \tau < \infty$. We consider vector-valued functions $u : [0, \tau] \to B$ and explain first some notions such as measurability and integrability. Most of the material in this subsection can be found in [93, 208, 222].

Definition 2.128. *Let u and s be vector-valued functions.*

(i) $s : [0, \tau] \to B$ *is called simple (or step function) if it is of the form*

$$s(t) = \sum_{i=1}^{m} \chi_{E_i}(t) u_i, \quad 0 \le t \le \tau,$$

where each E_i is a Lebesgue measurable subset of the interval $[0, \tau]$, $u_i \in B$ ($i = 1, \ldots, m$), and χ_{E_i} is the characteristic function of E_i.

(ii) $u : [0, \tau] \to B$ *is strongly measurable if a sequence (s_k) of simple functions $s_k : [0, \tau] \to B$ exists such that $s_k(t) \to u(t)$ as $k \to \infty$, for a.e. $t \in [0, \tau]$.*

(iii) $u : [0, \tau] \to B$ *is weakly measurable if for each $u^* \in B^*$ the mapping $t \to \langle u^*, u(t) \rangle$ is Lebesgue measurable.*

(iv) $u : [0, \tau] \to B$ *is almost separably valued if a subset $N \subset [0, \tau]$ of zero measure exists such that the set $\{u(t) : t \in [0, \tau] \backslash N\}$ is a separable subset of B.*

Theorem 2.129 (Pettis). *The function $u : [0, \tau] \to B$ is strongly measurable if and only if u is weakly measurable and almost separably valued.*

Definition 2.130. *The integral of vector-valued functions is defined as follows:*

(i) *The integral of the simple function $s(t) = \sum_{i=1}^{m} \chi_{E_i}(t) u_i$ is defined by*

$$\int_0^\tau s(t)\, dt = \sum_{i=1}^{m} \operatorname{meas}(E_i)\, u_i.$$

(ii) *The vector-valued function $u : [0, \tau] \to B$ is called integrable if a sequence (s_k) of simple functions exists such that*

$$\int_0^\tau \| s_k(t) - u(t) \|\, dt \to 0 \quad as \ k \to \infty.$$

(iii) *If $u : [0, \tau] \to B$ is integrable, its integral is defined by*

$$\int_0^\tau u(t)\, dt = \lim_{k \to \infty} \int_0^\tau s_k(t)\, dt.$$

Theorem 2.131. *The function* $u : [0, \tau] \to B$ *is integrable if and only if* u *is strongly measurable and* $t \to \|u(t)\|$ *is integrable. Furthermore, one has*

$$\left\| \int_0^\tau u(t) \, dt \right\| \le \int_0^\tau \|u(t)\| \, dt, \quad and \quad \left\langle u^*, \int_0^\tau u(t) \, dt \right\rangle = \int_0^\tau \langle u^*, u(t) \rangle \, dt,$$

for each $u^* \in B^*$.

Definition 2.132. *Let* $1 \le p \le \infty$. *We denote by* $L^p(0, \tau; B)$ *the space of (equivalent classes of) measurable functions* $u : [0, \tau] \to B$ *such that* $\|u(\cdot)\|$ *belongs to* $L^p(0, \tau; \mathbb{R})$ *with*

$$\|u\|_{L^p(0,\tau;B)} = \left(\int_0^\tau \|u(t)\|^p \, dt \right)^{1/p} \quad for \ 1 \le p < \infty,$$

$$\|u\|_{L^\infty(0,\tau;B)} = \operatorname*{ess\,sup}_{0 \le t \le \tau} \|u(t)\| < \infty.$$

The space $C([0, \tau]; B)$ *comprises of all continuous functions* $u : [0, \tau] \to B$ *with*

$$\|u\|_{C([0,\tau];B)} = \max_{0 \le t \le \tau} \|u(t)\| < \infty.$$

Theorem 2.133. *Let* B *and* Y *be Banach spaces. Then we have the following results:*

(i) $L^p(0, \tau; B)$ *with* $1 \le p \le \infty$ *and the norm given by Definition 2.132 is a Banach space.*

(ii) $C([0, \tau]; B)$ *is dense in* $L^p(0, \tau; B)$ *for* $1 \le p < \infty$, *and the embedding* $C([0, \tau]; B) \subset L^p(0, \tau; B)$ *is continuous.*

(iii) *If* B *is a Hilbert space with scalar product* $(\cdot, \cdot)_B$, *then* $L^2(0, \tau; B)$ *is also a Hilbert space with the scalar product*

$$(u, v) = \int_0^\tau (u(t), v(t))_B \, dt.$$

(iv) $L^p(0, \tau; B)$ *is separable if* B *is separable and* $1 \le p < \infty$.

(v) $L^p(0, \tau; B)$ *is uniformly (strictly) convex in the case where* B *is uniformly (strictly) convex and* $1 < p < \infty$.

(vi)   *If the embedding* $B \subset Y$ *is continuous, then the embedding*

$$L^r(0, \tau; B) \subset L^q(0, \tau; Y), \quad 1 \le q \le r \le \infty,$$

is also continuous.

(vii) *Let* B *be a reflexive and separable Banach space, and let* $1 < p < \infty$, $1/p + 1/q = 1$. *Then* $X = L^p(0, \tau; B)$ *is also reflexive and separable, and its dual space* X^* *is norm-isomorphic to* $L^q(0, \tau; B^*)$. *Therefore,* X^* *and* $L^q(0, \tau; B^*)$ *may be identified. The duality pairing* $\langle \cdot, \cdot \rangle_X$ *between* X *and its dual* X^* *can be written as*

$$\langle v, u \rangle_X = \int_0^\tau \langle v(t), u(t) \rangle_B \, dt \quad for \ all \ u \in X, \ v \in X^*.$$

Remark 2.134. We usually drop the subscripts X and B in $\langle v, u \rangle_X$ and $\langle v(t), u(t) \rangle_B$, respectively, because from the context, the type of duality pairing is clear.

2.4.3 Evolution Triple and Generalized Derivative

The material of this subsection is mainly taken from [208, 222].

Definition 2.135 (Evolution Triple). *A triple (V, H, V^*) is called an evolution triple if the following properties hold:*

(i)   *V is a real, separable, and reflexive Banach space, and H is a real, separable Hilbert space endowed with the scalar product $(\cdot, \cdot)$.*
(ii) *The embedding $V \subset H$ is continuous, and V is dense in H.*
(iii) *Identifying H with its dual H^* by the Riesz map, we then have $H \subset V^*$ with the equation*

$$\langle h, v \rangle_V = (h, v) \quad for \ \ h \in H \subset V^*, \ v \in V.$$

Remark 2.136. As V is reflexive and V is dense in H, the space H^* is dense in V^*, and hence, H is dense in V^*. It is a simple consequence of Proposition 2.25 in Sect. 2.1.2 applied to the embedding operator $i : V \to H$.

Example 2.137. Let $\Omega \subset \mathbb{R}^N$ be a bounded domain with Lipschitz boundary $\partial\Omega$, and let V be a closed subspace of $W^{1,p}(\Omega)$ with $2 \leq p < \infty$ such that $W_0^{1,p}(\Omega) \subset V \subset W^{1,p}(\Omega)$. Then (V, H, V^*) with $H = L^2(\Omega)$ is an evolution triple with all embeddings being, in addition, compact.

Definition 2.138. *Let Y, Z be Banach spaces, and $u \in L^1(0, \tau; Y)$ and $w \in L^1(0, \tau; Z)$. Then, the function w is called the generalized derivative of the function u in $(0, \tau)$ iff the following relation holds:*

$$\int_0^\tau \varphi'(t) u(t) \, dt = -\int_0^\tau \varphi(t) w(t) \, dt \quad for \ all \ \ \varphi \in C_0^\infty(0, \tau).$$

We write $w = u'$.

Theorem 2.139. *Let $V \subset H \subset V^*$ be an evolution triple, and let $1 \leq p, q \leq \infty$, $0 < \tau < \infty$. Let $u \in L^p(0, \tau; V)$; then the generalized derivative $u' \in L^q(0, \tau; V^*)$ exists iff there is a function $w \in L^q(0, \tau; V^*)$ such that*

$$\int_0^\tau (u(t), v)_H \varphi'(t) \, dt = -\int_0^\tau \langle w(t), v \rangle_V \varphi(t) \, dt$$

for all $v \in V$ and all $\varphi \in C_0^\infty(0, \tau)$. The generalized derivative u' is uniquely defined and $u' = w$.

Definition 2.140. *Let V be a real, separable, and reflexive Banach space, and let $X = L^p(0, \tau; V)$, $1 < p < \infty$. A space W is defined by*

$$W = \{u \in X : u' \in X^*\},$$

where u' is the generalized derivative, and $X^* = L^q(0, \tau; V^*)$, $1/p + 1/q = 1$.

Theorem 2.141 (Lions–Aubin). *Let B_0, B, B_1 be reflexive Banach spaces with $B_0 \subset B \subset B_1$, and assume $B_0 \subset B$ is compactly and $B \subset B_1$ is continuously embedded. Let $1 < p < \infty$, $1 < q < \infty$, and define $\mathcal{W}$ by*

$$\mathcal{W} = \{u \in L^p(0, \tau; B_0) : u' \in L^q(0, \tau; B_1)\}.$$

Then $\mathcal{W} \subset L^p(0, \tau; B)$ is compactly embedded.

Example 2.142. Let $\Omega \subset \mathbb{R}^N$ be a bounded domain with Lipschitz boundary $\partial\Omega$. As $W^{1,p}(\Omega) \subset L^p(\Omega)$ is compactly embedded, and $L^p(\Omega) \subset W^{1,p}(\Omega)^*$ is continuously embedded for $2 \le p < \infty$, Theorem 2.141 can be applied by setting $B_0 = W^{1,p}(\Omega)$, $B = L^p(\Omega)$ and $B_1 = W^{1,p}(\Omega)^*$, $2 \le p < \infty$. Thus, W defined in Definition 2.140, i.e.,

$$W = \{u \in L^p(0, \tau; W^{1,p}(\Omega)) : u' \in L^q(0, \tau; W^{1,p}(\Omega)^*)\},$$

is compactly embedded in $L^p(0, \tau; L^p(\Omega)) \equiv L^p(Q)$, where $Q = \Omega \times (0, \tau)$.

Let $\Omega \subset \mathbb{R}^N$ be as in Example 2.142, and $\Gamma = \partial\Omega \times (0, \tau)$. If $u \in X = L^p(0, \tau; W^{1,p}(\Omega))$, then for a.e. $t \in (0, \tau)$ the function $t \mapsto \gamma u(t) \in L^p(\partial\Omega)$ is well defined, where $\gamma : W^{1,p}(\Omega) \to L^p(\partial\Omega)$ denotes the trace operator (see Theorem 2.75). In view of Theorem 2.133 (vi) and the continuity of $\gamma : W^{1,p}(\Omega) \to L^p(\partial\Omega)$, we get that $t \mapsto \gamma u(t)$ belongs to $L^p(0, \tau; L^p(\partial\Omega)) \equiv L^p(\Gamma)$. If we denote the mapping that assigns $u \in X$ to the vector-valued function $t \mapsto \gamma u(t)$ again by γ, then it follows that $\gamma : X \to L^p(\Gamma)$ is linear and continuous. Moreover, as the trace operator $\gamma : W^{1,p}(\Omega) \to L^p(\partial\Omega)$ is even compact, we obtain the following result.

Proposition 2.143. *Let $\Omega \subset \mathbb{R}^N$ be a bounded domain with Lipschitz boundary $\partial\Omega$, and let $X = L^p(0, \tau; W^{1,p}(\Omega))$ with $2 \le p < \infty$. Then the trace operator $\gamma : W \to L^p(\Gamma)$ is compact.*

Proof: We apply Theorem 2.141. To this end, let $B_0 = W^{1,p}(\Omega)$, $B = W^{1-\varepsilon,p}(\Omega)$, and $B_1 = B_0^*$. As $B_0 \subset B$ is compactly embedded for any $\varepsilon \in (0, 1)$, and $B \subset B_1$ is continuously embedded, from Theorem 2.141, it follows that $W \subset L^p(0, \tau; W^{1-\varepsilon,p}(\Omega))$ is compactly embedded. If we select ε such that $0 < \varepsilon < 1 - 1/p$, then $\gamma : W^{1-\varepsilon,p}(\Omega) \to W^{1-\varepsilon-1/p,p}(\partial\Omega)$ is linear and continuous, and thus $\gamma : L^p(0, \tau; W^{1-\varepsilon,p}(\Omega)) \to L^p(0, \tau; W^{1-\varepsilon-1/p,p}(\Omega)) \subset L^p(\Gamma)$ is linear and continuous, which due to the compact embedding of $W \subset L^p(0, \tau; W^{1-\varepsilon,p}(\Omega))$ completes the proof. $\square$

Theorem 2.144. *Let $V \subset H \subset V^*$ be an evolution triple, and let $1 < p < \infty$, $1/p + 1/q = 1$, $0 < \tau < \infty$. Then the following hold:*

(i)   *The space W defined in Definition 2.140 is a real, separable, and reflexive Banach space with the norm*

$$\|u\|_W = \|u\|_X + \|u'\|_{X^*}.$$

(ii) *The embedding $W \subset C([0, \tau]; H)$ is continuous.*

(iii) *For all $u, v \in W$ and arbitrary t, s with $0 \le s \le t \le \tau$, the following generalized integration by parts formula holds:*

$$(u(t), v(t))_H - (u(s), v(s))_H = \int_s^t \langle u'(\zeta), v(\zeta) \rangle_V + \langle v'(\zeta), u(\zeta) \rangle_V \, d\zeta.$$

Remark 2.145. The integration by parts formula is equivalent to

$$\frac{d}{dt}(u(t), v(t))_H = \langle u'(t), v(t) \rangle_V + \langle v'(t), u(t) \rangle_V \quad \text{for a.e. } t \in (0, \tau).$$

In particular, for $u = v$, we obtain

$$\frac{d}{dt}\|u(t)\|_H^2 = 2\langle u'(t), u(t) \rangle_V,$$

which implies

$$\int_s^t \langle u'(\zeta), u(\zeta) \rangle_V \, d\zeta = \frac{1}{2}(\|u(t)\|_H^2 - \|u(s)\|_H^2). \tag{2.14}$$

In case that $V = W^{1,p}(\Omega)$, $2 \le p < \infty$, and $H = L^2(\Omega)$, we obtain the following generalization of formula (2.14), which will be useful for obtaining comparison principles in evolutionary problems.

Lemma 2.146. *Let $X = L^p(0, \tau; W^{1,p}(\Omega))$ with $2 \le p < \infty$ and $W = \{u \in X : u' \in X^*\}$, where $\Omega \subset \mathbb{R}^N$ is a bounded domain with Lipschitz boundary $\partial\Omega$. Let $\theta : \mathbb{R} \to \mathbb{R}$ be continuous and piecewise continuously differentiable with $\theta' \in L^\infty(\mathbb{R})$, and $\theta(0) = 0$, and let Θ denote the primitive of θ defined by*

$$\Theta(r) = \int_0^r \theta(s) \, ds.$$

Then, for $w \in W$, the following formula holds:

$$\int_r^s \langle w'(t), \theta(w(t)) \rangle \, dt = \int_\Omega \Theta(w(s)) \, dx - \int_\Omega \Theta(w(r)) \, dx, \tag{2.15}$$

for a.e. $0 \le r < s \le \tau$.

Proof: The proof makes use of density arguments and the generalized chain rule for Sobolev functions (see Lemma 2.84). Note first that in view of the assumptions on θ and Lemma 2.84, the composed function $\theta(w)$ is in X for $w \in W$. The space $C^1([0,\tau]; C^1(\overline{\Omega}))$ of smooth functions is dense in W (cf. [222, Chap. 23]). Let $w \in W$ be given. Then there is a sequence $(w_n) \subset C^1([0,\tau]; C^1(\overline{\Omega}))$ with $w_n \to w$ as $n \to \infty$. For the smooth functions w_n, we have

$$\int_r^s \langle w_n'(t), \theta(w_n(t)) \rangle \, dt = \int_r^s \int_\Omega w_n'(x,t) \theta(w_n(x,t)) \, dx dt$$

$$= \int_r^s \int_\Omega \frac{\partial}{\partial t} \Big(\Theta(w_n(x,t)) \Big) \, dx dt$$

$$= \int_\Omega \Big(\Theta(w_n(x,s)) - \Theta(w_n(x,r)) \Big) \, dx. \quad (2.16)$$

The assumptions on θ imply that θ is Lipschitz continuous, and thus, it follows that for some subsequence of (w_n) (again denoted by (w_n)),

$$\theta(w_n) \to \theta(w) \quad \text{in } X, \quad (2.17)$$

and due to the continuous embedding $W \subset C([0,\tau]; L^2(\Omega))$, one gets for all $t \in [0,\tau]$

$$\Theta(w_n(t)) \to \Theta(w(t)) \quad \text{in } L^2(\Omega). \quad (2.18)$$

By using (2.17), (2.18), we may pass to the limit in (2.16) for some subsequence, which completes the proof. □

Example 2.147. Let $\theta(s) = s$. Then θ trivially satisfies all assumptions of Lemma 2.146, and the primitive Θ is given by $\Theta(s) = (1/2)s^2$, and thus, formula (2.15) becomes

$$\int_r^s \langle w'(t), w(t) \rangle \, dt = \frac{1}{2} \int_\Omega (w(s))^2 \, dx - \frac{1}{2} \int_\Omega (w(r))^2 \, dx$$

$$= \frac{1}{2}(\|w(s)\|_H^2 - \|w(r)\|_H^2), \quad (2.19)$$

for all $0 \le r < s \le \tau$, where $H = L^2(\Omega)$, which is formula (2.14.)

The following example will play a crucial rule in obtaining comparison results.

Example 2.148. If $\theta(s) = s^+ = \max\{s, 0\}$, then its primitive can easily be seen to be $\Theta(s) = (1/2)(s^+)^2$, and thus, for $w \in W$, we get the formula

$$\int_r^s \langle w'(t), (w(t))^+ \rangle \, dt = \frac{1}{2}(\|(w(s))^+\|_H^2 - \|(w(r))^+\|_H^2). \quad (2.20)$$

2.4.4 Existence Results for Evolution Equations

The material of this subsection is mainly based on results obtained in [19, 20, 208]; see also [152, 222].

Let $V \subset H \subset V^*$ be an evolution triple, and let $X = L^p(0, \tau; V)$, X^* and W be the spaces of vector-valued functions as defined in Sect. 2.4.3 with $1 < p < \infty$, $1/p + 1/q = 1$, and $0 < \tau < \infty$. We provide an existence result for the evolution equation

$$u \in W: \quad u'(t) + A(t)u(t) = f(t),\ 0 < t < \tau, \quad u(0) = 0, \qquad (2.21)$$

where $f \in X^*$ is given and $A(t) : V \to V^*$ is some operator specified later. Without loss of generality, homogeneous initial values have been assumed, because inhomogeneous initial values can be transformed to homogeneous ones by translation. The generalized derivative $Lu = u'$ restricted to the subset

$$D(L) = \{u \in X : u' \in X^* \ \text{and}\ u(0) = 0\} = \{u \in W : u(0) = 0\}$$

defines a linear operator $L : D(L) \to X^*$ given by

$$\langle Lu, v \rangle = \int_0^\tau \langle u'(t), v(t) \rangle\, dt \quad \text{for all}\ v \in X.$$

The operator L has the following properties.

Lemma 2.149. *Let $V \subset H \subset V^*$ be an evolution triple, and let $X = L^p(0, \tau; V)$, where $1 < p < \infty$. Then the operator $L : D(L) \subset X \to X^*$ is densely defined, closed, and maximal monotone.*

Proof: First we note that the set M defined by

$$M = \{u \in C^1([0, \tau]; V) : u(0) = 0\}$$

satisfies $M \subset D(L)$ and $\overline{M} = X$, which shows that $\overline{D(L)} = X$, and thus, L is densely defined. Due to the continuous embedding $W \subset C([0, \tau]; H)$, it follows that $D(L)$ is closed in W, and thus, L is closed. From formula (2.19), we get

$$\langle Lu, u \rangle = \int_0^\tau \langle Lu(t), u(t) \rangle\, dt = \frac{1}{2}(\|u(\tau)\|_H^2 - \|u(0)\|_H^2) = \frac{1}{2}\|u(\tau)\|_H^2 \geq 0,$$

$$(2.22)$$

which shows that L is monotone. To prove that L is maximal monotone, we make use of the characterization of single-valued maximal monotone operators (see Sect. 2.3.3). To this end, suppose $(v, w) \in X \times X^*$ and

$$\langle w - Lu, v - u \rangle \geq 0 \quad \text{for all}\ u \in D(L). \qquad (2.23)$$

We need to show that $v \in D(L)$ and $w = Lv = v'$. Let u be chosen as

$$u = \varphi z, \quad \text{with } \varphi \in C_0^\infty(0, \tau) \text{ and } z \in V.$$

Obviously, $u \in D(L)$ with $u' = \varphi' z$, and $\langle Lu, u \rangle = 0$ in view of (2.22). Due to inequality (2.23), we then have

$$0 \le \langle w, v \rangle - \int_0^\tau \langle \varphi'(t)v(t) + \varphi(t)w(t), z \rangle_V \, dt \quad \text{for all } z \in V, \qquad (2.24)$$

which implies

$$\int_0^\tau \langle \varphi'(t)v(t) + \varphi(t)w(t), z \rangle_V \, dt = 0 \quad \text{for all } \varphi \in C_0^\infty(0, \tau),$$

and thus, $v' = w$. It remains to show that $v \in D(L)$. Again by applying formula (2.22) with u replaced by $v - u$, we obtain

$$0 \le \langle v' - u', v - u \rangle = \frac{1}{2}(\|v(\tau) - u(\tau)\|_H^2 - \|v(0) - u(0)\|_H^2). \qquad (2.25)$$

To complete the proof, we only need to show that $v(0) = 0$. To this end, choose a sequence $(v_n) \subset V$ with $\tau v_n \to v(\tau)$ in H (note that V is dense in H) and specialize $u(t) = t v_n$. Then $u \in D(L)$, and from (2.25), one obtains

$$0 \le \frac{1}{2}(\|v(\tau) - \tau v_n\|_H^2 - \|v(0)\|_H^2),$$

which by passing to the limit as $n \to \infty$ results in $v(0) = 0$. $\qquad \square$

Remark 2.150. With only slight modifications one can prove that $L : D(L) \subset X \to X^*$ defined by

$$Lu = u' : \quad D(L) = \{u \in X : u' \in X^* \text{ and } u(0) = u(\tau)\}$$

is a densely defined, closed, and maximal monotone operator (cf. [222, Proposition 32.10]).

Now we state the following conditions on the time-dependent operators $A(t) : V \to V^*$:

(H1) $\|A(t)u\|_{V^*} \le c_0\Big(\|u\|_V^{p-1} + k_0(t)\Big)$ for all $u \in V$ and $t \in [0, \tau]$ with some positive constant c_0 and $k_0 \in L^q(0, \tau)$.
(H2) $A(t) : V \to V^*$ is demicontinuous for each $t \in [0, \tau]$.
(H3) The function $t \to \langle A(t)u, v \rangle$ is measurable on $(0, \tau)$ for all $u, v \in V$.
(H4) $\langle A(t)u, u \rangle \ge c_1(\|u\|_V^p - k_1(t))$ for all $u \in V$ and $t \in [0, \tau]$ with some constant $c_1 > 0$ and some function $k_1 \in L^1(0, \tau)$.

Define an operator $\hat{A}$ related with $A(t)$ by

$$\hat{A}(u)(t) = A(t)u(t), \quad t \in [0, \tau], \qquad (2.26)$$

which may be considered as the associated Nemytskij operator generated by
the operator-valued function $t \mapsto A(t)$. Thus, problem (2.21) corresponds to
the following one:

$$u \in D(L) : Lu + \hat{A}(u) = f \quad \text{in } X^*. \tag{2.27}$$

Definition 2.151. *Let $D(L)$ be equipped with the graph norm; that is,*

$$\|u\|_L = \|u\|_X + \|Lu\|_{X^*}.$$

*The operator $\hat{A} : X \to X^*$ is called pseudomonotone with respect to the graph
norm topology of $D(L)$ (or pseudomonotone w.r.t. $D(L)$ for short); if for any
sequence $(u_n) \in D(L)$ satisfying*

$$u_n \rightharpoonup u \text{ in } X, \ Lu_n \rightharpoonup Lu \text{ in } X^*, \ and \ \limsup_{n \to \infty} \langle \hat{A}(u_n), u_n - u \rangle \leq 0,$$

it follows that

$$\hat{A}(u_n) \rightharpoonup \hat{A}(u) \text{ in } X^* \ and \ \langle \hat{A}(u_n), u_n \rangle \to \langle \hat{A}(u), u \rangle.$$

In an obvious similar way, the (S_+)-condition with respect to $D(L)$ is defined.

For the following surjectivity result, which yields the existence for problem
(2.27), we refer to [19, 152].

Theorem 2.152. *Let $L : D(L) \subset X \to X^*$ be as given above, and let $\hat{A} :
X \to X^*$ defined by (2.26) be bounded, demicontinuous, and pseudomonotone
w.r.t. $D(L)$. If $\hat{A}$ is coercive, then $(L + \hat{A})(D(L)) = X^*$; that is, $L + \hat{A}$ is
surjective.*

The next result shows that certain properties of the operators $A(t)$ are trans-
fered to its Nemytskij operator $\hat{A}$; cf. [20].

Theorem 2.153. *Let hypotheses (H1)–(H4) be satisfied. Then we have the
following results:*

(i)   *If $A(t) : V \to V^*$ is pseudomonotone for all $t \in [0, \tau]$, then $\hat{A} : X \to X^*$
      is pseudomonotone with respect to $D(L)$ according to Definition 2.151.*
(ii) *If $A(t) : V \to V^*$ has the (S_+)-property for all $t \in [0, \tau]$, then $\hat{A} : X \to X^*$
      has the (S_+)-property with respect to $D(L)$.*
(iii) *Hypotheses (H1) and (H3) imply that $\hat{A} : X \to X^*$ is bounded.*
(iv) *Hypotheses (H1)–(H3) imply that $\hat{A} : X \to X^*$ is demicontinuous.*
(v)   *Hypothesis (H4) implies that $\hat{A} : X \to X^*$ is coercive.*

2.4.5 Multivalued Evolution Equations

In this section, we briefly recall a general surjectivity result for multivalued operators in a real reflexive Banach space X, which allows us to deal with multivalued evolution equations in the form

$$u \in X : \quad u' + A(u) \ni f \text{ in } X^*, \quad u(0) = u_0. \tag{2.28}$$

To this end, we introduce first the notion of a multivalued pseudomonotone operator with respect to the graph norm topology of the domain $D(L)$ (w.r.t. $D(L)$ for short) of some linear, closed, densely defined, and maximal monotone operator $L : D(L) \subset X \to X^*$.

Definition 2.154. *Let* $L : D(L) \subset X \to X^*$ *be a linear, closed, densely defined, and maximal monotone operator. The operator* $A : X \to 2^{X^*}$ *is called pseudomonotone w.r.t. $D(L)$ if the following conditions are satisfied:*

(i)   *The set $A(u)$ is nonempty, bounded, closed, and convex for all $u \in X$.*
(ii) *A is upper semicontinuous from each finite-dimensional subspace of X to the weak topology of X^*.*
(iii) *If $(u_n) \subset D(L)$ with $u_n \rightharpoonup u$ in X, $Lu_n \rightharpoonup Lu$ in X^*, $u_n^* \in A(u_n)$ with $u_n^* \rightharpoonup u^*$ in X^*, and $\limsup\langle u_n^*, u_n - u \rangle \leq 0$, then $u^* \in A(u)$ and $\langle u_n^*, u_n \rangle \to \langle u^*, u \rangle$.*

Definition 2.155. *The operator $A : X \to 2^{X^*}$ is called coercive iff either the domain $D(A)$ of A is bounded or $D(A)$ is unbounded and*

$$\frac{\inf\{\langle v^*, v \rangle : v^* \in A(v)\}}{\|v\|_X} \to +\infty \quad as \ \|v\|_X \to \infty, \ v \in D(A).$$

The following surjectivity result can be found in [79, Theorem 1.3.73, p. 62].

Theorem 2.156. *Let X be a real reflexive, strictly convex Banach space with dual space X^*, and let $L : D(L) \subset X \to X^*$ be a linear, closed, densely defined, and maximal monotone operator. If the multivalued operator $A : X \to 2^{X^*}$ is pseudomonotone w.r.t. $D(L)$, bounded, and coercive, then $L + A$ is surjective; i.e., $(L + A)(D(L)) = X^*$.*

Consider the multivalued evolution equation

$$u \in X : \quad u' + A(u) \ni f \text{ in } X^*, \quad u(0) = 0, \tag{2.29}$$

where

$$X = L^p(0, \tau; V), \quad 1 < p < \infty,$$

and $V \subset H \subset V^*$ is an evolution triple with V being strictly convex. As earlier, we define the operator L by

$$Lu = u', \quad \text{with } D(L) = \{u \in W : u(0) = 0\} \tag{2.30}$$

(for W see Definition 2.140). Thus, problem (2.29) can equivalently be written in the form

$$u \in D(L): \quad u' + A(u) \ni f \text{ in } X^*. \tag{2.31}$$

Corollary 2.157. *If the multivalued operator $A : X \to 2^{X^*}$ in (2.29) is pseudomonotone w.r.t. $D(L)$, bounded, and coercive, then problem (2.29) has at least one solution.*

Proof: In view of Theorem 2.133, the Banach space X is reflexive and strictly convex. The operator $L : D(L) \subset X \to X^*$ given by (2.30) is densely defined, linear, closed, and maximal monotone (see Lemma 2.149). Thus, the assertion follows from Theorem 2.156.   □

2.5 Nonsmooth Analysis

The area of nonsmooth analysis is closely related with the development of a critical point theory for nondifferentiable functions, in particular, for locally Lipschitz continuous functions based on Clarke's generalized gradient. It provides an appropriate mathematical framework to extend the classic critical point theory for C^1-functionals in a natural way, and to meet specific needs in applications, such as in nonsmooth mechanics and engineering. In this section, we provide basic facts and results of nonsmooth analysis to such an extent as it will be needed in the study of the problems we shall be investigating in this book.

2.5.1 Clarke's Generalized Gradient

Throughout this section, X stands for a real Banach space endowed with the norm $\| \cdot \|$. The dual space of X is denoted X^*, and the notation $\langle \cdot, \cdot \rangle$ means the duality pairing between X^* and X.

We recall the following well-known definition.

Definition 2.158. *A functional $f : X \to \mathbb{R}$ is said to be locally Lipschitz if for every point $x \in X$ a neighborhood V of x in X and a constant $K > 0$ exist such that*

$$|f(y) - f(z)| \leq K\|y - z\|, \quad \forall\, y, z \in V.$$

Example 2.159. A convex and continuous function $f : X \to \mathbb{R}$ is locally Lipschitz. More generally, a convex function $f : X \to \mathbb{R}$, which is bounded above on a neighborhood of some point is locally Lipschitz (see [68, p. 34]).

Example 2.160. A functional $f : X \to \mathbb{R}$, which is Lipschitz continuous on bounded subsets of X is locally Lipschitz. The converse assertion is not generally true. For instance, consider the next situation (given through [203]). On the Hilbert space ℓ^2, let the function $f : \ell^2 \to \mathbb{R}$ be defined by

$$f(x) = \sup_{n \geq 0}(2n|x_n| - n), \quad \forall\, x \in \ell^2,$$

where x_n are the components of x. The function f is convex, continuous, and not bounded on the bounded sets. Indeed, f is defined on ℓ^2 because for any $x \in \ell^2$, the set

$$\{n : 2n|x_n| - n \geq 0\} = \left\{ n : |x_n| \geq \frac{1}{2} \right\}$$

is finite. The function f is convex because it is the upper hull of the convex functions f_n on ℓ^2 given by $f_n(x) = 2n|x_n| - n$. We note that f is zero on the ball centered at 0 and radius $\frac{1}{2}$ because $0 = f_0(x) \leq f(x)$ and $2|x_n| \leq 1$ if $\|x\| < \frac{1}{2}$. Being bounded on a nonempty open set, the function f is continuous. Finally, it is seen that $f(e_n) = n$, where e_n is the n-th vector of the canonical basis of ℓ^2. It turns out that the function f is not bounded from above on the unit sphere in ℓ^2. Consequently, the function f is not Lipschitz continuous on bounded subsets, but as pointed out in Example 2.159, f is locally Lipschitz.

The classic theory of differentiability does not work in the case of locally Lipschitz functions. However, a suitable subdifferential calculus approach has been developed by Clarke [68]. Here we give a brief introduction. Further details can be found in [68, 43, 79, 103, 173].

Definition 2.161. *Let $f : X \to \mathbb{R}$ be a locally Lipschitz function, and fix two points $u, v \in X$. The generalized directional derivative of f at u in the direction v is defined as follows:*

$$f^\circ(u; v) = \limsup_{\substack{x \to u \\ t \downarrow 0}} \frac{f(x + tv) - f(x)}{t}.$$

As f is locally Lipschitz, it is clear that $f^\circ(u; v) \in \mathbb{R}$.

Proposition 2.162. *If $f : X \to \mathbb{R}$ is a locally Lipschitz function, then the following holds:*

(i) *The function $f^\circ(u; \cdot) : X \to \mathbb{R}$ is subadditive, positively homogeneous, and satisfies the inequality*

$$|f^\circ(u; v)| \leq K\|v\|, \quad \forall\, v \in X,$$

where $K > 0$ is the Lipschitz constant of f near the point $u \in X$.

(ii) *$f^\circ(u; -v) = (-f)^\circ(u; v), \quad \forall\, v \in X$.*

(iii) *The function $(u, v) \in X \times X \mapsto f^\circ(u; v) \in \mathbb{R}$ is upper semicontinuous.*

Proof: The result follows directly from Definition 2.161.   □

The next definition focuses on the case where $f^o(u; v)$ reduces to the usual directional derivative

$$f'(u; v) = \lim_{t \downarrow 0} \frac{f(u + tv) - f(u)}{t}.$$

Definition 2.163. *A locally Lipschitz function $f : X \to \mathbb{R}$ is said to be regular at a point $u \in X$ if*

(i) *the directional derivative $f'(u; v)$ exists, for every $v \in X$.*
(ii) *$f^o(u; v) = f'(u; v)$, $\forall\, v \in X$.*

Significant classes of regular functions are given in the following examples.

Example 2.164. If the function $f : X \to \mathbb{R}$ is strictly differentiable, that is, for all $u \in X$, $f'(u) \in X^*$ exists such that

$$\lim_{\substack{w \to u \\ t \downarrow 0}} \frac{f(w + tv) - f(w)}{t} = \langle f'(u), v \rangle, \quad \forall\, v \in X,$$

where the convergence is uniform for v in compact sets, then f is locally Lipschitz and regular in the sense of Definition 2.163. In particular, if $f : X \to \mathbb{R}$ is a continuously differentiable function, then f is strictly differentiable, so it is locally Lipschitz and regular.

Example 2.165. A convex and continuous function $f : X \to \mathbb{R}$ is regular.

On the basis of Definition 2.161, one introduces the main notion in this section.

Definition 2.166. *The generalized gradient of a locally Lipschitz functional $f : X \to \mathbb{R}$ at a point $u \in X$ is the subset of X^* defined by*

$$\partial f(u) = \{\zeta \in X^* : f^o(u; v) \geq \langle \zeta, v \rangle, \ \forall\, v \in X\}.$$

By using the Hahn–Banach theorem (see [24, p. 1]), it follows $\partial f(u) \neq \emptyset$.

Example 2.167. If $f : X \to \mathbb{R}$ is a locally Lipschitz function that is Gâteaux differentiable and regular at the point $u \in X$, then one has $\partial f(u) = \{D_G f(u)\}$, where $D_G f(u)$ denotes the Gâteaux differential of f at u. Indeed, as f is Gâteaux differentiable and regular at u, we may write

$$\langle D_G f(u), v \rangle = f'(u; v) = f^o(u; v), \quad \forall\, v \in X,$$

that implies $D_G f(u) \in \partial f(u)$. Conversely, if $\zeta \in \partial f(u)$, from Definitions 2.166 and 2.163 in conjunction with the assumption that f is Gâteaux differentiable at u, it turns out that

$$\langle \zeta, v \rangle \leq f^o(u; v) = f'(u; v) = \langle D_G f(u), v \rangle, \quad \forall\, v \in X,$$

so $\zeta = D_G f(u)$.

Example 2.168. If $f : X \to \mathbb{R}$ is continuously differentiable, then $\partial f(u) = \{f'(u)\}$ for all $u \in X$, where $f'(u)$ denotes the Fréchet differential of f at u. It is a direct consequence of Example 2.167.

Example 2.169. If $f : X \to \mathbb{R}$ is convex and continuous, then the generalized gradient $\partial f(u)$ coincides with the subdifferential of f at u in the sense of convex analysis. It follows from Examples 2.159 and 2.165.

Remark 2.170. It is seen from Definition 2.166, Example 2.169, and Proposition 2.162(i) that the generalized gradient of a locally Lipschitz functional $f : X \to \mathbb{R}$ at a point $u \in X$ is given by

$$\partial f(u) = \partial(f^\circ(u; \cdot))(0),$$

where in the right-hand side, the subdifferential in the sense of convex analysis is written.

The next proposition presents some important properties of generalized gradients.

Proposition 2.171. *Let $f : X \to \mathbb{R}$ be a locally Lipschitz function. Then for any $u \in X$, the following properties hold:*

(i) $\partial f(u)$ *is a convex, weak*-compact subset of X^* and*

$$\|\zeta\|_{X^*} \leq K, \quad \forall\, \zeta \in \partial f(u),$$

 where $K > 0$ is the Lipschitz constant of f near u.
(ii) $f^\circ(u; v) = \max\{\langle \zeta, v \rangle : \zeta \in \partial f(u)\}, \forall\, v \in X$.
(iii) *The mapping $u \mapsto \partial f(u)$ is weak*-closed from X into X^*.*
(iv) *The mapping $u \mapsto \partial f(u)$ is upper semicontinuous from X into X^*, where X^* is equipped with the weak*-topology.*

Proof: As for (i) and (ii), one applies Definitions 2.161 and 2.166, and as for (iv), see [68]. To see (iii), let $(u_n) \subset X$ satisfy $u_n \to u$ in X, and let $\zeta_n \in \partial f(u_n)$ with $\zeta_n \rightharpoonup^* \zeta$ in X^*. We need to show that $\zeta \in \partial f(u)$. By Definition 2.166, we have $\langle \zeta_n, v \rangle \leq f^\circ(u_n; v)$ for all $v \in X$, which from the weak*-convergence of (ζ_n) and the upper semicontinuity of the function $x \mapsto f^\circ(x; v)$ according to Proposition 2.162(iii) implies

$$\langle \zeta, v \rangle \leq \limsup_{n \to \infty} f^\circ(u_n; v) \leq f^\circ(u; v) \text{ for all } v \in X,$$

and thus, $\zeta \in \partial f(u)$.   □

Remark 2.172. The definitions and results given here are applicable to a locally Lipschitz function $f : U \to \mathbb{R}$ on a nonempty, open subset U of the Banach space X.

In the final part of this subsection, we present, for the sake of providing more information on the development of generalized differentiation theory in variational analysis, some basic elements of another subdifferential calculus for nonsmooth functionals, namely the one introduced by Mordukhovich ([164], [165]). The subsequent chapters of the book will not make use of this theory because we need calculus rules and specific properties related to Clarke's concept of generalized gradient for locally Lipschitz functionals in the sense of Definition 2.166 as will be given in the next subsection devoted to calculus, but we consider that it is useful to outline here the subdifferentiation approach in [164], [165] (see also [22]).

Given a nonempty subset S of a Banach space X and a point $u \in S$, it is introduced, for every number $\varepsilon \geq 0$, the set of ε- normals to S at u as the subset of X^* equal to

$$\hat{N}_\varepsilon(u; S) = \{\zeta \in X^* : \limsup_{\substack{w \to u \\ w \in S}} \frac{\langle \zeta, w - u \rangle}{\|w - u\|} \leq \varepsilon\}.$$

The basic normal cone $N(u; S)$ to S at u is defined by

$$N(u; S) = \limsup_{\substack{w \to u, w \in S \\ \varepsilon \downarrow 0}} \hat{N}_\varepsilon(w; S),$$

where in the right-hand side, the sequential Painlevé–Kuratowski upper limit is written. Explicitly, this means that

$$N(u; S) = \{\zeta \in X^* : \text{ there are sequences } w_k \to u \text{ with } w_k \in S, \varepsilon_k \downarrow 0,$$

$$\zeta_k \rightharpoonup^* \zeta \text{ with } \zeta_k \in \hat{N}_{\varepsilon_k}(w_k; S) \text{ for all } k\}.$$

It is shown in [165, Theorem 2.9] that in the case where X is an Asplund space (i.e., every separable subspace of X has a separable dual), the formula of $N(u; S)$ results in

$$N(u; S) = \limsup_{w \to u, w \in S} \hat{N}_0(w; S).$$

Now we are in a position to introduce the notion of subdifferential of an extended real-valued function $f : X \to [-\infty, +\infty]$ at $u \in X$ with $f(u) \in \mathbb{R}$ as follows:

$$\hat{\partial} f(u) := \{\zeta \in X^* : (\zeta, -1) \in N((u, f(u)); \text{epi}(f))\}.$$

If $f : X \to \mathbb{R}$ is locally Lipschitz and X is an Asplund space, the relationship between the above subdifferential $\hat{\partial} f(u)$ and Clarke's generalized gradient in the sense of Definition 2.166 is expressed by the following formula:

$$\partial f(u) = \text{cl}^* \text{co} \, \hat{\partial} f(u), \quad \text{for all } u \in X$$

(see [165, Theorem 8.11]), where the notation cl*co stands for the convex closure in the weak* topology on the space X^*.

2.5.2 Some Calculus

This subsection is devoted to the basic calculus rules with generalized gradients.

Proposition 2.173. *Let $f : X \to \mathbb{R}$ be a locally Lipschitz function, let $\lambda \in \mathbb{R}$, and let $u \in X$. Then the following formula holds:*

$$\partial(\lambda f)(u) = \lambda \partial f(u).$$

In particular, one has

$$\partial(-f)(u) = -\partial f(u).$$

Proof: If $\lambda = 0$, the property is obvious. If $\lambda > 0$, we have

$$\zeta \in \partial(\lambda f)(u) \iff \langle \frac{1}{\lambda} \zeta, v \rangle \le \frac{1}{\lambda}(\lambda f)^o(u; v) = f^o(u; v), \quad \forall\, v \in X$$
$$\iff \zeta \in \lambda \partial f(u).$$

If $\lambda < 0$, we have $\zeta \in \partial(\lambda f)(u) \iff$

$$\langle \frac{1}{\lambda} \zeta, v \rangle = -\frac{1}{\lambda}\langle \zeta, -v \rangle \le -\frac{1}{\lambda}(\lambda f)^o(u; -v) = -\frac{1}{\lambda}(-\lambda f)^o(u; v)$$
$$= f^o(u; v), \quad \forall\, v \in X$$

$\iff \zeta \in \lambda \partial f(u)$, where Proposition 2.162(ii) has been used. $\qquad \square$

Proposition 2.174. *Let $f, g : X \to \mathbb{R}$ be locally Lipschitz functions. Then for every $u \in X$, the following inclusion holds:*

$$\partial(f + g)(u) \subset \partial f(u) + \partial g(u).$$

If, in addition, the functions f and g are regular at the point $u \in X$, then the above inclusion becomes an equality, and $f + g$ is regular at u.

Proof: Let $\zeta \in \partial(f + g)(u)$. Definition 2.166 ensures

$$\langle \zeta, v \rangle \le f^o(u; v) + g^o(u; v), \quad \forall\, v \in X. \tag{2.32}$$

Arguing by contradiction, let us admit that $\zeta \notin \partial f(u) + \partial g(u)$. Then, by separation in the space X^* endowed with the w^*-topology, $w \in X$ exists such that

$$\langle \zeta, w \rangle > \max\{\langle z, w \rangle \,:\, z \in \partial f(u) + \partial g(u)\}$$
$$= \max\{\langle z_1, w \rangle \,:\, z_1 \in \partial f(u)\} + \max\{\langle z_2, w \rangle \,:\, z_2 \in \partial g(u)\}$$
$$= f^o(u; w) + g^o(u; w),$$

where Proposition 2.171(ii) has been employed. It contradicts (2.32), which proves the first assertion in Proposition 2.174.

Suppose now that f and g are regular at u in X. Then, by means of Definition 2.166 for every $\zeta \in \partial f(u) + \partial g(u)$, we have

$$
\begin{aligned}
\langle \zeta, v \rangle &\leq f^\circ(u; v) + g^\circ(u; v) \\
&= f'(u; v) + g'(u; v) \\
&= (f + g)'(u; v) \\
&\leq (f + g)^\circ(u; v)
\end{aligned}
$$

for all $v \in X$. We conclude $\zeta \in \partial(f + g)(u)$, which completes the proof.   □

Remark 2.175. The inclusion of Proposition 2.174 becomes an equality also when at least one of the two locally Lipschitz functions is strictly differentiable.

We state a useful necessary condition of optimality in the case of locally Lipschitz functions.

Proposition 2.176. *If $u \in X$ is a local minimum or maximum point for the locally Lipschitz function $f : X \to \mathbb{R}$, then $0 \in \partial f(u)$.*

Proof: We may assume that u is a local minimum (if u is a local maximum, we can argue with $-f$). Then we obtain that $f^\circ(u; v) \geq 0$, $\forall\, v \in X$, which is equivalent to $0 \in \partial f(u)$.   □

The result below presents the mean value property for locally Lipschitz functionals due to Lebourg [148].

Theorem 2.177. *Let $f : X \to \mathbb{R}$ be a locally Lipschitz function. Then for all $x, y \in X$, $u = x + t_0(y - x)$ with $0 < t_0 < 1$, and $\zeta \in \partial f(u)$ exist, such that*

$$ f(y) - f(x) = \langle \zeta, y - x \rangle. $$

Proof: Consider the function $\theta : [0, 1] \to \mathbb{R}$ defined by

$$ \theta(t) = f(x + t(y - x)) + t[f(x) - f(y)], \quad \forall\, t \in [0, 1]. $$

The continuity of θ combined with the equalities $\theta(0) = \theta(1) = f(x)$ yields a point $t_0 \in (0, 1)$ where θ assumes the minimum or maximum. By Proposition 2.176, we find that

$$ 0 \in \partial\theta(t_0) \subset \langle \partial f(x + t_0(y - x)), y - x \rangle + [f(x) - f(y)]. $$

The conclusion of Theorem 2.177 follows.   □

Another important result in the calculus with generalized gradients is the chain rule.

Theorem 2.178. *Let $F : X \to Y$ be a continuously differentiable mapping between the Banach spaces X, Y, and let $g : Y \to \mathbb{R}$ be a locally Lipschitz* *function. Then the function* $g \circ F : X \to \mathbb{R}$ *is locally Lipschitz, and for any point* $u \in X$, *the formula holds:*

$$\partial(g \circ F)(u) \subset \partial g(F(u)) \circ DF(u), \tag{2.33}$$

in the sense that every element $z \in \partial(g \circ F)(u)$ *can be expressed as*

$$z = DF(u)^* \zeta, \quad \text{for some } \zeta \in \partial g(F(u)),$$

where $DF(u)^*$ *denotes the adjoint operator associated with the Fréchet differential* $DF(u)$ *of* F *at* u. *If, in addition,* F *maps every neighborhood of* u *onto a dense subset of a neighborhood of* $F(u)$, *then (2.33) is satisfied with equality.*

Proof: The mean value theorem for the continuously differentiable mapping F readily yields that $g \circ F$ is locally Lipschitz. According to Proposition 2.171 (ii), inclusion (2.33) is equivalent to the inequality

$$(g \circ F)^\circ(u; v) \leq \max\{\langle z, DF(u)v \rangle : z \in \partial g(F(u))\}$$
$$= g^\circ(F(u); DF(u)v), \quad \forall\, v \in X. \tag{2.34}$$

Fix $w, v \in X$ and $t > 0$. Applying Theorem 2.177 ensures the existence of $t_0, t_1 \in (0, 1)$ and $\zeta \in \partial g(F(w) + t_0(F(w + tv) - F(w)))$ such that

$$g \circ F(w + tv) - g \circ F(w) = \langle \zeta, F(w + tv) - F(w) \rangle = t \langle \zeta, DF(w + t_1 tv)v \rangle.$$

Dividing by t, then letting $w \to u$ in X and $t \to 0$, and taking into account that the multifunction ∂g is upper semicontinuous from X to X^* endowed with the w^*-topology [cf. Proposition 2.171(iv)], we obtain (2.34). Assuming now that F maps an arbitrary neighborhood of u onto a dense subset of a neighborhood of $F(u)$ implies

$$g^\circ(F(u); DF(u)v) = \limsup_{\substack{x \to u \\ t \downarrow 0}} \frac{g(F(x) + tDF(u)v) - g(F(x))}{t}$$

$$= \limsup_{\substack{x \to u \\ t \downarrow 0}} \frac{g(F(x + tv)) - g(F(x))}{t}$$

$$= (g \circ F)^\circ(u; v), \quad \forall\, v \in X.$$

Therefore, (2.34) holds with equality, so the same is true for (2.33). $\qquad\square$

Corollary 2.179. *Under the assumptions of the first part of Theorem 2.178, if* g *(or* $-g$) *is regular at* $F(u)$, *then* $g \circ F$ *(or* $-g \circ F$) *is regular at* u *and equality holds in (2.33).*

Proof: As $\partial(-g)(F(u)) = -\partial g(F(u))$, it is sufficient to suppose that g is regular at $F(u)$. It turns out that

$$g^o(F(u); DF(u)v) = g'(F(u); DF(u)v)$$

$$= \lim_{t \downarrow 0} \frac{g(F(u) + tDF(u)v) - g(F(u))}{t}$$

$$= \lim_{t \downarrow 0} \left[\frac{g(F(u) + tDF(u)v) - g(F(u + tv))}{t} \right.$$

$$\left. + \frac{g(F(u + tv)) - g(F(u))}{t} \right]$$

$$= (g \circ F)'(u; v) \leq (g \circ F)^o(u; v), \quad \forall \, v \in X.$$

Consequently, we have equality in (2.34), so equality holds in (2.33).   □

Corollary 2.180. *If a linear continuous embedding* $i : X \to Y$ *of the Banach space* X *into a Banach space* Y *exists, then for every locally Lipschitz function* $g : Y \to \mathbb{R}$, *we have*

$$\partial(g \circ i)(u) \subset i^* \partial g(i(u)), \quad \forall \, u \in X.$$

If, in addition, $i(X)$ *is dense in* Y, *then*

$$\partial(g \circ i)(u) = i^* \partial g(i(u)), \quad \forall \, u \in X.$$

Proof: One applies Theorem 2.178 for $F = i$.   □

Finally, we give Aubin–Clarke's Theorem [9] of subdifferentiation under the integral sign.

Let numbers $m \geq 1$, $1 < p < +\infty$, and let T be a positive complete measure space with $|T| < \infty$, where $|T|$ stands for the measure of T. Let $j : T \times \mathbb{R}^m \to \mathbb{R}$ be a function such that $j(\cdot, y) : T \to \mathbb{R}$ is measurable whenever $y \in \mathbb{R}^m$, and satisfies either

$$|j(x, y_1) - j(x, y_2)| \leq k(x)|y_1 - y_2|, \quad \text{a.a. } x \in T, \ \forall \, y_1, y_2 \in \mathbb{R}^m, \quad (2.35)$$

with a function $k \in L^q(T)$ and $1/p + 1/q = 1$, or, $j(x, \cdot) : \mathbb{R}^m \to \mathbb{R}$ is locally Lipschitz for almost all $x \in T$ and there are a constant $c > 0$ and a function $h \in L^q(T)$ such that

$$|z| \leq h(x) + c|y|^{p-1}, \quad \text{a.a. } x \in T, \quad \forall \, y \in \mathbb{R}^m, \ \forall \, z \in \partial_y j(x, y). \quad (2.36)$$

The notation $\partial_y j(x, y)$ in (2.36) means the generalized gradient of j with respect to the second variable $y \in \mathbb{R}^m$; i.e., $\partial_y j(x, y) = \partial j(x, \cdot)(y)$. We introduce the functional $J : L^p(T; \mathbb{R}^m) \to \mathbb{R}$ by

$$J(v) = \int_T j(x, v(x)) dx, \quad \forall \, v \in L^p(T; \mathbb{R}^m). \quad (2.37)$$

Theorem 2.181 (Aubin-Clarke's Theorem). *Under assumption (2.35) or (2.36), one has that the functional $J : L^p(T; \mathbb{R}^m) \to \mathbb{R}$ in (2.37) is Lipschitz continuous on the bounded subsets of $L^p(T; \mathbb{R}^m)$ and its generalized gradient satisfies*

$$\partial J(u) \subset \{w \in L^q(T; \mathbb{R}^m) : \ w(x) \in \partial_y j(x, u(x)) \ \text{for a.e.} \ x \in T\}. \quad (2.38)$$

Moreover, if $j(x, \cdot)$ is regular at $u(x)$ for almost all $x \in T$, then J is regular at u and (2.38) holds with equality.

Proof: Using Hölder's inequality in conjunction with (2.35) or (2.36), one verifies easily that J is Lipschitz continuous on bounded subsets of $L^p(T; \mathbb{R}^m)$. Definition 2.161 ensures that the map $x \mapsto j_y^o(x, u(x); v(x))$ is measurable on T [see the arguments given in the proof of Theorem 2.7.2 in Clarke [68] related with the superpositional measurability of $s \mapsto j_y^o(\cdot, s; 1)$], where the subscript y indicates that the generalized directional derivative j^o is taken with respect to the second variable. Furthermore, by assumption (2.35) or (2.36), it is known that this function is integrable. Let us check the inequality

$$J^o(u; v) \leq \int_T j_y^o(x, u(x); v(x)) dx, \quad \forall \, u, v \in L^p(T; \mathbb{R}^m). \quad (2.39)$$

If (2.35) is assumed, then Fatou's lemma leads directly to (2.39). In the case where (2.36) is admitted, Theorem 2.177 enables us to write

$$\frac{j(x, u(x) + \lambda v(x)) - j(x, u(x))}{\lambda} = \langle \zeta_x, v(x) \rangle,$$

with $\zeta_x \in \partial j(x, u^*(x))$ for some $u^*(x)$ lying on the open segment in $\mathbb{R}^m$ with endpoints $u(x)$ and $u(x) + \lambda v(x)$. Then Fatou's lemma implies (2.39). Notice that the application of Fatou's lemma is possible because of the growth condition in (2.36). In view of (2.39), any $z \in \partial J(u)$ belongs to the subdifferential at 0 of the convex function on $L^p(T; \mathbb{R}^m)$ given by

$$v \in L^p(T; \mathbb{R}^m) \mapsto \int_T j_y^o(x, u(x); v(x)) dx \in \mathbb{R}.$$

The subdifferentiation under the integral for the convex integrands (see [79]) and Remark 2.170 allow us to conclude that (2.38) holds. Finally, assume further that $j(x, \cdot)$ is regular at $u(x)$ for almost all $x \in T$. Then, under either assumption (2.35) or (2.36), we may apply Fatou's lemma to get

$$\liminf_{\lambda \downarrow 0} \frac{1}{\lambda}(J(u + \lambda v) - J(u)) \geq \int_T j_y'(x, u(x); v(x)) dx$$

$$= \int_T j_y^0(x, u(x); v(x)) dx, \quad \forall \, v \in L^p(T; \mathbb{R}^m).$$

Combining with (2.39), it follows that the directional derivative $J'(u; v)$ exists and $J'(u; v) = J^o(u; v)$ for every $v \in L^p(T; \mathbb{R}^m)$, thus we obtain the regularity of J at u, as well as the equality

$$J^o(u; v) = J'(u; v) = \int_T j'_y(x, u(x); v(x)) dx, \quad \forall \, v \in L^p(T; \mathbb{R}^m).$$

Thereby, due to the regularity assumption for $j(x, \cdot)$, it is seen that (2.38) becomes an equality. $\square$

2.5.3 Critical Point Theory

In this subsection, we present basic elements of a general critical point theory for nonsmooth functionals $I : X \to \mathbb{R} \cup \{+\infty\}$ on a real Banach space X verifying the structural hypothesis

(H) $I = \Phi + \Psi$, with $\Phi : X \to \mathbb{R}$ locally Lipschitz and $\Psi : X \to \mathbb{R} \cup \{+\infty\}$ convex, lower semicontinuous, and proper (i.e., $\not\equiv +\infty$).

For more details and developments, we refer to the works [102], [103, Chap. 4], [156], [171, Chap. 3], [173, Chap. 2].

Definition 2.182. *An element $u \in X$ is called a critical point of the functional $I : X \to \mathbb{R} \cup \{+\infty\}$ satisfying (H) if*

$$\Phi^o(u; v - u) + \Psi(v) - \Psi(u) \geq 0 \quad \forall \, v \in X, \tag{2.40}$$

where the notation $\Phi^o(u; \cdot)$ means the generalized directional derivative of Φ at u (see Definition 2.161).

Definition 2.182 can be expressed equivalently as follows.

Proposition 2.183. *An element $u \in X$ is a critical point of the functional $I : X \to \mathbb{R} \cup \{+\infty\}$ satisfying (H) if and only if $u \in D(\partial \Psi)$ and*

$$0 \in \partial \Phi(u) + \partial \Psi(u), \tag{2.41}$$

where the notations $\partial \Phi(u)$ and $\partial \Psi(u)$ stand for the generalized gradient of Φ at u and the subdifferential (in the sense of convex analysis) of Ψ at u, respectively, whereas $D(\partial \Psi)$ denotes the domain of the subdifferential $\partial \Psi$; i.e., $D(\partial \Psi) = \{x \in X : \partial \Psi(x) \neq \emptyset\}$.

Proof: Assume that $u \in X$ satisfies relation (2.40), or equivalently,

$$\Phi^o(u; w) + \Psi(w + u) - \Psi(u) \geq 0 \quad \forall \, w \in X.$$

It follows that 0 is a minimum point of the convex function

$$w \mapsto \Phi^o(u; w) + \Psi(w + u) - \Psi(u),$$

so $u \in D(\partial \Psi)$, and by using the subdifferential calculus for convex functions,

$$0 \in \partial(\Phi^o(u; \cdot) + \Psi(\cdot + u) - \Psi(u))(0) = \partial(\Phi^o(u; \cdot))(0) + \partial \Psi(u) = \partial \Phi(u) + \partial \Psi(u)$$

(see the last part of the statement of Proposition 2.174). Conversely, if (2.41) is satisfied, $\xi \in \partial\Phi(u)$ and $\eta \in \partial\Psi(u)$ exist such that $0 = \xi + \eta$ in X^*. Taking into account Definition 2.166 and because $\eta \in \partial\Psi(u)$, we derive

$$\Phi^o(u; v - u) + \Psi(v) - \Psi(u) \geq \langle \xi, v - u \rangle + \langle \eta, v - u \rangle = \langle \xi + \eta, v - u \rangle = 0$$

for all $v \in X$. $\square$

Corollary 2.184. *Let $\Phi : X \to \mathbb{R}$ be a locally Lipschitz function, and let K be a nonempty, closed, convex subset of X. Denote by $I_K : X \to \mathbb{R} \cup \{+\infty\}$ the indicator function of K; i.e., $I_K(x) = 0$ whenever $x \in K$ and $I_K = +\infty$ otherwise. Then $u \in X$ is a critical point of $\Phi + I_K$ if and only if $u \in K$ and $0 \in \partial\Phi(u) + N_K(u)$, where $N_K(u) = \{\eta \in X^* : \langle \eta, v - u \rangle \leq 0, \forall v \in K\}$ is the normal cone of K at u.*

Proof: One applies Proposition 2.183 for $\Psi = I_K$. $\square$

The examples below illustrate the concept of critical point introduced in Definition 2.182.

Example 2.185. Every local minimum $u \in X$ of a nonsmooth functional $I : X \to \mathbb{R} \cup \{+\infty\}$ satisfying (H) with $I(u) < +\infty$ is a critical point in the sense of Definition 2.182. Indeed, if $u \in X$ with $I(u) < +\infty$ is a local minimum of I, then, by convexity of Ψ, for any $v \in X$ and a small $t > 0$, we have

$$0 \leq I(u + t(v - u)) - I(u) \leq \Phi(u + t(v - u)) - \Phi(u) + t(\Psi(v) - \Psi(u)).$$

Dividing by t and letting $t \to 0^+$, we deduce that u fulfills Definition 2.182.

Example 2.186. Every minimum $u \in X$ of $\Phi|_K$ with $\Phi : X \to \mathbb{R}$ locally Lipschitz and a nonempty, closed, convex subset $K \subset X$ is a critical point of $\Phi + I_K$ in the sense of Definition 2.182. Indeed, if u is a minimum of Φ on K, then $u \in K$ and

$$\inf_X (\Phi + I_K) = (\Phi + I_K)(u) = \Phi(u),$$

and Example 2.185 leads to the desired conclusion.

Example 2.187. Every local maximum $u \in X$ of a nonsmooth functional $I : X \to \mathbb{R} \cup \{+\infty\}$ satisfying (H) with $I(u) < +\infty$ is a critical point in the sense of Definition 2.182. Indeed, under the given hypotheses, u is in the interior of the effective domain of Ψ, and thus, Ψ is Lipschitz continuous near u. Actually, $I = \Phi + \Psi$ is Lipschitz continuous near u and Proposition 2.174 yields

$$0 \in \partial I(u) = \partial(\Phi + \Psi)(u) \subset \partial\Phi(u) + \partial\Psi(u),$$

where $\partial\Phi(u)$ is the generalized gradient of Φ and $\partial\Psi(u)$ is the subdifferential of Ψ in the sense of convex analysis (see Example 2.169). According to Proposition 2.183, u is a critical point of I.

Example 2.188. Let $\Phi : X \to \mathbb{R}$ be a locally Lipschitz function. Setting $\Psi = 0$ in (H), we see by Definition 2.182 that $u \in X$ is a critical point of Φ if and only if $0 \in \partial\Phi(u)$. Therefore, in this case, Definition 2.182 reduces to the definition of Chang [64] for a critical point of a locally Lipschitz function. In particular, if $\Phi \in C^1(X)$ and $\Psi = 0$ in (H), one obtains the notion of critical point in the smooth critical point theory.

Example 2.189. Consider in assumption (H) that $\Phi \in C^1(X)$ and $\Psi : X \to \mathbb{R} \cup \{+\infty\}$ is convex, lower semicontinuous, and proper. Notice that the functional $I = \Phi + \Psi : X \to \mathbb{R} \cup \{+\infty\}$ complies with hypothesis (H). Then, according to Definition 2.182, $u \in X$ is a critical point of $I = \Phi + \Psi$ if and only if

$$\langle \Phi'(u), v - u \rangle + \Psi(v) - \Psi(u) \geq 0 \quad \forall\, v \in X;$$

i.e., $-\Phi'(u) \in \partial\Psi(u)$. Consequently, in this case, Definition 2.182 reduces to the definition of critical point as given by Szulkin [211].

We present the Palais–Smale condition for the class of nonsmooth functionals satisfying the structural hypothesis (H).

Definition 2.190. *The functional* $I = \Phi + \Psi : X \to \mathbb{R} \cup \{+\infty\}$ *in* (H) *is said to satisfy the Palais–Smale condition (for short,* (PS)) *if every sequence* $(u_n) \subset X$ *such that* $(I(u_n))$ *is bounded in* $\mathbb{R}$ *and*

$$\Phi^\circ(u_n; v - u_n) + \Psi(v) - \Psi(u_n) \geq -\varepsilon_n \|v - u_n\|, \quad \forall\, v \in X,$$

for a sequence (ε_n) *with* $\varepsilon_n \downarrow 0$, *contains a strongly convergent subsequence.*

Example 2.191. If $\Phi \in C^1(X)$ and $\Psi : X \to \mathbb{R} \cup \{+\infty\}$ is convex, lower semicontinuous, and proper, then Definition 2.190 coincides with the (PS) condition in the sense of Szulkin [211]. In particular, if $\Phi \in C^1(X)$ and $\Psi = 0$, then Definition 2.190 is the usual smooth (PS) condition.

We need the following result from [211].

Lemma 2.192. *Let* $\chi : X \to \mathbb{R} \cup \{+\infty\}$ *be a lower semicontinuous, convex function with* $\chi(0) = 0$. *If*

$$\chi(x) \geq -\|x\|, \quad \forall\, x \in X,$$

then $z \in X^*$ *exists such that* $\|z\|_{X^*} \leq 1$ *and*

$$\chi(x) \geq \langle z, x \rangle, \quad \forall\, x \in X.$$

Proof: Consider the following convex subsets A and B of $X \times \mathbb{R}$:

$$A = \{(x,t) \in X \times \mathbb{R} : \|x\| < -t\} \text{ and } B = \{(x,t) \in X \times \mathbb{R} : \chi(x) \leq t\}.$$

Notice that A is an open set, and due to the condition $\chi(x) \geq -\|x\|$, one has $A \cap B = \emptyset$. A well-known separation result (see [24, p. 5]) yields the existence of $\alpha, \beta \in \mathbb{R}$ and $w \in X^*$ such that $(w, \alpha) \neq (0, 0)$,

$$\langle w, x \rangle - \alpha t \geq \beta, \quad \forall\, (x, t) \in \bar{A}$$

and

$$\langle w, x \rangle - \alpha t \leq \beta, \quad \forall\, (x, t) \in B.$$

We see that $\beta = 0$ because $(0, 0) \in \bar{A} \cap B$. Set $t = -\|x\|$ in the first inequality above. It follows that $\langle w, x \rangle \geq -\alpha\|x\|$, $\forall\, x \in X$, which implies $\alpha > 0$ and $\|w\|_{X^*} \leq \alpha$. Set $z = \alpha^{-1}w$. Using $t = \chi(x)$, we deduce that $\langle z, x \rangle \leq \chi(x)$, $\forall\, x \in X$. As $\|w\|_{X^*} \leq \alpha$, we obtain $\|z\|_{X^*} \leq 1$. $\qquad \square$

The following result establishes the equivalence between Definition 2.190 with $\Psi = 0$ and the (PS) condition in the sense of Chang [64].

Proposition 2.193. *A locally Lipschitz function $\Phi : X \to \mathbb{R}$ satisfies the (PS) condition in the sense of Definition 2.190 if and only if Φ verifies the Palais–Smale condition as defined in [64].*

Proof: Assume that the locally Lipschitz function $\Phi : X \to \mathbb{R}$ satisfies the (PS) condition formulated in Definition 2.190. Let a sequence $(u_n) \subset X$ with $\Phi(u_n)$ bounded and for which

$$\lambda(u_n) = \inf_{w \in \partial\Phi(u_n)} \|w\|_{X^*} \to 0 \quad \text{as } n \to \infty.$$

It is known from Proposition 2.171 (i) that an element $z_n \in \partial\Phi(u_n)$ can be found such that $\lambda(u_n) = \|z_n\|_{X^*}$. As

$$\Phi^o(u_n; v) \geq \langle z_n, v \rangle \geq -\|z_n\|_{X^*}\|v\|, \quad \forall\, v \in X,$$

the inequality in Definition 2.190 (with $\Psi = 0$) is verified with $\varepsilon_n = \|z_n\|$. It implies that (u_n) possesses a convergent subsequence, which ensures the Palais–Smale condition in the sense of [64].

Conversely, we suppose that Φ verifies the Palais–Smale condition in the sense of [64]. Let (u_n) be a sequence as in Definition 2.190. We can apply Lemma 2.192 for $\chi = \frac{1}{\varepsilon_n}\Phi^o(u_n; \cdot)$, which gives an element $w_n \in X^*$ with $\|w_n\|_{X^*} \leq 1$ and

$$\frac{1}{\varepsilon_n}\Phi^o(u_n; x) \geq \langle w_n, x \rangle, \ \forall\, x \in X.$$

It follows that $\varepsilon_n w_n \in \partial\Phi(u_n)$ and $\varepsilon_n w_n \to 0$ in X^* as $n \to \infty$. According to the Palais–Smale condition in [64], we have that (u_n) contains a convergent subsequence, so the (PS) condition in the sense of Definition 2.190 holds. $\qquad \square$

2.5.4 Linking Theorem

The objective of this subsection is to provide sufficient conditions for the existence of critical points in the setting of functionals of type (H) in Sect. 2.5.3. We start with the following minimization result.

Theorem 2.194. *Assume that the function* $I = \Phi + \Psi : X \to \mathbb{R} \cup \{+\infty\}$ *satisfies hypothesis* (H), *is bounded from below, and verifies the* (PS) *condition. Then* $u \in X$ *exists such that* $I(u) = \inf_X I \in \mathbb{R}$ *and* u *is a critical point of* I *in the sense of Definition 2.182.*

Proof: Denote $m = \inf_X I \in \mathbb{R}$. We find a (minimizing) sequence $(u_n) \subset X$ such that

$$I(u_n) < m + \varepsilon_n^2,$$

for a sequence (ε_n) of positive numbers, with $\varepsilon_n \downarrow 0$. Applying Ekeland's variational principle (cf. [91]) to the function I, a sequence $(v_n) \subset X$ exists such that

$$I(v_n) < m + \varepsilon_n^2$$

and

$$I(v) \geq I(v_n) - \varepsilon_n \|v_n - v\|, \quad \forall\, v \in X,\ \forall\, n \in \mathbb{N}.$$

Setting $v = (1-t)v_n + tw$ in the above inequality, for arbitrary $0 < t < 1$ and $w \in X$, we obtain

$$\Phi((1-t)v_n + tw) + \Psi((1-t)v_n + tw)$$
$$\geq \Phi(v_n) + \Psi(v_n) - \varepsilon_n t\|w - v_n\|, \quad \forall\, w \in X,\ \forall\, t \in (0,1).$$

The convexity of $\Psi : X \to \mathbb{R} \cup \{+\infty\}$ yields

$$\Phi((1-t)v_n + tw) - t\Psi(v_n) + t\Psi(w)$$
$$\geq \Phi(v_n) - \varepsilon_n t\|w - v_n\|, \quad \forall\, w \in X,\ \forall\, t \in (0,1).$$

Dividing by t and letting $t \downarrow 0$, we deduce that for all $w \in X$, one has

$$\Phi^o(v_n; w - v_n) + \Psi(w) - \Psi(v_n)$$
$$\geq \limsup_{t \downarrow 0} \frac{1}{t}(\Phi(v_n + t(w - v_n)) - \Phi(v_n)) + \Psi(w) - \Psi(v_n) \geq -\varepsilon_n\|w - v_n\|.$$

On the other hand, we have $\Phi(v_n) + \Psi(v_n) \to m$ as $n \to \infty$. Then the (PS) condition (see Definition 2.190) implies that along a relabelled subsequence $v_n \to u$ in X, for some $u \in X$. The lower semicontinuity of I yields $I(u) \leq \liminf_{n \to \infty} I(v_n) \leq m$, so $I(u) = m$. Making use of Example 2.185, we derive that u is a critical point of I. $\qquad\square$

We now focus on the existence of critical points for functionals of type (H) that are not obtained by minimization, thus, saddle-points. The subsequent minimax principle makes use of the notion of linking as given in [88].

Definition 2.195. *Let S be a nonempty closed subset of the Banach space X, and let Q be a compact topological submanifold of X with nonempty boundary ∂Q (in the sense of manifolds with boundary). We say that S and Q link if $S \cap \partial Q = \emptyset$ and $f(Q) \cap S \neq \emptyset$ whenever $f \in \Gamma$, where*

$$\Gamma := \{f \in C(Q, X) : f|_{\partial Q} = \mathrm{id}_{\partial Q}\}.$$

Example 2.196. Let $X = E \times \mathbb{R}$, with a Banach space E, and let $0 < \rho < r$. The sets $S = E \times \{\rho\}$ and $Q = \{(0, tr) \in E \times \mathbb{R} : t \in [0, 1]\}$ link.

The following result given in [171, Chap. 3] provides critical points of saddle-point type for nonsmooth functionals having the structure in (H).

Theorem 2.197. *Let the functional $I : X \to \mathbb{R} \cup \{+\infty\}$ satisfy assumptions (H) and (PS). Let S and Q link in the sense of Definition 2.195. Assume further that*

$$\sup_Q I \in \mathbb{R}, \ b := \inf_S I \in \mathbb{R}, \ a := \sup_{\partial Q} I < b.$$

Then the number

$$c := \inf_{f \in \Gamma} \sup_{x \in Q} I(f(x)),$$

with Γ in Definition 2.195, is a critical value of I; that is, there is a critical point u of I in the sense of Definition 2.182 and $I(u) = c$. Moreover, $c \geq b$.

Proof: The inequality $c \geq b$ is a direct consequence of linking property in Definition 2.195. Arguing by contradiction we assume that c is not a critical value of I. Applying the deformation result in Theorem 3.1 in [171], with $\bar{\varepsilon} = c - a > 0$, we get an $\varepsilon \in (0, \bar{\varepsilon})$ as stated therein. Define Γ_1 as the set of all continuous mappings $\varphi : Q \to X$ such that

$$\varphi(\partial Q) \subset \left\{x \in X : I(x) \leq c - \frac{\varepsilon}{2}\right\}$$

and $\varphi|_{\partial Q}$, $\mathrm{id}_{\partial Q}$ are homotopic maps from ∂Q into $\{x \in X : I(x) \leq c - \frac{\varepsilon}{4}\}$. We have $\mathrm{id}_Q \in \Gamma_1$. Using the definitions of c and Γ_1, we obtain

$$c = \inf_{\varphi \in \Gamma_1} \sup_{x \in Q} I(\varphi(x)). \tag{2.42}$$

It is seen that Γ_1 is a closed subset of the Banach space $C(Q; X)$ with respect to the uniform norm $\|\varphi\| = \sup_{x \in Q} \|\varphi(x)\|$. Consider the lower semicontinuous functional $\Pi : \Gamma_1 \to \mathbb{R} \cup \{+\infty\}$ defined by

$$\Pi(\varphi) = \sup_{x \in Q} I(\varphi(x)), \quad \forall \varphi \in \Gamma_1.$$

Taking into account (2.42), Ekeland's variational principle [91] applied to the function Π on Γ_1 yields a $\varphi \in \Gamma_1$ satisfying $c \leq \Pi(\varphi) \leq c + \varepsilon$ and

$$\Pi(\psi) - \Pi(\varphi) \geq -\varepsilon \|\psi - \varphi\|, \quad \forall \, \psi \in \Gamma_1. \tag{2.43}$$

Theorem 3.1 in [171] provides a deformation $h_\varphi : W \times [0, \bar{s}] \to X$ corresponding to the compact set $A = \varphi(Q)$, where W is a closed neighborhood of A in X and $\bar{s}$ is a positive number, satisfying

$$\|v - h_\varphi(v, s)\| \leq s, \quad \forall \, v \in W, \, \forall \, s \in [0, \bar{s}], \tag{2.44}$$

$$\sup_{v \in A} I(h_\varphi(v, s)) - \sup_{v \in A} I(v) \leq -2\varepsilon s, \quad \forall \, s \in [0, \bar{s}]. \tag{2.45}$$

Let us show that for $\bar{s} > 0$ small enough, we have

$$h_\varphi(\varphi(\cdot), s) \in \Gamma_1, \quad \forall \, s \in [0, \bar{s}]. \tag{2.46}$$

To prove (2.46), it suffices to note that $h_\varphi(\varphi(\cdot), s)|_{\partial Q}$ and $\varphi|_{\partial Q}$ are homotopic maps from ∂Q into $\{x \in X : I(x) \leq c - \frac{\varepsilon}{2}\}$. Such a homotopy is $(x, t) \mapsto h_\varphi(\varphi(x), ts)$ as can be seen from (2.44) and (2.46). It follows from (2.45), (2.46), (2.43), and (2.44) that

$$-2\varepsilon s \geq \Pi(h_\varphi(\varphi(\cdot), s)) - \Pi(\varphi)$$
$$\geq -\varepsilon \|h_\varphi(\varphi(\cdot), s) - \varphi\| \geq -\varepsilon s, \quad \forall \, 0 \leq s \leq \bar{s}.$$

This contradiction proves that our initial assumption that c is not a critical value of I is not possible, which completes the proof. $\qquad \square$

Corollary 2.198. *Let E be a Banach space, let $\Phi : E \times \mathbb{R} \to \mathbb{R}$ be locally Lipschitz, and let $\Psi : E \times \mathbb{R} \to \mathbb{R} \cup \{+\infty\}$ be proper, convex, and lower semicontinuous. Suppose that the function*

$$F = \Phi + \Psi : E \times \mathbb{R} \to \mathbb{R} \cup \{+\infty\}$$

satisfies the (PS) condition and positive numbers ρ and r exist with $\rho < r$ such that $F(0, 0) \leq 0$, $F(0, r) \leq 0$, $0 < \inf_{v \in E} F(v, \rho) < +\infty$. Then

$$c = \inf\{\sup_{t \in [0,1]} F(g(t)) : g \in C([0, 1], E \times \mathbb{R}), \, g(0) = (0, 0), \, g(1) = (0, r)\}$$

is a critical value of F, and we have the estimate

$$\inf_{v \in E} F(v, \rho) \leq c \leq \sup_{t \in [0,1]} F(0, tr).$$

Proof: Apply Theorem 2.197 with $X = E \times \mathbb{R}$ and $I = F$ using the linking in Example 2.196. The last inequality above is obtained by taking the path $g(t) = (0, tr)$ for all $t \in [0, 1]$. $\qquad \square$

Remark 2.199. In the general setting of nonsmooth functionals verifying hypothesis (H), Theorem 2.197 incorporates important minimax results in the critical point theory such as the mountain-pass theorem, saddle-point theorem, and generalized mountain-pass theorem.

Remark 2.200. A basic hypothesis in Theorem 2.197 is that $b > a$. Under the situation of linking in Theorem 2.197, we then have $c > a$. As from the expression of the minimax value c it is seen that always $c \geq a$, the situation that is not covered by Theorem 2.197 is the so-called liming case $c = a$. The specific situation $c = a$ is treated in [156].

3

Variational Equations

This chapter deals with existence and comparison results for weak solutions of nonlinear elliptic and parabolic problems. The ideas and methods developed here will also be useful in the treatment of nonsmooth variational problems in later chapters. Section 3.1 deals with semilinear elliptic Dirichlet boundary value problems and may be considered as a preparatory section for Sect. 3.2 and Sect. 3.3, where general quasilinear elliptic and parabolic problems are treated. The purpose of Sect. 3.1 is to emphasize the basic ideas and to present various approaches without overburdening the presentation with too many technicalities. As an application of the general results of Sect. 3.2 combined with critical point theory, the existence of multiple and sign-changing solutions is proved in Sect. 3.4. Finally, in Sect. 3.5, the concept of sub-supersolutions is extended to some nonstandard elliptic boundary value problem, which in the one-space dimensional and semilinear case reduces to a second-order ordinary differential equation subject to periodic boundary conditions. The chapter concludes with bibliographical notes and further applications and extensions of the theory developed in the preceeding sections.

3.1 Semilinear Elliptic Equations

In this section, the basic ideas and methods to prove existence and comparison results will be demonstrated with the help of the following simple semilinear elliptic boundary value problem (BVP, for short):

$$-\Delta u + g(u) = f \quad \text{in} \quad \Omega, \quad u = 0 \quad \text{on} \quad \partial\Omega, \tag{3.1}$$

where $\Omega \subset \mathbb{R}^N$ is a bounded domain with Lipschitz boundary $\partial\Omega$, and $g : \mathbb{R} \to \mathbb{R}$ is a continuous function. Let $V = W^{1,2}(\Omega)$, $V_0 = W_0^{1,2}(\Omega)$, and assume $f \in V_0^*$. If G denotes the Nemytskij operator related with g by

$$G(u)(x) = g(u(x)),$$

then u is a called a weak solution of the BVP (3.1) if $G(u) \in L^2(\Omega)$ ($\subset V_0^*$) and u satisfies

$$u \in V_0 : \quad -\Delta u + G(u) = f \quad \text{in } V_0^*, \tag{3.2}$$

which is equivalent to

$$u \in V_0 : \quad \int_\Omega (\nabla u \nabla \varphi + G(u)\varphi) \, dx = \langle f, \varphi \rangle \quad \text{for all } \varphi \in V_0. \tag{3.3}$$

3.1.1 Comparison Principle

We first introduce the notion of (weak) sub- and supersolution for the BVP (3.1).

Definition 3.1. *The function $\underline{u} \in V$ is called a subsolution of (3.1) if $G(\underline{u}) \in L^2(\Omega)$, $\underline{u} \leq 0$ on $\partial \Omega$, and*

$$\int_\Omega (\nabla \underline{u} \nabla \varphi + G(\underline{u})\varphi) \, dx \leq \langle f, \varphi \rangle \quad \text{for all } \varphi \in V_0 \cap L_+^2(\Omega). \tag{3.4}$$

Similarly, we have the following definition for a supersolution.

Definition 3.2. *The function $\bar{u} \in V$ is called a supersolution of (3.1) if $G(\bar{u}) \in L^2(\Omega)$, $\bar{u} \geq 0$ on $\partial \Omega$, and*

$$\int_\Omega (\nabla \bar{u} \nabla \varphi + G(\bar{u})\varphi) \, dx \geq \langle f, \varphi \rangle \quad \text{for all } \varphi \in V_0 \cap L_+^2(\Omega). \tag{3.5}$$

Remark 3.3. The condition $\underline{u} \leq 0$ on $\partial \Omega$ means that $\gamma(\underline{u}) \leq 0$ in $L^2(\partial \Omega)$, where $\gamma : V \to L^2(\partial \Omega)$ denotes the trace operator (see Sect. 2.2.2). Inequality (3.4) is equivalent to the following inequality in V_0^*:

$$-\Delta \underline{u} + G(\underline{u}) \leq f,$$

where the order relation is generated by the dual-order cone of V_0^*. Similar statements hold for $\bar{u}$.

We are going to prove the following existence and comparison result.

Theorem 3.4. *Let $\underline{u}$ and $\bar{u}$ be sub- and supersolutions of (3.1), respectively, that satisfy $\underline{u} \leq \bar{u}$, and assume a local growth condition for g in the form*

$$|g(v(x))| \leq k(x) \quad \text{for all } v \in [\underline{u}, \bar{u}], \tag{3.6}$$

where $k \in L_+^2(\Omega)$. Then solutions of the BVP (3.1) exist within the ordered interval $[\underline{u}, \bar{u}]$.

Proof: By the growth condition (3.6), the Nemytskij operator G provides a continuous and bounded mapping from the ordered interval $[\underline{u}, \bar{u}] \subset L^2(\Omega)$ into $L^2(\Omega)$. As $L^2(\Omega)$ is continuously embedded into V_0^*, we may consider $G(u) \in L^2(\Omega)$ as an element of V_0^* given by

$$\langle G(u), \varphi \rangle = \int_\Omega G(u)(x) \, \varphi(x) \, dx, \quad \varphi \in V_0.$$

The Laplacian $-\Delta : V_0 \to V_0^*$ given by

$$\langle -\Delta u, \varphi \rangle = \int_\Omega \nabla u \nabla \varphi \, dx, \quad \varphi \in V_0,$$

defines a strongly monotone, bounded, and continuous mapping (see Lemma 2.111). We next consider the following auxiliary truncated BVP:

$$u \in V_0 : \quad -\Delta u + (G \circ T)u = f \quad \text{in } V_0^*, \tag{3.7}$$

where T is the truncation operator related with the given ordered pair of sub- and supersolutions, which is given by

$$Tu(x) = \begin{cases} \bar{u}(x) & \text{if} \quad u(x) > \bar{u}(x), \\ u(x) & \text{if} \quad \underline{u}(x) \leq u(x) \leq \bar{u}(x), \\ \underline{u}(x) & \text{if} \quad u(x) < \underline{u}(x). \end{cases}$$

Due to Lemma 2.89, the operator T is, in particular, bounded and continuous from from $L^2(\Omega)$ into $[\underline{u}, \bar{u}] \subset L^2(\Omega)$, which implies that the composition $G \circ T : L^2(\Omega) \to L^2(\Omega)$ is continuous and uniformly bounded with

$$\|(G \circ T)v\|_2 \leq \|k\|_2 \quad \text{for all} \ \ v \in L^2(\Omega), \tag{3.8}$$

where $\| \cdot \|_2$ denotes the norm in $L^2(\Omega)$. Due to the compact embedding of $V_0 \subset L^2(\Omega)$, we infer that $G \circ T : V_0 \to L^2(\Omega) \subset V_0^*$ is strongly continuous and, thus, in particular, pseudomonotone (see Lemma 2.98). As $-\Delta : V_0 \to V_0^*$ is strongly monotone, continuous, and bounded, it is also pseudomonotone, and thus by Lemma 2.98, we obtain that

$$-\Delta + G \circ T : V_0 \to V_0^* \tag{3.9}$$

is a pseudomonotone, bounded, and continuous operator. By Theorem 2.99, the BVP (3.7) possesses a solution provided that the operator $-\Delta + G \circ T : V_0 \to V_0^*$ is coercive. The latter, however, follows easily from the uniform boundedness (3.8) and the strong monotonicity of $-\Delta$, which yields

$$\langle -\Delta u + (G \circ T)u, u \rangle \geq c \|u\|_{V_0}^2 - \|k\|_2 \|u\|_2 \geq c \|u\|_{V_0}^2 - \|k\|_2 \|u\|_{V_0},$$

where we have used that $u \mapsto (\int_\Omega |\nabla u|^2 \, dx)^{1/2}$ defines an equivalent norm in V_0. This process completes the existence proof for the auxiliary BVP (3.7).

The assertion of the theorem is proved provided we can show that there is a solution u of the auxiliary problem (3.7), which satisfies $\underline{u} \leq u \leq \bar{u}$; since then we have $Tu = u$, and thus, u must be also a solution of the original problem (3.1). In fact we are going to show that any solution of (3.7) is contained in the ordered interval $[\underline{u}, \bar{u}]$. Let u be any solution of (3.7), which is equivalent to

$$\int_\Omega (\nabla u \nabla \varphi + (G \circ T)u\, \varphi)\, dx = \langle f, \varphi \rangle \quad \text{for all} \ \ \varphi \in V_0. \tag{3.10}$$

We first show that $u \leq \bar{u}$. Subtracting the inequality (3.5) for the supersolution from (3.10), we obtain for all $\varphi \in V_0 \cap L_+^2(\Omega)$ the inequality

$$\int_\Omega \nabla (u - \bar{u}) \nabla \varphi\, dx + \int_\Omega ((G \circ T)u - G(\bar{u}))\, \varphi\, dx \leq 0. \tag{3.11}$$

Testing (3.11) with $\varphi = (u - \bar{u})^+ \in V_0 \cap L_+^2(\Omega)$ and observing that

$$\int_\Omega ((G \circ T)u - G(\bar{u}))\, (u - \bar{u})^+\, dx = 0,$$

we get

$$0 \leq \int_\Omega |\nabla(u - \bar{u})^+|^2\, dx = \int_\Omega \nabla(u - \bar{u}) \nabla(u - \bar{u})^+\, dx \leq 0,$$

and thus $\|(u - \bar{u})^+\|_{V_0} = 0$, which implies $(u - \bar{u})^+ = 0$; i.e., $u \leq \bar{u}$. The proof for the inequality $\underline{u} \leq u$ can be done in an obvious similar way. This process completes the proof of the theorem.   □

Remark 3.5. The existence proof for the auxiliary BVP (3.7) can also be done in a more elementary way by using Lax–Milgram's Theorem (see Corollary 2.102) and Schauder's fixed point theorem (see Theorem 2.2), because here we have the special situation that V_0 is a Hilbert space and

$$\langle -\Delta u, \varphi \rangle = \int_\Omega \nabla u \nabla \varphi\, dx, \quad u, \varphi \in V_0,$$

defines a coercive and bounded bilinear form in V_0.

3.1.2 Directed and Compact Solution Set

We denote by $\mathcal{S}$ the set of all solutions of the BVP (3.1) within the ordered interval $[\underline{u}, \bar{u}]$. The main goal of this section is to show that $\mathcal{S}$ is a directed and compact set in V_0. The directedness will be seen as an immediate consequence of the following generalized version of Theorem 3.4.

Theorem 3.6. *Let $\underline{u}_1, \ldots, \underline{u}_k$ and $\bar{u}_1, \ldots, \bar{u}_m$ with $k, m \in \mathbb{N}$ be sub- and supersolutions of (3.1), respectively, such that*

$$\underline{u} = \max\{\underline{u}_1, \ldots, \underline{u}_k\} \leq \bar{u} = \min\{\bar{u}_1, \ldots, \bar{u}_m\}, \qquad (3.12)$$

and assume a local growth condition for g in the form

$$|g(v(x))| \leq k(x) \quad \text{for all} \ \ v \in [\min\{\underline{u}_1, \ldots, \underline{u}_k\}, \max\{\bar{u}_1, \ldots, \bar{u}_m\}], \qquad (3.13)$$

where $k \in L^2_+(\Omega)$. Then solutions of the BVP (3.1) exist within the ordered interval $[\underline{u}, \bar{u}]$.

Proof: The proof follows the idea of the proof of Theorem 3.4 and is based on the consideration of a suitably constructed auxiliary problem, which is now more involved and which reads as follows:

$$u \in V_0: \quad -\Delta u + P(u) = f \quad \text{in} \ V_0^*, \qquad (3.14)$$

where the operator P is given by

$$P(u) = (G \circ T)u + \sum_{i=1}^{m} |(G \circ T^i)u - (G \circ T)u| - \sum_{j=1}^{k} |(G \circ T_j)u - (G \circ T)u|, \qquad (3.15)$$

where the truncation operators T_j, T^i, and T are defined as follows:

$$Tu(x) = \begin{cases} \bar{u}(x) & \text{if} \ u(x) > \bar{u}(x), \\ u(x) & \text{if} \ \underline{u}(x) \leq u(x) \leq \bar{u}(x), \\ \underline{u}(x) & \text{if} \ u(x) < \underline{u}(x), \end{cases}$$

$$T_j u(x) = \begin{cases} \underline{u}_j(x) & \text{if} \ u(x) < \underline{u}_j(x), \\ u(x) & \text{if} \ \underline{u}_j(x) \leq u(x) \leq \bar{u}(x), \\ \bar{u}(x) & \text{if} \ u(x) > \bar{u}(x), \end{cases}$$

$$T^i u(x) = \begin{cases} \underline{u}(x) & \text{if} \ u(x) < \underline{u}(x), \\ u(x) & \text{if} \ \underline{u}(x) \leq u(x) \leq \bar{u}_i(x), \\ \bar{u}_i(x) & \text{if} \ u(x) > \bar{u}_i(x), \end{cases}$$

for $1 \leq i \leq m, 1 \leq j \leq k, x \in \Omega$. The operators $G \circ T$, $G \circ T_j$, $G \circ T^i$ stand for the compositions of the Nemytskij operator G and the truncation operators T, T_j, T^i, respectively, and we have

$$\langle |(G \circ T^i)u - (G \circ T)u|, v \rangle = \int_\Omega |g(\cdot, T^i u) - g(\cdot, Tu)| \, v \, dx$$

as well as

$$\langle |(G \circ T_j)u - (G \circ T)u|, v \rangle = \int_\Omega |g(\cdot, T_j u) - g(\cdot, Tu)| \, v \, dx$$

for all $u, v \in V_0$. As T_j, T^i, $T : L^2(\Omega) \to L^2(\Omega)$ are bounded and continuous, it follows from the compact embedding $V_0 \subset L^2(\Omega)$ and in view of the growth condition imposed on g that $P : V_0 \to L^2(\Omega) \subset V_0^*$ is uniformly bounded and completely continuous. The same arguments as in the proof of Theorem 3.4 apply to ensure that

$$-\Delta + P : V_0 \to V_0^*$$

is pseudomonotone, bounded, continuous, and coercive, which implies that $-\Delta + P : V_0 \to V_0^*$ is surjective and thus solutions of the auxiliary BVP (3.14) exist. The proof of the theorem is accomplished provided any solution u of (3.14) can be shown to satisfy

$$\underline{u}_j \leq u \leq \bar{u}_i, \quad 1 \leq i \leq m, \ 1 \leq j \leq k. \tag{3.16}$$

As a result, u satisfies also $\underline{u} \leq u \leq \bar{u}$, which finally results in $Tu = u$, $T_j u = u$, $T^i u = u$, and thus $P(u) = G(u)$ showing that u is a solution of the original problem (3.1) within $[\underline{u}, \bar{u}]$.

Let us first show that any solution u of (3.14) satisfies $u \leq \bar{u}_l$ for $l \in \{1, \ldots, m\}$ fixed. By assumption, $\bar{u}_l$ is a supersolution; i.e., $\bar{u}_l \geq 0$ on $\partial\Omega$ and

$$-\Delta \bar{u}_l + G(\bar{u}_l) \geq f \tag{3.17}$$

with respect to the dual-order cone of V_0^*. Subtracting (3.17) from (3.14), we obtain

$$-\Delta(u - \bar{u}_l) + P(u) - G(\bar{u}_l) \leq 0,$$

which is equivalent to

$$\int_\Omega \nabla(u - \bar{u}_l)\nabla\varphi \, dx + \int_\Omega (P(u) - G(\bar{u}_l))\varphi \, dx \leq 0 \tag{3.18}$$

for all $\varphi \in V_0 \cap L_+^2(\Omega)$. Taking the special test function $\varphi = (u - \bar{u}_l)^+$ in (3.18), we obtain for the first term on the left-hand side of (3.18)

$$\int_\Omega \nabla(u - \bar{u}_l)\nabla(u - \bar{u}_l)^+ \, dx = \int_\Omega |\nabla(u - \bar{u}_l)^+|^2 \, dx \geq 0. \tag{3.19}$$

Let us consider the second term on the left-hand side of (3.18); i.e.,

$$\int_\Omega (P(u) - G(\bar{u}_l))(u - \bar{u}_l)^+ \, dx = \int_{\{u > \bar{u}_l\}} (P(u) - G(\bar{u}_l))(u - \bar{u}_l) \, dx, \tag{3.20}$$

where $\{u > \bar{u}_l\} = \{x \in \Omega : u(x) > \bar{u}_l(x)\}$. As for the estimate of the right-hand side of (3.20), we note that $\bar{u}_l \geq \bar{u} \geq \underline{u} \geq \underline{u}_j$, which yields by taking into account the definition of the truncation operators that $T_j u(x) = \bar{u}(x) = Tu(x)$ for $x \in \{u > \bar{u}_l\}$ and all $j = 1, \ldots, k$, and thus,

$$\int_{\{u > \bar{u}_l\}} \sum_{j=1}^k |(G \circ T_j)u - (G \circ T)u|(u - \bar{u}_l) \, dx = 0.$$

By means of the last equation and taking into account that $T^l u(x) = \bar{u}_l(x)$ for $x \in \{u > \bar{u}_l\}$, we obtain the following estimate:

$$\int_\Omega (P(u) - G(\bar{u}_l))(u - \bar{u}_l)^+ \, dx$$

$$= \int_{\{u > \bar{u}_l\}} (P(u) - G(\bar{u}_l))(u - \bar{u}_l) \, dx$$

$$= \int_{\{u > \bar{u}_l\}} \left[((G \circ T)u - G(\bar{u}_l))(u - \bar{u}_l) + \sum_{i=1}^m |(G \circ T^i)u - (G \circ T)u|(u - \bar{u}_l) \right] dx$$

$$= \int_{\{u > \bar{u}_l\}} (G(\bar{u}) - G(\bar{u}_l) + |G(\bar{u}_l) - G(\bar{u})|)(u - \bar{u}_l) \, dx$$

$$+ \int_{\{u > \bar{u}_l\}} \sum_{i \neq l} |(G \circ T^i)u - G(\bar{u})|(u - \bar{u}_l) \, dx \geq 0. \tag{3.21}$$

Testing (3.18) with $\varphi = (u - \bar{u}_l)^+$, we get with the help of (3.19) and (3.21) the inequality

$$0 \leq \int_\Omega |\nabla (u - \bar{u}_l)^+|^2 \, dx \leq 0,$$

which implies $(u - \bar{u}_l)^+ = 0$, and hence, it follows $u \leq \bar{u}_l$ for any $l \in \{1, \ldots, m\}$. The proof of the inequalities $\underline{u}_j \leq u$ for $j = 1, \ldots, k$ can be done in a similar way. Thus, inequalities (3.16) are satisfied, which completes the proof of the theorem. □

As an immediate consequence of Theorem 3.6, we get the following corollary.

Corollary 3.7 (Directedness). *The solution set S of the BVP (3.1) is directed.*

Proof: Let $u_i \in S$, $i = 1, 2$. As any solution u_i is, in particular, also a subsolution, Theorem 3.6 ensures the existence of a solution u of the BVP (3.1) within the ordered interval $[\max\{u_1, u_2\}, \bar{u}]$. To this end, we only need to specialize $m = 2$ and $k = 1$ with $\bar{u}_1 = \bar{u}$. Thus, there is a solution u of the BVP (3.1) satisfying $\underline{u} \leq u_i \leq u \leq \bar{u}$, which means that S is upward directed. As any $u \in S$ is also a supersolution, one can show in just the same way that S is downward directed, which shows the directedness. □

An alternative method to prove directedness of the solution set S is based on the following result.

Theorem 3.8. *If u_1 and u_2 are subsolutions of the BVP (3.1) and if the Nemytskij operator $G : [\min\{u_1, u_2\}, \max\{u_1, u_2\}] \to L^2(\Omega)$ is well defined, then $\max\{u_1, u_2\}$ is a subsolution. Analogously, if u_1 and u_2 are supersolutions of the BVP (3.1) with the same assumption on the Nemytskij operator G, then $\min\{u_1, u_2\}$ is a supersolution.*

Proof: Let u_1 and u_2 be subsolutions; i.e., we have $u_k \leq 0$ on $\partial\Omega$ and

$$\int_\Omega (\nabla u_k \nabla \varphi + G(u_k)\varphi)\, dx \leq \langle f, \varphi \rangle \quad \text{for all} \quad \varphi \in V_0 \cap L_+^2(\Omega), \qquad (3.22)$$

and $k = 1, 2$. Denote $u = \max\{u_1, u_2\}$. Then according to Sect. 2.2.3, we have $u \in V$ and $u \leq 0$ on $\partial\Omega$, and thus for u being a subsolution, we need to verify the following inequality:

$$\int_\Omega (\nabla u \nabla \varphi + G(u)\varphi)\, dx \leq \langle f, \varphi \rangle \quad \text{for all} \quad \varphi \in V_0 \cap L_+^2(\Omega), \qquad (3.23)$$

where

$$G(u)(x) = \begin{cases} G(u_1)(x) & \text{if} \quad x \in \{u_1 \geq u_2\}, \\ G(u_2)(x) & \text{if} \quad x \in \{u_2 > u_1\}. \end{cases}$$

For any $\varepsilon > 0$, we introduce the nondecreasing, piecewise differentiable function $\theta_\varepsilon : \mathbb{R} \to \mathbb{R}$ given by

$$\theta_\varepsilon(s) = \begin{cases} 0 & \text{if} \ s \leq 0, \\ \frac{1}{\varepsilon} s & \text{if} \ 0 < s < \varepsilon, \\ 1 & \text{if} \ s \geq \varepsilon. \end{cases} \qquad (3.24)$$

Furthermore, if $\mathcal{D}_+$ denotes the following set of nonnegative smooth functions:

$$\mathcal{D}_+ = \{\psi \in C_0^\infty(\Omega) : \psi \geq 0 \ \text{in} \ \Omega\},$$

then its closure in V coincides with $V_0 \cap L_+^2(\Omega)$. Now we apply special test functions to (3.22). With $\psi \in \mathcal{D}_+$, we take in case $k = 1$ the test function

$$\varphi = \psi\,(1 - \theta_\varepsilon(u_2 - u_1)) \in V_0 \cap L_+^2(\Omega),$$

and in case $k = 2$, we take

$$\varphi = \psi\,\theta_\varepsilon(u_2 - u_1) \in V_0 \cap L_+^2(\Omega).$$

Adding the resulting inequalities, we obtain

$$\int_\Omega \Big(\nabla u_1 \nabla(\psi\,(1 - \theta_\varepsilon(u_2 - u_1))) + \nabla u_2 \nabla(\psi\,\theta_\varepsilon(u_2 - u_1)) \Big)\, dx$$

$$+ \int_\Omega \Big(G(u_1)\psi\,(1 - \theta_\varepsilon(u_2 - u_1)) + G(u_2)\psi\,\theta_\varepsilon(u_2 - u_1) \Big)\, dx$$

$$\leq \langle f, \psi \rangle, \qquad (3.25)$$

which yields

$$\int_\Omega \Big(\nabla u_1 \nabla \psi + \nabla(u_2 - u_1)\nabla(\psi\,\theta_\varepsilon(u_2 - u_1)) \Big)\, dx$$

$$+ \int_{\Omega} \Big(G(u_1)\psi + (G(u_2) - G(u_1)) \, \psi \, \theta_\varepsilon(u_2 - u_1) \Big) dx$$

$$\leq \langle f, \psi \rangle. \tag{3.26}$$

By means of the chain rule (see Sect. 2.3.3), we get

$$\nabla(\psi \, \theta_\varepsilon(u_2 - u_1)) = \nabla\psi \, \theta_\varepsilon(u_2 - u_1) + \psi \, \theta_\varepsilon'(u_2 - u_1)\nabla(u_2 - u_1),$$

which gives the following estimate for the first term on the left-hand side of (3.26):

$$\int_{\Omega} \Big(\nabla u_1 \nabla\psi + \nabla(u_2 - u_1)\nabla(\psi \, \theta_\varepsilon(u_2 - u_1)) \Big) dx$$

$$= \int_{\Omega} \Big(\nabla u_1 \nabla\psi + |\nabla(u_2 - u_1)|^2 \psi \, \theta_\varepsilon'(u_2 - u_1)) \Big) dx$$

$$+ \int_{\Omega} \nabla(u_2 - u_1)\nabla\psi \, \theta_\varepsilon(u_2 - u_1) \, dx$$

$$\geq \int_{\Omega} \Big(\nabla u_1 \nabla\psi + \nabla(u_2 - u_1)\nabla\psi \, \theta_\varepsilon(u_2 - u_1) \Big) dx. \tag{3.27}$$

From (3.26) and (3.27), we obtain for any $\psi \in \mathcal{D}_+$ and $\varepsilon > 0$,

$$\int_{\Omega} \Big(\nabla u_1 \nabla\psi + \nabla(u_2 - u_1)\nabla\psi \, \theta_\varepsilon(u_2 - u_1) \Big) dx$$

$$+ \int_{\Omega} \Big(G(u_1)\psi + (G(u_2) - G(u_1)) \, \psi \, \theta_\varepsilon(u_2 - u_1) \Big) dx$$

$$\leq \langle f, \psi \rangle.$$

Applying Lebesgue's dominated convergence theorem (see Theorem 2.63) and taking into account that

$$\theta_\varepsilon(u_2 - u_1) \to \chi_{\{u_2 - u_1 > 0\}} \quad \text{as} \quad \varepsilon \to 0,$$

where $\chi_{\{u_2 - u_1 > 0\}}$ is the characteristic function of the set $\{u_2 - u_1 > 0\} = \{x \in \Omega : u_2(x) - u_1(x) > 0\}$, we finally get

$$\int_{\Omega} \Big(\nabla u_1 + \nabla(u_2 - u_1)\chi_{\{u_2 - u_1 > 0\}} \Big) \nabla\psi \, dx$$

$$+ \int_{\Omega} \Big(G(u_1) + (G(u_2) - G(u_1))\chi_{\{u_2 - u_1 > 0\}} \Big) \psi \, dx$$

$$\leq \langle f, \psi \rangle. \tag{3.28}$$

Inequality (3.28) is equivalent with

$$\int_{\Omega} (\nabla u \nabla\psi + G(u)\psi) \, dx \leq \langle f, \psi \rangle \quad \text{for all} \quad \psi \in \mathcal{D}_+, \tag{3.29}$$

with $u = \max\{u_1, u_2\}$, and thus, u is a subsolution, because $\mathcal{D}_+$ is dense in $V_0 \cap L^2_+(\Omega)$. As for the proof of the second part of the theorem, which consists in showing that if u_1 and u_2 are supersolutions, then $\min\{u_1, u_2\}$ is a supersolution, we can proceed in just the same way. This process completes the proof of the theorem. $\square$

Remark 3.9. With the help of Theorem 3.8 and Theorem 3.4, we can provide an alternative to prove the directedness of the solution set $\mathcal{S}$ as follows. If $u_1, u_2 \in \mathcal{S}$, then u_1 and u_2 are, in particular, subsolutions of the BVP (3.1), which in view of Theorem 3.8 implies that $\max\{u_1, u_2\} \in [\underline{u}, \bar{u}]$ is a subsolution. Therefore, we may apply Theorem 3.4, which yields the existence of a solution of the BVP (3.1) within the ordered interval $[\max\{u_1, u_2\}, \bar{u}]$; i.e., a $u \in \mathcal{S}$ exists such that $u \geq u_1$ and $u \geq u_2$, which implies that $\mathcal{S}$ is upward directed. As $u_1, u_2 \in \mathcal{S}$, are also, in particular, supersolutions of the BVP (3.1), from Theorem 3.8, we infer that $\min\{u_1, u_2\} \in [\underline{u}, \bar{u}]$ is a supersolution, and hence, Theorem 3.4 implies the existence of a solution of the BVP (3.1) within the ordered interval $[\underline{u}, \min\{u_1, u_2\}]$; i.e., a $u \in \mathcal{S}$ exists such that $u \leq u_1$ and $u \leq u_2$, which implies that $\mathcal{S}$ is downward directed. Thus, $\mathcal{S}$ is both upward and downward directed; i.e., $\mathcal{S}$ is directed.

Theorem 3.10 (Compactness). *The solution set $\mathcal{S}$ is compact in V_0.*

Proof: Let $(u_n) \subset \mathcal{S}$, i.e., $u_n \in [\underline{u}, \bar{u}]$ and u_n is a solution of the BVP (3.1), which means

$$u_n \in V_0 : \quad \int_\Omega (\nabla u_n \nabla \varphi + G(u_n)\varphi)\, dx = \langle f, \varphi \rangle \quad \text{for all} \quad \varphi \in V_0. \qquad (3.30)$$

Testing (3.30) with $\varphi = u_n$ and noting that

$$\|G(u_n)\|_2 \leq c \quad \text{for all} \quad n \in \mathbb{N},$$

we see that $\|u_n\|_{V_0} \leq c$ for all $n \in \mathbb{N}$, and thus, a subsequence (u_k) of (u_n) exists such that

$$u_k \rightharpoonup u \text{ in } V_0, \quad u_k \to u \text{ in } L^2(\Omega). \qquad (3.31)$$

Replacing n in (3.30) by k and applying the convergence properties (3.31), we see that the limit u satisfies

$$u \in V_0 : \quad \int_\Omega (\nabla u \nabla \varphi + G(u)\varphi)\, dx = \langle f, \varphi \rangle \quad \text{for all} \quad \varphi \in V_0, \qquad (3.32)$$

which implies that the limit u belongs to $\mathcal{S}$. We are going to show that (u_k) is not only weakly convergent to u in V_0 but also strongly convergent. To this end, we subtract (3.32) from (3.30) with n replaced by k and obtain by using the test function $\varphi = u_k - u$ the following relation:

$$\int_\Omega |\nabla(u_k - u)|^2\, dx = \int_\Omega (G(u) - G(u_k))(u_k - u)\, dx. \qquad (3.33)$$

As $u_k \to u$ in $L^2(\Omega)$ implies $G(u_k) \to G(u)$ in $L^2(\Omega)$, the right-hand side of (3.33) tends to zero, which shows that $u_k \to u$ in V_0, completing the proof. $\square$

3.1.3 Extremal Solutions

As in Sect. 3.1.2, we denote by $\mathcal{S}$ the set of all solutions of the BVP (3.1) within the ordered interval $[\underline{u}, \bar{u}]$. The main goal of this section is to show that $\mathcal{S}$ has extremal elements; i.e., the greatest and smallest element in $\mathcal{S}$ exists.

Theorem 3.11. *The solution set $\mathcal{S}$ possesses extremal elements; i.e., there is a greatest solution u^* and a smallest solution u_* of the BVP (3.1) within $[\underline{u}, \bar{u}]$.*

Proof: The main tools used in the proof are Corollary 3.7 and Theorem 3.10. We focus on the existence of the greatest element of $\mathcal{S}$.

As V_0 is separable, it follows that $\mathcal{S} \subset V_0$ is separable, so a countable, dense subset $Z = \{z_n : n \in \mathbb{N}\}$ of $\mathcal{S}$ exists. By Corollary 3.7, $\mathcal{S}$ is, in particular, upward directed, so we can construct an increasing sequence $(u_n) \subset \mathcal{S}$ as follows. Let $u_1 = z_1$. Select $u_{n+1} \in \mathcal{S}$ such that

$$\max\{z_n, u_n\} \leq u_{n+1} \leq \bar{u}.$$

The existence of u_{n+1} is due to Corollary 3.7. From the compactness of $\mathcal{S}$ according to Theorem 3.10, a subsequence of (u_n) exists, denoted again (u_n), and an element $u \in \mathcal{S}$ such that $u_n \to u$ in V_0, and $u_n(x) \to u(x)$ a.e. in Ω. This last property of (u_n) combined with its increasing monotonicity implies that the entire sequence is convergent in V_0, and moreover, $u = \sup_n u_n$. By construction, we see that

$$\max\{z_1, z_2, \ldots, z_n\} \leq u_{n+1} \leq u, \quad \text{for all } n \in \mathbb{N},$$

thus, $Z \subset [\underline{u}, u]$. As the interval $[\underline{u}, u]$ is closed in V_0, we infer

$$\mathcal{S} \subset \overline{Z} \subset \overline{[\underline{u}, u]} = [\underline{u}, u],$$

which in conjunction with $u \in \mathcal{S}$ ensures that $u = u^*$ is the greatest solution of the BVP (3.1).

The existence of the smallest solution u_* of the BVP (3.1) can be proved in a similar way using the fact that $\mathcal{S}$ is also downward directed.   □

Remark 3.12. An alternative proof of the greatest and smallest elements of $\mathcal{S}$, which is based on Zorn's lemma (see Theorem 2.62) is as follows. Again we focus on the existence of the greatest element of $\mathcal{S}$. To this end, we first show that $\mathcal{S}$ possesses maximal elements (see Definition 2.61) by applying Zorn's lemma. Therefore, let $\mathcal{C} \subset \mathcal{S}$ be a well-ordered chain. By Theorem 3.10, $\mathcal{S}$ is compact, and thus $\mathcal{C}$ is bounded in V_0, which implies the existence of an increasing sequence $(u_n) \subset \mathcal{C}$ satisfying

$$u_n \to \sup \mathcal{C} \quad \text{in } L^2(\Omega),$$

$$u_n \rightharpoonup \sup \mathcal{C} \quad \text{in} \quad V_0.$$

The compactness of $\mathcal{S}$ implies that $u = \sup \mathcal{C}$ belongs to $\mathcal{S}$, which shows that any well-ordered chain $\mathcal{C}$ of $\mathcal{S}$ has an upper bound in $\mathcal{S}$. Now Zorn's lemma can be applied, which yields the existence of a maximal element $\hat{u} \in \mathcal{S}$. As $\mathcal{S}$ is, in particular, upward directed, the maximal element of $\mathcal{S}$ is uniquely defined and must be the greatest one; i.e., $\hat{u} = u^*$. To see the uniqueness of the maximal element, assume there are maximal elements $\hat{u}_1$ and $\hat{u}_2$ with $\hat{u}_1 \neq \hat{u}_2$. Because $\mathcal{S}$ is upward directed, an element $u \in \mathcal{S}$ exists such that

$$\max\{\hat{u}_1, \hat{u}_2\} \leq u \leq \bar{u}. \tag{3.34}$$

However, (3.34) implies that $u = \hat{u}_1$ and $u = \hat{u}_2$, which contradicts $\hat{u}_1 \neq \hat{u}_2$. This process completes the alternative proof of extremal elements.

We finally study some order-preserving property of the extremal solutions of the BVP (3.1). First, consider the case $g = 0$. In this case, the BVP (3.1) is uniquely solvable, and the unique solution is order preserving with respect to the right-hand side f, which means that the solution operator of the linear BVP

$$u \in V_0 : \quad -\Delta u = f \quad \text{in} \quad V_0^*, \tag{3.35}$$

denoted by $(-\Delta)^{-1} : V_0^* \to V_0$ is isotone; i.e., $(-\Delta)^{-1}$ satisfies

$$f_1 \leq f_2 \implies u_1 \leq u_2, \tag{3.36}$$

where $u_k = (-\Delta)^{-1} f_k$, $k = 1, 2$. The uniqueness result for (3.35) follows from the fact that $-\Delta : V_0 \to V_0^*$ is a strongly monotone, linear, continuous, and bounded operator, and property (3.36) can be seen from the following simple calculation. Let $f_1 \leq f_2$ in V_0^*, which means

$$\langle f_1, \varphi \rangle \leq \langle f_2, \varphi \rangle \quad \text{for all} \quad \varphi \in V_0 \cap L_+^2(\Omega),$$

and let $u_k = (-\Delta)^{-1} f_k$, $k = 1, 2$ denote the corresponding unique solutions. By subtraction we obtain

$$-\Delta(u_1 - u_2) = f_1 - f_2 \leq 0 \quad \text{in} \quad V_0^*, \tag{3.37}$$

which yields with the nonnegative test function $\varphi = (u_1 - u_2)^+$, the following inequality:

$$0 \leq \|\nabla(u_1 - u_2)^+\|_2^2 = \int_\Omega \nabla(u_1 - u_2)\nabla(u_1 - u_2)^+ \, dx \leq 0,$$

and thus, $(u_1 - u_2)^+ = 0$; i.e., $u_1 \leq u_2$.

However, in general, BVP (3.1) is not uniquely solvable, and therefore, the above arguments are not applicable. But still an order-preserving property can be shown to hold for the extremal solutions. Consider the BVP (3.1) with right-hand sides f_k, $k = 1, 2$; i.e.,

$$u \in V_0 : \quad -\Delta u + G(u) = f_k \quad \text{in} \quad V_0^*, \tag{3.38}$$

and assume that $\underline{u}$ and $\bar{u}$ are sub- and supersolutions of both problems (3.38) with $\underline{u} \leq \bar{u}$.

Theorem 3.13. *Let $u_k^* \in [\underline{u}, \bar{u}]$ denote the greatest solution of the BVP (3.38) with right-hand side f_k. If $f_1 \leq f_2$, then $u_1^* \leq u_2^*$, and similarly, $u_{1*} \leq u_{2*}$ for the smallest solutions u_{k*} of BVP (3.38).*

Proof: To prove $u_1^* \leq u_2^*$, let us consider the BVP

$$u \in V_0 : \quad -\Delta u + G(u) = f_1 \quad \text{in} \quad V_0^*. \tag{3.39}$$

In view of $f_1 \leq f_2$, any solution of (3.39) is a subsolution of the BVP

$$u \in V_0 : \quad -\Delta u + G(u) = f_2 \quad \text{in} \quad V_0^*, \tag{3.40}$$

and thus, in particular, the greatest solution u_1^* of (3.39) is a subsolution of (3.40) as well. Therefore, u_1^* and $\bar{u}$ is an ordered pair of sub- and supersolutions of the BVP (3.40), which by applying Theorem 3.4 implies the existence of a solution u of (3.40) satisfying

$$u_1^* \leq u \leq \bar{u}. \tag{3.41}$$

As u_2^* is the greatest solution of the BVP (3.40) within $[\underline{u}, \bar{u}]$, it follows $u \leq u_2^*$, and thus, we finally get $u_1^* \leq u_2^*$. The proof of $u_{1*} \leq u_{2*}$ can be done analogously. $\qquad\square$

3.2 Quasilinear Elliptic Equations

Let $\Omega \subset \mathbb{R}^N$ be as in Sect. 3.1. In this section, we extend the results of Sect. 3.1 to the following quasilinear elliptic BVP:

$$-\sum_{i=1}^{N} \frac{\partial}{\partial x_i} a_i(\cdot, u, \nabla u) + g(\cdot, u, \nabla u) = f \quad \text{in} \quad \Omega, \quad u = 0 \quad \text{on} \quad \partial\Omega. \tag{3.42}$$

In the study of (3.42), we assume for the coefficient functions $a_i : \Omega \times \mathbb{R} \times \mathbb{R}^N \to \mathbb{R}$, $i = 1, \ldots, N$, the so-called Leray–Lions conditions (H1)–(H3) formulated in Sect. 2.3.2, and $g : \Omega \times \mathbb{R} \times \mathbb{R}^N \to \mathbb{R}$ is assumed to be a Carathéodory function; i.e.,

$$x \mapsto g(x, s, \xi) \quad \text{is measurable in } \Omega \text{ for all } (s, \xi) \in \mathbb{R} \times \mathbb{R}^N,$$
$$(s, \xi) \mapsto g(x, s, \xi) \quad \text{is continuous in } \mathbb{R} \times \mathbb{R}^N \text{ for a.e. } x \in \Omega.$$

Set $V = W^{1,p}(\Omega)$ and $V_0 = W_0^{1,p}(\Omega)$. Let $a : V \times V \to \mathbb{R}$ be the semilinear form given by

$$a(u, v) = \int_\Omega \sum_{i=1}^N a_i(\cdot, u, \nabla u) \frac{\partial v}{\partial x_i} \, dx, \qquad (3.43)$$

which is well defined for any $(u, v) \in V \times V$ in view of (H1). By means of (3.43), we introduce an operator A through

$$\langle Au, \varphi \rangle = a(u, \varphi) \quad \text{for all} \ \ \varphi \in V_0, \qquad (3.44)$$

which is well defined for any $u \in V$; i.e., $Au \in V_0^*$, and $A : V \to V_0^*$ is continuous and bounded in view of the Carathéodory and growth condition (H1) imposed on a_i, $i = 1, \ldots, N$.

Let G denote the Nemytskij operator related to g by

$$G(u)(x) = g(x, u(x), \nabla u(x)),$$

and suppose that $f \in V_0^*$.

Definition 3.14. *The function* $u \in V_0$ *is called a (weak) solution of the BVP (3.42) if* $G(u) \in L^q(\Omega) \subset V_0^*$ *and* u *satisfies the equation*

$$Au + G(u) = f \quad \text{in} \ \ V_0^*. \qquad (3.45)$$

Note q denotes the conjugate Hölder exponent to p; i.e., $1/p + 1/q = 1$ and equation (3.45) is equivalent to

$$a(u, \varphi) + \int_\Omega G(u) \varphi \, dx = \langle f, \varphi \rangle \quad \text{for all} \ \ \varphi \in V_0. \qquad (3.46)$$

3.2.1 Comparison Principle

Let hypotheses (H1)–(H3) formulated in Sect. 2.3.2 be fulfilled throughout this section. In a similar way as in Sect. 3.1.1, we introduce the notion of (weak) sub- and supersolution.

Definition 3.15. *The function* $\underline{u} \in V$ *is called a subsolution of the BVP (3.42) if* $G(\underline{u}) \in L^q(\Omega)$, $\underline{u} \leq 0$ *on* $\partial\Omega$, *and*

$$a(\underline{u}, \varphi) + \int_\Omega G(\underline{u}) \varphi \, dx \leq \langle f, \varphi \rangle \quad \text{for all} \ \ \varphi \in V_0 \cap L_+^p(\Omega). \qquad (3.47)$$

By reversing the inequality sign in Definition 3.15, we get the following definition for the supersolution.

Definition 3.16. *The function $\bar{u} \in V$ is called a supersolution of the BVP (3.42) if $G(\bar{u}) \in L^q(\Omega)$, $\bar{u} \geq 0$ on $\partial\Omega$, and*

$$a(\bar{u}, \varphi) + \int_\Omega G(\bar{u})\,\varphi\,dx \geq \langle f, \varphi \rangle \quad \text{for all } \varphi \in V_0 \cap L_+^p(\Omega). \tag{3.48}$$

The following comparison principle is a classic result that can be found, e.g., in [83].

Theorem 3.17. *Let $\underline{u}$ and $\bar{u}$ be sub- and supersolutions of the BVP (3.42), respectively, that satisfy $\underline{u} \leq \bar{u}$; assume (H1)–(H3) and the following local growth condition for g :*

$$|g(x, s, \xi)| \leq k_1(x) + c\,|\xi|^{p-1} \tag{3.49}$$

for a.e. $x \in \Omega$, for all $\xi \in \mathbb{R}^N$, and for all $s \in [\underline{u}(x), \bar{u}(x)]$, where $k_1 \in L_+^q(\Omega)$ and $c > 0$. Then solutions of the BVP (3.42) exist within the ordered interval $[\underline{u}, \bar{u}]$.

Proof: Let T be the truncation operator introduced in the proof of Theorem 3.4. According to Lemma 2.89, we have that $T : V \to V$ is continuous and bounded. As we are interested in the existence of solutions of the BVP (3.42) within the interval $[\underline{u}, \bar{u}]$, we consider first the following auxiliary truncated BVP:

$$u \in V_0 : \quad A_T u + \lambda B(u) + (G \circ T)u = f \quad \text{in } V_0^*, \tag{3.50}$$

where the operator $A_T : V_0 \to V_0^*$ is defined by

$$\langle A_T u, \varphi \rangle = a_T(u, \varphi) = \int_\Omega \sum_{i=1}^N a_i(\cdot, Tu, \nabla u)\frac{\partial \varphi}{\partial x_i}\,dx \quad \text{for all } \varphi \in V_0. \tag{3.51}$$

The parameter $\lambda > 0$ will be specified later, and B is the Nemytskij operator generated by the following cutoff function $b : \Omega \times \mathbb{R} \to \mathbb{R}$ related to the ordered pair of sub- and supersolutions, and given by

$$b(x, s) = \begin{cases} (s - \bar{u}(x))^{p-1} & \text{if } s > \bar{u}(x), \\ 0 & \text{if } \underline{u}(x) \leq s \leq \bar{u}(x), \\ -(\underline{u}(x) - s)^{p-1} & \text{if } s < \underline{u}(x). \end{cases} \tag{3.52}$$

We readily verify that b is a Carathéodory function satisfying the growth condition

$$|b(x, s)| \leq k_2(x) + c_3\,|s|^{p-1} \tag{3.53}$$

for a.e. $x \in \Omega$, for all $s \in \mathbb{R}$, with some function $k_2 \in L_+^q(\Omega)$ and a constant $c_3 \geq 0$. Moreover, one has the following estimate:

$$\int_{\Omega} b(x, u(x))\, u(x)\, dx \geq c_4 \, \|u\|_p^p - c_5, \quad \forall\, u \in L^p(\Omega), \qquad (3.54)$$

where c_4 and c_5 are some positive constants and $\|\cdot\|_p$ denotes the norm in $L^p(\Omega)$. In view of (3.53), the Nemytskij operator $B : L^p(\Omega) \to L^q(\Omega)$ is continuous and bounded, and thus due to the compact embedding $V_0 \subset L^p(\Omega)$, it follows that $B : V_0 \to V_0^*$ is completely continuous. Hypotheses (H1)–(H3), the growth condition (3.49), and the continuity of $T : V \to V$ imply that $A_T + G \circ T : V_0 \to V_0^*$ is bounded, continuous, and pseudomonotone due to Theorem 2.109. As $B : V_0 \to V_0^*$ is bounded and completely continuous, it follows that

$A_T + \lambda B + G \circ T : V_0 \to V_0^*$ is bounded, continuous, and pseudomonotone.

By the Main Theorem on pseudomonotone operators (see Theorem 2.99), the auxiliary BVP (3.50) possesses solutions provided that $A_T + \lambda B + G \circ T : V_0 \to V_0^*$ is coercive, which will be shown next. From (3.54), we get

$$\langle B(u), u \rangle = \int_{\Omega} B(u) u \, dx \geq c_4 \, \|u\|_p^p - c_5. \qquad (3.55)$$

The growth condition (3.49) in conjunction with Young's inequality implies the estimate

$$\left| \int_{\Omega} (G \circ T)(u)\, u \, dx \right| \leq \int_{\Omega} \left(|k_1 u| + c\, |\nabla u|^{p-1} |u| \right) dx$$
$$\leq \|k_1\|_q \|u\|_p + c \, \|\nabla u\|_p^{p-1} \|u\|_p$$
$$\leq \varepsilon \, \|\nabla u\|_p^p + c(\varepsilon) \|u\|_p^p + c, \qquad (3.56)$$

for any $\varepsilon > 0$, where $c > 0$ is some generic constant and $c(\varepsilon)$ is some positive constant depending only on ε. By means of hypothesis (H3), we obtain

$$\langle A_T u, u \rangle \geq \nu \, \|\nabla u\|_p^p - \|k\|_1. \qquad (3.57)$$

Hence, from (3.55)–(3.57), we get

$$\langle (A_T + \lambda B + G \circ T) u, u \rangle \geq (\nu - \varepsilon) \, \|\nabla u\|_p^p + (\lambda c - c(\varepsilon)) \|u\|_p^p - c, \qquad (3.58)$$

which yields the coercivity of the operator $A_T + \lambda B + G \circ T : V_0 \to V_0^*$ by choosing $\varepsilon < \nu$ and λ sufficiently large such that $\lambda c - c(\varepsilon) > 0$.

The assertion of the theorem is proved provided we can show that any solution of the auxiliary BVP (3.50) is contained in the interval $[\underline{u}, \bar{u}]$ formed by the sub- and supersolution; i.e., we are going to prove now that $\underline{u} \leq u \leq \bar{u}$, for any solution u of (3.50).

Let us show that $u \leq \bar{u}$. Taking in (3.50) and (3.48) the nonnegative test function $\varphi = (u - \bar{u})^+ \in V_0 \cap L^p(\Omega)$ and subtracting (3.48) from (3.50) we obtain

$$\langle A_T u - A\bar{u} + \lambda B(u) + (G \circ T)u - G(\bar{u}), (u - \bar{u})^+\rangle \le 0. \qquad (3.59)$$

By definition of A_T given in (3.51) and using hypothesis (H2), we have

$$\langle A_T u - A\bar{u}, (u - \bar{u})^+\rangle$$
$$= a_T(u, (u - \bar{u})^+) - a(\bar{u}, (u - \bar{u})^+)$$
$$= \int_\Omega \sum_{i=1}^N \Big(a_i(\cdot, Tu, \nabla u) - a_i(\cdot, \bar{u}, \nabla \bar{u})\Big) \frac{\partial(u - \bar{u})^+}{\partial x_i}\, dx$$
$$= \int_{\{u > \bar{u}\}} \sum_{i=1}^N \Big(a_i(\cdot, \bar{u}, \nabla u) - a_i(\cdot, \bar{u}, \nabla \bar{u})\Big) \frac{\partial(u - \bar{u})}{\partial x_i}\, dx \ge 0, \qquad (3.60)$$

where $\{u > \bar{u}\} = \{x \in \Omega : u(x) > \bar{u}(x)\}$. By definition of the truncation operator T, the third term on the left-hand side of (3.59) results in

$$\langle (G \circ T)u - G(\bar{u}), (u - \bar{u})^+\rangle$$
$$= \int_\Omega \Big(g(\cdot, Tu, \nabla Tu) - g(\cdot, \bar{u}, \nabla \bar{u})\Big)(u - \bar{u})^+\, dx$$
$$= \int_{\{u > \bar{u}\}} \Big(g(\cdot, \bar{u}, \nabla \bar{u}) - g(\cdot, \bar{u}, \nabla \bar{u})\Big)(u - \bar{u})\, dx = 0. \qquad (3.61)$$

Applying the definition of the cutoff function b, we get for the second term on the left-hand side of (3.59)

$$\langle \lambda B(u), (u - \bar{u})^+\rangle = \int_{\{u > \bar{u}\}} (u - \bar{u})^p\, dx = \lambda \, \|(u - \bar{u})^+\|_p^p, \qquad (3.62)$$

and hence, it follows from (3.59)–(3.62)

$$\|(u - \bar{u})^+\|_p^p \le 0,$$

which implies $(u - \bar{u})^+ = 0$; i.e., $u \le \bar{u}$. The proof of the inequality $\underline{u} \le u$ can be done in an obvious similar way, which completes the proof of the theorem.
□

3.2.2 Directed and Compact Solution Set

Theorem 3.17 of the previous section shows that the set S of all solutions of the BVP (3.42) within the ordered interval $[\underline{u}, \bar{u}]$ is nonempty. In this section, we prove directedness and compactness results for S. Unlike the corresponding results in the semilinear case, the proofs in the quasilinear case are much more involved. Although for the compactness of S the same hypotheses as in Theorem 3.17 are sufficient, we will see that the proof of the directedness of S requires some additional assumption, which is basically caused by the dependence of the coefficients $a_i(\cdot, u, \nabla u)$ on u.

Theorem 3.18. *Let the hypotheses of Theorem 3.17 be fulfilled. Then the solution set $\mathcal{S}$ is compact.*

Proof: Let $(u_n) \subset \mathcal{S}$ be any sequence. We have to show that there is a subsequence of (u_n), which is strongly convergent in V_0 to some $u \in \mathcal{S}$. By definition of $\mathcal{S}$, we have $u_n \in [\underline{u}, \bar{u}]$, and

$$u_n \in V_0: \quad \langle Au_n + G(u_n), \varphi \rangle = \langle f, \varphi \rangle \quad \text{for all} \quad \varphi \in V_0. \tag{3.63}$$

Taking in (3.63) the special test function $\varphi = u_n$, we get by hypothesis (H3) for the first term on the left-hand side the estimate

$$\langle Au_n, u_n \rangle \geq \nu \|\nabla u_n\|_p^p - \|k\|_1, \tag{3.64}$$

and by means of the growth condition (3.49) on g in conjunction with Young's inequality, the following estimate for the second term on the left-hand side of (3.63) is readily verified:

$$\left| \int_\Omega G(u_n) u_n \, dx \right| \leq \int_\Omega \left(|k_1 u_n| + c |\nabla u_n|^{p-1} |u| \right) dx$$
$$\leq \|k_1\|_q \|u_n\|_p + c \|\nabla u_n\|_p^{p-1} \|u_n\|_p$$
$$\leq \varepsilon \|\nabla u_n\|_p^p + c(\varepsilon), \tag{3.65}$$

for any $\varepsilon > 0$, where $c(\varepsilon)$ is some positive constant depending only on ε. For (3.65), we have used the fact that (u_n) is bounded in $L^p(\Omega)$ due to $u_n \in [\underline{u}, \bar{u}]$. For the right-hand side of (3.63) with $\varphi = u_n$, we obtain by using Young's inequality the estimate

$$|\langle f, u_n \rangle| \leq \|f\|_{V_0^*} \|u_n\|_{V_0} \leq c(\delta, f) + \delta \|\nabla u_n\|_p^p, \tag{3.66}$$

for any $\delta > 0$, where $c(\delta, f) > 0$ is some constant depending only on δ and the norm of f. Thus, from (3.63) with $\varphi = u_n$ and (3.64)–(3.66), we obtain the estimate

$$(\nu - \varepsilon - \delta) \|\nabla u_n\|_p^p \leq c(\varepsilon, \delta, f) \quad \text{for all} \ n \in \mathbb{N}, \tag{3.67}$$

which shows the boundedness of the sequence (u_n) in V_0 when selecting ε and δ sufficiently small such that $\varepsilon + \delta < \nu$. Thus, a subsequence of (u_n) exists, denoted by (u_k) such that

$$u_k \rightharpoonup u \text{ in } V_0, \quad u_k \to u \text{ in } L^p(\Omega). \tag{3.68}$$

Replacing n by k in (3.63) and taking $\varphi = u_k - u$, we obtain

$$\langle Au_k, u_k - u \rangle = \int_\Omega G(u_k)(u - u_k) \, dx + \langle f, u_k - u \rangle. \tag{3.69}$$

Passing to the lim sup in (3.69) as $k \to \infty$ and taking into account the boundedness of (u_k) in V_0 as well as the convergence properties (3.68), we get

$$\limsup_{k} \langle Au_k, u_k - u \rangle \leq 0,$$

which due to the (S_+)-property of the operator A (see Theorem 2.109) implies the strong convergence $u_k \to u$ in V_0. To complete the proof we only need to show that the limit u belongs to S. This process however, follows immediately from (3.63) by replacing n by k and passing to the limit as $k \to \infty$. $\qquad \square$

To show that the solution set S of the BVP (3.42) is a directed set, the following additional assumption on the coefficients $a_i : \Omega \times \mathbb{R} \times \mathbb{R}^N \to \mathbb{R}$ is required.

(H4) Modulus of Continuity Condition: Let a function $k_3 \in L_+^q(\Omega)$ and a function $\omega : \mathbb{R}_+ \to \mathbb{R}_+$ exist such that

$$|a_i(x, s, \xi) - a_i(x, s', \xi)| \leq [k_3(x) + |s|^{p-1} + |s'|^{p-1} + |\xi|^{p-1}] \omega(|s - s'|),$$

holds for a.e. $x \in \Omega$, for all $s, s' \in \mathbb{R}$ and for all $\xi \in \mathbb{R}^N$, where $\omega : \mathbb{R}_+ \to \mathbb{R}_+$ is a continuous function with the property

$$\int_{0+} \frac{dr}{\omega(r)} = +\infty. \tag{3.70}$$

Remark 3.19. Equation (3.70) means that for every $\varepsilon > 0$, the integral taken over $[0, \varepsilon]$ diverges; i.e.,

$$\int_0^\varepsilon \frac{dr}{\omega(r)} = +\infty.$$

Hypothesis (H4) includes, for example, $\omega(r) = c\,r$, $\forall\, r \geq 0$, i.e., a Lipschitz condition of the coefficients $a_i(x, s, \xi)$ with respect to s.

In the semilinear case, the directedness of the solution set S is an immediate consequence of either Theorem 3.6 or Theorem 3.8. Both theorems can be extended to the quasilinear case considered in this section under the assumptions of Theorem 3.17 and the additional hypothesis (H4). As the extension of Theorem 3.6 to the quasilinear case has been treated in great detail in [43], we provide in the following theorem the extension of Theorem 3.8.

Theorem 3.20. *Assume hypotheses (H1)–(H4), and let u_1 and u_2 be subsolutions of the BVP (3.42) such that the Nemytskij operator*

$$G : [\min\{u_1, u_2\}, \max\{u_1, u_2\}] \to L^q(\Omega)$$

is well defined. Then $\max\{u_1, u_2\}$ is a subsolution of the BVP (3.42). Analogously, if u_1 and u_2 are supersolutions of the BVP (3.42) with the same assumption on the Nemytskij operator G, then $\min\{u_1, u_2\}$ is a supersolution.

Proof: We consider first the case that u_1 and u_2 are subsolutions; i.e., we have $u_k \leq 0$ on $\partial \Omega$ and

$$a(u_k, \varphi) + \int_\Omega G(u_k)\varphi \, dx \leq \langle f, \varphi \rangle \quad \text{for all} \quad \varphi \in V_0 \cap L_+^p(\Omega), \qquad (3.71)$$

and $k = 1, 2$. Denote $u = \max\{u_1, u_2\}$; then according to Sect. 2.2.3, we have $u \in V$ and $u \leq 0$ on $\partial \Omega$, and

$$\nabla u(x) = \begin{cases} \nabla u_1(x) & \text{if} \quad u_1(x) \geq u_2(x), \\ \nabla u_2(x) & \text{if} \quad u_2(x) \geq u_1(x). \end{cases}$$

Thus, for u being a subsolution, we need to verify the following inequality:

$$a(u, \varphi) + \int_\Omega G(u)\varphi \, dx \leq \langle f, \varphi \rangle \quad \text{for all} \quad \varphi \in V_0 \cap L_+^p(\Omega), \qquad (3.72)$$

where

$$G(u)(x) = \begin{cases} G(u_1)(x) & \text{if} \quad x \in \{u_1 \geq u_2\}, \\ G(u_2)(x) & \text{if} \quad x \in \{u_2 > u_1\}. \end{cases}$$

In view of hypothesis (H4), for any fixed $\varepsilon > 0$, a $\delta(\varepsilon) \in (0, \varepsilon)$ exists such that

$$\int_{\delta(\varepsilon)}^\varepsilon \frac{1}{w(r)} \, dr = 1.$$

This property allows us to introduce the function $\theta_\varepsilon : \mathbb{R} \to \mathbb{R}_+$, which is defined by

$$\theta_\varepsilon(s) = \begin{cases} 0 & \text{if} \quad s < \delta(\varepsilon), \\ \int_{\delta(\varepsilon)}^s \frac{1}{w(r)} \, dr & \text{if} \quad \delta(\varepsilon) \leq s \leq \varepsilon, \\ 1 & \text{if} \quad s > \varepsilon. \end{cases} \qquad (3.73)$$

Obviously, for each $\varepsilon > 0$, the function θ_ε is continuous, piecewise differentiable, and the derivative is nonnegative and bounded. Thus, the function θ_ε is Lipschitz continuous and nondecreasing, and moreover, it satisfies

$$\theta_\varepsilon \to \chi_{\{s > 0\}} \quad \text{as } \varepsilon \to 0,$$

where $\chi_{\{s > 0\}}$ is the characteristic function of the set $\{s > 0\} = \{s \in \mathbb{R} : s > 0\}$. In addition, one has

$$\theta_\varepsilon'(s) = \begin{cases} \dfrac{1}{w(s)} & \text{if} \quad \delta(\varepsilon) < s < \varepsilon, \\ 0 & \text{if} \quad s \notin [\delta(\varepsilon), \varepsilon]. \end{cases} \qquad (3.74)$$

As in the proof of Theorem 3.8, let us introduce the set $\mathcal{D}_+$ of nonnegative smooth functions; i.e.,

$$\mathcal{D}_+ = \{\psi \in C_0^\infty(\Omega) : \psi \geq 0\}.$$

In case that $k = 1$, we take the special test function

$$\varphi = \psi\,(1 - \theta_\varepsilon(u_2 - u_1)) \in V_0 \cap L_+^p(\Omega),$$

where $\psi \in \mathcal{D}_+$, and in case $k = 2$, we take

$$\varphi = \psi\,\theta_\varepsilon(u_2 - u_1) \in V_0 \cap L_+^p(\Omega).$$

Adding the resulting inequalities, we obtain

$$a(u_1, \psi) + a(u_2, \psi\,\theta_\varepsilon(u_2 - u_1)) - a(u_1, \psi\,\theta_\varepsilon(u_2 - u_1))$$

$$+ \int_\Omega G(u_1)\,\psi\,dx + \int_\Omega (G(u_2) - G(u_1))\,\psi\,\theta_\varepsilon(u_2 - u_1))\,dx$$

$$\leq \langle f, \psi \rangle. \tag{3.75}$$

We discuss next the terms depending on ε. To this end, we need the partial derivative $\partial/\partial x_i$ of $\psi\,\theta_\varepsilon(u_2 - u_1)$, which can be calculated by applying the generalized chain rule (see Lemma 2.84) as follows:

$$\frac{\partial}{\partial x_i}(\psi\,\theta_\varepsilon(u_2 - u_1)) = \frac{\partial \psi}{\partial x_i}\theta_\varepsilon(u_2 - u_1) + \psi\,\theta_\varepsilon'(u_2 - u_1)\frac{\partial(u_2 - u_1)}{\partial x_i}. \tag{3.76}$$

By means of (3.76), we first get

$$a(u_2, \psi\,\theta_\varepsilon(u_2 - u_1)) - a(u_1, \psi\,\theta_\varepsilon(u_2 - u_1))$$

$$= \int_\Omega \sum_{i=1}^N \Big(a_i(\cdot, u_2, \nabla u_2) - a_i(\cdot, u_1, \nabla u_1)\Big)\frac{\partial(\psi\,\theta_\varepsilon(u_2 - u_1))}{\partial x_i}\,dx$$

$$= \int_\Omega \sum_{i=1}^N \Big(a_i(\cdot, u_2, \nabla u_2) - a_i(\cdot, u_1, \nabla u_1)\Big)\psi\,\theta_\varepsilon'(u_2 - u_1)\frac{\partial(u_2 - u_1)}{\partial x_i}\,dx$$

$$+ \int_\Omega \sum_{i=1}^N \Big(a_i(\cdot, u_2, \nabla u_2) - a_i(\cdot, u_1, \nabla u_1)\Big)\theta_\varepsilon(u_2 - u_1)\frac{\partial \psi}{\partial x_i}\,dx,$$

which yields by applying hypotheses (H2) and (H4) as well as the properties of $\theta_\varepsilon(u_2 - u_1)$ the following estimate:

$$a(u_2, \psi\,\theta_\varepsilon(u_2 - u_1)) - a(u_1, \psi\,\theta_\varepsilon(u_2 - u_1))$$

$$= \int_\Omega \sum_{i=1}^N \Big(a_i(\cdot, u_2, \nabla u_2) - a_i(\cdot, u_2, \nabla u_1)\Big)\psi\,\theta_\varepsilon'(u_2 - u_1)\frac{\partial(u_2 - u_1)}{\partial x_i}\,dx$$

$$+ \int_\Omega \sum_{i=1}^N \Big(a_i(\cdot, u_2, \nabla u_1) - a_i(\cdot, u_1, \nabla u_1)\Big)\psi\,\theta_\varepsilon'(u_2 - u_1)\frac{\partial(u_2 - u_1)}{\partial x_i}\,dx$$

$$+ \int_\Omega \sum_{i=1}^N \Big(a_i(\cdot, u_2, \nabla u_2) - a_i(\cdot, u_1, \nabla u_1) \Big) \theta_\varepsilon (u_2 - u_1) \frac{\partial \psi}{\partial x_i} \, dx$$

$$\geq \int_\Omega \sum_{i=1}^N \Big(a_i(\cdot, u_2, \nabla u_2) - a_i(\cdot, u_1, \nabla u_1) \Big) \theta_\varepsilon (u_2 - u_1) \frac{\partial \psi}{\partial x_i} \, dx$$

$$- N \int_{\{\delta(\varepsilon) < u_2 - u_1 < \varepsilon\}} \Big(k_3 + |u_1|^{p-1} + |u_2|^{p-1} + |\nabla u_1|^{p-1} \Big) \times$$

$$\times \psi \, |\nabla (u_2 - u_1)| \, dx. \tag{3.77}$$

For the last term on the right-hand side of (3.77), we have

$$\int_{\{\delta(\varepsilon) < u_2 - u_1 < \varepsilon\}} \Big(k_3 + |u_1|^{p-1} + |u_2|^{p-1} + |\nabla u_1|^{p-1} \Big) \psi \, |\nabla (u_2 - u_1)| \, dx \to 0 \tag{3.78}$$

as $\varepsilon \to 0$. By means of Lebesgue's dominated convergence theorem (see Theorem 2.63), the first term on the right-hand side of (3.77) yields as $\varepsilon \to 0$ the relation

$$\lim_{\varepsilon \to 0} \int_\Omega \sum_{i=1}^N \Big(a_i(\cdot, u_2, \nabla u_2) - a_i(\cdot, u_1, \nabla u_1) \Big) \theta_\varepsilon (u_2 - u_1) \frac{\partial \psi}{\partial x_i} \, dx$$

$$= \int_\Omega \sum_{i=1}^N \Big(a_i(\cdot, u_2, \nabla u_2) - a_i(\cdot, u_1, \nabla u_1) \Big) \chi_{\{u_2 > u_1\}} \frac{\partial \psi}{\partial x_i} \, dx, \tag{3.79}$$

and thus from (3.78) and (3.79) we get

$$\liminf_{\varepsilon \to 0} [a(u_2, \psi \, \theta_\varepsilon (u_2 - u_1)) - a(u_1, \psi \, \theta_\varepsilon (u_2 - u_1))]$$

$$\geq \int_\Omega \sum_{i=1}^N \Big(a_i(\cdot, u_2, \nabla u_2) - a_i(\cdot, u_1, \nabla u_1) \Big) \chi_{\{u_2 > u_1\}} \frac{\partial \psi}{\partial x_i} \, dx. \tag{3.80}$$

Again by Lebesgue's dominated convergence theorem, we have

$$\lim_{\varepsilon \to 0} \int_\Omega (G(u_2) - G(u_1)) \, \psi \, \theta_\varepsilon (u_2 - u_1)) \, dx$$

$$= \int_\Omega (G(u_2) - G(u_1)) \, \chi_{\{u_2 > u_1\}} \psi \, dx. \tag{3.81}$$

Finally, from (3.75), (3.80), and (3.81), we obtain

$$\int_\Omega \sum_{i=1}^N a_i(\cdot, u_1, \nabla u_1) \frac{\partial \psi}{\partial x_i} \, dx$$

$$+ \int_\Omega \sum_{i=1}^N \Big(a_i(\cdot, u_2, \nabla u_2) - a_i(\cdot, u_1, \nabla u_1) \Big) \chi_{\{u_2 > u_1\}} \frac{\partial \psi}{\partial x_i} \, dx$$

$$+ \int_\Omega \Big(G(u_1) + (G(u_2) - G(u_1))\,\chi_{\{u_2 > u_1\}}\Big)\psi\,dx$$
$$\leq \langle f, \psi \rangle, \tag{3.82}$$

which is nothing else as

$$a(u, \psi) + \int_\Omega G(u)\psi\,dx \leq \langle f, \psi \rangle \quad \text{for all} \quad \psi \in \mathcal{D}_+. \tag{3.83}$$

As $\mathcal{D}_+$ is dense in $V_0 \cap L^p_+(\Omega)$ from (3.83), we obtain (3.72), which proves that $u = \max\{u_1, u_2\}$ is a subsolution whenever u_1 and u_2 are subsolutions. That the minimum of two supersolutions becomes a supersolution can be proved analogously. This process completes the proof of the theorem. □

The following corollary is an immediate consequence of Theorem 3.20.

Corollary 3.21 (Directedness). *Assume hypotheses (H1)–(H4), and let the growth condition (3.49) be satisfied with respect to the ordered interval $[\underline{u}, \bar{u}]$. Then the solution set $\mathcal{S}$ of the BVP (3.42) is directed.*

Proof: The same arguments as in the proof of Corollary 3.7 apply. □

3.2.3 Extremal Solutions

Let $\mathcal{S}$ be the same as in Sect. 3.2.2. The following extremality result can easily be proved by means of the directedness and compactness results of the previous section.

Theorem 3.22. *Under the hypotheses of Corollary 3.21, the solution set $\mathcal{S}$ possesses extremal elements; i.e., there is a greatest solution u^* and a smallest solution u_* of the quasilinear BVP (3.42) within the ordered interval $[\underline{u}, \bar{u}]$ of sub- and supersolutions.*

Proof: The main tools used in the proof are Theorem 3.18, Theorem 3.20, and Corollary 3.21. We focus on the existence of the greatest element of $\mathcal{S}$.

As V_0 is separable, we have that $\mathcal{S} \subset V_0$ is separable, so a countable, dense subset $Z = \{z_n : n \in \mathbb{N}\}$ of $\mathcal{S}$ exists. By Corollary 3.21, $\mathcal{S}$ is upward directed, so we can construct an increasing sequence $(u_n) \subset \mathcal{S}$ as follows. Let $u_1 = z_1$. Select $u_{n+1} \in \mathcal{S}$ such that

$$\max\{z_n, u_n\} \leq u_{n+1} \leq \bar{u}.$$

The existence of such an $u_{n+1} \in \mathcal{S}$ follows from Corollary 3.21. By induction we get an increasing sequence $(u_n) \subset \mathcal{S}$, which converges to $u = \sup_n u_n$ in $L^p(\Omega)$ because the order cone $L^p_+(\Omega)$ is a fully regular order cone, (see [111]). As $\mathcal{S}$ is a compact subset of V_0, a subsequence of (u_n) exists, which converges in V_0, and whose limit belongs to $\mathcal{S}$. As $u_n \to u$ in $L^p(\Omega)$, all convergent subsequences must have the same limit, and thus, the entire increasing sequence (u_n) satisfies $u_n \to u \in \mathcal{S}$ strongly in V_0. By construction, we see that

$$\max\{z_1, z_2, \ldots, z_n\} \leq u_{n+1} \leq u, \quad \text{for all } n \in \mathbb{N},$$

and thus, u is an upper bound for Z, which implies $Z \subset [\underline{u}, u]$. As the interval $[\underline{u}, u]$ is closed in V_0 and Z is dense in $\mathcal{S}$, we infer

$$\mathcal{S} \subset \overline{Z} \subset \overline{[\underline{u}, u]} = [\underline{u}, u],$$

which, because $u \in \mathcal{S}$, ensures that $u = u^*$ is the greatest element of $\mathcal{S}$; i.e., u is the greatest solution of the BVP (3.42) within $[\underline{u}, \bar{u}]$. The existence of the smallest solution u_* of the BVP (3.42) can be proved in a similar way by using the fact that $\mathcal{S}$ is also downward directed. $\square$

The extremality result due to Theorem 3.22 allows us to extend the order-preserving property of the extremal solutions in the semilinear case to the quasilinear case. Consider the BVP (3.42) with right-hand sides $f_k \in V_0^*$, $k = 1, 2$; i.e.,

$$u \in V_0 : \quad Au + G(u) = f_k \quad \text{in } V_0^*. \tag{3.84}$$

Let $\underline{u}$ and $\bar{u}$ be an ordered pair of sub- and supersolutions for both problems (3.84), and denote by u_{k*} and u_k^* the corresponding smallest and greatest solutions of (3.84), respectively, within $[\underline{u}, \bar{u}]$. With these notations, the following result holds.

Theorem 3.23. *If $f_1 \leq f_2$, then $u_1^* \leq u_2^*$, and similarly, $u_{1*} \leq u_{2*}$.*

Proof: The proof follows the same idea as for the proof of Theorem 3.13. $\square$

Example 3.24. Let us consider the following special case of the BVP (3.42):

$$-\Delta_p u + g(u) = f \quad \text{in } \Omega, \quad u = 0 \quad \text{on } \partial\Omega, \tag{3.85}$$

where $\Delta_p u = \text{div}\,(|\nabla u|^{p-2}\nabla u)$ is the p-Laplacian, and $f \in V_0^*$ is given. Let $\underline{u}$ and $\bar{u}$ be an ordered pair of sub- and supersolutions, and assume that $g : \mathbb{R} \to \mathbb{R}$ is a continuous function that satisfies the growth condition:

$$|g(s)| \leq c\,(1 + |s|^{p-1}) \quad \text{for all } s \in \mathbb{R}, \tag{3.86}$$

where c is some positive constant.

Corollary 3.25. *The BVP (3.85) possesses solutions within the ordered interval $[\underline{u}, \bar{u}]$, and the solution set $\mathcal{S}$ of all solutions of (3.85) in $[\underline{u}, \bar{u}]$ is compact in V_0 and has extremal elements that depend monotonically on the right-hand side f.*

Proof: In the special case of the p-Laplacian, the coefficients a_i, $i = 1, \ldots, N$, defining the differential operator are given by

$$a_i(x, s, \xi) = |\xi|^{p-2}\xi_i.$$

As a_i do not depend on x and s, hypothesis (H4) is trivially satisfied. Hypotheses (H1) and (H3) are easily seen to be true, and hypothesis (H2) follows from the monotonicity inequality of Sect. 2.2.4. For any $s \in [\underline{u}(x), \bar{u}(x)]$, one has

$$|s| \leq |\underline{u}(x)| + |\bar{u}(x)|,$$

and thus, the growth condition (3.49) of Theorem 3.17 follows from (3.86) with

$$k_1(x) = c \left(1 + (|\underline{u}(x)| + |\bar{u}(x)|)^{p-1} \right).$$

Note $k_1 \in L^q(\Omega)$. Therefore, Theorem 3.17, Theorem 3.8–Theorem 3.11, and Corollary 3.21 are applicable, which completes the proof of the corollary. □

Remark 3.26. Note that the "global" growth condition (3.86) is more restrictive than the "local" one given by (3.49) of Theorem 3.17. Of course, also for Corollary 3.25 to hold, it is enough to assume only a local growth condition for g with respect to the ordered interval of sub- and supersolutions; i.e.,

$$|g(s)| \leq k_1(x) \quad \text{for all} \ \ s \in [\underline{u}(x), \bar{u}(x)].$$

Remark 3.27. Theorem 3.17–Theorem 3.23 and Corollary 3.21 remain true if the hypotheses (H2) and (H3) of Sect. 2.3.2 imposed on the coefficients a_i, $i = 1, \ldots, N$, are replaced by the following strong ellipticity:

$$\sum_{i=1}^{N} (a_i(x, s, \xi) - a_i(x, s, \xi'))(\xi_i - \xi_i') \geq \mu |\xi - \xi'|^p, \qquad (3.87)$$

where $\mu > 0$ (for a.e. $x \in \Omega$, for all $s \in \mathbb{R}$, and for all $\xi, \xi' \in \mathbb{R}^N$), whereas condition (3.70) of hypothesis (H4) is replaced by

$$\int_{0+} \frac{dr}{\omega^q(r)} = +\infty, \qquad (3.88)$$

with q being the Hölder conjugate to p. Condition (3.88) allows us to deal with coefficients $a_i(x, s, \xi)$, which satisfy, e.g., a Hölder condition with respect to s. For a more detailed analysis in this case, we refer to [43].

3.3 Quasilinear Parabolic Equations

In this section, we shall use the notations and results provided in Sect. 2.4. Let $\Omega \subset \mathbb{R}^N$ be a bounded domain with Lipschitz boundary $\partial\Omega$, $Q = \Omega \times (0, \tau)$, and $\Gamma = \partial\Omega \times (0, \tau)$, with $\tau > 0$. Our goal is to prove comparison, extremality, and compactness results for the following quasilinear initial-boundary value problem (IBVP, for short):

$$u_t + Au + g(\cdot, \cdot, u, \nabla u) = f \quad \text{in} \ \ Q,$$

$$u = 0 \text{ in } \Omega \times \{0\}, \quad u = 0 \text{ on } \Gamma, \tag{3.89}$$

where A is assumed to be a second-order quasilinear differential operator in divergence form given by

$$Au(x,t) = -\sum_{i=1}^{N} \frac{\partial}{\partial x_i} a_i(x, t, u(x,t), \nabla u(x,t)). \tag{3.90}$$

We impose the following hypotheses of Leray–Lions type on the coefficients $a_i : Q \times \mathbb{R} \times \mathbb{R}^N \to \mathbb{R}$, $i = 1, \ldots, N$.

(A1) Carathéodory and Growth Condition: Each $a_i(x,t,s,\xi)$ satisfies Carathéodory conditions, i.e., is measurable in $(x,t) \in Q$ for all $(s,\xi) \in \mathbb{R} \times \mathbb{R}^N$ and continuous in (s,ξ) for a.e. $(x,t) \in Q$. A constant $c_0 > 0$ and a function $k_0 \in L^q(Q)$ exist so that

$$|a_i(x,t,s,\xi)| \leq k_0(x,t) + c_0(|s|^{p-1} + |\xi|^{p-1})$$

for a.e. $(x,t) \in Q$ and for all $(s,\xi) \in \mathbb{R} \times \mathbb{R}^N$, with $|\xi|$ denoting the Eucleadian norm of the vector $\xi \in \mathbb{R}^N$.

(A2) Monotonicity Type Condition: The coefficients a_i satisfy a monotonicity condition with respect to ξ in the form

$$\sum_{i=1}^{N} (a_i(x,t,s,\xi) - a_i(x,t,s,\xi'))(\xi_i - \xi_i') > 0$$

for a.e. $(x,t) \in Q$, for all $s \in \mathbb{R}$, and for all $\xi, \xi' \in \mathbb{R}^N$ with $\xi \neq \xi'$.

(A3) Coercivity Type Condition:

$$\sum_{i=1}^{N} a_i(x,t,s,\xi)\xi_i \geq \nu|\xi|^p - k(x,t)$$

for a.e. $(x,t) \in Q$, for all $s \in \mathbb{R}$, and for all $\xi \in \mathbb{R}^N$ with some constant $\nu > 0$ and some function $k \in L^1(Q)$.

As the coefficients a_i are not necessarily differentiable, the IBVP (3.89) has to be understood in an appropriate generalized sense, a motivation of which has already been given in Sect. 2.4.1 for a simpler IBVP. As in the previous section, let $V = W^{1,p}(\Omega)$ and $V_0 = W_0^{1,p}(\Omega)$. We assume throughout this section

$$2 \leq p < \infty,$$

and set $H = L^2(\Omega)$. Then $V \subset H \subset V^*$ (respectively, $V_0 \subset H \subset V_0^*$) forms an evolution triple with all embeddings being continuous, dense, and compact, (see Sect. 2.4.3). We introduce the spaces $X = L^p(0,\tau;V)$ and $X_0 = L^p(0,\tau;V_0)$ of vector-valued functions, and we define

$$W = \{u \in X : u_t \in X^*\}, \quad (\text{respectively,} \ W_0 = \{u \in X_0 : u_t \in X_0^*\}),$$

where the derivative $u_t = \partial u/\partial t = u'$ is understood in the sense of vector-valued distributions (see Sect. 2.4). Let $a : X \times X \to \mathbb{R}$ be the semilinear form related to the differential operator A and given by

$$a(u,v) = \int_Q \sum_{i=1}^N a_i(\cdot, \cdot, u, \nabla u) \frac{\partial v}{\partial x_i} \, dx dt; \qquad (3.91)$$

then due to (A1), a is well defined for any $(u,v) \in X \times X$. Again by (A1), for fixed $u \in X$ the mapping $\varphi \mapsto a(u, \varphi)$ is linear and continuous on X (respectively, on X_0). With the help of (3.91), we introduce an operator (again denoted by A) defined by

$$\langle Au, \varphi \rangle = a(u, \varphi) \quad \text{for all} \ \varphi \in X_0, \qquad (3.92)$$

where $\langle \cdot, \cdot \rangle$ is the duality pairing between X_0^* and X_0. From (3.92) for any $u \in X$, we have $Au \in X_0^*$, and in view of (A1), the operator $A : X \to X_0^*$ is continuous and bounded (respectively, $A : X_0 \to X_0^*$ is continuous and bounded).

Let us agree to use the notation $\langle \cdot, \cdot \rangle$ for any of the dual pairings between X and X^*, X_0 and X_0^*, V and V^*, and V_0 and V_0^*. For example, with $f \in X_0^*, u \in X_0$,

$$\langle f, u \rangle = \int_0^\tau \langle f(t), u(t) \rangle \, dt.$$

Furthermore, a natural partial ordering in $L^p(Q)$ is defined by $u \le w$ if and only if $w - u$ belongs to the positive cone $L_+^p(Q)$ of all nonnegative elements of $L^p(Q)$. It induces a corresponding partial ordering also in the subspace W of $L^p(Q)$, and if $u, w \in W$ with $u \le w$, then

$$[u, w] = \{v \in W : u \le v \le w\}$$

denotes the ordered interval formed by u and w.

As for the lower order terms g of the parabolic equation, we assume the following hypothesis.

(H) The function $g : Q \times \mathbb{R} \times \mathbb{R}^N \to \mathbb{R}$ is a Carathéodory function that satisfies the growth condition

$$|g(x, t, s, \xi)| \le k_1(x, t) + c_1 |\xi|^{p-1} \qquad (3.93)$$

for a.e. $(x, t) \in Q$, for all $\xi \in \mathbb{R}^N$, and for all $s \in [\underline{v}(x, t), \bar{v}(x, t)]$, where $k_1 \in L_+^q(Q)$, c_1 is some positive constant, and $[\underline{v}, \bar{v}]$ is some ordered interval specified later.

Denote by G the Nemytskij operator related to g, i.e.,

$$G(u)(x, t) = g(x, t, u(x, t), \nabla u(x, t)),$$

and assume the right-hand side f of the parabolic equation to be an element of X_0^*. Then a generalized solution of the IBVP (3.89) is defined as follows.

Definition 3.28. *The function $u \in W_0$ is a solution of the IBVP (3.89) if $G(u) \in L^q(Q)$, $u(\cdot, 0) = 0$ in Ω, and*

$$u' + Au + G(u) = f \quad in \ \ X_0^*. \tag{3.94}$$

Remark 3.29. We have assumed homogeneous initial and boundary conditions in the IBVP (3.89) only for the sake of simplicity. Without loss of generality, nonhomogeneous initial and boundary conditions of the form

$$u = \psi \ \ in \ \ \Omega \times \{0\}, \quad u = h \ \ on \ \ \Gamma \tag{3.95}$$

can be treated as well, where $\psi \in H$, and $h = \gamma(w)$ is the trace of a function $w \in W$. The corresponding nonhomogeneous IBVP will be reduced to the homogeneous case by translation. To this end, let $\hat{u} \in W$ be any function that satisfies the initial and boundary data (3.95); i.e.,

$$\hat{u} = \psi \ \ in \ \ \Omega \times \{0\}, \quad \hat{u} = h \ \ on \ \ \Gamma.$$

The existence of such a function $\hat{u}$ will be demonstrated in the next section. Consider now the nonhomogeneous IBVP

$$
\begin{aligned}
u_t + Au + g(\cdot, \cdot, u, \nabla u) &= f \quad in \ \ Q, \\
u = \psi \ \ in \ \ \Omega \times \{0\}, \quad u &= h \ \ on \ \ \Gamma,
\end{aligned}
\tag{3.96}
$$

and perform the translation

$$u = \hat{u} + v,$$

which yields the following homogeneous IBVP in v:

$$
\begin{aligned}
v_t + \hat{A}v + \hat{g}(\cdot, \cdot, v, \nabla v) &= \hat{f} \quad in \ \ Q, \\
v = 0 \ \ in \ \ \Omega \times \{0\}, \quad v &= 0 \ \ on \ \ \Gamma,
\end{aligned}
\tag{3.97}
$$

where the coefficients $\hat{a}_i$ of the transformed operator $\hat{A}$, $\hat{g}$, and $\hat{f}$ are given by

$$
\begin{aligned}
\hat{a}_i(x, t, s, \xi) &= a_i(x, t, s + \hat{u}(x, t), \xi + \nabla\hat{u}(x, t)), \\
\hat{g}(x, t, s, \xi) &= g(x, t, s + \hat{u}(x, t), \xi + \nabla\hat{u}(x, t)), \\
\hat{f} &= f - \hat{u}_t.
\end{aligned}
\tag{3.98}
$$

To verify that the transformed IBVP (3.97) is of the same structure as the original homogeneous IBVP (3.89), we need to show that the data $\hat{a}_i$, $\hat{g}$, and $\hat{f}$ preserve the regularity and structure hypotheses imposed on a_i, g, and f, respectively. First, note that $\hat{u}_t \in X^* \subset X_0^*$, and thus, $\hat{f} \in X_0^*$. One readily verifies that $\hat{g}$ is a Carathéodory function satisfying the growth condition (3.93) with possibly a different function k_1 and constant c_1, and which is true with respect to the shifted ordered interval $[\underline{u} - \hat{u}, \bar{u} - \hat{u}]$. The coefficients $\hat{a}_i$ are Carathéodory functions as well, and they satisfy apparently a similar growth condition as in (A1). As a_i satisfy (A2), we obtain

$$\sum_{i=1}^{N}(\hat{a}_i(x,t,s,\xi) - \hat{a}_i(x,t,s,\xi'))(\xi_i - \xi_i')$$

$$= \sum_{i=1}^{N}\left(\hat{a}_i(x,t,s,\xi) - \hat{a}_i(x,t,s,\xi')\right)\left((\xi_i + \hat{u}_{x_i}(x,t)) - (\xi_i' + \hat{u}_{x_i}(x,t))\right) > 0$$

for a.e. $(x,t) \in Q$, for all $s \in \mathbb{R}$, and for all $\xi, \xi' \in \mathbb{R}^N$ with $\xi \neq \xi'$, which shows that $\hat{a}_i$ satisfy (A2) as well. Consider now $\sum_{i=1}^{N} \hat{a}_i(x,t,s,\xi)\xi_i$. For its estimate below, we use the following elementary inequality. There is a positive constant c such that

$$|\xi|^p \leq (|\xi + \eta| + |\eta|)^p \leq c\left(|\xi + \eta|^p + |\eta|^p\right) \quad \text{for all } \xi, \eta \in \mathbb{R}^N. \tag{3.99}$$

Applying (A3) satisfied by a_i, we obtain

$$\sum_{i=1}^{N} \hat{a}_i(x,t,s,\xi)\xi_i = \sum_{i=1}^{N} a_i(x,t,s + \hat{u}(x,t), \xi + \nabla\hat{u}(x,t))\xi_i$$

$$\geq \nu\,|\xi + \nabla\hat{u}(x,t)|^p - k(x,t)$$

$$- \sum_{i=1}^{N} a_i(x,t,s + \hat{u}(x,t), \xi + \nabla\hat{u}(x,t))\hat{u}_{x_i}(x,t). \tag{3.100}$$

By means of the growth condition of (A1) and Young's inequality, we get for the last term the estimate

$$\left|\sum_{i=1}^{N} a_i(x,t,s + \hat{u}(x,t), \xi + \nabla\hat{u}(x,t))\hat{u}_{x_i}(x,t)\right|$$

$$\leq \frac{\varepsilon}{c_0}|k_0(x,t)|^q + \varepsilon\,|s + \hat{u}(x,t)|^p + \varepsilon\,|\xi + \nabla\hat{u}(x,t)|^p$$

$$+ c(\varepsilon)|\nabla\hat{u}(x,t)|^p \tag{3.101}$$

for every $\varepsilon > 0$, where $c(\varepsilon)$ is a positive constant only depending on ε. Taking into account the inequality [see (3.99]

$$|\xi + \eta|^p \geq \frac{1}{c}|\xi|^p - |\eta|^p \quad \text{for all } \xi, \eta \in \mathbb{R}^N,$$

and (3.100) and (3.101), we get

$$\sum_{i=1}^{N} \hat{a}_i(x,t,s,\xi)\xi_i \geq \frac{\nu - \varepsilon}{c}|\xi|^p - \frac{\varepsilon}{c}|s|^p - \hat{k}_\varepsilon(x,t), \tag{3.102}$$

where $\hat{k}_\varepsilon \in L^1(Q)$. If a_i do not depend on s, the above estimates yield

$$\sum_{i=1}^{N} \hat{a}_i(x,t,\xi)\xi_i \geq \frac{\nu - \varepsilon}{c} |\xi|^p - \hat{k}_\varepsilon(x,t) \qquad (3.103)$$

for any $\varepsilon > 0$, which shows that in this case, for $\varepsilon < \nu$, also hypothesis (A3) is satisfied. Therefore, in the special case that $a_i(x,t,\xi)$ do not depend on s, we have justified that homogeneous initial and boundary values can be assumed without loss of generality. In the general case, (A3) is not necessarily true for the transformed problem due to the term $-\frac{\varepsilon}{c} |s|^p$ on the right-hand side of (3.102). However, as we will see later, the comparison technique to be developed here, which is based on sub- and supersolutions, turns out to be flexible enough to compensate also this drawback, so that also in the general case, we are allowed to deal with the homogeneous IBVP without loss of generality.

3.3.1 Parabolic Equation with p-Laplacian

Let us consider the following IBVP:

$$u_t - \Delta_p u = f \quad \text{in} \ \ Q,$$
$$u = \psi \ \ \text{in} \ \ \Omega \times \{0\}, \quad u = 0 \ \ \text{on} \ \ \Gamma, \qquad (3.104)$$

where $\Delta_p u = \operatorname{div}(|\nabla u|^{p-2}\nabla u)$ is the p-Laplacian, $f \in X_0^*$, and $\psi \in H$. Problem (3.104) is a special case of the IBVP (3.89) and arises from (3.89) by setting $g = 0$ and $a_i(x,t,s,\xi) = |\xi|^{p-2}\xi_i$. According to Definition 3.28, a function u is a solution of the IBVP (3.104) if u satisfies

$$u \in W_0, \ u(x,0) = \psi(x) \ \text{ and } \ u' - \Delta_p u = f \quad \text{in} \ \ X_0^*. \qquad (3.105)$$

Lemma 3.30. *The IBVP (3.104) (respectively, (3.105) has a unique solution.*

Proof: The operator $-\Delta_p : X_0 \to X_0^*$ given by

$$\langle -\Delta u, \varphi \rangle = \int_Q |\nabla u|^{p-2}\nabla u \nabla \varphi \, dxdt \quad \text{for all } \varphi \in X_0$$

can easily be seen to be a monotone, continuous, bounded, and coercive operator. Hence, the existence of a unique solution of the IBVP (3.105) follows by applying [222, Theorem 30.A]. □

With the help of Lemma 3.30 we can prove the existence of a function $\hat{u} \in W$ satisfying given nonhomogeneous initial and boundary data as it was assumed in Remark 3.29 of the previous section.

Corollary 3.31. *Let $\psi \in H$ and $h = \gamma(w)$, where $\gamma(w)$ denotes the trace of a function $w \in W$ on Γ. Then a function $\hat{u} \in W$ exists with*

$$\hat{u} = \psi \ \ \text{in} \ \ \Omega \times \{0\}, \quad \hat{u} = \gamma(w) \ \ \text{on} \ \ \Gamma.$$

Proof: As $W \subset C([0, \tau]; H)$ is continuously embedded, we have $w(\cdot, 0) \in H$. The IBVP (3.104) with ψ replaced by $\psi - w(\cdot, 0) \in H$ has a unique solution $u \in W_0$, and thus, the function $\hat{u}$ defined by $\hat{u} = w + u$ satisfies

$$\hat{u} \in W, \quad \gamma(\hat{u}) = \gamma(w) \quad \text{on} \quad \Gamma, \quad \hat{u}(x, 0) = w(x, 0) + u(x, 0) = \psi(x),$$

which proves the corollary. $\qquad\qquad\Box$

Definition 3.32. *A function* $\underline{u} \in W$ *is called a subsolution of the IBVP (3.104) if* $\underline{u}(x, 0) \leq \psi(x)$ *for* $x \in \Omega$, $\underline{u} \leq 0$ *on* Γ, *and*

$$\langle \underline{u}' - \Delta_p \underline{u}, \varphi \rangle \leq \langle f, \varphi \rangle \quad \text{for all} \quad \varphi \in X_0 \cap L_+^p(Q).$$

Similarly, $\bar{u}$ is a supersolution of the IBVP (3.104) if the reversed inequalities in Definition 3.32 hold with $\underline{u}$ replaced by $\bar{u}$.

Lemma 3.33. *Let* $\underline{u}$ *and* $\bar{u}$ *be sub- and supersolutions of the IBVP (3.104). Then* $\underline{u} \leq \bar{u}$ *and the unique solution* u *of (3.104) satisfies* $u \in [\underline{u}, \bar{u}]$.

Proof: Subtracting the corresponding inequalities satisfied by the sub- and supersolution, respectively, we get

$$\langle \underline{u}' - \bar{u}', \varphi \rangle + \int_Q \left(|\nabla \underline{u}|^{p-2} \nabla \underline{u} - |\nabla \bar{u}|^{p-2} \nabla \bar{u} \right) \nabla \varphi \, dx dt \leq 0 \qquad (3.106)$$

for all $\varphi \in X_0 \cap L_+^p(Q)$. In particular, $\varphi = (\underline{u} - \bar{u})^+$ is an admissible test function for inequality (3.106), because $\gamma((\underline{u} - \bar{u})^+) = 0$ on Γ, and thus, $(\underline{u} - \bar{u})^+ \in X_0 \cap L_+^p(Q)$. Moreover, as $(\underline{u} - \bar{u})^+(x, 0) = 0$, we see from Example 2.148 that

$$\langle (\underline{u} - \bar{u})', (\underline{u} - \bar{u})^+ \rangle = \frac{1}{2} \|(\underline{u} - \bar{u})^+(\cdot, \tau)\|_H^2. \qquad (3.107)$$

As $2 \leq p < \infty$, we obtain for the integral on the left-hand side of (3.106) the following estimate:

$$\int_Q \left(|\nabla \underline{u}|^{p-2} \nabla \underline{u} - |\nabla \bar{u}|^{p-2} \nabla \bar{u} \right) \nabla (\underline{u} - \bar{u})^+ \, dx dt$$

$$\geq c \int_Q |\nabla (\underline{u} - \bar{u})^+|^p \, dx dt \geq \hat{c} \, \|(\underline{u} - \bar{u})^+\|_{X_0}^p. \qquad (3.108)$$

Testing (3.106) with $\varphi = (\underline{u} - \bar{u})^+$ and applying (3.107) and (3.108) results in

$$\|(\underline{u} - \bar{u})^+\|_{X_0} = 0,$$

which implies $(\underline{u} - \bar{u})^+ = 0$; i.e., $\underline{u} \leq \bar{u}$. Finally, because the unique solution u of the IBVP (3.104) is both a subsolution and a supersolution, it follows that $u \in [\underline{u}, \bar{u}]$. $\qquad\qquad\Box$

3.3.2 Comparison Principle for Quasilinear Equations

The main goal of this section is to prove a comparison result for the general IBVP (3.89) based on the notions of sub- and supersolutions, which are defined as follows.

Definition 3.34. *The function $\underline{u} \in W$ is called a subsolution of the IBVP (3.89) if $G(\underline{u}) \in L^q(Q)$, $\underline{u} \leq 0$ in $\Omega \times \{0\}$, $\underline{u} \leq 0$ on Γ, and*

$$\underline{u}' + A\underline{u} + G(\underline{u}) \leq f \quad \text{in } X_0^*,$$

which is equivalent to

$$\langle \underline{u}', \varphi \rangle + a(\underline{u}, \varphi) + \int_Q G(\underline{u})\,\varphi\,dxdt \leq \langle f, \varphi \rangle \quad \text{for all } \varphi \in X_0 \cap L_+^p(Q).$$

Similarly, we define a supersolution as follows.

Definition 3.35. *The function $\bar{u} \in W$ is called a supersolution of the IBVP (3.89) if $G(\bar{u}) \in L^q(Q)$, $\bar{u} \geq 0$ in $\Omega \times \{0\}$, $\bar{u} \geq 0$ on Γ, and*

$$\bar{u}' + A\bar{u} + G(\bar{u}) \geq f \quad \text{in } X_0^*,$$

which is equivalent to

$$\langle \bar{u}', \varphi \rangle + a(\bar{u}, \varphi) + \int_Q G(\bar{u})\,\varphi\,dxdt \geq \langle f, \varphi \rangle \quad \text{for all } \varphi \in X_0 \cap L_+^p(Q).$$

In preparation of our main result, we provide first an existence result for an associated auxiliary truncated IBVP of the form

$$u_t + A_T u + (G \circ T)u + \lambda B(u) = f \quad \text{in } Q,$$
$$u = 0 \quad \text{in } \Omega \times \{0\}, \quad u = 0 \quad \text{on } \Gamma, \tag{3.109}$$

where T is the truncation operator related with an ordered pair of sub- and supersolutions given by

$$Tu(x,t) = \begin{cases} \bar{u}(x,t) & \text{if } u(x,t) > \bar{u}(x,t), \\ u(x,t) & \text{if } \underline{u}(x,t) \leq u(x,t) \leq \bar{u}(x,t), \\ \underline{u}(x,t) & \text{if } u(x,t) < \underline{u}(x,t), \end{cases}$$

and $G \circ T$ denotes the composition of the operators G and T. The operator A_T is defined by

$$\langle A_T u, \varphi \rangle = a_T(u, \varphi) = \int_Q \sum_{i=1}^N a_i(\cdot, \cdot, Tu, \nabla u) \frac{\partial \varphi}{\partial x_i}\,dxdt \quad \text{for all } \varphi \in X_0. \tag{3.110}$$

In just the same way as in the elliptic case, the parameter $\lambda > 0$ will be specified later, and B stands for the Nemytskij operator generated by the cutoff function $b : Q \times \mathbb{R} \to \mathbb{R}$, which is related to the ordered pair of sub- and supersolutions, and given by

$$b(x,t,s) = \begin{cases} (s - \bar{u}(x,t))^{p-1} & \text{if } s > \bar{u}(x,t), \\ 0 & \text{if } \underline{u}(x,t) \leq s \leq \bar{u}(x,t), \\ -(\underline{u}(x,t) - s)^{p-1} & \text{if } s < \underline{u}(x,t). \end{cases} \tag{3.111}$$

Again we have that b is a Carathéodory function satisfying

$$|b(x,t,s)| \leq k_2(x,t) + c_2 |s|^{p-1} \tag{3.112}$$

for a.e. $(x,t) \in Q$, for all $s \in \mathbb{R}$, with some function $k_2 \in L_+^q(Q)$ and a constant $c_2 \geq 0$, and

$$\int_Q b(x,t,u(x,t))\,u(x,t)\,dxdt \geq c_3 \|u\|_p^p - c_4 \quad \text{for all } u \in L^p(Q), \tag{3.113}$$

where c_3 and c_4 are some positive constants. To transform the IBVP (3.109) into an equivalent operator equation, we introduce the operator $L = \partial/\partial t$ with domain $D(L)$ given by

$$D(L) = \{u \in X_0 : u_t \in X_0^* \text{ and } u(\cdot,0) = 0 \text{ in } \Omega\}.$$

Then $L : D(L) \subset X_0 \to X_0^*$ is a closed, densely defined, and maximal monotone operator (see Lemma 2.149), and generalized or weak solutions of the IBVP (3.109) are solutions of the following operator equation:

$$u \in D(L): \quad Lu + A_T u + (G \circ T)u + \lambda B(u) = f \quad \text{in } X_0^*. \tag{3.114}$$

Lemma 3.36. *Assume hypotheses (A1)–(A3), and let hypothesis (H) be satisfied with respect to an ordered pair of sub- and supersolutions. If $\lambda > 0$ is sufficiently large, then the IBVP (3.109) possesses solutions.*

Proof: We note first that the truncation operator $T : X \to X$ is continuous and bounded (see Sect. 2.2.3 or [84]). Hypothesis (H) implies that the operator $G : [\underline{u}, \bar{u}] \subset X \to L^q(Q)$ is continuous and bounded, and thus, the composition $G \circ T : X_0 \to L^q(Q) \subset X_0^*$ is continuous and bounded. In view of (A1), the operator $A_T : X_0 \to X_0^*$ is continuous and bounded. The Carathéodory and growth condition (3.112) of b imply that $B : L^p(Q) \to L^q(Q)$ is continuous and bounded, which due to the continuous embedding $X_0 \subset L^p(Q)$ shows that $B : X_0 \to X_0^*$ is continuous and bounded. Moreover, because of the compact embedding $W_0 \subset L^p(Q)$, it follows that $B : X_0 \to X_0^*$ is even completely continuous w.r.t $D(L)$. Hypotheses (A1), (A2), and (H) are sufficient to prove that $A_T + G \circ T + \lambda B : X_0 \to X_0^*$ is pseudomonotone w.r.t. $D(L)$. For a proof, we refer to [152] and [174]. So far we know that the operator $A_T + G \circ T + \lambda B : X_0 \to X_0^*$ is continuous, bounded, and pseudomonotone w.r.t. $D(L)$.

According to Theorem 2.152, the operator $L + A_T + G \circ T + \lambda B : D(L) \to X_0^*$ is surjective; i.e., the IBVP (3.114) has a solution, provided $A_T + G \circ T + \lambda B : X_0 \to X_0^*$ is, in addition, coercive. This result will be shown next. From (A3) and (3.113), it follows that

$$\langle A_T u + \lambda B(u), u \rangle \geq \nu \, \|\nabla u\|_p^p + \lambda c_3 \|u\|_p^p - \|k\|_1 - \lambda c_4. \tag{3.115}$$

By means of (H) in conjunction with

$$|\nabla Tu| \leq |\nabla \underline{u}| + |\nabla \bar{u}| + |\nabla u|,$$

we obtain an estimate in the form

$$|\langle (G \circ T)u, u \rangle| = \left| \int_Q (G \circ T)(u)\, u \, dx dt \right| \leq \int_Q (k_4 + c_5 |\nabla u|^{p-1}) |u| \, dx dt,$$

for some constant $c_5 > 0$ and $k_4 \in L_+^q(Q)$, which by applying Young's inequality yields the estimate

$$|\langle (G \circ T)u, u \rangle| \leq \varepsilon \, \|\nabla u\|_p^p + c(\varepsilon)(\|u\|_p^p + 1) \tag{3.116}$$

for any $\varepsilon > 0$. With (3.115), and (3.116) we finally get the estimate

$$\langle (A_T + G \circ T + \lambda B)u, u \rangle \geq (\nu - \varepsilon) \|\nabla u\|_p^p + \left(\lambda c_3 - c(\varepsilon) \right) \|u\|_p^p - \hat{c}(\varepsilon). \tag{3.117}$$

If we select $\varepsilon < \nu$, then for λ sufficiently large, we get $\lambda c_3 - c(\varepsilon) > 0$, which proves the coercivity of $A_T + G \circ T + \lambda B : X_0 \to X_0^*$ in view of (3.117). This process completes the proof.   □

Now we can prove the following comparison principle.

Theorem 3.37. *Let $\underline{u}$ and $\bar{u}$ be sub-and supersolutions of the IBVP (3.89) satisfying $\underline{u} \leq \bar{u}$, and assume the hypotheses of Lemma 3.36. Then the IBVP (3.89) has solutions within the interval $[\underline{u}, \bar{u}]$.*

Proof: First we note that by using the operator $L = \partial / \partial t$ introduced above, we can rewrite the notion of solution of the IBVP (3.89) given in Definition 3.28 as the following operator equation:

$$u \in D(L): \quad Lu + Au + G(u) = f \quad \text{in } X_0^*, \tag{3.118}$$

where $G(u) \in L^q(Q)$. The proof of the theorem is based on the auxiliary truncated IBVP (3.109); i.e.,

$$u \in D(L): \quad Lu + A_T u + (G \circ T)u + \lambda B(u) = f \quad \text{in } X_0^*.$$

The existence of solutions of (3.109) is ensured by Lemma 3.36. The proof of the theorem is achieved if we only know that any solution u of the auxiliary problem (3.109) is in fact contained in the interval $[\underline{u}, \bar{u}]$, because then $Tu = u$ and $B(u) = 0$; that is, u must be a solution of the original IBVP (3.89). Let us show first $u \leq \bar{u}$.

According to Definition 3.35, the function $\bar{u} \in W$ is a supersolution of the IBVP (3.89) if $G(\bar{u}) \in L^q(Q)$, $\bar{u} \geq 0$ in $\Omega \times \{0\}$, $\bar{u} \geq 0$ on Γ, and

$$\bar{u}' + A\bar{u} + G(\bar{u}) \geq f \quad \text{in } X_0^*, \tag{3.119}$$

where inequality (3.119) has to be considered in the weak sense, i.e., with respect to the dual-order cone. From the initial and boundary conditions for $\bar{u}$, and in view of the lattice structure of X, we get

$$(u - \bar{u})^+ \in X_0 \cap L_+^p(Q) \quad \text{and} \quad (u - \bar{u})^+(x, 0) = 0 \quad \text{for } x \in \Omega. \tag{3.120}$$

Testing (3.109) and (3.119) with $\varphi = (u - \bar{u})^+$ and subtracting (3.119) from (3.109), we arrive at

$$\langle (u - \bar{u})', (u - \bar{u})^+ \rangle + \langle A_T u - A\bar{u}, (u - \bar{u})^+ \rangle$$
$$+ \int_Q \left((G \circ T)u - G(\bar{u}) \right) (u - \bar{u})^+ \, dx dt$$
$$+ \lambda \int_Q B(u) (u - \bar{u})^+ \, dx dt \leq 0. \tag{3.121}$$

In view of Example 2.148 and (3.120), we obtain for the first term on the left-hand side of (3.121)

$$\langle (u - \bar{u})', (u - \bar{u})^+ \rangle = \frac{1}{2} \| (u - \bar{u})^+(\cdot, \tau) \|_H^2. \tag{3.122}$$

By definition of A_T and hypothesis (A2), the second term of (3.121) yields

$$\langle A_T u - A\bar{u}, (u - \bar{u})^+ \rangle$$
$$= \int_Q \sum_{i=1}^N \left(a_i(\cdot, \cdot, Tu, \nabla u) - a_i(\cdot, \cdot, \bar{u}, \nabla \bar{u}) \right) \frac{\partial (u - \bar{u})^+}{\partial x_i} \, dx dt$$
$$= \int_{\{u > \bar{u}\}} \sum_{i=1}^N \left(a_i(\cdot, \cdot, \bar{u}, \nabla u) - a_i(\cdot, \cdot, \bar{u}, \nabla \bar{u}) \right) \frac{\partial (u - \bar{u})}{\partial x_i} \, dx dt$$
$$\geq 0. \tag{3.123}$$

For the third term we get

$$\int_Q \left((G \circ T)u - G(\bar{u}) \right) (u - \bar{u})^+ \, dx dt$$
$$= \int_{\{u > \bar{u}\}} \left(g(\cdot, \cdot, \bar{u}, \nabla \bar{u}) - g(\cdot, \cdot, \bar{u}, \nabla \bar{u}) \right) (u - \bar{u}) \, dx dt$$
$$= 0, \tag{3.124}$$

and finally, by definition of the cutoff function b, the last term on the left-hand side of (3.121) becomes

$$\lambda \int_Q B(u)\,(u - \bar{u})^+ \,dxdt = \lambda \int_{\{u > \bar{u}\}} (u - \bar{u})^p \,dxdt = \lambda \,\|(u - \bar{u})^+\|_p^p. \quad (3.125)$$

Thus, (3.121)–(3.125) yields

$$0 \le \lambda \,\|(u - \bar{u})^+\|_p^p \le 0,$$

which implies $(u - \bar{u})^+ = 0$; i.e., $u \le \bar{u}$ in Q. The proof of the inequality $\underline{u} \le u$ can be done analogously, which shows that any solution u of the auxiliary problem (3.109) is contained in the interval $[\underline{u}, \bar{u}]$ completing the proof of the theorem. $\square$

Remark 3.38. Notice that the role played by the term $\lambda\,B(u)$ of the auxiliary truncated IBVP (3.109) is twofold. On the one hand, we see from the proof of Lemma 3.36 that $\lambda\,B(u)$ provides a coercivity generating term if λ is chosen sufficiently large. On the other hand, it allows for the comparison of the solutions of the auxiliary problem with the sub- and supersolutions, which finally proves our comparison principle. It is this term that can be used, in addition, to compensate the term $-(\varepsilon/c)\,|s|^p$ we mentioned in Remark 3.29 that arises when transforming inhomogeneous initial and boundary data to homogeneous ones in the general case of the operator A.

3.3.3 Directed and Compact Solution Set

Let us again denote the set of all solutions of the IBVP (3.89) within the ordered interval $[\underline{u}, \bar{u}]$ by $\mathcal{S}$. We are going to show that $\mathcal{S}$ is a directed set, which is compact in W_0. For the latter, the following preliminary result will be useful.

Lemma 3.39. *Let the operator A given by (3.90) satisfy hypotheses (A1)–(A3). Then $A : X_0 \to X_0^*$ is bounded, continuous, and pseudomonotone w.r.t. $D(L)$, and it has the (S_+)-property w.r.t. $D(L)$.*

Proof: The proof is a consequence of Theorem 2.153, where the (S_+)-property w.r.t. $D(L)$ means: If $(u_n) \subset W_0$ satisfies $u_n \rightharpoonup u$ in X_0, $Lu_n \rightharpoonup Lu$ in X_0^*, and

$$\limsup_{n \to \infty} \langle Au_n, u_n - u \rangle \le 0,$$

then $u_n \to u$ in X_0. $\square$

Theorem 3.40 (Compactness). *Under the hypotheses of Theorem 3.37, the solution set $\mathcal{S}$ is compact in W_0.*

Proof: First note that $S \subset [\underline{u}, \bar{u}]$, which shows that S is bounded in $L^p(Q)$. As $u \in S$ satisfies

$$u \in D(L): \quad Lu + Au + G(u) = f \quad \text{in } X_0^*,$$

we get from (A3) and

$$\langle Lu, u \rangle = \frac{1}{2} \|u(\cdot, \tau)\|_H^2 \geq 0,$$

$$\nu \|\nabla u\|_p^p \leq \|k\|_1 + \|f\|_{X_0^*} \|u\|_{X_0} + |\langle G(u), u \rangle|. \qquad (3.126)$$

By means of (H) and taking into account that S is bounded in $L^p(Q)$, the last term on the right-hand side of (3.126) can be estimated as follows:

$$|\langle G(u), u \rangle| \leq \varepsilon \|\nabla u\|_p^p + c(\varepsilon), \qquad (3.127)$$

for any $\varepsilon > 0$, where $c(\varepsilon)$ is some constant only depending on ε. Thus, (3.126) and (3.127) yield

$$(\nu - \varepsilon) \|\nabla u\|_p^p \leq c(\varepsilon) + \|f\|_{X_0^*} \|u\|_{X_0} \quad \text{for all } u \in S, \qquad (3.128)$$

which by selecting $\varepsilon < \nu$ implies the boundedness of S in X_0. As $A : X_0 \to X_0^*$ and $G : X_0 \cap S \to X_0^*$ are bounded operators, from

$$Lu = f - Au - G(u),$$

we see that

$$\|Lu\|_{X_0^*} \leq c \quad \text{for all } u \in S,$$

which in view of (3.128) implies that

$$\|u\|_{W_0} \leq c \quad \text{for all } u \in S. \qquad (3.129)$$

Let $(u_n) \subset S$ be any sequence. Then (u_n) is bounded in W_0, and thus, by the reflexivity of W_0, a subsequence (u_k) exists that is weakly convergent in W_0 to u; i.e., $u_k \rightharpoonup u$ in X_0, $u_k' \rightharpoonup u'$ in X_0^*. As $D(L) \subset W_0$ is convex and closed, it follows that $D(L)$ is weakly closed, and thus, $u \in D(L)$. Therefore, we have $u_k \rightharpoonup u$ in X_0 and $Lu_k \rightharpoonup Lu$ in X_0^*. As solutions of the IBVP (3.89), u_k satisfy

$$u_k \in D(L): \quad \langle Lu_k + Au_k + G(u_k), u_k - u \rangle = \langle f, u_k - u \rangle. \qquad (3.130)$$

Using

$$\langle Lu_k, u_k - u \rangle = \langle Lu_k - Lu, u_k - u \rangle + \langle Lu, u_k - u \rangle \geq \langle Lu, u_k - u \rangle,$$

we get from (3.130)

$$\langle Au_k, u_k - u \rangle \leq \langle Lu, u - u_k \rangle + \langle f, u_k - u \rangle + \int_Q |G(u_k)(u_k - u)| \, dxdt,$$

$$(3.131)$$

where the last term on the right-hand side of inequality (3.131) can be estimated by applying hypothesis (H) as follows:

$$\int_Q |G(u_k)(u_k - u)| \, dxdt \leq c \, (1 + \|\nabla u_k\|_p^{p-1}) \|u_k - u\|_p. \qquad (3.132)$$

The compact embedding $W_0 \subset L^p(Q)$ implies $u_k \to u$ in $L^p(Q)$, and thus, from (3.131), (3.132), and the boundedness of (u_k) in W_0, we obtain

$$\limsup_{k \to \infty} \langle Au_k, u_k - u \rangle \leq 0, \qquad (3.133)$$

which in view of the (S_+)-property of A according to Lemma 3.39 shows that (u_k) is strongly convergent in X_0; i.e., $u_k \to u$ in X_0. The strong convergence of (u_k) in X_0 and the weak convergence $Lu_k \rightharpoonup Lu$ allow us to pass to the limit in

$$Lu_k + Au_k + G(u_k) = f \quad \text{in } X_0^*,$$

as $k \to \infty$, which shows that $u \in \mathcal{S}$. Moreover, by the strong convergence $u_k \to u$ in X_0 and the continuity of the operators $A : X_0 \to X_0^*$ and $G : X_0 \to X_0^*$, we have

$$\|Lu_k - Lu\|_{X_0^*} \leq \|G(u_k) - G(u)\|_{X_0^*} + \|Au_k - Au\|_{X_0^*} \to 0,$$

as $k \to \infty$, which proves the strong convergence of (u_k) to u in W_0.   □

To show the directedness of the solution set $\mathcal{S}$, the following additional assumption on the coefficients $a_i : Q \times \mathbb{R} \times \mathbb{R}^N \to \mathbb{R}$, $i = 1, \ldots, N$, is needed.

(A4) **Modulus of Continuity Condition:** Let a function $k_3 \in L_+^q(Q)$ and a function $\omega : \mathbb{R}_+ \to \mathbb{R}_+$ exist such that

$$|a_i(x,t,s,\xi) - a_i(x,t,s',\xi)| \leq [k_3(x,t) + |s|^{p-1} + |s'|^{p-1} + |\xi|^{p-1}] \, \omega(|s - s'|),$$

holds for a.e. $(x,t) \in Q$, for all $s, s' \in \mathbb{R}$ and for all $\xi \in \mathbb{R}^N$, where $\omega : \mathbb{R}_+ \to \mathbb{R}_+$ is a continuous function with the property

$$\int_{0+} \frac{dr}{\omega(r)} = +\infty. \qquad (3.134)$$

Note that (A4) is empty if $a_i(x,t,s,\xi)$, $i = 1, \ldots, N$, do not depend on s. Also, Remark 3.19 holds correspondingly. The directedness of $\mathcal{S}$ will be seen to be an immediate consequence of the following generalized comparison result.

Theorem 3.41. *Let hypotheses (A1)–(A4) be satisfied. Let $\underline{u}_1, \ldots, \underline{u}_k$ and $\bar{u}_1, \ldots, \bar{u}_m$ with $k, m \in \mathbb{N}$ be sub- and supersolutions of the IBVP (3.89), respectively, such that*

$$\underline{u} = \max\{\underline{u}_1, \ldots, \underline{u}_k\} \leq \bar{u} = \min\{\bar{u}_1, \ldots, \bar{u}_m\}, \tag{3.135}$$

and assume hypothesis (H) on g to be satisfied with respect to the ordered interval $[\underline{v}, \bar{v}]$, where

$$\underline{v} = \min\{\underline{u}_1, \ldots, \underline{u}_k\}, \quad \bar{v} = \max\{\bar{u}_1, \ldots, \bar{u}_m\}. \tag{3.136}$$

Then solutions of the IBVP (3.89) exist within the ordered interval $[\underline{u}, \bar{u}]$.

Proof: The following auxiliary truncated IBVP plays a key role in the proof:

$$u \in D(L) : Lu + Au + P(u) + \lambda B(u) = f \quad \text{in } X_0^*, \tag{3.137}$$

where B is as above the Nemytskij operator generated by the cutoff function b given by (3.111), and $\lambda > 0$ is a free parameter to be specified later. The operator P is given by

$$P(u) = (G \circ T)u + \sum_{i=1}^{m} |(G \circ T^i)u - (G \circ T)u| - \sum_{j=1}^{k} |(G \circ T_j)u - (G \circ T)u|, \tag{3.138}$$

where the truncation operators T_j, T^i, and T are defined in a similar way as in Sect. 3.1.2; i.e.,

$$Tu(x,t) = \begin{cases} \bar{u}(x,t) & \text{if } u(x,t) > \bar{u}(x,t), \\ u(x,t) & \text{if } \underline{u}(x,t) \leq u(x,t) \leq \bar{u}(x,t), \\ \underline{u}(x,t) & \text{if } u(x,t) < \underline{u}(x,t); \end{cases}$$

$$T_j u(x,t) = \begin{cases} \underline{u}_j(x,t) & \text{if } u(x,t) < \underline{u}_j(x,t), \\ u(x,t) & \text{if } \underline{u}_j(x,t) \leq u(x,t) \leq \bar{u}(x,t), \\ \bar{u}(x,t) & \text{if } u(x,t) > \bar{u}(x,t); \end{cases}$$

$$T^i u(x,t) = \begin{cases} \underline{u}(x,t) & \text{if } u(x,t) < \underline{u}(x,t), \\ u(x,t) & \text{if } \underline{u}(x,t) \leq u(x,t) \leq \bar{u}_i(x,t), \\ \bar{u}_i(x,t) & \text{if } u(x,t) > \bar{u}_i(x,t); \end{cases}$$

for $1 \leq i \leq m, 1 \leq j \leq k$, $(x,t) \in Q$. The operators $G \circ T$, $G \circ T_j$, $G \circ T^i$ stand for the compositions of the Nemytskij operator G and the truncation operators T, T_j, and T^i, respectively. Furthermore, we have

$$\langle |(G \circ T^i)u - (G \circ T)u|, v \rangle = \int_Q |g(\cdot, \cdot, T^i u, \nabla T^i u) - g(\cdot, \cdot, Tu, \nabla Tu)| \, v \, dx dt$$

as well as

$$\langle |(G \circ T_j)u - (G \circ T)u|, v \rangle = \int_Q |g(\cdot, \cdot, T_j u, \nabla T_j u) - g(\cdot, \cdot, Tu, \nabla Tu)| \, v \, dx dt,$$

for all $u, v \in X_0$. As T_j, T^i, and $T : X_0 \to X_0$ are bounded and continuous, the operator $P : X_0 \to L^q(Q) \subset X_0^*$ preserves the structure and mapping properties of $G \circ T$. We may apply similar arguments as in the proof of Lemma 3.36 to ensure the existence of solutions of the auxiliary IBVP (3.137). The proof of the theorem is completed provided we can show that any solution u of (3.137) satisfies

$$\underline{u}_j \le u \le \bar{u}_i, \quad 1 \le i \le m, \ 1 \le j \le k, \tag{3.139}$$

because then u satisfies also $\underline{u} \le u \le \bar{u}$, which finally results in $Tu = u$, $T_j u = u$, $T^i u = u$, and thus, $P(u) = G(u)$, as well as $B(u) = 0$. It shows that u is a solution of the original problem IBVP (3.89), which lies in $[\underline{u}, \bar{u}]$.

Let us first show that any solution u of the auxiliary problem (3.137) satisfies $u \le \bar{u}_l$ for $l \in \{1, \ldots, m\}$ fixed. By assumption, $\bar{u}_l$ is a supersolution; i.e., we have $\bar{u}_l \ge 0$ in $\Omega \times \{0\}$, $\bar{u}_l \ge 0$ on Γ, and

$$\bar{u}_l' + A\bar{u}_l + G(\bar{u}_l) \ge f \quad \text{in } X_0^*. \tag{3.140}$$

Subtracting (3.140) from (3.137), we get

$$u' - \bar{u}_l' + Au - A\bar{u}_l + P(u) - G(\bar{u}_l) \le 0 \quad \text{in } X_0^*. \tag{3.141}$$

Let $\theta_\varepsilon : \mathbb{R} \to \mathbb{R}_+$ be the Lipschitz continuous and nondecreasing function introduced in Sect. 3.2.2 and defined by (3.73); i.e.,

$$\theta_\varepsilon(s) = \begin{cases} 0 & \text{if } s < \delta(\varepsilon), \\ \displaystyle\int_{\delta(\varepsilon)}^s \frac{1}{\omega(r)} \, dr & \text{if } \delta(\varepsilon) \le s \le \varepsilon, \\ 1 & \text{if } s > \varepsilon. \end{cases}$$

Then the composed function $\theta_\varepsilon(u - \bar{u}_l) : Q \to \mathbb{R}_+$ is in $X_0 \cap L_+^p(Q)$. Let Θ_ε be the primitive of the function θ_ε defined by

$$\Theta_\varepsilon(r) = \int_0^r \theta_\varepsilon(s) \, ds;$$

then in view of Lemma 2.146, one has

$$\langle (u - \bar{u}_l)', \theta_\varepsilon(u - \bar{u}_l) \rangle = \int_\Omega \Theta_\varepsilon(u - \bar{u}_l)(x, \tau) \, dx \ge 0, \tag{3.142}$$

because $\Theta_\varepsilon(u - \bar{u}_l)(x, 0) = 0$ for $x \in \Omega$. The partial derivative $\partial/\partial x_i$ of $\theta_\varepsilon(u - \bar{u}_l)$ yields

$$\frac{\partial}{\partial x_i} \theta_\varepsilon(u - \bar{u}_l) = \theta_\varepsilon'(u - \bar{u}_l) \frac{\partial(u - \bar{u}_l)}{\partial x_i}. \tag{3.143}$$

By means of (3.143), (A2), and (A4), we derive the following estimate:

$$\langle Au - A(\bar{u}_l), \theta_\varepsilon(u - \bar{u}_l) \rangle$$

$$= \int_Q \sum_{i=1}^N \left(a_i(x, t, u, \nabla u) - a_i(x, t, \bar{u}_l, \nabla \bar{u}_l) \right) \frac{\partial}{\partial x_i} \theta_\varepsilon(u - \bar{u}_l) \, dx \, dt$$

$$\geq \int_Q \sum_{i=1}^N \left(a_i(x, t, u, \nabla u) - a_i(x, t, u, \nabla \bar{u}_l) \right) \frac{\partial(u - \bar{u}_l)}{\partial x_i} \theta'_\varepsilon(u - \bar{u}_l) \, dx \, dt$$

$$- N \int_Q (k_3 + |u|^{p-1} + |\bar{u}_l|^{p-1} + |\nabla \bar{u}_l|^{p-1}) |\nabla(u - \bar{u}_l)| \, dx \, dt$$

$$\geq -N \int_{\{\delta(\varepsilon) < u - \bar{u}_l < \varepsilon\}} h \, |\nabla(u - \bar{u}_l)| \, dx \, dt, \tag{3.144}$$

where $h = k_3 + |u|^{p-1} + |\bar{u}_l|^{p-1} + |\nabla \bar{u}_l|^{p-1} \in L^q(Q)$. The term on the right-hand side of (3.144) tends to zero as $\varepsilon \to 0$. By using

$$\theta_\varepsilon \to \chi_{\{s>0\}} \quad \text{as } \varepsilon \to 0,$$

where $\chi_{\{s>0\}}$ is the characteristic function of the set $\{s > 0\} = \{s \in \mathbb{R} : s > 0\}$, and applying Lebesgue's dominated convergence theorem, it follows that

$$\lim_{\varepsilon \to 0} \int_Q B(u) \, \theta_\varepsilon(u - \bar{u}_l) \, dx \, dt = \int_Q B(u) \, \chi_{\{u - \bar{u}_l > 0\}} \, dx \, dt \tag{3.145}$$

and

$$\lim_{\varepsilon \to 0} \int_Q (P(u) - G(\bar{u}_l)) \, \theta_\varepsilon(u - \bar{u}_l) \, dx \, dt = \int_Q (P(u) - G(\bar{u}_l)) \, \chi_{\{u - \bar{u}_l > 0\}} \, dx \, dt. \tag{3.146}$$

By definition of the operator B, the right-hand side of (3.145) can be estimated in the following way:

$$\int_Q B(u) \, \chi_{\{u - \bar{u}_l > 0\}} \, dx \, dt = \int_{\{u > \bar{u}_l\}} (u - \bar{u})^{p-1} \, dx \, dt \geq \int_{\{u > \bar{u}_l\}} (u - \bar{u}_l)^{p-1} \, dx \, dt. \tag{3.147}$$

Applying similar arguments as in the estimate (3.21), we have for the right-hand side of (3.146) the estimate

$$\int_Q (P(u) - G(\bar{u}_l)) \, \chi_{\{u - \bar{u}_l > 0\}} \, dx \, dt \geq 0. \tag{3.148}$$

Finally, testing inequality (3.141) with $\theta_\varepsilon(u - \bar{u}_l) \in X_0 \cap L^p_+(Q)$ and using (3.142), as well as (3.144)–(3.148), one obtains as $\varepsilon \to 0$

$$0 \leq \int_Q [(u - \bar{u}_l)^+]^{p-1}\, dxdt = \int_{\{u > \bar{u}_l\}} (u - \bar{u}_l)^{p-1} dx\, dt \leq 0, \qquad (3.149)$$

which implies $(u - \bar{u}_l)^+ = 0$; i.e., $u \leq \bar{u}_l$, for any $l \in \{1, \ldots, m\}$. The proof of the inequalities $\underline{u}_j \leq u$ for $j = 1, \ldots, k$ can be done in a similar way. Thus, inequalities (3.139) are satisfied, which completes the proof of the theorem. $\qquad \square$

Corollary 3.42 (Directedness). *Assume hypotheses (A1)–(A4), and let (H) be satisfied with respect to the ordered interval $[\underline{u}, \bar{u}]$ of sub- and supersolution. Then the solution set S of the IBVP (3.89) is directed.*

Proof: Taking into account that any solution of the IBVP (3.89) is a subsolution and a supersolution as well, the assertion follows from Theorem 3.41. $\quad \square$

3.3.4 Extremal Solutions

With the help of the compactness and directedness results given by Theorem 3.40, Theorem 3.41, and Corollary 3.42, the following extremality property of the solution set S can be shown in a similar way as the corresponding result in the elliptic case (see Theorem 3.22).

Theorem 3.43 (Extremal Solutions). *Under the assumptions of Corollary 3.42, the solution set S of the IBVP (3.89) has extremal elements; i.e., there is a greatest solution u^* and a smallest solution u_* of the IBVP (3.89) within the ordered interval $[\underline{u}, \bar{u}]$ of sub- and supersolutions.*

Theorem 3.43 in conjunction with the comparison principle formulated in Theorem 3.37 will allow us to verify an order-preserving property of extremal solutions. To this end, we consider the IBVP (3.89) with right-hand sides $f_k \in X_0^*$, $k = 1, 2$; i.e.,

$$u \in D(L): \quad Lu + Au + G(u) = f_k \quad \text{in } X_0^*. \qquad (3.150)$$

Let u_{k*} and u_k^* denote the corresponding smallest and greatest solutions of (3.150) with respect to a common ordered interval $[\underline{u}, \bar{u}]$ of sub- and supersolutions. The order-preserving property of the extremal solutions is given in the next theorem.

Theorem 3.44. *Assume the hypotheses of Corollary 3.42. If $f_1 \leq f_2$, then $u_1^* \leq u_2^*$, and $u_{1*} \leq u_{2*}$.*

Proof: Let us show: $u_1^* \leq u_2^*$. Consider the IBVP

$$u \in D(L): \quad Lu + Au + G(u) = f_1 \quad \text{in } X_0^*. \qquad (3.151)$$

Then the greatest solution u_1^* of (3.151) is readily seen to be a subsolution of the IBVP

$$u \in D(L): \quad Lu + Au + G(u) = f_2 \quad \text{in } X_0^*, \tag{3.152}$$

because $f_1 \leq f_2$. As u_1^* and $\bar{u}$ form a pair of sub- and supersolutions of (3.152), a solution u of (3.152) exists within the interval $[u_1^*, \bar{u}]$. Because u_2^* is the greatest solution of (3.152) within $[\underline{u}, \bar{u}]$, we get $u_1^* \leq u \leq u_2^*$, which proves the assertion. Similarly, $u_{1*} \leq u_{2*}$ can be shown. $\qquad \square$

Remark 3.45. Theorem 3.37–Theorem 3.44 and Corollary 3.42 remain true if the hypotheses (A2) and (A3) imposed on the coefficients a_i, $i = 1, \ldots, N$, are replaced by the following strong ellipticity:

$$\sum_{i=1}^{N} (a_i(x,t,s,\xi) - a_i(x,s,t,\xi'))(\xi_i - \xi_i') \geq \mu \, |\xi - \xi'|^p, \tag{3.153}$$

where $\mu > 0$ (for a.e. $(x,t) \in Q$, for all $s \in \mathbb{R}$, and for all $\xi, \xi' \in \mathbb{R}^N$), and condition (3.134) of hypothesis (A4) is replaced by

$$\int_{0+} \frac{dr}{\omega^q(r)} = +\infty, \tag{3.154}$$

with q being the Hölder conjugate to p. Condition (3.154) allows us to deal with coefficients $a_i(x,t,s,\xi)$, which satisfy, e.g., a Hölder condition with respect to s. For a more detailed analysis in this case, we refer to [43].

3.4 Sign-Changing Solutions via Fučik Spectrum

This section deals with the existence of sign-changing and multiple solutions for a class of nonlinear elliptic Dirichlet-problems involving the p-Laplacian. Let $\Omega \subset \mathbb{R}^N$, $N \geq 1$ be a bounded domain with smooth boundary $\partial\Omega$, and let $V_0 = W_0^{1,p}(\Omega)$ and $V = W^{1,p}(\Omega)$. We consider the quasilinear elliptic boundary value problem (BVP for short)

$$u \in V_0: \quad -\Delta_p u = f(x,u) \quad \text{in } V_0^*, \tag{3.155}$$

where $\Delta_p u = \text{div}\,(|\nabla u|^{p-2}\nabla u)$ is the p-Laplacian, $1 < p < \infty$. Although there are many existence and multiplicity results for (3.155) in the literature (see [70, 71, 74, 87, 119, 186]), only a few papers deal with sign-changing solutions, such as [221, 220, 118]. The approach in [118] and [221, 220] is based, among others on the calculation of critical groups and the construction of pseudo-gradient vector field in V_0, respectively. The approach suggested here is different and relies on a combined use of the results on extremal solutions provided in Sect. 3.2 and the variational characterization of the Fučik spectrum. The results presented here are based on the paper in [61].

3.4.1 Introduction

The existence of sign-changing and multiple solutions for (3.155) will be studied under the following assumptions on the right-hand side f.

(H1) $f : \Omega \times \mathbb{R} \to \mathbb{R}$ is a Carathéodory function satisfying the growth condition

$$|f(x,t)| \le c\,(1 + |t|^{p-1}), \qquad (3.156)$$

for a.e. $x \in \Omega$, and for all $t \in \mathbb{R}$, and f is assumed to be of the form

$$f(x,t) = a\,(t^+)^{p-1} - b\,(t^-)^{p-1} + g(x,t), \qquad (3.157)$$

with

$$\lim_{t \to 0} \frac{g(x,t)}{|t|^{p-1}} = 0 \quad \text{uniformly in } x, \qquad (3.158)$$

where $(a,b) \in \mathbb{R}^2$, and $t^+ = \max\{t,0\}$, and $t^- = \max\{-t,0\}$.

Definition 3.46. *The set Σ_p of those points $(a,b) \in \mathbb{R}^2$ for which the asymptotic problem*

$$u \in V_0 : \quad -\Delta_p\, u = a\,(u^+)^{p-1} - b\,(u^-)^{p-1} \quad in \ \ V_0^* \qquad (3.159)$$

has a nontrivial solution is called the Fučik spectrum of the p-Laplacian on Ω.

The Fučik spectrum was introduced in the semilinear case $p = 2$ by Dancer [73] and Fučik [96] who recognized its significance for the solvability of problems with jumping nonlinearities. In the semilinear ordinary differential equation (ODE) case, $p = 2$, $N = 1$, Fučik [96] showed that Σ_2 consists of a sequence of hyperbolic-like curves passing through the points (λ_l, λ_l), where $(\lambda_l)_{l \in \mathbb{N}}$ are the eigenvalues of $-d^2/dx^2$, with one or two curves going through each point. Drábek [86] has recently shown that Σ_p has this same general shape for all $p > 1$ in the ODE case.

In the partial differential equation (PDE) case, $N \ge 2$, much of the work to date on Σ_p has been done for the semilinear case $p = 2$. It is now known that Σ_2 consists, at least locally, of curves emanating from the points (λ_l, λ_l) (see [72, 73, 77, 96, 157]). Schechter [204] has shown that Σ_2 contains two continuous and strictly decreasing curves through (λ_l, λ_l), which may coincide, such that the points in the square $(\lambda_{l-1}, \lambda_{l+1})^2$ that are either below the lower curve or above the upper curve are not in Σ_2, whereas the points between them may or may not belong to Σ_2 when they do not coincide.

In the quasilinear PDE case $p \ne 2$, $N \ge 2$, it is known that the first eigenvalue λ_1 of $-\Delta_p$ is positive, simple, and admits a positive eigenfunction φ_1 (see Lindqvist [151]), so Σ_p clearly contains the two lines $\lambda_1 \times \mathbb{R}$ and $\mathbb{R} \times \lambda_1$. In addition, $\sigma(-\Delta_p)$ has an unbounded sequence of variational eigenvalues (λ_l)

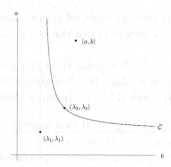

Fig. 3.1. Fučik Spectrum

satisfying a standard min-max characterization, and Σ_p contains the corresponding sequence of points (λ_l, λ_l). A first nontrivial curve $\mathcal{C}$ in Σ_p through (λ_2, λ_2) asymptotic to $\lambda_1 \times \mathbb{R}$ and $\mathbb{R} \times \lambda_1$ at infinity was recently constructed and variationally characterized by a mountain-pass procedure by Cuesta, de Figueiredo, and Gossez [71] (see Fig. 3.1).

More recently, unbounded sequences of curves (analogous to the lower and upper curves of Schechter) have been constructed and variationally characterized by min-max procedures by Micheletti and Pistoia [161] for $p \geq 2$ and by Perera [185] for all $p > 1$.

The main goal of this section is to identify the set of points (a, b) relative to the Fučik spectrum that ensure the existence of sign-changing solutions of (3.155). More precisely, assuming the existence of a positive supersolution $\bar{u}$ and a negative subsolution $\underline{u}$ of (3.155) and (a, b) located above the curve $\mathcal{C}$, we prove the existence of at least three nontrivial solutions within the order interval $[\underline{u}, \bar{u}]$: a positive solution, a negative solution, and a sign-changing solution.

3.4.2 Preliminaries

As usual we denote the norm in V_0 and $L^p(\Omega)$ by $\|\cdot\|_{V_0}$ and $\|\cdot\|_p$, respectively. Consider the boundary value problem

$$u \in V_0 : \quad -\Delta_p u = h \quad \text{in } V_0^*. \tag{3.160}$$

Besides the hypothesis (H1), we will assume the following hypotheses to hold throughout the rest of Sect. 3.4.

(H2) A positive supersolution $\bar{u}$ and a negative subsolution $\underline{u}$ of (3.155) exist, and the point $(a, b) \in \mathbb{R}^2$ is above the curve $\mathcal{C}$ of the Fučik spectrum (see Fig. 3.1).

(H3) Any solution u of (3.160) with $h \in L^\infty(\Omega)$ belongs to $C^1(\overline{\Omega})$.

Remark 3.47. (i) Assuming the existence of super- and subsolutions as in hypothesis (H2) is a weaker assumption than the usual condition on the jumping nonlinearity at infinity.

(ii) By (H3), we impose $C^1(\overline{\Omega})$-regularity of the solution of (3.160). As for regularity results up to the boundary ($C^{1,\alpha}$-regularity), we refer to Giaquinta and Giusti [98], Liu and Barrett [153], Lieberman [150], or Giuffré [100].

Lemma 3.48. *If $u \geq$ (respectively, $\leq$) 0 is a solution of (3.155), then either $u >$ (respectively, $<$) 0 or $u \equiv 0$. Moreover, if $u > 0$, then there is an $\varepsilon > 0$ such that $u \geq \varepsilon \varphi_1$, where φ_1 is the positive eigenfunction that belongs to the first eigenvalue of $-\Delta_p$.*

Proof: First we note that by the results of Anane [6] and di Benedetto [85] any solution u of (3.155) belongs to $L^\infty(\Omega) \cap C^1(\Omega)$, and thus, the right-hand side of (3.155) yields a function $h \in L^\infty(\Omega)$, which by (H3) implies that $u \in C^1(\overline{\Omega})$. If $u \geq 0$ is a solution of (3.155) that is not identically zero, then by means of the Harnack inequality (Trudinger [216, Theorem 1.1]), u must be positive in Ω. For $\varrho > 0$, let $\Omega_\varrho = \{x \in \overline{\Omega} : \text{dist}(x, \partial\Omega) \leq \varrho\}$. Then for ϱ sufficiently small, we have $f(x, u(x)) \geq 0$ for all $x \in \Omega_\varrho$ by (H1) and (H2). This result allows us to apply the strong maximum principle from Vázquez [217] to get the strict inequality $(\partial u/\partial\nu)(x) > 0$ for all $x \in \partial\Omega$, where ν is the interior normal at x. The eigenfunction φ_1 of the first eigenvalue of $-\Delta_p$ is positive, is of class $C^{1,\alpha}(\overline{\Omega})$ for $\alpha \in (0,1)$, and satisfies $(\partial\varphi_1/\partial\nu)(x) > 0$ (see [6] and [151]). Therefore, for ε sufficiently small, we obtain $u \geq \varepsilon \varphi_1$ in $\overline{\Omega}$. $\qquad\square$

Lemma 3.49. *Given a bounded sequence $(u_n) \subset V_0$ and a sequence of positive reals (ε_n) with $\varepsilon_n \to 0$ as $n \to \infty$; then for a subsequence,*

$$\frac{1}{\varepsilon_n^{p-1}} \int_\Omega |g(x, \varepsilon_n u_n(x))|\, dx \to 0 \quad \text{as } n \to \infty. \tag{3.161}$$

Furthermore, if G is the primitive of g, i.e., $G(x, t) = \displaystyle\int_0^t g(x, s)\, ds$, then

$$\frac{1}{\varepsilon_n^p} \int_\Omega |G(x, \varepsilon_n u_n(x))|\, dx \to 0 \quad \text{as } n \to \infty \tag{3.162}$$

for a subsequence.

Proof: Passing to a subsequence [again denoted by (u_n)], we may assume that $u_n \to u$ a.e. and in $L^p(\Omega)$. By Egoroff's theorem (Theorem 2.66), for any $\mu > 0$, there is a measurable subset Ω_μ of Ω such that $|\Omega \setminus \Omega_\mu| \leq \mu$ and $u_n \to u$ uniformly on Ω_μ. Thus, $\varepsilon_n u_n \to 0$ a.e. in Ω_μ. We have

$$\frac{1}{\varepsilon_n^{p-1}} \int_\Omega |g(x, \varepsilon_n u_n(x))|\, dx$$

$$= \int_{\Omega_\mu} \frac{|g(x, \varepsilon_n u_n(x))|}{\varepsilon_n^{p-1} |u_n(x)|^{p-1}} |u_n(x)|^{p-1} \, dx$$

$$+ \int_{\Omega \setminus \Omega_\mu} \frac{|g(x, \varepsilon_n u_n(x))|}{\varepsilon_n^{p-1} |u_n(x)|^{p-1}} |u_n(x)|^{p-1} \, dx. \qquad (3.163)$$

By (3.156)–(3.158), it follows that

$$\frac{|g(x, t)|}{|t|^{p-1}} \leq C. \qquad (3.164)$$

The first integral on the right hand-side of (3.163) tends to zero by the asymptotic behavior (3.158) of g, (3.164), and Lebesgue's dominated convergence theorem (observe that the integrand is majorized by $C \left(|u(x)| + \delta \right)^{p-1}$ for any $\delta > 0$ because the uniform convergence in Ω_μ). The second integral is bounded by

$$C |\Omega \setminus \Omega_\mu|^{\frac{1}{p}} \|u_n\|_p^{\frac{p-1}{p}} \leq C \mu^{\frac{1}{p}} \to 0 \quad \text{as } \mu \to 0,$$

which proves (3.161). Observing that the elementary inequality

$$|G(x, \varepsilon_n u_n(x))| \leq \varepsilon_n |u_n(x)| |g(x, \tau_n(x) \varepsilon_n u_n(x))|$$

holds, where $0 < \tau_n(x) \leq 1$, which yields

$$\frac{1}{\varepsilon_n^p} \int_\Omega |G(x, \varepsilon_n u_n(x))| \, dx \leq \int_\Omega \frac{|g(x, \tau_n(x) \varepsilon_n u_n(x))|}{\tau_n(x)^{p-1} \varepsilon_n^{p-1} |u_n(x)|^{p-1}} |u_n(x)|^p \, dx,$$

we see that (3.162) follows similarly. $\qquad \square$

Lemma 3.50. *Problem (3.155) has a positive solution $u > 0$ within the order interval $[0, \overline{u}]$ and a negative solution $u < 0$ within the order interval $[\underline{u}, 0]$.*

Proof: In the proof, we focus on the existence of a positive solution only, because the existence of a negative solution can be shown in a similar way.

As is well known, solutions of (3.155) are the critical points of the smooth functional

$$\Phi(u) = \int_\Omega \left(|\nabla u|^p - p \, F(x, u) \right) dx, \quad u \in V_0, \qquad (3.165)$$

where $F(x, t) = \int_0^t f(x, s) \, ds$. Let $\overline{f}$ be the following truncated nonlinearity:

$$\overline{f}(x, t) = \begin{cases} 0 & \text{if } t \leq 0, \\ f(x, t) & \text{if } 0 < t < \overline{u}(x), \\ f(x, \overline{u}(x)) & \text{if } t \geq \overline{u}(x), \end{cases} \qquad (3.166)$$

and $\overline{F}$ its associated primitive given by

$$\overline{F}(x,t) = \int_0^t \overline{f}(x,s)\, ds.$$

Consider the functional

$$\overline{\Phi}(u) = \int_\Omega \left(|\nabla u|^p - p\,\overline{F}(x,u) \right) dx$$

whose critical points are the solutions of the auxiliary boundary value problem

$$u \in V_0: \quad -\Delta_p u = \overline{f}(x,u) \quad \text{in } V_0^*. \tag{3.167}$$

Obviously, $\overline{\Phi}: V_0 \to \mathbb{R}$ is bounded from below, weakly lower semicontinuous, and coercive. Thus, there is a global minimizer; so a critical point u of $\overline{\Phi}$, which is a solution of (3.167), i.e.,

$$0 = \langle \overline{\Phi}'(u), \varphi \rangle = \int_\Omega \left(|\nabla u|^{p-2}\nabla u \nabla \varphi - \overline{f}(x,u)\,\varphi \right) dx. \tag{3.168}$$

We will show that this global minimizer is in fact a positive solution of (3.155) within $[0, \overline{u}]$. Taking in (3.168) the special test function $\varphi = u^- = \max\{-u, 0\}$, we get in view of the definition of $\overline{f}$ the equation

$$0 = \int_\Omega \left(|\nabla u|^{p-2}\nabla u \nabla u^- - \overline{f}(x,u)\,u^- \right) dx = \|u^-\|_{V_0}^p,$$

which shows $u^- = 0$, and thus, $u \ge 0$. As $\overline{u}$ is a supersolution, $\varphi = (u - \overline{u})^+ \in V_0 \cap L_+^p(\Omega)$, so by definition of the supersolution and (3.168), we obtain

$$0 \ge \int_\Omega \Big[\left(|\nabla u|^{p-2}\nabla u - |\nabla \overline{u}|^{p-2}\nabla \overline{u} \right) \nabla (u - \overline{u})^+$$

$$- \left(\overline{f}(x,u) - f(x,\overline{u}) \right)(u - \overline{u})^+ \Big] dx$$

$$= \int_{\{u > \overline{u}\}} \left(|\nabla u|^{p-2}\nabla u - |\nabla \overline{u}|^{p-2}\nabla \overline{u} \right)(\nabla u - \nabla \overline{u})\, dx \ge 0,$$

which implies that $\nabla (u - \overline{u})^+ = 0$, and thus, $u \le \overline{u}$. This result shows that the global minimizer u of the functional $\overline{\Phi}$ satisfies $u \in [0, \overline{u}]$, and thus, u is a solution of (3.155) because of the definition of $\overline{f}$. As $a > \lambda_1$, we get by hypothesis (H1) that

$$\overline{\Phi}(\varepsilon\,\varphi_1) < 0, \quad \varepsilon > 0 \text{ small.}$$

As u is a global minimizer of $\overline{\Phi}$, it follows that $\overline{\Phi}(u) \le \overline{\Phi}(\varepsilon\,\varphi) < 0$, and thus, in view of Lemma 3.48, u must be a positive solution of (3.155).   □

Definition 3.51. *A solution u_+ is called the smallest positive solution of (3.155) if any other positive solution u of problem (3.155) satisfies $u \ge u_+$. Similarly, u_- is the greatest negative solution of (3.155) if any other negative solution u satisfies $u \le u_-$.*

Lemma 3.52. *Problem (3.155) has a smallest positive solution u_+ and a greatest negative solution u_-.*

Proof: We are going to prove the existence of the smallest positive solution only, because the proof of the existence of the greatest negative solution is analogous. In view of Lemma 3.50, a positive solution $u \in [0, \bar{u}]$ exists, and applying Lemma 3.48, there is a $\varepsilon > 0$ small enough such that $\varepsilon\,\varphi_1 \leq u$, where φ_1 is the positive eigenfunction that belongs to the first eigenvalue λ_1 of $-\Delta_p$. As $a > \lambda_1$, one readily verifies that $\varepsilon\,\varphi_1$ is a subsolution of problem (3.155) for sufficiently small $\varepsilon > 0$. Thus, there is a $\varepsilon_0 > 0$ such that $\varepsilon_0\,\varphi_1$ and $\bar{u}$ form an ordered pair of sub- and supersolutions. Applying Theorem 3.22 on the existence of extremal solutions for general quasilinear elliptic problems, we obtain the existence of a smallest and greatest solution of (3.155) with respect to the order interval $[\varepsilon_0\,\varphi_1, \bar{u}]$. We denote the smallest solution within this interval by u_0. Now let (ε_n) be a decreasing sequence with $\varepsilon_n \to 0$ as $n \to \infty$, and denote by u_n the corresponding smallest solution of (3.155) with respect to the order interval $[\varepsilon_n\varphi_1, \bar{u}]$. Then obviously (u_n) is a decreasing sequence of smallest positive solutions of (3.155), which converges to its nonnegative pointwise limit u_* in $L^p(\Omega)$. We will show that u_* is in fact the smallest positive solution; i.e., $u_* = u_+$. First we verify that u_* is a solution of (3.155). As the u_n are solutions of (3.155), we get from (3.155) with test function u_n the equation

$$\|u_n\|_{V_0}^p = \int_\Omega |\nabla u_n|^{p-2}\,\nabla u_n \nabla u_n\, dx = \int_\Omega f(x, u_n)\, u_n\, dx,$$

which by the growth condition (H1) and the boundedness in $L^p(\Omega)$ of the sequence (u_n) implies its boundedness in V_0; i.e., $\|u_n\|_{V_0} \leq c$. Thus, a subsequence weakly convergent in V_0 exists, and because of the strong convergence of (u_n) in $L^p(\Omega)$, even the entire sequence is weakly convergent in V_0 with weak limit u_*. From (3.155) with the test function $u_n - u_*$, we obtain

$$\langle -\Delta_p u_n, u_n - u_* \rangle = \int_\Omega |\nabla u_n|^{p-2}\,\nabla u_n \nabla(u_n - u_*)\, dx$$
$$= \int_\Omega f(x, u_n)\, (u_n - u_*)\, dx,$$

which implies that

$$\limsup_n \langle -\Delta_p u_n, u_n - u_* \rangle \leq 0. \tag{3.169}$$

The weak convergence of (u_n) and (3.169) along with the (S_+)-property of the operator $-\Delta_p$ (see Lemma 2.111) yield its strong convergence in V_0. This process allows the passage to the limit in (3.155) with u replaced by u_n, and hence, u_* is a solution of problem (3.155). To show that $u_* > 0$, our argument is by contradiction. Suppose $u_* = 0$; that is, $u_n \to 0$ in V_0. As $u_n > 0$, we may consider $\tilde{u}_n = u_n/\|u_n\|_{V_0}$, which satisfies

$$\int_\Omega |\nabla \tilde{u}_n|^{p-2}\, \nabla \tilde{u}_n \nabla\, \varphi\, dx = \int_\Omega \left[a\, \tilde{u}_n^{p-1} + \frac{g(x, u_n)}{\|u_n\|_{V_0}^{p-1}} \right] \varphi\, dx. \tag{3.170}$$

By definition $\|\tilde{u}_n\|_{V_0} = 1$, so there is a subsequence $(\tilde{u}_n)$ that converges weakly in V_0 and strongly in $L^p(\Omega)$ to $\tilde{u}$ due to the compact embedding of $V_0 \subset L^p(\Omega)$. Taking in (3.170) as special test function $\varphi = \tilde{u}_n - \tilde{u}$, we get for the right-hand side of (3.170)

$$\int_\Omega \left[a\, \tilde{u}_n^{p-1} + \frac{g(x, u_n)}{|u_n(x)|^{p-1}}\, |\tilde{u}_n(x)|^{p-1} \right] (\tilde{u}_n - \tilde{u})\, dx \to 0,$$

as $n \to \infty$, because the terms in parentheses are $L^q(\Omega)$-bounded. Hence, (3.170) implies that

$$\limsup_n \langle -\Delta_p \tilde{u}_n, \tilde{u}_n - \tilde{u} \rangle \le 0,$$

which because of the S_+-property of $-\Delta_p$ implies the strong convergence of $\tilde{u}_n \to \tilde{u}$ in V_0. Moreover, the second integral term on the right-hand side of (3.170) converges to zero by Lemma 3.49, so we may pass to the limit to get

$$\int_\Omega |\nabla \tilde{u}|^{p-2}\, \nabla \tilde{u} \nabla\, \varphi\, dx = \int_\Omega a\, \tilde{u}^{p-1}\, \varphi\, dx \quad \text{for all}\ \ \varphi \in C_0^\infty(\Omega);$$

i.e., $\tilde{u}$ satisfies the boundary value problem

$$\tilde{u} \in V_0: \quad -\Delta_p \tilde{u} = a\, \tilde{u}^{p-1} \quad \text{in}\ V_0^*. \tag{3.171}$$

As $\|\tilde{u}_n\|_{V_0} = 1$ and $\tilde{u}_n > 0$, by Lemma 3.48, we have the same properties for $\tilde{u}$, which, however, contradicts that a nontrivial solution of (3.171) changes sign. So far we have shown that the limit u_* of the least solutions $u_n \in [\varepsilon_n \varphi_1, \bar{u}]$ is a positive solution of (3.155). Finally, to prove that u_* is the smallest positive solution, let w be any positive solution of (3.155). Then by Lemma 3.48, there is a $\varepsilon_n > 0$ for n sufficiently large such that $\varepsilon_n \varphi_1 \le w$, which by definition of the sequence of smallest solutions (u_n) yields $u_* \le u_n \le w$ (for n sufficiently large), which proves that $u_* = u_+$ is in fact the smallest positive one. $\square$

3.4.3 Main Result

The main result of Sect. 3.4 reads as follows.

Theorem 3.53. *Let hypotheses (H1)–(H3) be satisfied. Then the BVP (3.155) has at least three nontrivial solutions: a positive solution, a negative solution, and a sign-changing solution.*

Proof: We introduce the cutoff function $\widetilde{f}_+$ by

$$\widetilde{f}_+(x,t) = \begin{cases} 0 & \text{if } t \leq 0, \\ f(x,t) & \text{if } 0 < t < u_+(x), \\ f(x,u_+(x)) & \text{if } t \geq u_+(x), \end{cases}$$

with its primitive $\widetilde{F}_+$ given by

$$\widetilde{F}_+(x,t) = \int_0^t \widetilde{f}_+(x,s)\,ds.$$

Consider the functional

$$\widetilde{\Phi}_+(u) = \int_\Omega \left(|\nabla u|^p - p\widetilde{F}_+(x,u) \right)\,dx.$$

Arguments similar to those in the proof of Lemma 3.50 show that critical points of $\widetilde{\Phi}_+$ are solutions of the BVP (3.155) in the order interval $[0, u_+]$, so 0 and u_+ are the only critical points of $\widetilde{\Phi}_+$ by Lemmas 3.48 and 3.52. Now, $\widetilde{\Phi}_+$ is bounded from below and coercive, and

$$\widetilde{\Phi}_+(\varepsilon\,\varphi_1) < 0, \quad \text{for } \varepsilon > 0 \text{ small},$$

because $a > \lambda_1$, so $\widetilde{\Phi}_+$ has a global minimizer at a negative critical level. It follows that u_+ is the (strict) global minimizer of $\widetilde{\Phi}_+$ and $\widetilde{\Phi}_+(u_+) < 0$.

Now let

$$\widetilde{f}(x,t) = \begin{cases} f(x,u_-(x)) & \text{if } t \leq u_-(x), \\ f(x,t) & \text{if } u_-(x) < t < u_+(x), \\ f(x,u_+(x)) & \text{if } t \geq u_+(x), \end{cases}$$

with its primitive

$$\widetilde{F}(x,t) = \int_0^t \widetilde{f}(x,s)\,ds,$$

and the associated functional $\widetilde{\Phi}$ given by

$$\widetilde{\Phi}(u) = \int_\Omega \left(|\nabla u|^p - p\widetilde{F}(x,u) \right)\,dx.$$

As before, critical points of $\widetilde{\Phi}$ are solutions of the BVP (3.155) in the order interval $[u_-, u_+]$, so it follows from Lemma 3.48 and Lemma 3.52 that any nontrivial critical point of $\widetilde{\Phi}$ other than $u_\pm$ is a sign-changing solution. We are going to prove this latter assertion by using the following auxiliary result.

Lemma 3.54. *The solutions $u_\pm$ are strict local minimizers of $\widetilde{\Phi}$, and it holds that $\widetilde{\Phi}(u_\pm) < 0$.*

Proof: We only consider u_+ as the argument for u_- is similar. Suppose that there is a sequence $u_j \to u_+$ in V_0, $u_j \neq u_+$ with $\widetilde{\Phi}(u_j) \leq \widetilde{\Phi}(u_+)$. By (3.156) and (3.158), we have

$$|\widetilde{F}(x,t)| \leq C\,|t|^p,$$

which implies that

$$\widetilde{\Phi}(u_j) = \int_{\Omega} \left(|\nabla u_j^+|^p - p\,\widetilde{F}(x, u_j^+)\right)\,dx + \int_{\Omega} \left(|\nabla u_j^-|^p - p\,\widetilde{F}(x, -u_j^-)\right)\,dx$$

$$\geq \widetilde{\Phi}_+(u_j^+) + \|u_j^-\|_{V_0}^p - C\,\|u_j^-\|_p^p.$$

If $u_j^- = 0$, then $u_j^+ \neq u_+$ and

$$\widetilde{\Phi}_+(u_j^+) \leq \widetilde{\Phi}(u_j) \leq \widetilde{\Phi}(u_+) = \widetilde{\Phi}_+(u_+),$$

which contradicts that u_+ is the unique global minimizer of $\widetilde{\Phi}_+$; so $u_j^- \neq 0$. We will show that

$$\|u_j^-\|_{V_0}^p > C\,\|u_j^-\|_p^p, \quad j \text{ large}. \tag{3.172}$$

Assume for the moment the last inequality, then we have the contradiction $\widetilde{\Phi}_+(u_j^+) < \widetilde{\Phi}_+(u_+)$. To see that (3.172) holds, we first note that the measure of the set $\Omega_j = \{x \in \Omega : u_j(x) < 0\}$ goes to zero. To see this result, given $\varepsilon > 0$, take a compact subset Ω^ε of Ω such that $|\Omega \setminus \Omega^\varepsilon| < \varepsilon$ and let $\Omega_j^\varepsilon = \Omega^\varepsilon \cap \Omega_j$. Then

$$\|u_j - u_+\|_p^p \geq \int_{\Omega_j^\varepsilon} |u_j - u_+|^p\,dx \geq \int_{\Omega_j^\varepsilon} u_+^p\,dx \geq c^p\,|\Omega_j^\varepsilon| \tag{3.173}$$

where $c = \min_{\Omega^\varepsilon} u_+ > 0$. Thus, in view of (3.173), we get $|\Omega_j^\varepsilon| \to 0$. As $\Omega_j \subset \Omega_j^\varepsilon \cup (\Omega \setminus \Omega^\varepsilon)$ and $\varepsilon > 0$ is arbitrary, we see that $|\Omega_j| \to 0$ as $j \to \infty$.

If (3.172) does not hold, setting $\widetilde{u}_j = u_j^- / \|u_j^-\|_p$, it follows that $\|\widetilde{u}_j\|_{V_0}$ is bounded for some subsequence, so $\widetilde{u}_j \to \widetilde{u}$ in $L^p(\Omega)$ and a.e. in Ω for a further subsequence, where $\|\widetilde{u}\|_p = 1$ and $\widetilde{u} \geq 0$. But then $\Omega_\mu = \{x \in \Omega : \widetilde{u}(x) \geq \mu\}$ has positive measure for all sufficiently small $\mu > 0$ and

$$\|\widetilde{u}_j - \widetilde{u}\|_p^p \geq \int_{\Omega_\mu \setminus \Omega_j} |\widetilde{u}_j - \widetilde{u}|^p\,dx = \int_{\Omega_\mu \setminus \Omega_j} \widetilde{u}^p\,dx \geq \mu^p\,(|\Omega_\mu| - |\Omega_j|). \tag{3.174}$$

As the right-hand side of (3.174) tends to $\mu^p\,|\Omega_\mu| > 0$ as $j \to \infty$, we get a contradiction. $\square$

By means of Lemma 3.54, we will show that $\widetilde{\Phi}$ has a nontrivial critical point other than $u_\pm$, which completes the proof of our main result. We note first that a standard deformation argument ensures the existence of a mountain-pass point u_1 at the critical value

$$c = \inf_{\pi \in \Pi} \max_{u \in \pi([-1,1])} \widetilde{\Phi}(u) > \widetilde{\Phi}(u_\pm), \tag{3.175}$$

where $\Pi = \{\pi \in C([-1,1]; V_0) : \pi(\pm 1) = u_\pm\}$ is the class of paths joining $u_\pm$. To show that $u_1 \neq 0$, we will construct a path that lies in $\widetilde{\Phi}^0 = \{u \in V_0 : \widetilde{\Phi}(u) < 0\}$.

First we show that, for all sufficiently small $\varepsilon > 0$, $\pm \varepsilon \varphi_1$ can be joined by a path π_ε in $\widetilde{\Phi}^0$. Note that $\widetilde{\Phi}$ can be rewritten in the form

$$\widetilde{\Phi}(u) = I_{(a,b)}(u) - \int_\Omega \widetilde{G}(x, u)\, dx,$$

where

$$I_{(a,b)}(u) = \int_\Omega \left(|\nabla u|^p - a\,(u^+)^p - b\,(u^-)^p \right) dx$$

is the functional associated with (3.159) and

$$\widetilde{G}(x, t) = p\,\widetilde{F}(x, t) - a\,(t^+)^p - b\,(t^-)^p = o(|t|^p) \quad \text{as } t \to 0.$$

As the tuple (a, b) is above the curve $\mathcal{C}$ of the Fučik spectrum, there is a path π_0 in

$$\left\{ u \in V_0 : I_{(a,b)}(u) < 0,\ \|u\|_p = 1 \right\}$$

joining $\pm \varphi_1$ by the construction of $\mathcal{C}$ according to [71]. For $u \in \pi_0([-1, 1])$, we have

$$\widetilde{\Phi}(\varepsilon u) \leq \varepsilon^p \left[\max I_{(a,b)}(\pi_0([-1, 1])) + \int_\Omega \frac{|\widetilde{G}(x, \varepsilon u)|}{\varepsilon^p}\, dx \right], \tag{3.176}$$

and the last integral on the right-hand side of (3.176) goes to 0 uniformly on the compact set $\pi_0([-1, 1])$ as $\varepsilon \to 0$ by Lemma 3.49; so we can take $\pi_\varepsilon = \varepsilon\,\pi_0$.

We complete the proof by showing that $\pm \varepsilon \varphi_1$ and $u_\pm$ can be joined by paths in $\widetilde{\Phi}^0$. Again we only consider $\varepsilon \varphi_1$ and u_+. Setting $\alpha = \inf \widetilde{\Phi}_+ = \widetilde{\Phi}_+(u_+)$ and $\beta = \widetilde{\Phi}_+(\varepsilon \varphi_1) = \widetilde{\Phi}(\varepsilon \varphi_1) < 0$, by the second deformation lemma (see Chang [63]), the sublevel set $\widetilde{\Phi}_+^\alpha = \left\{ u \in V_0 : \widetilde{\Phi}_+(u) \leq \alpha \right\} = \{u_+\}$ is a strong deformation retract of $\widetilde{\Phi}_+^\beta$; that is, there is an $\eta \in C([0, 1] \times \widetilde{\Phi}_+^\beta, \widetilde{\Phi}_+^\beta)$ such that

(i) $\eta(0, u) = u$ for all $u \in \widetilde{\Phi}_+^\beta$.
(ii) $\eta(t, u_+) = u_+$ for all $t \in [0, 1]$.
(iii) $\eta(1, u) = u_+$ for all $u \in \widetilde{\Phi}_+^\beta$.

In particular, $\pi = \eta(\cdot, \varepsilon \varphi_1)$ is a path in $\widetilde{\Phi}_+^\beta$ joining $\varepsilon \varphi_1$ and u_+. Now the path π_+ defined by $\pi_+(t) = \pi(t)^+$ also joins $\varepsilon \varphi_1$ and u_+, and

$$\widetilde{\Phi}(\pi_+(t)) = \widetilde{\Phi}_+(\pi(t)) - \int_\Omega |\nabla \pi(t)^-|^p \leq \beta < 0,$$

which completes the proof of Theorem 3.53. $\square$

3.5 Quasilinear Elliptic Problems of Periodic Type

In this section, we are interested in a variational sub-supersolution approach to a quasilinear elliptic boundary value problem that, in the one-space dimensional and semilinear case, is a boundary value problem for a second-order scalar ordinary differential equation subject to *periodic boundary conditions*. The latter problem was first studied by Hans Knobloch [125] and later by many other authors using various kinds of nonlinear analysis methods (see [205], [158], [126], [107]). The continuation and extension of the one-space dimensional case to the higher-space dimensional case presented here is based on a recent result due to V. Le and K. Schmitt [147].

3.5.1 Problem Setting

Let $\Omega \subset \mathbb{R}^N$ be a bounded domain with smooth boundary $\partial\Omega$. We consider the following boundary value problem (BVP, for short):

$$-\text{div}[a(x, \nabla u)] + f(x, u) = 0, \ x \in \Omega, \tag{3.177}$$

$$u(x) = \text{constant}, \ x \in \partial\Omega, \tag{3.178}$$

$$\int_{\partial\Omega} a(x, \nabla u)\,\nu\,dS = 0, \tag{3.179}$$

where ν denotes the outward unit normal on $\partial\Omega$, and

$$\text{div}[a(x, \nabla u)] = \sum_{i=1}^{N} \frac{\partial}{\partial x_i} a_i(x, \nabla u).$$

(Note that in condition (3.178), it is understood that the trace $\gamma(u)$ of u is a constant function, with the constant not being fixed.) We assume that the coefficients $a_i : \Omega \times \mathbb{R}^N \to \mathbb{R}$ are Carathéodory functions satisfying conditions (H1)–(H3) of Sect. 2.3.2, which can be reformulated in terms of the vector function $a = (a_1, \ldots, a_N)$ as follows.

(H1) A constant $c_0 > 0$ and a function $k_0 \in L^q(\Omega)$ exist so that

$$|a(x, \xi)| \le k_0(x) + c_0 |\xi|^{p-1}, \tag{3.180}$$

for a.e. $x \in \Omega$, and for all $\xi \in \mathbb{R}^N$, where p and q are Hölder conjugate reals with $1 < p < \infty$, $1/p + 1/q = 1$.

(H2) $a(x, \xi)$ is monotone in ξ; that is,

$$[a(x, \xi) - a(x, \xi')](\xi - \xi') > 0 \tag{3.181}$$

for a.e. $x \in \Omega$, and for all $\xi, \xi' \in \mathbb{R}^N$ with $\xi \ne \xi'$.

(H3) a satisfies the following coercivity property: $k \in L^1(\Omega)$ and $\mu > 0$ exist
such that

$$a(x, \xi)\, \xi \geq \mu\, |\xi|^p - k(x) \tag{3.182}$$

for a.e. $x \in \Omega$, and for all $\xi \in \mathbb{R}^N$.

Remark 3.55. When $N = 1$ and $\Omega = (c, d)$ is some bounded interval, the
boundary condition (3.178)–(3.179) becomes the following boundary condition
on (c, d):

$$u(c) = u(d),\ a(c, u'(c)) = a(d, u'(d)),$$

which, when $a(x, v) = v$ corresponds to the usual set of periodic boundary
conditions,

$$u(c) = u(d),\ u'(c) = u'(d).$$

Example 3.56. The p-Laplacian is a prototype of the operator a above; i.e.,

$$a(x, \nabla u) = |\nabla u|^{p-2}\nabla u,\ 1 < p < \infty.$$

It is easy to check that a satisfies conditions (H1)–(H3) above. In this case,
the boundary condition (3.179) becomes

$$\int_{\partial\Omega} |\nabla u|^{p-2}\frac{\partial u}{\partial \nu}\, dS = 0.$$

Assume that $f : \Omega \times \mathbb{R} \to \mathbb{R}$ is a Carathéodory function with some appropriate growth condition to be specified later. As in previous sections, we
denote by $V = W^{1,p}(\Omega)$, and define an operator $A : V \to V^*$ via the following
semilinear form:

$$\langle Au, v \rangle = \int_\Omega a(x, \nabla u)\, \nabla v\, dx \quad \text{for all } u, v \in V.$$

Let F denote the Nemytskij operator related to f. If $F(u) \in L^q(\Omega)$ for $u \in V$,
then F defines a mapping (which is again denoted by F) from V into V^* by

$$\langle F(u), v \rangle = \int_\Omega F(u)\, v\, dx.$$

Hypotheses (H1)–(H3) imply that A is continuous, bounded, monotone, and
satisfies

$$\langle Au, u \rangle \geq \mu\|\nabla u\|_p^p - \|k\|_1, \quad \text{for all } u \in V, \tag{3.183}$$

where $\|\cdot\|_r$ stands for the norm in $L^r(\Omega)$ for $1 \leq r < \infty$. Let

$$V_c = \{u \in V : u|_{\partial\Omega} = \text{constant}\}.$$

Then V_c is a closed subspace of V, and thus, a reflexive Banach space with
the restricted norm of V. The weak (variational) formulation of the boundary
value problem (3.177)–(3.179) is given as follows.

Definition 3.57. *The function $u \in V_c$ is called a (weak) solution of the BVP (3.177)–(3.179) if the following variational equality holds:*

$$\int_\Omega a(x, \nabla u) \nabla v \, dx + \int_\Omega f(x, u) \, v \, dx = 0 \quad \text{for all } v \in V_c. \tag{3.184}$$

Let us justify that Definition 3.57 provides an appropriate notion of weak solution. Note that if u satisfies (3.177)–(3.179) and $v \in V_c$, then

$$
\begin{aligned}
0 &= -\int_\Omega \operatorname{div} a(x, \nabla u) \, v \, dx + \int_\Omega f(x, u) \, v \, dx \\
&= \int_\Omega a(x, \nabla u) \, \nabla v \, dx - (v|_{\partial\Omega}) \int_{\partial\Omega} a(x, \nabla u) \, \nu \, dS + \int_\Omega f(x, u) \, v \, dx \\
&= \int_\Omega a(x, \nabla u) \, \nabla v \, dx + \int_\Omega f(x, u) \, v \, dx.
\end{aligned}
$$

Hence, we have (3.184). Conversely, if $u \in V_c$ is a solution of (3.184), then by choosing $v \in C_0^\infty(\Omega) \subset V_c$ in (3.184) and applying the Divergence theorem as above, we see that (3.177) holds. Choosing $v = 1$ in (3.184), we have $\int_\Omega f(x, u) dx = 0$. On the other hand, integrating (3.177) over Ω and using once more the Divergence theorem yield

$$0 = -\int_\Omega \operatorname{div} a(x, \nabla u) dx + \int_\Omega f(x, u) dx = -\int_{\partial\Omega} a(x, \nabla u) \, \nu \, dS.$$

Hence, we have the boundary condition (3.179).

3.5.2 Sub-Supersolutions

We will study the existence of solutions of (3.184) by first defining appropriate concepts of sub- and supersolutions.

Definition 3.58. *A function $\underline{u}$ (respectively, $\bar{u}$) in V_c is called a subsolution (respectively, supersolution) of (3.184) if*

$$\int_\Omega a(x, \nabla \underline{u}) \, \nabla v \, dx + \int_\Omega f(x, \underline{u}) v \, dx \leq 0 \quad (\text{respectively}, \geq 0), \tag{3.185}$$

for all $v \in V_c \cap L_+^p(\Omega)$.

Remark 3.59. In case of the Laplacian, i.e., $a(x, \nabla u) = \nabla u$ and $p = 2$, or when $N = 1$ (ordinary differential equation case), the above definition of sub- and supersolutions is the variational form of those given in [206], without imposing additional smoothness assumptions.

As is the case with solutions satisfying additional smoothness conditions, sub- and supersolutions, when smooth enough, satisfy additional boundary conditions. Let us see this in the case of the p-Laplacian. For assume that $\alpha \in V_c \cap W^{2,p}(\Omega)$ satisfies (cf. (17) of [206]):

$$\int_\Omega |\nabla\alpha|^{p-2}\nabla\alpha\nabla\phi\,dx + \int_\Omega f(x,\alpha)\phi\,dx \le 0 \qquad (3.186)$$

for all $\phi \in C_0^\infty(\Omega)$ with $\phi \ge 0$, and

$$\int_{\partial\Omega} |\nabla\alpha|^{p-2}\nabla\alpha\,\nu\,dS \le 0. \qquad (3.187)$$

As $\alpha \in W^{2,p}(\Omega)$, the divergence theorem implies that

$$\int_\Omega [-\operatorname{div}\left(|\nabla\alpha|^{p-2}\nabla\alpha\right) + f(x,\alpha)]\,\phi\,dx \le 0$$

for all $\phi \in C_0^\infty(\Omega)$, $\phi \ge 0$; i.e. (in the sense of distributions),

$$-\operatorname{div}\left(|\nabla\alpha|^{p-2}\nabla\alpha\right) + f(x,\alpha) \le 0 \quad \text{a.e. on } \Omega. \qquad (3.188)$$

Let $v \in V_c \cap L_+^p(\Omega)$. It follows from (3.188) that

$$0 \ge \int_\Omega [-\operatorname{div}\left(|\nabla\alpha|^{p-2}\nabla\alpha\right) + f(x,\alpha)]\,v\,dx$$
$$= \int_\Omega |\nabla\alpha|^{p-2}\nabla\alpha\,\nabla v\,dx - \int_{\partial\Omega} |\nabla\alpha|^{p-2}\frac{\partial\alpha}{\partial\nu}\,v\,dS + \int_\Omega f(x,\alpha)\,v\,dx.$$

Hence,

$$\int_\Omega |\nabla\alpha|^{p-2}\nabla\alpha\,\nabla v\,dx + \int_\Omega f(x,\alpha)\,v\,dx \le (v|_{\partial\Omega}) \int_{\partial\Omega} |\nabla\alpha|^{p-2}\frac{\partial\alpha}{\partial\nu}\,dS \le 0;$$

that is, α satisfies (3.185). Conversely, assume $\alpha \in V_c \cap W^{2,p}(\Omega)$ satisfies (3.185). As $C_0^\infty(\Omega) \subset V_c$, we have (3.186). To prove that α satisfies (3.187), we choose a sequence (Ω_n) of subdomains of Ω such that

$$\overline{\Omega_n} \subset \Omega_{n+1}, \text{ for all } n, \text{ and } \Omega = \bigcup_{n=1}^\infty \Omega_n. \qquad (3.189)$$

For each $n \in \mathbb{N}$, choose $\phi_n \in C_0^\infty(\Omega)$ such that $0 \le \phi_n(x) \le 1$ for all $x \in \Omega$, and $\phi_n(x) = 1$ for all $x \in \Omega_n$. Let $v_n = 1 - \phi_n$ $(n \in \mathbb{N})$. Then $v_n \in V_c$, $v_n = 1$ on $\partial\Omega$, and $0 \le v_n \le 1$ on Ω. Letting $v = v_n$ in (3.185), we get

$$0 \ge \int_\Omega |\nabla\alpha|^{p-2}\nabla\alpha\,\nabla v_n\,dx + \int_\Omega f(x,\alpha)\,v_n\,dx$$
$$= \int_\Omega [-\operatorname{div}\left(|\nabla\alpha|^{p-2}\nabla\alpha\right) + f(x,\alpha)]\,v_n\,dx + \int_{\partial\Omega} |\nabla\alpha|^{p-2}\frac{\partial\alpha}{\partial\nu}\,v_n\,dS$$

$$= \int_\Omega \left[-\mathrm{div}\left(|\nabla\alpha|^{p-2}\nabla\alpha\right) + f(x,\alpha)\right] v_n\, dx + \int_{\partial\Omega} |\nabla\alpha|^{p-2}\frac{\partial\alpha}{\partial\nu}\, dS.$$
$$(3.190)$$

Because $v_n = 0$ on Ω_n, from (3.189) and the dominated convergence theorem, we obtain

$$\lim_{n\to\infty} \int_\Omega \left[-\mathrm{div}\left(|\nabla\alpha|^{p-2}\nabla\alpha\right) + f(x,\alpha)\right] v_n\, dx = 0.$$

Letting $n \to \infty$ in (3.190), we obtain

$$\int_{\partial\Omega} |\nabla\alpha|^{p-2}\frac{\partial\alpha}{\partial\nu}\, dS \leq 0,$$

which is (3.187).

3.5.3 Existence Result

The main result of Sect. 3.5 is the following theorem.

Theorem 3.60. *Assume a pair of sub- and supersolution $\underline{u}$ and $\bar{u}$ of (3.184) exists such that $\underline{u} \leq \bar{u}$ and that f satisfies the following growth condition:*

$$|f(x,u)| \leq k_1(x), \tag{3.191}$$

for a.e. $x \in \Omega$, for all $u \in [\underline{u}(x), \bar{u}(x)]$, with $k_1 \in L_+^q(\Omega)$. Then, (3.184) has a solution $u \in V_c$ such that $\underline{u} \leq u \leq \bar{u}$.

Proof: As earlier in this chapter, we introduce the truncation operator T and the cutoff function $b : \Omega \times \mathbb{R} \to \mathbb{R}$ related to the given sub- and supersolutions by

$$Tu(x) = \begin{cases} \bar{u}(x) & \text{if } u(x) > \bar{u}(x), \\ u(x) & \text{if } \underline{u}(x) \leq u(x) \leq \bar{u}(x), \\ \underline{u}(x) & \text{if } u(x) < \underline{u}(x), \end{cases}$$

and

$$b(x,s) = \begin{cases} (s - \bar{u}(x))^{p-1} & \text{if } s > \bar{u}(x), \\ 0 & \text{if } \underline{u}(x) \leq s \leq \bar{u}(x), \\ -(\underline{u}(x) - s)^{p-1} & \text{if } s < \underline{u}(x), \end{cases} \tag{3.192}$$

and note that b is a Carathéodory function satisfying

$$|b(x,s)| \leq k_2(x) + c_2 |s|^{p-1} \tag{3.193}$$

for a.e. $x \in \Omega$, for all $s \in \mathbb{R}$, with some function $k_2 \in L_+^q(\Omega)$ and a constant $c_2 > 0$. Therefore, the operator $B : V \to V^*$ given by

$$\langle Bu, v \rangle = \int_\Omega b(x, u) v \, dx \quad \text{for all } v \in V,$$

is well defined, completely continuous, and bounded. Moreover, there are $c_3, c_4 > 0$ such that

$$\langle Bu, u \rangle \geq c_3 \|u\|_p^p - c_4 \quad \text{for all } u \in V. \tag{3.194}$$

Let us consider the following variational equation in V_c:

$$u \in V_c : \quad \langle Au + B(u) + (F \circ T)u, v \rangle = 0 \quad \text{for all } v \in V_c. \tag{3.195}$$

It follows from (3.191) that $F \circ T : V \to V^*$ is well defined and completely continuous. Because A is monotone, it follows that $A + B + F \circ T$ is pseudo-monotone. Next, let us show that $A + B + F \circ T$ is coercive on V in the following sense:

$$\lim_{\|u\|_V \to \infty} \frac{\langle Au + B(u) + (F \circ T)u, u \rangle}{\|u\|_V} = +\infty. \tag{3.196}$$

In fact, from the definition of the truncation operator T and (3.191),

$$|\langle (F \circ T)u, u \rangle| = \left| \int_\Omega f(x, Tu) u \, dx \right| \leq \int_\Omega k_1 |u| \, dx \leq \|k_1\|_q \|u\|_p. \tag{3.197}$$

Combining (3.197) with (3.194) and (3.183), we get

$$\begin{aligned}
\langle &Au + B(u) + (F \circ T)u, u \rangle \\
&\geq \mu \|\nabla u\|_p^p - \|k\|_1 + c_3 \|u\|_p^p - c_4 - \|k_1\|_q \|u\|_p \\
&\geq \min\{\mu, c_3\}(\|u\|_p^p + \|\nabla u\|_p^p) - \|k_1\|_q \|u\|_V - \|k\|_1 - c_4 \\
&= c_5 \|u\|_V^p - c_6 \|u\|_V - c_7, \quad \text{for all } u \in V,
\end{aligned}$$

with $c_5, c_6, c_7 > 0$. Because $p > 1$, this estimate implies (3.196).

As V_c is a closed subspace of V, the existence of solutions of (3.195) follows from Theorem 2.99 of Sect. 2.3. Assume that u is any solution of (3.195). To complete the proof of the theorem, we only need to prove that

$$\underline{u} \leq u \leq \bar{u} \text{ a.e. in } \Omega, \tag{3.198}$$

because then $Tu = u$, $B(u) = 0$, and thus, u is also a solution of (3.184). Let us verify the first inequality in (3.198). As $u, \underline{u} \in V$, we have $(\underline{u} - u)^+ \in V$ from the lattice structure of V (see Sect. 2.2.3). Moreover, because the traces $\gamma(u) = u|_{\partial\Omega}$ and $\gamma(\underline{u}) = \underline{u}|_{\partial\Omega}$ are constants, we have

$$\gamma((\underline{u} - u)^+) = (\gamma(\underline{u}) - \gamma(u))^+ = \text{constant},$$

(see Sect. 2.2.3); i.e.,

$$(\underline{u} - u)^+ \in V_c. \tag{3.199}$$

Choosing $v = (\underline{u} - u)^+$ in (3.195), we obtain

$$\int_\Omega a(x, \nabla u)\, \nabla(\underline{u} - u)^+ \, dx + \int_\Omega [b(x, u) + f(x, Tu)](\underline{u} - u)^+ \, dx = 0. \tag{3.200}$$

On the other hand, taking $v = (\underline{u} - u)^+ (\geq 0)$ as a special test function in inequality (3.185) gives us

$$\int_\Omega a(x, \nabla \underline{u})\, \nabla(\underline{u} - u)^+ \, dx + \int_\Omega f(x, \underline{u})(\underline{u} - u)^+ \, dx \leq 0. \tag{3.201}$$

Subtracting (3.200) from (3.201) yields

$$\int_\Omega [a(x, \nabla \underline{u}) - a(x, \nabla u)]\, \nabla(\underline{u} - u)^+ \, dx + \int_\Omega [f(x, \underline{u}) - f(x, Tu)](\underline{u} - u)^+ \, dx$$

$$\leq \int_\Omega b(x, u)(\underline{u} - u)^+ \, dx. \tag{3.202}$$

By means of (3.181) and Example 2.86, we have

$$\int_\Omega [a(x, \nabla \underline{u}) - a(x, \nabla u)]\, \nabla[(\underline{u} - u)^+]\, dx$$

$$= \int_{\{\underline{u} > u\}} [a(x, \nabla \underline{u}) - a(x, \nabla u)]\, (\nabla \underline{u} - \nabla u)\, dx$$

$$\geq 0, \tag{3.203}$$

where $\{\underline{u} > u\} = \{x \in \Omega : \underline{u}(x) > u(x)\}$. From the definition of Tu, we have $Tu(x) = \underline{u}(x)$ on $\{\underline{u} > u\}$, and thus,

$$\int_\Omega [f(x, \underline{u}) - f(x, Tu)](\underline{u} - u)^+ \, dx$$

$$= \int_{\{\underline{u} > u\}} [f(x, \underline{u}) - f(x, Tu)](\underline{u} - u)\, dx = 0. \tag{3.204}$$

Using (3.203) and (3.204) in (3.202), we obtain

$$0 \leq \int_\Omega b(x, u)(\underline{u} - u)^+ \, dx = -\int_{\{\underline{u} > u\}} (\underline{u} - u)^p \, dx \leq 0.$$

This result implies that

$$\int_{\{\underline{u} > u\}} (\underline{u} - u)^p \, dx = \int_\Omega [(\underline{u} - u)^+]^p \, dx = 0,$$

i.e., $(\underline{u} - u)^+ = 0$, which proves the first inequality in (3.198). The other inequality there is established in the same way, which completes the proof of the theorem.   □

Remark 3.61. By modifying the proof of Theorem 3.60 in the spirit of Theorem 3.6 or Theorem 3.41, we can extend Theorem 3.60 to the existence of solutions of (3.184) between a finite number of sub- and supersolutions. In fact, we can show that if $\underline{u}_1, \ldots, \underline{u}_k$ (respectively, $\bar{u}_1, \ldots, \bar{u}_m$) are subsolutions (respectively, supersolutions) of (3.184) such that

$$\max\{\underline{u}_1, \ldots, \underline{u}_k\} \leq \min\{\bar{u}_1, \ldots, \bar{u}_m\},$$

and that f satisfies an appropriate growth condition between these sub- and supersolutions, then a solution u of (3.184) exists such that

$$\max\{\underline{u}_1, \ldots, \underline{u}_k\} \leq u \leq \min\{\bar{u}_1, \ldots, \bar{u}_m\}.$$

Remark 3.62. We note that for the method of proof of Theorem 3.60 to work, the important property of the subspace V_c that was needed was that $u^+ \in V_c$ for any $u \in V_c$. We therefore see that Theorem 3.60 remains valid, if V_c is replaced by any subspace $\widetilde{V}$, which has this property (and, of course, the definitions of sub- and supersolutions are appropriately modified). This more general theorem, for example, contains the sub-supersolution existence result for boundary value problems subject to Neumann boundary conditions.

3.6 Notes and Comments

The sub-supersolution method was motivated by the well-known Perron arguments on sub- and superharmonic functions and was used in [4, 15, 159, 202] to study the solvability of nonlinear elliptic and parabolic problems in the classic sense. In these papers, it was also established the existence of extremal classic solutions. The sub-supersolution argument was later employed in [83, 84] to study the existence of weak solutions of quasilinear elliptic and parabolic variational equations. However, weak extremal solutions were not investigated in those works. The existence of weak extremal solutions was considered among others in [43, 75, 106, 133, 138, 146, 189] (see also the references in [43]). The concept of sub-supersolutions has been established for various different types of nonlinear elliptic and parabolic problems, such as quasilinear Dirichlet-periodic boundary value problems in [31], reaction-diffusion equations under nonlinear and nonlocal flux boundary conditions in [44, 49], and boundary value problems in unbounded domains in [32, 33, 42]. An extension of this method to quasilinear elliptic problems with a right-hand side f not in V_0^* but in $L^1(\Omega)$ has been considered in [62]. Nonlinear parabolic problems with $1 < p < 2$ were treated in [67, 92]. The sub-supersolution method has been proved to be a powerful and fruitful tool not only in the qualitative analysis for a wide range of nonlinear elliptic and parabolic problems, but also in their quantitative analysis. This method coupled with the monotone iteration has been proved to be an effective and flexible technique in the theoretical as well as the constructive analysis of semilinear elliptic and parabolic problems, which has extensively been discussed within the framework of classic solutions in the monograph by C. V. Pao, cf. [182]. The constructive aspect of the sub-subsolution method has been improved in a recent monograph by V. Lakshmikantham and S. Köksal [135] by combining it with the so-called quasilinearization method to get not only monotone but also rapidly convergent approximate solutions (see also [48, 49]). For these iteration methods to work, comparison principles of related linearized problems based on sub-supersolutions are the main tools.

4

Multivalued Variational Equations

The subject of this chapter is boundary value problems for quasilinear differential inclusions of elliptic and parabolic type whose governing multivalued terms are of Clarke's gradient type. We introduce concepts of sub- and supersolutions that are designed to obtain existence and comparison results and that generalize the notion of sub- and supersolutions of variational equations considered in Chap. 3 in a natural way. Thus, the least requirement of any notion of sub-supersolutions for inclusions is that to include the corresponding notion for equations as introduced in Chap. 3. In Sect. 4.1, we first provide some motivation for differential inclusions with the help of elementary examples and introduce the basic concept of sub- and supersolutions. Depending on the structure and growth assumptions imposed on the multivalued terms, the notion of sub- and supersolutions and the comparison principles related with them are further developed in Sect. 4.2, Sect. 4.3, and Sect. 4.5 for general quasilinear elliptic and parabolic inclusion problems. As an application of the theory presented in this chapter, an elliptic inclusion is considered whose multivalued term is given in Sect. 4.4 by the difference of Clarke's generalized gradient and the usual subdifferential. An alternative notion of sub-supersolution existing in the literature and its relation to the one introduced here is considered in Sect. 4.6. The chapter concludes with comments and further bibliographical notes.

4.1 Motivation and Introductory Examples

To motivate the study of differential inclusions, let us consider the following simple discontinuous semilinear BVP:

$$-\Delta u + g(u) = 1 \quad \text{in} \quad \Omega, \quad u = 0 \quad \text{on} \quad \partial\Omega, \tag{4.1}$$

where $\Omega \subset \mathbb{R}^N$ is a bounded domain with Lipschitz boundary $\partial\Omega$ and the nonlinearity $g : \mathbb{R} \to \mathbb{R}$ is assumed to be the Heaviside step function; i.e.,

$$g(s) = \begin{cases} 0 & \text{if } s \leq 0, \\ 1 & \text{if } s > 0. \end{cases} \tag{4.2}$$

For the BVP (4.1) with the nonlinearity (4.2), one readily verifies that the constant functions $\underline{u}(x) \equiv -c$ and $\bar{u}(x) \equiv c$ with $c > 0$ form an ordered pair of sub- and supersolutions, respectively. However, in Chap. 1, it has been proved that (4.1) does not possess any solution. Consequently, in general, the existence and comparison principle proved in Chap. 3 fails if the nonlinearities involved are discontinuous. Therefore, instead of the discontinuous BVP (4.1), we consider the following relaxed multivalued BVP:

$$-\Delta u + [g(u-0), g(u+0)] \ni 1 \quad \text{in } \Omega, \quad u = 0 \quad \text{on } \partial\Omega, \tag{4.3}$$

which arises from (4.1) by replacing the discontinuous nonlinearity g by the multifunction $s \mapsto [g(s-0), g(s+0)]$, with $g(s \pm 0)$ being the one-sided limits of g at s (see Fig. 1.2). Note that the multifunction in Fig. 1.2 is a maximal monotone graph in $\mathbb{R}^2$. In Sect. 4.1.1, we give some further motivation for studying the differential inclusion (4.3) and provide a notion of its solution, and in Sect. 4.1.2, a first comparison principle will be given. In Sect. 4.1.3, we present an existence and comparison result for a nonmonotone elliptic differential inclusion whose multifunction is of the form shown in Fig. 1.4. The main goal of Sect. 4.1 is to introduce basic ideas in the study of existence and comparison principles for differential inclusions with the help of simple model problems.

4.1.1 Motivation

Let $j : \mathbb{R} \to \mathbb{R}$ be the primitive of g vanishing at 0 for g in (4.2); i.e.,

$$j(s) = \int_0^s g(t)\, dt,$$

which yields $j(s) = s^+$ (see Fig. 1.3), and consider the minimization problem

$$u \in V_0 : \quad E(u) = \inf_{v \in V_0} E(v), \tag{4.4}$$

where $V_0 = W_0^{1,2}(\Omega)$ and E is the functional given by

$$E(v) = \frac{1}{2} \int_\Omega |\nabla v|^2\, dx + \int_\Omega (j(v) - v)\, dx, \quad v \in V_0. \tag{4.5}$$

The functional $E : V_0 \to \mathbb{R}$ is easily seen to be convex (even strictly convex) and continuous, and thus it is also weakly sequentially lower semicontinuous. Also, one readily verifies that E is coercive in the sense that E satisfies

$$E(v) \to +\infty \quad \text{as} \quad \|v\|_{V_0} \to +\infty.$$

Thus, we may apply Weierstrass' Theorem 2.53, which ensures the existence of solutions of problem (4.4). Moreover, because $E : V_0 \to \mathbb{R}$ is strictly convex, the minimum problem (4.4) has a uniquely defined solution. By using elementary facts from the calculus of convex analysis, solutions of (4.4) can equivalently be characterized as the critical points of the nonsmooth functional E; i.e., we have the following equivalence:

$$u \in V_0 : \quad E(u) = \inf_{v \in V_0} E(v) \iff 0 \in \partial E(u), \tag{4.6}$$

where $\partial E(u)$ is the subdifferential of E at u. By standard calculation, we obtain

$$0 \in \partial E(u) \iff \Delta u + 1 \in \partial (J \circ i)(u), \tag{4.7}$$

where $i : V_0 \to L^2(\Omega)$ is the embedding operator and $J : L^2(\Omega) \to \mathbb{R}$ is the following integral functional:

$$J(u) = \int_\Omega j(u(x)) \, dx,$$

which is convex and even Lipschitz continuous in view of $j(s) = s^+$. Because V_0 is dense in $L^2(\Omega)$, we may apply the chain rule given in Corollary 2.180 to evaluate the subdifferential of $J \circ i$ at u, which results in

$$\partial (J \circ i)(u) = i^* \partial J(i(u)), \quad u \in V_0, \tag{4.8}$$

where i^* is the adjoint operator to i. The subdifferential of convex functions and convex integral functionals has been characterized by Example 2.58 and Example 2.59, respectively, given in Sect. 2.1.3, and thus, we obtain for $v \in L^2(\Omega)$

$$v^* \in \partial J(v) \iff v^* \in L^2(\Omega) \text{ and } v^*(x) \in \partial j(v(x)) = [\underline{g}(v(x)), \bar{g}(v(x))],$$

where $\underline{g}(s) = g(s - 0)$ and $\bar{g}(s) = g(s + 0)$ are the left-sided and right-sided limits of g at s. As here g is the Heaviside function, the multifunction $s \mapsto \partial j(s)$ is the maximal monotone graph that arises from the Heaviside function by filling in the gap at the point of discontinuity (see Fig. 1.2). So far we have seen that the minimum problem (4.4) is equivalent to the variational inequality

$$u \in V_0 : \quad \langle -\Delta u - 1, v - u \rangle + (J \circ i)(v) - (J \circ i)(u) \geq 0, \quad \forall\, v \in V_0, \tag{4.9}$$

which is defined on the entire space V_0, and which in turn is equivalent to the Dirichlet problem of the differential inclusion

$$u \in V_0 : \quad -\Delta u + \partial j(u) \ni 1 \quad \text{in } V_0^*. \tag{4.10}$$

Taking into account the chain rule (4.8) and the characterization of the subdifferential of integral functionals as given above, the notion of (weak) solution of the inclusion problem (4.10) is as follows.

Definition 4.1. *The function $u \in V_0$ is a solution of the inclusion problem (4.10) if there is a $w \in L^2(\Omega) \subset V_0^*$ such that*

(i) $w(x) \in \partial j(u(x))$ *for a.e.* $x \in \Omega$.
(ii) $-\Delta u + w = 1$ *in* V_0^* $\Longleftrightarrow$ $\int_{\Omega} (\nabla u \nabla \varphi + (w - 1) \varphi) dx = 0, \ \forall \ \varphi \in V_0$.

We already know that problem (4.10) is uniquely solvable, and we readily observe that $u = 0$ is the unique solution, because $(u, w) \in V_0 \times L^2(\Omega)$ with $(u, w) = (0, 1)$ fulfills Definition 4.1. To summarize our considerations above, we have seen that on the one hand, the differential inclusion problem (4.10) may be considered as a relaxation of the discontinuous problem (4.1) that arises by replacing the discontinuous nonlinearity g by the associated multivalued nonlinearity $\partial j(s) = [\underline{g}(s), \bar{g}(s)]$, and on the other hand, the inclusion problem (4.10) may be interpreted as the Euler–Lagrange equation of the nonsmooth (locally Lipschitz and convex) functional E defined in (4.5) or as the necessary (in our case also sufficient) condition for critical points of E.

4.1.2 Comparison Principle: Subdifferential Case

Let us consider the differential inclusion problem (4.10), which according to the preceding section is uniquely solvable. Extending the notion of sub- and supersolution for (single-valued) BVPs in a natural way, we introduce the following notion for the multivalued problem (4.10).

Definition 4.2. *The function $\underline{u} \in V = W^{1,2}(\Omega)$ is a subsolution of the inclusion problem (4.10) if there is a $\underline{w} \in L^2(\Omega) \subset V_0^*$ such that*

(i) $\underline{u} \leq 0$ *on* $\partial\Omega$.
(ii) $\underline{w}(x) \in \partial j(\underline{u}(x))$ *for a.e.* $x \in \Omega$.
(iii) $-\Delta \underline{u} + \underline{w} \leq 1$ *in* V_0^*.

Note that inequality (iii) of Definition 4.2 has to be taken with respect to the order defined by the dual-order cone of V_0^*; i.e., that (iii) means

$$\int_{\Omega} (\nabla \underline{u} \nabla \varphi + (\underline{w} - 1) \varphi) dx \leq 0, \quad \text{for all } \varphi \in V_0 \cap L^2_+(\Omega).$$

Analogously, we define the supersolution as follows.

Definition 4.3. *The function $\bar{u} \in V = W^{1,2}(\Omega)$ is a supersolution of the inclusion problem (4.10) if there is a $\bar{w} \in L^2(\Omega) \subset V_0^*$ such that*

(i) $\bar{u} \geq 0$ *on* $\partial\Omega$.
(ii) $\bar{w}(x) \in \partial j(\bar{u}(x))$ *for a.e.* $x \in \Omega$.
(iii) $-\Delta \bar{u} + \bar{w} \geq 1$ *in* V_0^*.

Remark 4.4. The determination of sub- and supersolutions for the differential inclusion (4.10) can be reduced to finding sub- and supersolutions of an appropriately associated differential equation. For this purpose, let $\tilde{g} : \mathbb{R} \to \mathbb{R}$ be any single-valued selection of the maximal monotone graph ∂j; i.e., $\tilde{g}(s) \in \partial j(s)$ for all $s \in \mathbb{R}$, and consider the following (single-valued) BVP:

$$-\Delta u + \tilde{g}(u) = 1 \quad \text{in} \quad \Omega, \quad u = 0 \quad \text{on} \quad \partial\Omega. \tag{4.11}$$

Even though the BVP (4.11) may not have any solution, its subsolutions (supersolutions) are subsolutions (supersolutions) of the inclusion (4.10). To see this, let $\underline{u} \in V$ be a subsolution of (4.11) in the sense of Chap. 3; i.e., $\underline{u} \le 0$ on $\partial\Omega$, $\tilde{g}(\underline{u}) \in L^2(\Omega)$, and the inequality

$$-\Delta \underline{u} + \tilde{g}(\underline{u}) \le 1 \quad \text{in} \quad V_0^*$$

holds. By setting $\underline{w} = \tilde{g}(\underline{u})$, then obviously $\underline{u}$ and $\underline{w}$ satisfy all conditions of Definition 4.2. Similarly we show that any supersolution of (4.11) is a supersolution of (4.10).

Lemma 4.5. *Let $\underline{u}$ and $\bar{u}$ be sub- and supersolutions, respectively, of problem (4.10). Then $\underline{u} \le \bar{u}$ in Ω.*

Proof: From Definition 4.2 and Definition 4.3, we obtain $\underline{u} - \bar{u} \le 0$ on $\partial\Omega$ and

$$\int_\Omega \nabla(\underline{u} - \bar{u})\nabla\varphi \, dx + \int_\Omega (\underline{w} - \bar{w}) \, \varphi \, dx \le 0 \quad \text{for all } \varphi \in V_0 \cap L^2_+(\Omega), \tag{4.12}$$

where $\underline{w}(x) \in \partial j(\underline{u}(x))$ and $\bar{w}(x) \in \partial j(\bar{u}(x))$ for a.e. $x \in \Omega$. Taking $\varphi = (\underline{u} - \bar{u})^+$ as a special test function in (4.12), we get

$$\int_\Omega |\nabla(\underline{u} - \bar{u})^+|^2 \, dx + \int_{\{\underline{u} > \bar{u}\}} (\underline{w} - \bar{w})(\underline{u} - \bar{u}) \, dx \le 0. \tag{4.13}$$

As $s \mapsto \partial j(s)$ is a maximal monotone graph, the second integral on the left-hand side of (4.13) is nonnegative, and thus from (4.13), we obtain

$$\int_\Omega |\nabla(\underline{u} - \bar{u})^+|^2 \, dx = 0,$$

which implies $(\underline{u} - \bar{u})^+ = 0$; i.e., $\underline{u} \le \bar{u}$.   □

Remark 4.6. As any solution of the inclusion (4.10) is both a sub- and supersolution, Lemma 4.5 provides an alternative to prove the uniqueness of solution for (4.10). However, it should be noted that if the Laplacian is replaced by a more general (not necessarily strictly monotone) elliptic operator, sub- and supersolutions need not be order related to each other.

The definitions and results regarding the differential inclusion problem (4.10) can easily be extended to the following more general BVP:

$$u \in V_0 : \quad -\Delta u + \partial j(u) \ni f \quad \text{in } V_0^*, \tag{4.14}$$

where $f \in V_0^*$, and $j : \mathbb{R} \to \mathbb{R}$ is a convex function whose subdifferential $\partial j : \mathbb{R} \to 2^{\mathbb{R}} \setminus \{\emptyset\}$ satisfies a linear growth condition; i.e., there is a constant $c \geq 0$ such that

$$|\eta| \leq c(1 + |s|), \quad \eta \in \partial j(s), \tag{4.15}$$

for all $s \in \mathbb{R}$. The following existence and comparison principle of the subdifferential inclusion (4.14) can be obtained by an almost straightforward application of methods and results used in the treatment of the special problem (4.10).

Lemma 4.7. *Let $j : \mathbb{R} \to \mathbb{R}$ be a convex function whose subdifferential $\partial j : \mathbb{R} \to 2^{\mathbb{R}} \setminus \{\emptyset\}$ satisfies (4.15). Then the the BVP (4.14) enjoys the following properties:*

(i) *The BVP (4.14) is uniquely solvable.*
(ii) *If $\underline{u}$ and $\bar{u}$ are sub- and supersolutions of (4.14), then $\underline{u} \leq \bar{u}$.*
(iii) *Let u_i be the unique solution of (4.14) with the right-hand side $f_i \in V_0^*$, $i = 1, 2$. If $f_1 \leq f_2$, then $u_1 \leq u_2$. In other words, the operator $-\Delta + \partial j : V_0 \to V_0^*$ is inverse monotone increasing.*

Proof: Ad (i). As j is convex on the real line and its subdifferential satisfies a linear growth condition, it follows that the integral functional $J : L^2(\Omega) \to \mathbb{R}$ given by

$$J(v) = \int_{\Omega} j(v(x)) \, dx, \quad v \in L^2(\Omega),$$

is well defined, convex, and locally Lipschitz continuous. Moreover, because j is bounded below by an affine function and Ω is a bounded domain, we get the following estimate:

$$J(v) \geq -c(1 + |v|_2), \quad v \in L^2(\Omega), \tag{4.16}$$

where c is some positive constant and $|\cdot|_2$ denotes the norm in $L^2(\Omega)$. As in the special case above, one easily verifies that solutions of the BVP (4.14) are the critical points of the (nonsmooth) functional

$$E(v) = \frac{1}{2} \int_{\Omega} |\nabla v|^2 \, dx + (J \circ i)(v) - \langle f, v \rangle, \quad v \in V_0, \tag{4.17}$$

where $i : V_0 \to L^2(\Omega)$ is the embedding operator. The functional E is strictly convex, and in view of (4.16), it is also coercive. Therefore, the only critical point of E is its minimum point u satisfying

$$u \in V_0 : \quad E(u) = \inf_{v \in V_0} E(v),$$

whose existence is ensured by Weierstrass' Theorem 2.53.

Ad (ii). The proof can be done in just the same way as for Lemma 4.5.

Ad (iii). The unique solution u_i with right-hand side f_i satisfies

$$\int_\Omega \nabla u_i \nabla \varphi \, dx + \int_\Omega w_i \varphi \, dx = \langle f_i, \varphi \rangle, \quad \text{for all } \varphi \in V_0, \qquad (4.18)$$

where $w_i \in L^2(\Omega)$ satisfies $w_i(x) \in \partial j(u_i(x))$ for a.e. $x \in \Omega$ and $i = 1, 2$. By subtraction, we get from (4.18) the relation

$$\int_\Omega \nabla(u_1 - u_2) \nabla \varphi \, dx + \int_\Omega (w_1 - w_2) \varphi \, dx \leq 0 \quad \text{for all } \varphi \in V_0 \cap L^2_+(\Omega).$$
$$(4.19)$$

Testing (4.19) with the nonnegative function $\varphi = (u_1 - u_2)^+$, one concludes in a similar way as in the proof of Lemma 4.5 that $(u_1 - u_2)^+ = 0$; i.e., $u_1 \leq u_2$.
$\square$

4.1.3 Comparison Principle: Clarke's Gradient Case

We now extend our considerations of the previous section to the BVP

$$u \in V_0 : \quad -\Delta u + \partial j(u) \ni f \quad \text{in } V_0^*, \qquad (4.20)$$

where $f \in V_0^*$, and $\partial j : \mathbb{R} \to 2^{\mathbb{R}} \setminus \{\emptyset\}$ is the generalized Clarke's gradient of a locally Lipschitz function $j : \mathbb{R} \to \mathbb{R}$ that is assumed to fulfill the following structure and growth conditions:

(H1) A constant $c_1 \geq 0$ exists such that

$$\eta_1 \leq \eta_2 + c_1(s_2 - s_1)$$

for all $\eta_i \in \partial j(s_i)$, $i = 1, 2$, and for all s_1, s_2 with $s_1 < s_2$.

(H2) A constant $c_2 \geq 0$ exists such that

$$|\eta| \leq c_2(1 + |s|), \quad \eta \in \partial j(s),$$

for all $s \in \mathbb{R}$.

A model for a generalized Clarke's gradient satisfying (H1) and (H2) is illustrated in Fig. 1.4. The graph of ∂j given in Fig. 1.4 can easily be seen to have a representation in the form

$$\partial j(s) = \partial j_1(s) - \partial j_2(s), \quad s \in \mathbb{R}, \qquad (4.21)$$

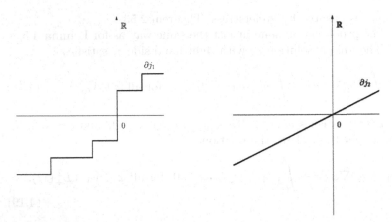

Fig. 4.1. Graph of ∂j_1 **Fig. 4.2.** Graph of ∂j_2

where ∂j_i, $i = 1, 2$, are the maximal monotone graphs shown in Fig. 4.1 and Fig. 4.2 with $\partial j_2(s) = \{c_1 s\}$, $c_1 > 0$, being a line having a positive slope; i.e., in this case, we have

$$\partial j(s) + c_1 s = \partial j_1(s), \quad s \in \mathbb{R}, \tag{4.22}$$

which implies (H1). The notion of sub- and supersolution for the BVP (4.20) that is similar to the one for the special BVP (4.10) is given by the following definition.

Definition 4.8. *The function $\underline{u}$ $(\bar{u}) \in V = W^{1,2}(\Omega)$ is a subsolution (supersolution) of the inclusion problem (4.20) if there is a function $\underline{w}$ $(\bar{w}) \in L^2(\Omega) \subset V_0^*$ such that*

(i) $\underline{u} \leq 0$ $(\bar{u} \geq 0)$ on $\partial \Omega$.
(ii) $\underline{w}(x) \in \partial j(\underline{u}(x))$ $(\bar{w}(x) \in \partial j(\bar{u}(x)))$ for a.e. $x \in \Omega$.
(iii) $-\Delta \underline{u} + \underline{w} \leq f$ $(-\Delta \bar{u} + \bar{w} \geq f)$ in V_0^*.

Note that problem (4.20), in general, is not uniquely solvable. To see this, consider for example the special case

$$j(s) = -\frac{\lambda}{2} s^2, \quad f = 0,$$

where λ may be any eigenvalue of the Laplacian. The function $s \mapsto -\frac{\lambda}{2} s^2$ is locally Lipschitz with $\partial j(s) = \{-\lambda s\}$, and thus, hypotheses (H1) and (H2) are trivially satisfied. However, in this case, the BVP (4.20), which reduces to the eigenvalue problem

$$u \in V_0 : \quad \Delta u + \lambda u = 0 \quad \text{in } V_0^*$$

is not uniquely solvable. Consequently, sub- and supersolutions of (4.20) need not be order related; i.e., Lemma 4.5 of the previous section no longer holds for inclusions of Clarke's gradient type. However, we are going to prove an existence and comparison principle for (4.20) under hypotheses (H1) and (H2), which basically asserts the existence of solutions of (4.20) within the interval formed by an ordered pair of sub- and supersolutions. Furthermore, we will show that the set of all solutions of (4.20) contained in the interval of sub- and supersolutions has extremal elements and is compact. The relatively simple structure of the BVP (4.20) allows for its treatment to use only very basic tools. The next two lemmas provide some preliminary results that will be used in the proof of the main existence and comparison principle of this section.

Lemma 4.9. *Let $s \mapsto \partial j(s)$ be the generalized Clarke's gradient of a locally Lipschitz function $j : \mathbb{R} \to \mathbb{R}$ satisfying (H1). Then the multifunction $s \mapsto \partial j(s) + c_1 s$ is a maximal monotone graph in $\mathbb{R}^2$.*

Proof: Hypothesis (H1) implies that the multifunction $s \mapsto \partial j(s) + c_1 s$ is monotone. To prove that its graph is maximal monotone, let $r, w \in \mathbb{R}$ such that

$$[w - (z + c_1 s)](r - s) \geq 0, \quad \forall s \in \mathbb{R}, \ \forall z \in \partial j(s). \tag{4.23}$$

We must show that $w \in \partial j(r) + c_1 r$. Assume $w \notin \partial j(r) + c_1 r$. Then $\lambda \in \mathbb{R}$, $\lambda \neq 0$ exists such that

$$\max_{\zeta \in \partial j(r)} (\zeta \lambda) < (w - c_1 r)\lambda. \tag{4.24}$$

Let $t_n \downarrow 0$ and $\gamma_n \in \partial j(r + t_n \lambda)$. By (4.23), we may write

$$[w - (\gamma_n + c_1(r + t_n \lambda))](r - (r + t_n \lambda)) \geq 0,$$

or equivalently,

$$\lambda[w - (\gamma_n + c_1(r + t_n \lambda))] \leq 0.$$

Along a relabeled subsequence, one has $\gamma_n \to \gamma$ with $\gamma \in \partial j(r)$ (because ∂j has closed graph in $\mathbb{R}^2$). Passing to the limit in the previous inequality, we get

$$\lambda[w - (\gamma + c_1 r)] \leq 0;$$

that is,

$$(w - c_1 r)\lambda \leq \gamma \lambda,$$

which contradicts (4.24). $\square$

Lemma 4.10. *Let $j : \mathbb{R} \to \mathbb{R}$ be a locally Lipschitz function satisfying (H1). Then $\hat{j} : \mathbb{R} \to \mathbb{R}$ defined by*

$$\hat{j}(s) = j(s) + \frac{c_1}{2} s^2$$

is convex and locally Lipschitz, and its subdifferential is given by

$$\partial \hat{j}(s) = \partial j(s) + c_1 s,$$

where ∂j is the generalized Clarke's gradient of j.

Proof: The function $\hat{j}(s) = j(s) + \frac{c_1}{2} s^2$ is locally Lipschitz, and its generalized Clarke's gradient is given by $\partial \hat{j}(s) = \partial j(s) + c_1 s$, which is a maximal monotone graph according to Lemma 4.9. To complete the proof, we need to show that $\hat{j}$ is convex. By means of Lebourg's Theorem (see Theorem 2.177) and using the maximal monotonicity of the multifunction $s \mapsto \partial \hat{j}(s)$, we readily verify that $\partial \hat{j} : \mathbb{R} \to 2^{\mathbb{R}} \setminus \{\emptyset\}$ is also maximal cyclic monotone, where $\partial \hat{j}$ is called cyclic monotone iff the inequality

$$s_1^*(s_1 - s_2) + s_2^*(s_2 - s_3) + \cdots + s_n^*(s_n - s_{n+1}) \geq 0 \qquad (4.25)$$

holds for all $s_i \in \mathbb{R}$ and $s_i^* \in \partial \hat{j}(s_i)$, $i = 1, \ldots, n$, and $n \in \mathbb{N}$, where we set $s_{n+1} = s_1$. In fact we have

$$\sum_{k=1}^{n} (\hat{j}(s_k) - \hat{j}(s_{k+1})) = 0. \qquad (4.26)$$

(Note that $s_{n+1} = s_1$.) By means of Lebourg's Theorem, we get

$$\hat{j}(s_k) - \hat{j}(s_{k+1}) = \xi_k^*(s_k - s_{k+1}), \quad \xi_k^* \in \partial \hat{j}(\xi_k), \qquad (4.27)$$

where $\xi_k = s_k + \theta_k(s_{k+1} - s_k)$ for some $\theta_k \in (0, 1)$. If $s_k^* \in \partial \hat{j}(s_k)$, then the following inequality holds:

$$(\xi_k^* - s_k^*)(s_k - s_{k+1}) \leq 0. \qquad (4.28)$$

Thus, from (4.26) to (4.28) along with the maximal monotonicity of $\partial \hat{j}$, we obtain (4.25).

A general result due to Rockafellar (see [222, Corollary 32.18] or [223, Theorem 47.5]) asserts that the maximal cyclic monotonicity of a multifunction $A : X \to 2^{X^*}$ on the real Banach space X is equivalent to the existence of a convex, proper, and lower semicontinuous function $f : X \to (-\infty, +\infty]$ satisfying $A = \partial f$. Applying this result to the special multifunction $\partial \hat{j}$ yields the existence of a convex function $\tilde{j} : \mathbb{R} \to \mathbb{R}$ defined on the entire real line (and thus, $\tilde{j}$ is even locally Lipschitz) such that for its subdifferential we have $\partial \tilde{j} = \partial \hat{j}$. As the function $\tilde{j}$ is uniquely determined to within a constant c, we infer

$$\hat{j}(s) = \tilde{j}(s) + c, \quad s \in \mathbb{R};$$

i.e., $\hat{j} : \mathbb{R} \to \mathbb{R}$ must be convex. $\square$

Now we are in a position to prove the following existence, comparison, and compactness result.

Theorem 4.11. *Let $\underline{u}$ and $\bar{u}$ be sub- and supersolutions of the differential inclusion problem (4.20) that satisfy $\underline{u} \leq \bar{u}$. Then under hypotheses (H1) and (H2), extremal solutions of (4.20) exist within the ordered interval $[\underline{u}, \bar{u}]$. Moreover, the solution set S of all solutions of (4.20) lying in $[\underline{u}, \bar{u}]$ is compact.*

Proof: First, we note that problem (4.20) is equivalent to the following one:

$$u \in V_0: \quad -\Delta u + \partial j(u) + c_1 u \ni f + c_1 u \quad \text{in } V_0^*, \tag{4.29}$$

where $c_1 \geq 0$ is the constant of (H1). With $\hat{j}$ introduced in Lemma 4.10, problem (4.20) is equivalent to

$$u \in V_0: \quad -\Delta u + \partial \hat{j}(u) \ni f + c_1 u \quad \text{in } V_0^*, \tag{4.30}$$

where $\partial \hat{j}$ is the maximal monotone graph of the convex function $\hat{j}$ (see Lemma 4.10) that satisfies a linear growth condition. The proof of the existence and comparison result of (4.20) or equivalently of (4.30) formulated in the theorem will be based on the following iteration:

$$u_{n+1} \in V_0: \quad -\Delta u_{n+1} + \partial \hat{j}(u_{n+1}) \ni f + c_1 u_n \quad \text{in } V_0^*. \tag{4.31}$$

We remark that if $u_n \in L^2(\Omega)$, then $f_n = f + c_1 u_n \in V_0^*$ and $u_{n+1} \in V_0$ is the uniquely defined solution of the inclusion (4.31), which is of subdifferential type. Let us start the iteration (4.31) with the given supersolution of (4.20); i.e., $u_0 = \bar{u}$. The first iterate u_1 is the unique solution of the inclusion

$$u \in V_0: \quad -\Delta u + \partial \hat{j}(u) \ni f + c_1 u_0 \quad \text{in } V_0^*. \tag{4.32}$$

As the supersolution u_0 of (4.20) is also a supersolution of (4.32), it follows by the comparison result for inclusions of subdifferential type according to Lemma 4.7 (ii) that $u_1 \leq u_0$. Assume $u_n \leq u_{n-1}$. The iterates u_{n+1} and u_n are the unique solutions of (4.32) with right-hand sides $f_n = f + c_1 u_n$ and $f_{n-1} = f + c_1 u_{n-1}$, respectively. In view of $u_n \leq u_{n-1}$, we have $f_n \leq f_{n-1}$, and thus by applying the comparison result due to Lemma 4.7 (iii), it follows that $u_{n+1} \leq u_n$. In a similar way by induction, we obtain $\underline{u} \leq u_n$ for all $n \in \mathbb{N}$, and hence, our iteration (4.31) above with $u_0 = \bar{u}$ yields a monotone decreasing sequence of iterates satisfying

$$\underline{u} \leq \cdots \leq u_{n+1} \leq u_n \leq \cdots \leq u_1 \leq u_0 = \bar{u}. \tag{4.33}$$

From (4.33), the a.e. pointwise limit of the monotone sequence (u_n) exists; i.e.,

$$u_n(x) \to u^*(x) \quad \text{for a.e. } x \in \Omega.$$

Obviously $u^* \in [\underline{u}, \bar{u}]$, and by applying Lebesque's dominated convergence theorem, we readily see that

$$u_n \to u^* \quad \text{in } L^2(\Omega).$$

We are going to show that u^* is the greatest solution of (4.20) within the ordered interval $[\underline{u}, \bar{u}]$. As solutions of (4.31), the iterates u_{n+1} satisfy

$$u_{n+1} \in V_0 : \quad -\Delta u_{n+1} + w_{n+1} = f + c_1 u_n \quad \text{in } V_0^*, \qquad (4.34)$$

where $w_{n+1} \in L^2(\Omega) \subset V_0^*$ and $w_{n+1}(x) \in \partial \hat{j}(u_{n+1}(x))$. As $\partial \hat{j}$ satisfies a linear growth condition and because the sequence (u_n) is bounded in $L^2(\Omega)$, it follows that (w_n) is bounded in $L^2(\Omega)$ as well, which in view of (4.34) implies the boundedness of (u_n) in V_0 and, thus, the existence of a weakly convergent subsequence. The compact embedding $V_0 \subset L^2(\Omega)$ in conjunction with the monotonicity of the sequence (u_n) yield the weak convergence of the entire sequence in V_0; i.e.,

$$u_n \rightharpoonup u^* \quad \text{in } V_0.$$

Testing (4.34) with $\varphi = u_{n+1} - u^*$, we obtain

$$\int_\Omega |\nabla(u_{n+1} - u^*)|^2 \, dx = \langle f + \Delta u^*, u_{n+1} - u^* \rangle$$
$$+ \int_\Omega (c_1 u_n - w_{n+1})(u_{n+1} - u^*) \, dx. \qquad (4.35)$$

As (u_n) is weakly convergent in V_0 and strongly convergent in $L^2(\Omega)$, and (w_n) is bounded in $L^2(\Omega)$, it follows that the right-hand side of (4.35) tends to zero, which implies the strong convergence of (u_n) in V_0. To show that the limit u^* is a solution of (4.20) or equivalently of (4.29), we note that the BVP (4.31) is equivalent to the variational inequality

$$\langle -\Delta u_{n+1} - f - c_1 u_n, v - u_{n+1} \rangle + (\hat{J} \circ i)(v) - (\hat{J} \circ i)(u_{n+1}) \geq 0 \qquad (4.36)$$

for all $v \in V_0$, where $\hat{J} : L^2(\Omega) \to \mathbb{R}$ is the convex and locally Lipschitz functional generated by the convex function $\hat{j} : \mathbb{R} \to \mathbb{R}$. Using the convergence properties of the iterates, we may pass to the limit as $n \to \infty$ in (4.36), which yields

$$\langle -\Delta u^* - f - c_1 u^*, v - u^* \rangle + (\hat{J} \circ i)(v) - (\hat{J} \circ i)(u^*) \geq 0$$

for all $v \in V_0$, which is equivalent to

$$\Delta u^* + f + c_1 u^* \in \partial(\hat{J} \circ i)(u^*);$$

i.e., a $w^* \in L^2(\Omega)$ exists such that $w^*(x) \in \partial \hat{j}(u^*(x))$ for a.e. $x \in \Omega$, and

$$\Delta u^* + f + c_1 u^* = w^* \quad \text{in } V_0^*. \qquad (4.37)$$

By Lemma 4.10, we have $\partial \hat{j}(s) = \partial j(s) + c_1 s$, which implies $w^*(x) - c_1 u^*(x) \in \partial j(u^*(x))$, and thus from (4.37), we obtain

$$-\Delta u^* + \eta^* = f \quad \text{in } V_0^*,$$

where $\eta^* = w^* - c_1 u^*$, which proves that u^* is a solution of (4.20) lying within $[\underline{u}, \bar{u}]$.

Next we show that u^* is the greatest solution of (4.20) in $[\underline{u}, \bar{u}]$. To this end, let u be any solution of (4.20) [or equivalently of (4.29)] in $[\underline{u}, \bar{u}]$. In particular, u is then a subsolution of (4.29), and using the comparison result of Lemma 4.7 (ii) and the induction principle, we get $u \leq u_n$ for all $n \in \mathbb{N}$, which proves $u \leq u^*$; i.e., u^* is the greatest solution.

Starting the iteration with the subsolution $u_0 = \underline{u}$, by similar arguments, we can prove the existence of the smallest solution u_* of (4.20) in $[\underline{u}, \bar{u}]$.

To complete the proof of the theorem, we are going to show that the solution set $\mathcal{S}$ of all solutions of (4.20) in $[\underline{u}, \bar{u}]$ is compact. Let $(u_n) \subset \mathcal{S}$ be any sequence. Obviously, (u_n) is bounded in $L^2(\Omega)$ and satisfies the BVP

$$u_n \in V_0: \quad -\Delta u_n + w_n = f \quad \text{in } V_0^*, \tag{4.38}$$

where $w_n \in L^2(\Omega)$ satisfy $w_n(x) \in \partial j(u_n(x))$ for a.e. $x \in \Omega$. In view of the linear growth of $s \mapsto \partial j(s)$, the sequence (w_n) is bounded in $L^2(\Omega)$, which according to (4.38) implies the boundedness of (u_n) in V_0. Thus, a subsequence (u_k) of (u_n) exists that is weakly convergent in V_0 and strongly convergent in $L^2(\Omega)$ to the limit u. Replacing n by k in (4.38), and testing the resulted equation with $\varphi = u_k - u \in V_0$, we obtain

$$\int_\Omega |\nabla(u_k - u)|^2 \, dx = \langle \Delta u + f, u_k - u \rangle - \int_\Omega w_k (u_k - u) \, dx \to 0,$$

which implies the strong convergence of (u_k) in V_0. In the same way as for u^* above, we can prove that the limit u is a solution of (4.20), which completes the proof of the theorem. □

4.2 Inclusions with Global Growth on Clarke's Gradient

As a model problem we consider the following quasilinear elliptic inclusion under homogeneous Dirichlet boundary condition governed by the p-Laplacian Δ_p in the form:

$$u \in V_0: \quad -\Delta_p u + f(u) + \partial j(u) \ni h \quad \text{in } V_0^*, \tag{4.39}$$

where $V_0 = W_0^{1,p}(\Omega)$, $1 < p < \infty$, $h \in V_0^*$, and Ω is as in Sect. 4.1. The multifunction $\partial j : \mathbb{R} \to 2^{\mathbb{R}} \setminus \{\emptyset\}$ is the generalized Clarke's gradient of a locally Lipschitz function $j : \mathbb{R} \to \mathbb{R}$, and $f : \mathbb{R} \to \mathbb{R}$ is assumed to be a continuous function. We impose the following hypotheses on ∂j and f.

(H1) A constant $c_1 \geq 0$ exists such that

$$\eta_1 \leq \eta_2 + c_1(s_2 - s_1)^{p-1}$$

for all $\eta_i \in \partial j(s_i)$, $i = 1, 2$, and for all s_1, s_2 with $s_1 < s_2$.

(H2) A constant $c_2 \geq 0$ exists such that

$$|\eta| \leq c_2(1 + |s|^{p-1}), \quad \eta \in \partial j(s)$$

for all $s \in \mathbb{R}$.

(H3) A constant $c_3 \geq 0$ exists such that

$$|f(s)| \leq c_3(1 + |s|^{p-1} \quad \text{for all } s \in \mathbb{R}.$$

Remark 4.12. Note that for both the generalized Clarke's gradient ∂j and f, we have assumed by (H2) and (H3) a global growth condition. We will see later how to deal with the more general case of local growth conditions. Rather than to treat the case of local growth conditions for the inclusion (4.39), we are going to consider this case in a general setting of quasilinear elliptic inclusions. It will be done in the next section. We remark also that the methodology to be developed for problem (4.39) can easily be extended to inclusions with general quasilinear elliptic operators. Only for simplifying our presentation and to emphasize the basic ideas, we have restricted ourselves in this section to the p-Laplacian and a lower order term f that is independent of the gradient.

The main goal of this section is to generalize the sub-supersolution method introduced in Sect. 4.1 to problem (4.39). It should be noted that the extension of the existence and comparison results obtained in Sect. 4.1 to problem (4.39) is by no means straightforward and requires much more involved tools.

Let us denote $V = W^{1,p}(\Omega)$, $V_0 = W_0^{1,p}(\Omega)$, and let q be the Hölder conjugate to p with $1 < p < \infty$. Throughout this section, we assume hypotheses (H1)–(H3). We first introduce the notion of solution and sub-supersolution for problem (4.39).

Definition 4.13. *The function $u \in V_0$ is called a solution of (4.39) if there is a function $w \in L^q(\Omega) \subset V_0^*$ such that*

(i) $w(x) \in \partial j(u(x))$ *for a.e. $x \in \Omega$.*
(ii) $\langle -\Delta_p u + f(u) + w, \varphi \rangle = \langle h, \varphi \rangle$ *for all $\varphi \in V_0$.*

Definition 4.14. *The function $\underline{u}$ $(\bar{u}) \in V$ is a subsolution (supersolution) of the inclusion problem (4.39) if there is a function $\underline{w}$ $(\bar{w}) \in L^q(\Omega) \subset V_0^*$ such that*

(i) $\underline{u} \leq 0$ $(\bar{u} \geq 0)$ *on $\partial\Omega$.*
(ii) $\underline{w}(x) \in \partial j(\underline{u}(x))$ $(\bar{w}(x) \in \partial j(\bar{u}(x)))$ *for a.e. $x \in \Omega$.*
(iii) $\langle -\Delta_p\underline{u} + f(\underline{u}) + \underline{w}, \varphi \rangle \leq \langle h, \varphi \rangle$ *for all $\varphi \in V_0 \cap L_+^p(\Omega)$.*
  $(\langle -\Delta_p\bar{u} + f(\bar{u}) + \bar{w}, \varphi \rangle \geq \langle h, \varphi \rangle$ *for all $\varphi \in V_0 \cap L_+^p(\Omega))$.*

Remark 4.15. By using the definition of the generalized Clarke's gradient ∂j (see Sect. 2.5), one readily sees that any solution of (4.39) is a solution of the following associated *hemivariational inequality*

$$u \in V_0: \quad \langle -\Delta_p u + f(u) - h, v - u \rangle + \int_\Omega j^0(u; v - u)\, dx \geq 0, \quad \forall\, v \in V_0,$$

$$(4.40)$$

where $j^0(s; r)$ denotes the generalized directional derivative of j at s in the direction r. The reverse is true only if j satisfies some additional regularity condition in the sense of Clarke (see Definition 2.163). Comparison principles for hemivariational inequalities will be studied in Chap. 6.

4.2.1 Preliminaries

We introduce the functional $J : L^p(\Omega) \to \mathbb{R}$ by

$$J(v) = \int_\Omega j(v(x))\, dx, \quad \forall\, v \in L^p(\Omega).$$

Using the growth condition (H2) and Lebourg's mean value theorem (see Theorem 2.177), we note that the functional J is well defined and Lipschitz continuous on bounded sets in $L^p(\Omega)$, thus locally Lipschitz. Moreover, the Aubin–Clarke theorem (see Theorem 2.181) ensures that, for each $u \in L^p(\Omega)$, we have

$$\xi \in \partial J(u) \Longrightarrow \xi \in L^q(\Omega) \text{ with } \xi(x) \in \partial j(u(x)) \text{ for a.e. } x \in \Omega.$$

Consider now the multivalued operator $\partial(J|_{V_0}) : V_0 \to 2^{V_0^*}$, where $J|_{V_0} : V_0 \to \mathbb{R}$ denotes the restriction of J to V_0, which can be expressed by $J|_{V_0} = J \circ i$ with $i : V_0 \to L^p(\Omega)$ being the embedding operator.

Lemma 4.16. *The multivalued operator* $\partial(J|_{V_0}) : V_0 \to 2^{V_0^*}$ *is bounded and pseudomonotone in the sense of Definition 2.120.*

Proof: By the chain rule (see Corollary 2.180), we have [note that V_0 is dense in $L^p(\Omega)$]

$$\partial(J|_{V_0})(v) = \partial(J \circ i)(v) = i^* \partial J(i(v)) \quad \text{for all } v \in V_0, \qquad (4.41)$$

where i^* denotes the adjoint operator to i. The growth condition (H2) ensures that $\partial(J|_{V_0})$ is bounded. As $J : L^p(\Omega) \to \mathbb{R}$ is locally Lipschitz, it follows that $\partial(J|_{V_0})(v)$ is nonempty for all $v \in V_0$, and from Proposition 2.171 that $\partial J(i(v))$ is a convex and weak*-compact subset of $L^q(\Omega) = (L^p(\Omega))^*$ satisfying

$$\|\zeta\|_q \leq K, \quad \forall\, \zeta \in \partial J(i(v)),$$

which implies that $\partial(J|_{V_0})(v) = i^* \partial J(i(v))$ is convex and closed in V_0^*. In view of Proposition 2.123, the lemma is proved provided $\partial(J|_{V_0}) : V_0 \to 2^{V_0^*}$ is a generalized pseudomonotone operator in the sense of Definition 2.121. To this end, let $v_n \rightharpoonup v$ in V_0 and $w_n \rightharpoonup w$ in V_0^* with $w_n \in \partial(J|_{V_0})(v_n)$. The compactness of the embedding $V_0 \subset L^p(\Omega)$ implies that we have $v_n \to v$

in $L^p(\Omega)$. Using the density of V_0 in $L^p(\Omega)$, we know that $w_n \in \partial J(v_n)$, so (w_n) is bounded in $L^q(\Omega)$. Thus, we obtain that $w_n \rightharpoonup w$ in $L^q(\Omega)$ along a subsequence. This result yields $w \in \partial J(v)$ because $u \mapsto \partial J(u)$ is weak closed from $L^p(\Omega)$ into $L^q(\Omega)$, and

$$\langle w_n, v_n \rangle_{V_0^*, V_0} = \langle w_n, v_n \rangle_{L^q(\Omega), L^p(\Omega)} \to \langle w, v \rangle_{L^q(\Omega), L^p(\Omega)} = \langle w, v \rangle_{V_0^*, V_0},$$

which proves the generalized pseudomonotonicity of the operator $\partial(J|_{V_0})$.   □

If F denotes the Nemytskij operator related with the continuous function f, then in view of (H3), $F : L^p(\Omega) \to L^q(\Omega)$ is continuous and bounded. Thus, by the compact embedding $i : V_0 \to L^p(\Omega)$, we have $F : V_0 \to V_0^*$ is bounded and completely continuous. As $-\Delta_p : V_0 \to V_0^*$ $(1 < p < \infty)$ is continuous, bounded, and strictly monotone, we obtain that $-\Delta_p + F : V_0 \to V_0^*$ is a continuous, bounded, and pseudomonotone operator. Taking Lemma 4.16 into account and applying Theorem 2.124, we get the following result.

Lemma 4.17. *The operator* $-\Delta_p + F + \partial(J|_{V_0}) : V_0 \to 2^{V_0^*}$ *is bounded and pseudomonotone.*

Let B be the Nemytskij operator introduced in (3.50), which is generated by the cutoff function $b : \Omega \times \mathbb{R} \to \mathbb{R}$ related to some ordered pair $\underline{u} \le \bar{u}$ in V as follows:

$$b(x, s) = \begin{cases} (s - \bar{u}(x))^{p-1} & \text{if } s > \bar{u}(x), \\ 0 & \text{if } \underline{u}(x) \le s \le \bar{u}(x), \\ -(\underline{u}(x) - s)^{p-1} & \text{if } s < \underline{u}(x). \end{cases} \tag{4.42}$$

We already know that $B : L^p(\Omega) \to L^q(\Omega)$ is bounded and continuous, and thus, $B : V_0 \to V_0^*$ is bounded and completely continuous (see Sect. 3.2.1).

Lemma 4.18. *The operator* $-\Delta_p + F + \lambda B + \partial(J|_{V_0}) : V_0 \to 2^{V_0^*}$ *is bounded, pseudomonotone, and coercive provided $\lambda > 0$ is sufficiently large.*

Proof: As $B : V_0 \to V_0^*$ is bounded and completely continuous, from Lemma 4.17, it immediately follows that $-\Delta_p + F + \lambda B + \partial(J|_{V_0}) : V_0 \to 2^{V_0^*}$ is bounded and pseudomonotone for any λ. Therefore, we only need to show its coercivity when λ is sufficiently large. To this end, let $w \in \partial(J|_{V_0})(v)$. By means of (H2), we get

$$|\langle w, v \rangle| \le \int_\Omega |w\, v|\, dx \le c_2 \int_\Omega (1 + |v|^{p-1})|v|\, dx \le C(\|v\|_p + \|v\|_p^p), \tag{4.43}$$

for some generic positive constant C, where $\|\cdot\|_p$ denotes the norm in $L^p(\Omega)$. From (H3), we obtain

$$|\langle F(v), v \rangle| \le \int_\Omega |f(v)\, v|\, dx \le C(\|v\|_p + \|v\|_p^p), \tag{4.44}$$

and the estimate (3.54) of the cutoff function yields

$$\langle B(v), v \rangle = \int_\Omega b(\cdot, v) \, v \, dx \geq c_4 \|v\|_p^p - c_5 \tag{4.45}$$

for some positive constants c_4 and c_5. Thus, from (4.43) to (4.45), we obtain the estimate

$$\langle -\Delta_p v + F(v) + \lambda B(v) + w, v \rangle$$
$$\geq \|\nabla v\|_p^p + (\lambda c_4 - 2C)\|v\|_p^p - 2C\|v\|_p - \lambda c_5. \tag{4.46}$$

Selecting $\lambda > 2C/c_4$, estimate (4.46) implies the coercivity.   □

Let us consider the following auxiliary problem associated with (4.39):

$$u \in V_0 : \quad -\Delta_p u + f(u) + \lambda b(\cdot, u) + \partial j(u) \ni h \ \text{ in } \ V_0^*. \tag{4.47}$$

An existence result for (4.47) is given by the following corollary.

Corollary 4.19. *If $\lambda > 0$ is sufficiently large, then problem (4.47) possesses solutions.*

Proof: According to Lemma 4.18, the operator $-\Delta_p + F + \lambda B + \partial(J|_{V_0}) : V_0 \to 2^{V_0^*}$ is bounded, pseudomonotone, and coercive. Therefore, by Theorem 2.125, $-\Delta_p + F + \lambda B + \partial(J|_{V_0})$ is surjective, i.e., range$(-\Delta_p + F + \lambda B + \partial(J|_{V_0})) = V_0^*$, which means that a $u \in V_0$ exists such that $h \in -\Delta_p u + F(u) + \lambda B(u) + \partial(J|_{V_0})(u)$; i.e., there is a $\eta \in \partial J(u)$ such that $\eta \in L^q(\Omega)$ with $\eta(x) \in \partial j(u(x))$ and

$$-\Delta_p u + F(u) + \lambda B(u) + \eta = h \ \text{ in } \ V_0^*,$$

which means that $u \in V_0$ is a solution of (4.47).   □

Lemma 4.20. *Let $(u_n) \subset V_0$ such that $u_n \rightharpoonup u$ in V_0. If $w_n \in \partial(J|_{V_0})(u_n) = \partial(J \circ i)(u_n) \subset V_0^*$, then $w_n = i^*(z_n)$ with $z_n \in \partial J(u_n) \subset L^q(\Omega)$, and a subsequence (w_k) of (w_n) exists with $w_k = i^*(z_k) \to w = i^*(z)$ in V_0^*, where $z_k \rightharpoonup z$ in $L^q(\Omega)$ with $z \in \partial J(u)$ and $w = i^*(z) \in \partial(J|_{V_0})(u)$. Moreover, the limit $z \in L^q(\Omega)$ satisfies $z(x) \in \partial j(u(x))$ for a.e. $x \in \Omega$.*

Proof: As $V_0 \subset L^p(\Omega)$ is densely embedded, the chain rule in Corollary 2.180 applies and yields $\partial(J \circ i)(u) = i^* \partial J(i(u)) = i^* \partial J(u)$ for all $u \in V_0$, where i^* denotes the adjoint operator to the embedding $i : V_0 \to L^p(\Omega)$ and $\partial J : L^p(\Omega) \to 2^{L^q(\Omega)}$ is Clarke's generalized gradient of the integral functional $J : L^p(\Omega) \to \mathbb{R}$. Therefore, $w_n \in \partial(J|_{V_0})(u_n)$ are of the form $w_n = i^*(z_n)$ with $z_n \in \partial J(u_n)$. The weak convergence of the sequence (u_n) in V_0 implies its strong convergence in $L^p(\Omega)$, which because of the local Lipschitz continuity of $J : L^p(\Omega) \to \mathbb{R}$ results in the boundedness of (z_n) in $L^q(\Omega)$, and thus, a subsequence (z_k) of (z_n) exists with $z_k \rightharpoonup z$ in $L^q(\Omega)$ and $w_k = i^*(z_k) \to w = i^*(z)$ in V_0^* because of the compactness of the adjoint operator

i^*. By Proposition 2.171, Clarke's generalized gradient $\partial J : L^p(\Omega) \to 2^{L^q(\Omega)}$ is weak* closed from $L^p(\Omega)$ into $L^q(\Omega)$, where $L^q(\Omega)$ is equipped with the weak* topology, which coincides with the weak topology from the reflexivity of $L^q(\Omega)$. Hence, it follows that $z \in \partial J(u)$ and $w = i^*z \in i^*\partial J(u) = \partial(J \circ i)(u)$. Moreover, in view of Theorem 2.181, we have $z(x) \in j(u(x))$ for a.e. $x \in \Omega$, which completes the proof. □

Remark 4.21. Because the embedding i and its adjoint operator i^* are injective, we usually identify $w = i^*(z) \in \partial(J \circ i)(u) \subset V_0^*$ with $z \in \partial J(u) \subset L^q(\Omega)$. This identification will be used in what follows. We note that the result of Lemma 4.20 has been applied already in the proof of Lemma 4.16.

4.2.2 Comparison and Compactness Results

The goal of this section is twofold. We first prove an existence, comparison, and compactness result for problem (4.39), and then we show that the set of all solutions within the ordered interval of sub- and supersolutions has extremal elements.

Theorem 4.22. *Let $\underline{u}$ and $\bar{u}$ be sub- and supersolutions of (4.39) such that $\underline{u} \leq \bar{u}$. Then problem (4.39) has solutions within the ordered interval $[\underline{u}, \bar{u}]$, and the set S of all solutions of (4.39) contained in $[\underline{u}, \bar{u}]$ is compact in V_0.*

Proof: The proof will be given in two steps.

Step 1: Existence and Comparison

Consider the following auxiliary truncated problem:

$$u \in V_0 : \quad -\Delta_p u + f(Tu) + \lambda b(\cdot, u) + \partial j(u) \ni h \quad \text{in } V_0^*, \tag{4.48}$$

where T is the usual truncation operator related to the given pair of sub- and supersolutions (see Sect. 3.1.1). Let $F \circ T$ be the composition of the Nemytskij operator F with the truncation T. As $T : L^p(\Omega) \to L^p(\Omega)$ is bounded and continuous, it follows that $F \circ T : L^p(\Omega) \to L^q(\Omega)$ is bounded and continuous as well, and thus, $F \circ T : V_0 \to V_0^*$ is bounded and completely continuous. Therefore, Corollary 4.19 can likewise be applied to problem (4.48), which proves the existence of solutions of (4.48) for $\lambda > 0$ sufficiently large. To complete the existence and comparison part of the proof, we only need to show that any solution u of (4.48) is contained in the ordered interval $[\underline{u}, \bar{u}]$, because then $Tu = u$ and $b(\cdot, u) = 0$, and thus, u must be a solution of (4.39) in $[\underline{u}, \bar{u}]$. Let us check that $u \leq \bar{u}$, where u is a solution of (4.48); i.e., $u \in V_0$ and

$$-\Delta_p u + f(Tu) + \lambda b(\cdot, u) + w = h \quad \text{in } V_0^*, \tag{4.49}$$

with $w \in L^q(\Omega)$ and $w(x) \in \partial j(u(x))$ for a.e. $x \in \Omega$. As $\bar{u}$ is a supersolution of (4.39), we have $\bar{u} \geq 0$ on $\partial\Omega$, and a function $\bar{w} \in L^q(\Omega)$ exists with $\bar{w}(x) \in \partial j(\bar{u}(x))$ for a.e. $x \in \Omega$ such that

$$-\Delta_p \bar{u} + f(\bar{u}) + \bar{w} \geq h \quad \text{in} \quad V_0^*. \tag{4.50}$$

Subtracting (4.50) from (4.49) and testing the resulted inequality with $(u - \bar{u})^+ \in V_0 \cap L_+^p(\Omega)$, we obtain

$$\langle -(\Delta_p u - \Delta_p \bar{u}) + f(Tu) - f(\bar{u}) + w - \bar{w} + \lambda b(\cdot, u), (u - \bar{u})^+ \rangle \leq 0. \tag{4.51}$$

For the terms on the left-hand side of (4.51), the following relations can easily be verified:

$$\langle -(\Delta_p u - \Delta_p \bar{u}), (u - \bar{u})^+ \rangle \geq 0, \tag{4.52}$$

$$\langle f(Tu) - f(\bar{u}), (u - \bar{u})^+ \rangle = \int_\Omega (f(Tu) - f(\bar{u})(u - \bar{u})^+ \, dx = 0, \tag{4.53}$$

$$\langle \lambda b(\cdot, u), (u - \bar{u})^+ \rangle = \lambda \int_\Omega [(u - \bar{u})^+]^p \, dx. \tag{4.54}$$

By means of (H1) and the properties of w and $\bar{w}$, we get

$$\langle w - \bar{w}, (u - \bar{u})^+ \rangle = \int_{\{u > \bar{u}\}} (w - \bar{w})(u - \bar{u}) \, dx \geq -c_1 \int_\Omega [(u - \bar{u})^+]^p \, dx. \tag{4.55}$$

Taking (4.52)–(4.55) into account from (4.51), we obtain the following estimate:

$$(\lambda - c_1) \int_\Omega [(u - \bar{u})^+]^p \, dx \leq 0. \tag{4.56}$$

If $\lambda > 0$ is chosen sufficiently large such that, in addition, $\lambda > c_1$, then from (4.56), we infer $(u - \bar{u})^+ = 0$; i.e., $u \leq \bar{u}$. The proof of $\underline{u} \leq u$ can be done in a similar way, which completes Step 1.

Step 2: Compactness

Let $\mathcal{S}$ denote the set of all solutions of (4.39) within the interval $[\underline{u}, \bar{u}]$, which is nonempty due to Step 1. Let $(u_n) \subset \mathcal{S}$; i.e., $u_n \in [\underline{u}, \bar{u}]$ and

$$-\Delta_p u_n + f(u_n) + w_n = h \quad \text{in} \quad V_0^*, \tag{4.57}$$

where $w_n \in L^q(\Omega)$ with $w_n(x) \in \partial j(u_n(x))$ for a.e. $x \in \Omega$. As (u_n) is bounded in $L^p(\Omega)$, it follows that both sequences (w_n) and $(f(u_n))$ are bounded in $L^q(\Omega)$ in view of (H2) and (H3), respectively. Thus, from (4.57), we immediately get the boundedness of (u_n) in V_0, which implies the existence of subsequences (u_k) of (u_n) and (w_k) of (w_n) satisfying

$$u_k \rightharpoonup u \quad \text{in} \quad V_0 \quad \text{and} \quad w_k \rightharpoonup w \quad \text{in} \quad L^q(\Omega). \tag{4.58}$$

The compact embedding $V_0 \subset L^p(\Omega)$ implies $u_k \to u$ in $L^p(\Omega)$, which together with (4.58) yields

$$\langle -\Delta_p u_k, u_k - u \rangle = \langle h, u_k - u \rangle - \int_{\Omega} (f(u_k) + w_k)(u_k - u)\, dx \to 0, \quad (4.59)$$

and thus,

$$\lim_{k\to\infty} \langle -\Delta_p u_k, u_k - u \rangle = 0. \tag{4.60}$$

Relation (4.60) in conjunction with the (S_+)-property of the p-Laplacian implies the strong convergence of the subsequence (u_k) in V_0; i.e., $u_k \to u$ in V_0. To complete the compactness proof, we only need to check that the limit u belongs to S. By (4.58), we have $w_k \rightharpoonup w$ in $L^q(\Omega)$, which implies $w_k \to w$ in V_0^* because of the compactness of the adjoint operator $i^* : L^q(\Omega) \to V_0^*$. Replacing n in (4.57) by k and taking into account the convergence properties of (u_k) and (w_k) as well as the continuity of $-\Delta_p : V_0 \to V_0^*$ and $F : V_0 \to V_0^*$ (F is even completely continuous), we may pass to the limit as $k \to \infty$, which yields

$$-\Delta_p u + f(u) + w = h \quad \text{in } V_0^*.$$

Obviously $u \in [\underline{u}, \bar{u}]$, and thus, the proof is complete provided w can be shown to satisfy $w(x) \in \partial j(u(x))$ for a.e. $x \in \Omega$, because then u belongs to S. In view of $w_k(x) \in \partial j(u_k(x))$, we obtain

$$\int_{\Omega} w_k(x)\varphi(x)\, dx \le \int_{\Omega} j^\circ(u_k(x); \varphi(x))\, dx, \quad \forall\, \varphi \in L^p(\Omega). \tag{4.61}$$

As (u_k) is, in particular, strongly convergent in $L^p(\Omega)$, a subsequence of (u_k) exists [again denoted by by (u_k)], which is a.e. pointwise convergent; i.e.,

$$u_k(x) \to u(x) \quad \text{for a.e. } x \in \Omega.$$

The a.e. pointwise convergence, the weak convergence of (w_k), and the upper semicontinuity of $(s, r) \mapsto j_1^\circ(s; r)$ allows us to pass to the limit in (4.61), which by applying Fatou's Lemma results in

$$\int_{\Omega} w(x)\varphi(x)\, dx \le \int_{\Omega} j^\circ(u(x); \varphi(x))\, dx, \quad \forall\, \varphi \in L^p(\Omega). \tag{4.62}$$

From the last inequality, we are going to deduce $w(x) \in \partial j(u(x))$. Note that (4.62), in particular, holds for all $\varphi \in L^p_+(\Omega)$, which because $r \mapsto j^\circ(s; r)$ is positively homogeneous yields

$$\int_{\Omega} w(x)\varphi(x)\, dx \le \int_{\Omega} j^\circ(u(x); 1)\, \varphi(x)\, dx, \quad \forall\, \varphi \in L^p_+(\Omega). \tag{4.63}$$

Clarke's generalized directional derivative j° satisfies (see Proposition 2.171)

$$j^\circ(s; r) = \max\{\zeta r : \zeta \in \partial j(s)\}, \tag{4.64}$$

which implies the existence of a function $w^* : \Omega \to \mathbb{R}$ such that

$$j^o(u(x); 1) = w^*(x) \quad \text{for a.e. } x \in \Omega, \tag{4.65}$$

where

$$w^*(x) = \max\{\zeta : \zeta \in \partial j(u(x))\}. \tag{4.66}$$

(Note: $\partial j(s)$ is a nonempty, convex, and compact subset of $\mathbb{R}$.) We next verify that $w^* \in L^q(\Omega)$. The function $s \mapsto j^o(s; 1)$ is upper semicontinuous, and due to (H2), it satisfies the growth condition

$$|j^o(s; 1)| \le c_2(1 + |s|^{p-1}). \tag{4.67}$$

By applying general approximation results for lower (respectively, upper) semicontinuous functions in Hilbert spaces (see [8]), a sequence of locally Lipschitz functions exists converging pointwise to j^o. It implies that $s \mapsto j^o(s; 1)$ is superpositionally measurable, which means that the function $x \mapsto j^o(u(x); 1)$ is measurable whenever $u : \Omega \to \mathbb{R}$ is measurable. Thus, in view of (4.65) and (4.67), we infer that $w^* \in L^q(\Omega)$, and (4.63) yields

$$\int_\Omega w(x)\varphi(x)\,dx \le \int_\Omega w^*(x)\,\varphi(x)\,dx, \quad \forall\, \varphi \in L^p_+(\Omega). \tag{4.68}$$

From (4.68), it follows that

$$w(x) \le w^*(x) \quad \text{for a.e. } x \in \Omega. \tag{4.69}$$

Testing (4.62) with nonpositive functions $\varphi = -\psi$ where $\psi \in L^p_+(\Omega)$, we get

$$-\int_\Omega w(x)\psi(x)\,dx \le \int_\Omega j^o(u(x); -1)\,\psi(x)\,dx, \quad \forall\, \psi \in L^p_+(\Omega). \tag{4.70}$$

Similar arguments as before apply to ensure the existence of a function $\varrho \in L^q(\Omega)$ such that

$$\varrho(x) = \max\{-\zeta : \zeta \in \partial j(u(x))\} = -\min\{\zeta : \zeta \in \partial j(u(x))\}. \tag{4.71}$$

Setting $w_* = -\varrho$, we obtain from (4.70)

$$-\int_\Omega w(x)\psi(x)\,dx \le -\int_\Omega w_*(x)\,\psi(x)\,dx, \quad \forall\, \psi \in L^p_+(\Omega),$$

and thus,

$$\int_\Omega w(x)\psi(x)\,dx \ge \int_\Omega w_*(x)\,\psi(x)\,dx, \quad \forall\, \psi \in L^p_+(\Omega), \tag{4.72}$$

which implies that

$$w(x) \geq w_*(x) \quad \text{for a.e. } x \in \Omega. \tag{4.73}$$

From (4.69) and (4.73), it follows that

$$w_*(x) \leq w(x) \leq w^*(x) \quad \text{for a.e. } x \in \Omega,$$

which in view of (4.66), (4.71), and $w_* = -\varrho$ results in $w(x) \in \partial j(u(x))$ as asserted. $\square$

Next we are going to show the existence of extremal solutions in $\mathcal{S}$. A crucial step toward this goal is to prove that $\mathcal{S}$ enjoys the property of directedness. This property, however, is an immediate consequence of the following lemma.

Lemma 4.23. *If $u_1, u_2 \in \mathcal{S}$, then $\max\{u_1, u_2\}$ is a subsolution and $\min\{u_1, u_2\}$ is a supersolution of the inclusion problem (4.39).*

Proof: We first prove that $u = \max\{u_1, u_2\}$ is a subsolution of (4.39) where $u_k, k = 1, 2$, are solutions of (4.39) within $[\underline{u}, \bar{u}]$; i.e.,

$$u_k \in V_0 : \quad \langle -\Delta_p u_k + f(u_k) + w_k, \varphi \rangle = \langle h, \varphi \rangle \quad \text{for all } \varphi \in V_0, \tag{4.74}$$

with $w_k \in L^q(\Omega)$ and $w_k(x) \in \partial j(u_k(x))$ for a.e. $x \in \Omega$. To this end, we introduce the function w by

$$w(x) = \begin{cases} w_1(x) & \text{if } x \in \{u_1 \geq u_2\}, \\ w_2(x) & \text{if } x \in \{u_2 > u_1\}. \end{cases}$$

Obviously $w \in L^q(\Omega)$ and $w(x) \in \partial j(u(x))$ for a.e. $x \in \Omega$. Note that $u \in V_0$, and thus, u is a subsolution provided that u satisfies the following inequality:

$$\langle -\Delta_p u + f(u) + w, \varphi \rangle \leq \langle h, \varphi \rangle \quad \text{for all } \varphi \in V_0 \cap L^p_+(\Omega). \tag{4.75}$$

To verify (4.75), we employ a special test function technique that has been developed in Sect. 3.1.2. For $k = 1$, we use in (4.74) the test function

$$\varphi = \psi \left(1 - \theta_\varepsilon(u_2 - u_1) \right) \in V_0 \cap L^p_+(\Omega),$$

and for $k = 2$, we use the test function

$$\varphi = \psi \, \theta_\varepsilon(u_2 - u_1) \in V_0 \cap L^p_+(\Omega),$$

where $\psi \in \mathcal{D}_+$ with

$$\mathcal{D}_+ = \{\psi \in C_0^\infty(\Omega) : \psi \geq 0 \text{ in } \Omega\},$$

and for any $\varepsilon > 0$, the nondecreasing, piecewise differentiable function $\theta_\varepsilon : \mathbb{R} \to \mathbb{R}$ is given by

$$\theta_\varepsilon(s) = \begin{cases} 0 & \text{if } s \leq 0, \\ \frac{1}{\varepsilon} s & \text{if } 0 < s < \varepsilon, \\ 1 & \text{if } s \geq \varepsilon. \end{cases}$$

Applying the special test functions given above to (4.74) and adding the resulted equations, we obtain

$$\int_\Omega |\nabla u_1|^{p-2} \nabla u_1 \nabla \psi \, dx$$

$$+ \int_\Omega (|\nabla u_2|^{p-2} \nabla u_2 - |\nabla u_1|^{p-2} \nabla u_1) \nabla(\psi \, \theta_\varepsilon(u_2 - u_1)) \, dx$$

$$+ \int_\Omega \Big(f(u_1)\psi + (f(u_2) - f(u_1)) \, \psi \, \theta_\varepsilon(u_2 - u_1) \Big) dx$$

$$+ \int_\Omega \Big(w_1 \psi + (w_2 - w_1) \, \psi \, \theta_\varepsilon(u_2 - u_1) \Big) dx$$

$$= \langle h, \psi \rangle. \tag{4.76}$$

By using

$$\nabla(\psi \, \theta_\varepsilon(u_2 - u_1)) = \nabla \psi \, \theta_\varepsilon(u_2 - u_1) + \psi \, \theta'_\varepsilon(u_2 - u_1)\nabla(u_2 - u_1),$$

we can estimate the second integral on the left-hand side of (4.76) as follows:

$$\int_\Omega (|\nabla u_2|^{p-2} \nabla u_2 - |\nabla u_1|^{p-2} \nabla u_1) \nabla(\psi \, \theta_\varepsilon(u_2 - u_1)) \, dx$$

$$\geq \int_\Omega (|\nabla u_2|^{p-2} \nabla u_2 - |\nabla u_1|^{p-2} \nabla u_1) \nabla \psi \, \theta_\varepsilon(u_2 - u_1) \, dx. \tag{4.77}$$

With (4.77), we get from (4.76) the inequality

$$\int_\Omega |\nabla u_1|^{p-2} \nabla u_1 \nabla \psi \, dx$$

$$+ \int_\Omega (|\nabla u_2|^{p-2} \nabla u_2 - |\nabla u_1|^{p-2} \nabla u_1) \nabla \psi \, \theta_\varepsilon(u_2 - u_1) \, dx$$

$$+ \int_\Omega \Big(f(u_1)\psi + (f(u_2) - f(u_1)) \, \psi \, \theta_\varepsilon(u_2 - u_1) \Big) dx$$

$$+ \int_\Omega \Big(w_1 \psi + (w_2 - w_1) \, \psi \, \theta_\varepsilon(u_2 - u_1) \Big) dx$$

$$\leq \langle h, \psi \rangle, \tag{4.78}$$

which by passing to the limit as $\varepsilon \to 0$ yields

$$\int_\Omega |\nabla u_1|^{p-2} \nabla u_1 \nabla \psi \, dx$$

$$+ \int_\Omega (|\nabla u_2|^{p-2} \nabla u_2 - |\nabla u_1|^{p-2} \nabla u_1) \chi_{\{u_2 - u_1 > 0\}} \nabla \psi \, dx$$

$$+ \int_\Omega \Big(f(u_1) + (f(u_2) - f(u_1)) \chi_{\{u_2 - u_1 > 0\}} \Big) \psi \, dx$$

$$+ \int_\Omega \Big(w_1 + (w_2 - w_1) \chi_{\{u_2 - u_1 > 0\}} \Big) \psi \, dx$$

$$\leq \langle h, \psi \rangle, \tag{4.79}$$

where

$$\theta_\varepsilon (u_2 - u_1) \to \chi_{\{u_2 - u_1 > 0\}} \quad \text{a.e. in } \Omega \text{ as } \quad \varepsilon \to 0,$$

and $\chi_{\{u_2 - u_1 > 0\}}$ is the characteristic function of the set $\{u_2 - u_1 > 0\} = \{x \in \Omega : u_2(x) - u_1(x) > 0\}$. Inequality (4.79) is, however, nothing else but inequality (4.75), which proves that $u = \max\{u_1, u_2\}$ is a subsolution. By analogous reasoning, one shows that $\min\{u_1, u_2\}$ is a supersolution that completes the proof. $\qquad\square$

As an immediate consequence of Lemma 4.23 and Theorem 4.22, we obtain the following corollary.

Corollary 4.24. *The solution set S of (4.39) is directed and possesses extremal elements.*

Proof: Let $u_1, u_2 \in S$. From Lemma 4.23, $\max\{u_1, u_2\} \in [\underline{u}, \bar{u}]$ is a subsolution, and thus, by Theorem 4.22, a solution of (4.39) exists within the interval $[\max\{u_1, u_2\}, \bar{u}]$, which shows that S is upward directed. In a similar way, one can prove that S is also downward directed. The existence proof of the greatest and smallest elements of S follows the same arguments as used in the proof of Theorem 3.11 and is based on the directedness and compactness of S. $\qquad\square$

Remark 4.25. Under the assumptions (H1) and (H2) on Clarke's generalized gradient, the results of this section can be extended to general quasilinear elliptic inclusions in the form

$$-\sum_{i=1}^N \frac{\partial}{\partial x_i} a_i(\cdot, u, \nabla u) + f(\cdot, u, \nabla u) + \partial j(\cdot, u) \ni h, \tag{4.80}$$

where $a_i : \Omega \times \mathbb{R} \times \mathbb{R}^N \to \mathbb{R}$ satisfy the usual Leray–Lions conditions and condition (H3) above is replaced by

$$|f(x, s, \xi)| \leq k(x) + c_3(|s|^{p-1} + |\xi|^{p-1})$$

with $k \in L^q(\Omega)$. The approach to treat boundary value problems for (4.80) is closely related to methods that will be developed in Chap. 6 for quasilinear hemivariational inequalities. Therefore, we refer to Chap. 6 for the general problem (4.80). Instead in the next section we are going to deal with the Dirichlet problem of quasilinear elliptic inclusions in the form (4.80) under local growth conditions on the governing nonlinearities and Clarke's generalized gradient whose treatment requires different tools.

4.3 Inclusions with Local Growth on Clarke's Gradient

Let $\Omega \subset \mathbb{R}^N$ be a bounded domain with Lipschitz boundary $\partial \Omega$. In this section, we consider the Dirichlet problem for the following general elliptic inclusion:

$$Au + Fu + \partial j(\cdot, u) \ni h \quad \text{in} \quad \Omega, \quad u = 0 \quad \text{on} \quad \partial \Omega, \qquad (4.81)$$

where A is a second-order quasilinear differential operator in divergence form of Leray–Lions type given by

$$Au(x) = -\sum_{i=1}^{N} \frac{\partial}{\partial x_i} a_i(x, u(x), \nabla u(x)),$$

and F is the Nemytskij operator of the lower order terms generated by a function $f : \Omega \times \mathbb{R} \times \mathbb{R}^N \to \mathbb{R}$ and defined by

$$Fu(x) = f(x, u(x), \nabla u(x)).$$

The function $j : \Omega \times \mathbb{R} \to \mathbb{R}$ is assumed to be the primitive vanishing at 0 of some locally bounded and Borel measurable function $g : \Omega \times \mathbb{R} \to \mathbb{R}$; i.e.,

$$j(x, s) = \int_0^s g(x, \tau) \, d\tau. \qquad (4.82)$$

Thus, $j(x, \cdot) : \mathbb{R} \to \mathbb{R}$ is locally Lipschitz and Clarke's generalized gradient $\partial j(x, \cdot) : \mathbb{R} \to 2^{\mathbb{R}} \setminus \{\emptyset\}$ of j with respect to its second argument exists, which is given by

$$\partial j(x, s) := \{\zeta \in \mathbb{R} : j^o(x, s; r) \geq \zeta r, \quad \forall \, r \in \mathbb{R}\}, \qquad (4.83)$$

where $j^o(x, s; r)$ denotes the generalized directional derivative of j at s in the direction r. As in the previous section, we set $V = W^{1,p}(\Omega)$ and $V_0 = W_0^{1,p}(\Omega)$, $1 < p < \infty$, and denote by V^* and V_0^* their corresponding dual spaces, respectively.

The main goal of this section is to prove comparison, extremality, and compactness results of the inclusion problem (4.81). Unlike in the previous section, only a local growth condition on Clarke's generalized gradient is imposed, which makes the treatment of problem (4.81) difficult, because now the integral functional J generated by $j(x, \cdot)$ is no longer locally Lipschitz continuous. To overcome this difficulty, the approach here is based on a regularization technique combined with appropriate truncation and special test function techniques.

4.3.1 Comparison Principle

We impose the following hypotheses of Leray–Lions type on the coefficient functions a_i, $i = 1, \ldots, N$, of the operator A.

(A1) Each $a_i : \Omega \times \mathbb{R} \times \mathbb{R}^N \to \mathbb{R}$ satisfies the Carathéodory conditions; i.e., $a_i(x, s, \xi)$ is measurable in $x \in \Omega$ for all $(s, \xi) \in \mathbb{R} \times \mathbb{R}^N$ and continuous in (s, ξ) for almost all $x \in \Omega$. A constant $c_0 > 0$ and a function $k_0 \in L^q(\Omega), 1/p + 1/q = 1$ exist, such that

$$|a_i(x, s, \xi)| \le k_0(x) + c_0(|s|^{p-1} + |\xi|^{p-1}),$$

for a.e. $x \in \Omega$ and for all $(s, \xi) \in \mathbb{R} \times \mathbb{R}^N$.

(A2) $\sum_{i=1}^{N}(a_i(x, s, \xi) - a_i(x, s, \xi'))(\xi_i - \xi'_i) > 0$ for a.e. $x \in \Omega$, for all $s \in \mathbb{R}$, and for all $\xi, \xi' \in \mathbb{R}^N$ with $\xi \ne \xi'$.

(A3) $\sum_{i=1}^{N} a_i(x, s, \xi)\xi_i \ge \nu|\xi|^p - k_1(x)$ for a.e. $x \in \Omega$, for all $s \in \mathbb{R}$, and for all $\xi \in \mathbb{R}^N$ with some constant $\nu > 0$ and some function $k_1 \in L^1(\Omega)$.

(A4) $|a_i(x, s, \xi) - a_i(x, s', \xi)| \le [k_2(x) + |s|^{p-1} + |s'|^{p-1} + |\xi|^{p-1}]\omega(|s - s'|)$, for some function $k_2 \in L^q(\Omega)$, for a.e. $x \in \Omega$, for all $s, s' \in \mathbb{R}$ and for all $\xi \in \mathbb{R}^N$, where $\omega : [0, \infty) \to [0, \infty)$ is a continuous function satisfying

$$\int_{0+} \frac{dr}{\omega(r)} = +\infty. \tag{4.84}$$

Remark 4.26. Hypothesis (A4) is a condition on the modulus of continuity and includes for example $\omega(r) = c\,r$, for all $r \ge 0$, with $c > 0$, i.e., a Lipschitz condition of the coefficients $a_i(x, s, \xi)$ with respect to s.

As a consequence of (A1), the semilinear form a associated with the operator A by

$$\langle Au, \varphi \rangle := a(u, \varphi) = \sum_{i=1}^{N} \int_{\Omega} a_i(x, u, \nabla u)\frac{\partial \varphi}{\partial x_i}\,dx, \quad \text{for all } \varphi \in V_0$$

is well defined for any $u \in V$, and the operator $A : V \to V_0^*$ is continuous and bounded. The notion of weak solution as well as of weak sub- and supersolutions of problem (4.81) reads as follows.

Definition 4.27. *A function $u \in V_0$ is a solution of the BVP (4.81) if $Fu \in L^q(\Omega)$ and if there is a function $v \in L^q(\Omega)$ such that*

(i) $v(x) \in \partial j(x, u(x))$ *for a.e. $x \in \Omega$.*

(ii) $\langle Au, \varphi \rangle + \int_{\Omega}(Fu(x) + v(x))\,\varphi(x)\,dx = \langle h, \varphi \rangle, \quad \forall\, \varphi \in V_0$.

Definition 4.28. *A function $\underline{u} \in V$ is called a subsolution of (4.81) if $F\underline{u} \in L^q(\Omega)$ and if there is a function $\underline{v} \in L^q(\Omega)$ such that*

(i) $\underline{u} \le 0$ *on $\partial\Omega$.*

(ii) $\underline{v}(x) \in \partial j(x, \underline{u}(x))$ *for a.e. $x \in \Omega$.*

(iii) $\langle A\underline{u}, \varphi \rangle + \int_{\Omega}(F\underline{u}(x) + \underline{v}(x))\,\varphi(x)\,dx \le \langle h, \varphi \rangle, \quad \forall\, \varphi \in V_0 \cap L_+^p(\Omega)$.

Similarly, a function $\bar{u} \in V$ is a supersolution of (4.81) if the reversed inequalities hold in Definition 4.28 with $\underline{u}$ and $\underline{v}$ replaced by $\bar{u}$ and $\bar{v}$, respectively.

Remark 4.29. One possibility to determine sub- and supersolutions of (4.81) is to replace the multivalued problem by the following single-valued one:

$$Au + Fu + \hat{g}(\cdot, u) = h, \quad \text{in } \Omega, \quad u = 0, \quad \text{on } \partial\Omega, \qquad (4.85)$$

where $\hat{g} : \Omega \times \mathbb{R} \to \mathbb{R}$ may be any single-valued selection of $\partial j(\cdot, u)$ such as the functions $\underline{g}$ and $\bar{g}$ given below. Then obviously any subsolution (supersolution) of (4.85) is also a subsolution (supersolution) of the inclusion (4.81) according to Definition 4.28.

Let $\underline{u}$ and $\bar{u}$ be sub- and supersolutions of (4.81) such that $\underline{u} \leq \bar{u}$. We assume the following local growth and structure conditions on g and f.

(H1) The function $g : \Omega \times \mathbb{R} \to \mathbb{R}$ satisfies:
    (i) g is Borel-measurable in $\Omega \times \mathbb{R}$, and $g(x, \cdot) : \mathbb{R} \to \mathbb{R}$ is locally bounded.
    (ii)   Constants $\alpha > 0$ and $c_1 \geq 0$ exist such that

$$g(x, s_1) \leq g(x, s_2) + c_1 (s_2 - s_1)^{p-1},$$

        for a.e. $x \in \Omega$, and for all s_1, s_2 with $\underline{u}(x) - \alpha \leq s_1 < s_2 \leq \bar{u}(x) + \alpha$.
    (iii)   There is a function $k_3 \in L_+^q(\Omega)$ such that

$$|g(x, s)| \leq k_3(x),$$

        for a.e. $x \in \Omega$, and for all $s \in [\underline{u}(x) - 2\alpha, \bar{u}(x) + 2\alpha]$, where α is as in (ii).
(H2) $f : \Omega \times \mathbb{R} \times \mathbb{R}^N \to \mathbb{R}$ is a Carathéodory function, and a function $k_4 \in L_+^q(\Omega)$ exists such that for some constant $c_2 \geq 0$, the following estimate holds:

$$|f(x, s, \xi)| \leq k_4(x, t) + c_2 |\xi|^{p-1},$$

for a.e. $x \in \Omega$, for all $\xi \in \mathbb{R}^N$, and for all $s \in [\underline{u}(x), \bar{u}(x)]$.

Remark 4.30. First we note that the global growth conditions in the preceding section imply the corresponding local growth conditions given here. Furthermore, from (H1)(i), Clarke's generalized gradient $\partial j(x, s)$ can be represented in the form

$$\partial j(x, s) = [\underline{g}(x, s), \bar{g}(x, s)], \quad \text{for all } s \in \mathbb{R}, \text{ and a.e. } x \in \Omega, \qquad (4.86)$$

where

$$\underline{g}(x, s) = \lim_{\delta \downarrow 0} \underline{g}_\delta(x, s), \quad \bar{g}(x, s) = \lim_{\delta \downarrow 0} \bar{g}_\delta(x, s),$$

with

$$\underline{g}_\delta(x, s) = \operatorname*{ess\,inf}_{|t-s|<\delta} g(x, t), \quad \bar{g}_\delta(x, s) = \operatorname*{ess\,sup}_{|t-s|<\delta} g(x, t).$$

Hypothesis (H1)(ii) implies that Clarke's generalized gradient $\partial j(x, s)$ fulfills the condition: For $\eta_i \in \partial j(x, s_i)$, $i = 1, 2$, we have

$$\eta_1 \le \eta_2 + c_1 \, (s_2 - s_1)^{p-1}, \tag{4.87}$$

for a.e. $x \in \Omega$, and for all s_1, s_2 with $\underline{u}(x) - \alpha \le s_1 < s_2 \le \bar{u}(x) + \alpha$. Condition (4.87) will play a crucial role in the proof of the comparison and extremality results.

In preparation for the comparison principle to be proved in this section, we introduce truncation mappings T and T_α related with the given sub- and supersolutions $\underline{u}$ and $\bar{u}$, respectively, as follows:

$$Tu(x) = \begin{cases} \bar{u}(x) & \text{if } u(x) > \bar{u}(x), \\ u(x) & \text{if } \underline{u}(x) \le u(x) \le \bar{u}(x), \\ \underline{u}(x) & \text{if } u(x,t) < \underline{u}(x), \end{cases}$$

and with α given in (H1)(ii), we define the truncation operator T_α by

$$T_\alpha u(x) = \begin{cases} \bar{u}(x) + \alpha & \text{if } u(x) > \bar{u}(x) + \alpha, \\ u(x) & \text{if } \underline{u}(x) - \alpha \le u(x) \le \bar{u}(x) + \alpha, \\ \underline{u}(x) - \alpha & \text{if } u(x) < \underline{u}(x) - \alpha. \end{cases}$$

It is known that these truncation operators T, and T_α are continuous and bounded from V into V (see Chap. 2). Furthermore, let $\rho : \mathbb{R} \to \mathbb{R}$ be a mollifier function; that is, $\rho \in C_0^\infty((-1,1))$, $\rho \ge 0$ and

$$\int_{-\infty}^{+\infty} \rho(s) \, ds = 1.$$

For any $\varepsilon > 0$, we define the regularization g^ε of g with respect to the second variable by the convolution; i.e.,

$$g^\varepsilon(x,s) = \frac{1}{\varepsilon} \int_{-\infty}^{+\infty} g(x, s - \varsigma) \rho\left(\frac{\varsigma}{\varepsilon}\right) d\varsigma. \tag{4.88}$$

From hypothesis (H1)(iii), it readily follows that for any $u \in [\underline{u}, \bar{u}]$ and for any $\varepsilon : 0 < \varepsilon < 2\alpha$, we get the estimate

$$|g^\varepsilon(x,t,u(x,t))| \le k_3(x). \tag{4.89}$$

Let G^ε denote the Nemytskij operator associated with g^ε. Then by means of the truncation operator T_α, we define the regularized truncated Nemytskij operator

$$G_\alpha^\varepsilon(u)(x) = (G^\varepsilon \circ T_\alpha)(u)(x) = g^\varepsilon(x, T_\alpha u(x)). \tag{4.90}$$

Finally, we use again the cutoff function $b : \Omega \times \mathbb{R} \to \mathbb{R}$ given by

$$b(x,s) = \begin{cases} (s - \bar{u}(x))^{p-1} & \text{if } s > \bar{u}(x), \\ 0 & \text{if } \underline{u}(x) \le s \le \bar{u}(x), \\ -(\underline{u}(x) - s)^{p-1} & \text{if } s < \underline{u}(x), \end{cases}$$

which is a Carathéodory function satisfying the growth condition

$$|b(x,s)| \le k_5(x) + c_3|s|^{p-1} \tag{4.91}$$

for a.e. $x \in \Omega$ and for all $s \in \mathbb{R}$, where $c_3 > 0$ and $k_5 \in L^q(\Omega)$. Moreover, we have the following estimate:

$$\int_\Omega b(x,u(x))\, u(x)\, dx \ge c_4\|u\|_p^p - c_5, \quad \forall\, u \in L^p(\Omega), \tag{4.92}$$

for some constants $c_4 > 0$ and $c_5 > 0$. Thus, by (4.91), the Nemytskij operator $B : L^p(\Omega) \to L^q(\Omega)$ defined by

$$Bu(x) = b(x,u(x))$$

is continuous and bounded.

The main result of this subsection is the following comparison principle.

Theorem 4.31. *Let hypotheses (A1)–(A3) and (H1), (H2) be satisfied. Then the inclusion problem (4.81) possesses at least one solution u within the ordered interval $[\underline{u}, \bar{u}]$ formed by the given sub- and supersolution $\underline{u}$ and $\bar{u}$, respectively.*

Proof: The proof will be given in four steps.

Step 1: Existence result for an auxiliary problem.

We introduce the following regularized truncated BVP:

$$(P_\varepsilon) \quad A_T u + (F \circ T)(u) + G_\alpha^\varepsilon(u) + \lambda\, B(u) = h, \quad \text{in } \Omega, \quad u = 0, \quad \text{on } \partial\Omega,$$

where $F \circ T$ is the composition of F and T, G_α^ε is given by (4.90), and $\lambda > 0$ is some constant to be specified later. The operator A_T is defined by

$$A_T u(x) = -\sum_{i=1}^N \frac{\partial}{\partial x_i} a_i(x, Tu(x), \nabla u(x)). \tag{4.93}$$

By hypotheses (A1)–(A3) and the continuity of the truncation operator $T : V \to V$, it follows that the operator $A_T : V_0 \to V_0^*$ is continuous, bounded, and pseudomonotone, and, moreover, the following estimate holds:

$$\langle A_T u, u \rangle = \int_\Omega \sum_{i=1}^N a_i(x, Tu, \nabla u)\, \frac{\partial u}{\partial x_i}\, dx \ge \nu\, \|\nabla u\|_p^p - \|k_1\|_1 \tag{4.94}$$

for all $u \in V_0$. The truncation operator $T : V_0 \to [\underline{u}, \bar{u}] \cap V_0$ is bounded and continuous, and from (H2), the Nemytskij operator F generated by f is bounded and continuous from $[\underline{u}, \bar{u}] \cap V_0$ into $L^q(\Omega) \subset V_0^*$. Thus, the composed operator $F \circ T : V_0 \to V_0^*$ given by

$$\langle (F \circ T)(u), \varphi \rangle = \int_{\Omega} f(\cdot, Tu(\cdot), \nabla Tu(\cdot)) \, \varphi \, dx \quad \text{for all} \quad \varphi \in V_0,$$

is bounded and continuous, and the following estimate holds:

$$|\langle (F \circ T)(u), u \rangle| \leq \left(\|k_4\|_q + c_2 \|\nabla u\|_p^{p-1} \right) \|u\|_p \tag{4.95}$$

for all $u \in V_0$. Hypotheses (A1), (A2), and (H2) imply that the sum $A_T + F \circ T : V_0 \to V_0^*$ is a pseudomonotone operator (cf. Theorem 2.109). The operators G_{α}^{ε}, $B : L^p(\Omega) \to L^q(\Omega) \subset V_0^*$ are continuous and bounded, and thus, G_{α}^{ε}, $B : V_0 \to V_0^*$ are completely continuous from the compact embedding $V_0 \subset L^p(\Omega)$. Hence, it follows that the operator

$$\mathcal{A} = A_T + F \circ T + G_{\alpha}^{\varepsilon} + \lambda B : V_0 \to V_0^*$$

is continuous, bounded, and pseudomonotone, and thus, by the main theorem on pseudomonotone operators (see Theorem 2.99), $\mathcal{A} : V_0 \to V_0^*$ is surjective provided that $\mathcal{A}$ is, in addition, coercive. By (H1)(iii), it follows that $G_{\alpha}^{\varepsilon} : L^p(\Omega) \to L^q(\Omega)$ is uniformly bounded, and we have

$$|\langle G_{\alpha}^{\varepsilon}(u), u \rangle| = \int_{\Omega} |G_{\alpha}^{\varepsilon}(u) \, u \, dx| \leq \|k_3\|_q \|u\|_p, \quad \text{for all} \quad u \in L^p(\Omega). \tag{4.96}$$

Young's inequality yields for any $\eta > 0$

$$\|\nabla u\|_p^{p-1} \|u\|_p \leq \eta \|\nabla u\|_p^p + C(\eta) \|u\|_p^p, \tag{4.97}$$

where $C(\eta) > 0$ is some constant depending only on η. Thus, from (4.92), (4.94), (4.95), (4.96), and (4.97) we obtain the estimate

$$\langle \mathcal{A}u, u \rangle \geq \nu \|\nabla u\|_p^p - \eta \|\nabla u\|_p^p - C(\eta) \|u\|_p^p + \lambda c_4 \|u\|_p^p$$
$$- \left(\|k_3\|_q + \|k_4\|_q \right) \|u\|_p - \lambda c_5 - \|k_1\|_1. \tag{4.98}$$

Selecting η small enough such that $\eta < \nu$, and choosing the parameter λ sufficiently large such that the inequality $\lambda c_4 - C(\eta) > 0$ holds, we see from (4.98) that $\mathcal{A}$ is coercive, which completes the existence proof of problem (P_{ε}).

Step 2: Convergence of some subsequence of solutions of (P_{ε}).

Let (ε_n) be a sequence such that $\varepsilon_n \in (0, \alpha)$ and $\varepsilon_n \to 0$ as $n \to \infty$. From the previous step, we know that for all n, problem (P_{ε_n}) has at least one solution, which will be denoted by u_n. The coercivity of the operator $\mathcal{A}$ implies that the sequence (u_n) of solutions of (P_{ε_n}) is bounded in V_0. We are going to prove the following convergence properties of some subsequence of (u_n), which is again denoted by (u_n):

(i) $u_n \rightharpoonup u$ (weakly) in V_0 as $n \to \infty$.
(ii) $u_n \to u$ (strongly) in $L^p(\Omega)$ as $n \to \infty$.

(iii) $G_\alpha^{\varepsilon_n}(u_n) \rightharpoonup v$ (weakly) in $L^q(\Omega)$ as $n \to \infty$, where $v(x) \in \partial j(x, (T_\alpha u)(x))$
for a.e. $x \in \Omega$.

Properties (i) and (ii) readily follow from the fact that V_0 is reflexive and
$V_0 \subset L^p(\Omega)$ is compactly embedded. The first part of (iii) is a consequence
of the boundedness of $(G_\alpha^{\varepsilon_n}(u_n))$ in $L^q(\Omega)$, which implies the existence of a
weakly convergent subsequence in $L^q(\Omega)$ with weak limit v. To complete the
proof for (iii), we need to show that $v(x) \in \partial j(x, (T_\alpha u)(x))$ for a.e $x \in \Omega$,
which in view of (4.86) is equivalent to

$$v(x) \in [\underline{g}(x, (T_\alpha u)(x)), \bar{g}(x, (T_\alpha u)(x))]. \tag{4.99}$$

The strong convergence of $u_n \to u$ in $L^p(\Omega)$ and the continuity of the truncation operator T_α imply the strong convergence of $T_\alpha u_n \to T_\alpha u$ in $L^p(\Omega)$. By
the definition of the regularization g^ε according to (4.88), we get the inequality

$$\underset{|s-(T_\alpha u_n)(x)|<\varepsilon_n}{\mathrm{ess\,inf}} g(x,s) \le g^{\varepsilon_n}(x, (T_\alpha u_n)(x)) \le \underset{|s-(T_\alpha u_n)(x)|<\varepsilon_n}{\mathrm{ess\,sup}} g(x,s). \tag{4.100}$$

The strong convergence of $(T_\alpha u_n)$ in $L^p(\Omega)$ implies the almost everywhere
convergence of some subsequence [again denoted by $(T_\alpha u_n)$]. By applying
Egoroff's theorem (see Theorem 2.66), for any $\delta > 0$, there is a measurable
subset $E \subset \Omega$ with Lebesgue measure $|E| < \delta$ such that

$$T_\alpha u_n \to T_\alpha u \quad \text{(uniformly) in } \Omega \setminus E.$$

Let $\varrho \in (0, \alpha)$ arbitrarily be given. Then from the uniform convergence of
$T_\alpha u_n$ in $\Omega \setminus E$ and because of $\varepsilon_n \to 0$, there is a $N(\varrho)$ such that for all
$n > N(\varrho)$ we have

$$0 < \varepsilon_n < \varrho/2 \quad \text{and} \quad |(T_\alpha u_n)(x) - (T_\alpha u)(x)| < \varrho/2, \quad \forall x \in \Omega \setminus E. \tag{4.101}$$

Thus, from (4.100) and (4.101), we get for $x \in \Omega \setminus E$ and $n > N(\varrho)$

$$g^{\varepsilon_n}(x, (T_\alpha u_n)(x)) \le \underset{|s-(T_\alpha u_n)(x)|<\varrho/2}{\mathrm{ess\,sup}} g(x,s) \le \underset{|s-(T_\alpha u)(x)|<\varrho}{\mathrm{ess\,sup}} g(x,s), \tag{4.102}$$

and similarly

$$g^{\varepsilon_n}(x, (T_\alpha u_n)(x)) \ge \underset{|s-(T_\alpha u_n)(x)|<\varrho/2}{\mathrm{ess\,inf}} g(x,s) \ge \underset{|s-(T_\alpha u)(x)|<\varrho}{\mathrm{ess\,inf}} g(x,s). \tag{4.103}$$

From (4.102) and (4.103), we obtain for all $n > N(\varrho)$

$$\underline{g}_\varrho(x, (T_\alpha u)(x)) \le g^{\varepsilon_n}(x, (T_\alpha u_n)(x)) \le \bar{g}_\varrho(x, (T_\alpha u)(x)) \tag{4.104}$$

for all $x \in \Omega \setminus E$. Let $\chi_{\{\Omega \setminus E\}}$ denote the characteristic function of the set $\Omega \setminus E$,
and let $\varphi \in L_+^p(\Omega)$. As $g^{\varepsilon_n}(\cdot, (T_\alpha u_n)(\cdot)) = G_\alpha^{\varepsilon_n}(u_n) \rightharpoonup v$ weakly in $L^q(\Omega)$, we

obtain from (4.104) by multiplying the inequality with $\varphi \chi_{\{\Omega \setminus E\}} \in L_+^p(\Omega)$, integrating over Ω, and passing to the limit as $n \to \infty$ the following inequality:

$$\int_\Omega \underline{g}_\varrho(x, (T_\alpha u)(x)) \, \varphi(x) \, \chi_{\{\Omega \setminus E\}}(x) \, dx \le \int_\Omega v(x) \, \varphi(x) \, \chi_{\{\Omega \setminus E\}}(x) \, dx$$

$$\le \int_\Omega \bar{g}_\varrho(x, (T_\alpha u)(x)) \, \varphi(x) \, \chi_{\{\Omega \setminus E\}}(x) \, dx,$$

which yields for any $\varphi \in L_+^p(\Omega \setminus E)$

$$\int_{\Omega \setminus E} \underline{g}_\varrho(x, (T_\alpha u)(x)) \, \varphi(x) \, dx \le \int_{\Omega \setminus E} v(x) \, \varphi(x) \, dx$$

$$\le \int_{\Omega \setminus E} \bar{g}_\varrho(x, (T_\alpha u)(x)) \, \varphi(x) \, dx. \tag{4.105}$$

Inequality (4.105) holds for any $\varrho \in (0, \alpha)$, and thus by applying Fatou's lemma (see Theorem 2.65), we get as $\varrho \to 0$

$$\int_{\Omega \setminus E} \underline{g}(x, (T_\alpha u)(x)) \, \varphi(x) \, dx \le \int_{\Omega \setminus E} v(x) \, \varphi(x) \, dx$$

$$\le \int_{\Omega \setminus E} \bar{g}(x, (T_\alpha u)(x)) \, \varphi(x) \, dx$$

for all $\varphi \in L_+^p(\Omega \setminus E)$, which yields

$$v(x) \in [\underline{g}(x, (T_\alpha u)(x)), \bar{g}(x, (T_\alpha u)(x))], \quad \text{for a.e. } x \in \Omega \setminus E. \tag{4.106}$$

As the Lebesgue measure $|E| < \delta$ and $\delta > 0$ may be arbitrarily small, the inclusion (4.106) must hold for a.e. $x \in \Omega$, which completes the proof of (iii); i.e., we have for a.e. $x \in \Omega$

$$v(x) \in [\underline{g}(x, (T_\alpha u)(x)), \bar{g}(x, (T_\alpha u)(x))] = \partial j(x, (T_\alpha u)(x)). \tag{4.107}$$

Step 3: Passage to the limit in (P_{ε_n}).

Let (u_n) be a sequence of solutions of (P_{ε_n}) for $\varepsilon_n \to 0$ with the convergence properties (i), (ii), and (iii) of the previous Step 2, i.e., u_n satisfies

$$(P_{\varepsilon_n}) \quad u_n \in V_0 : \quad \langle A_T u_n + (F \circ T) u_n + G_\alpha^{\varepsilon_n}(u_n) + \lambda B u_n, \varphi \rangle = \langle h, \varphi \rangle$$

for all $\varphi \in V_0$. Hypotheses (A1)–(A3), (H2), and the fact that the truncation operator $T : V_0 \to V_0 \cap [\underline{u}, \bar{u}]$ is continuous and bounded imply that $A_T + F \circ T : V_0 \to V_0^*$ possesses the so-called (S_+)-property (cf. Theorem 2.109), which means that

$$u_n \rightharpoonup u \text{ in } V_0 \text{ and } \limsup_{n \to \infty} \langle (A_T + F \circ T) u_n, u_n - u \rangle \le 0$$

implies $u_n \to u$ (strongly) in V_0. From (P_{ε_n}), we obtain with $\varphi = u_n - u$

$$
\begin{aligned}
&\langle (A_T + F \circ T)u_n, u_n - u \rangle \\
&= \langle G_\alpha^{\varepsilon_n}(u_n) + \lambda B u_n, u - u_n \rangle + \langle h, u_n - u \rangle \\
&= \int_\Omega \Big(G_\alpha^{\varepsilon_n}(u_n) + \lambda B u_n \Big)(u - u_n)\, dx + \langle h, u_n - u \rangle.
\end{aligned}
\tag{4.108}
$$

As $B : L^p(\Omega) \to L^q(\Omega)$ is continuous and bounded, and $G_\alpha^{\varepsilon_n}(u_n) \rightharpoonup v$ weakly in $L^q(\Omega)$, the right-hand side of (4.108) tends to zero if $u_n \rightharpoonup u$ in V_0, which from the (S_+)-property of $A_T + F \circ T$ implies the strong convergence of (u_n) in V_0. This result allows the passage to the limit as $n \to \infty$ in (P_{ε_n}), which yields the following problem (P_0) for the limit u:

(P_0) $u \in V_0 : \quad \langle A_T u + (F \circ T)u + v + \lambda B(u), \varphi \rangle = \langle h, \varphi \rangle$

for all $\varphi \in V_0$, where $v \in L^q(\Omega)$ satisfies $v(x) \in \partial j(x, (T_\alpha u)(x))$ for a.e. $x \in \Omega$.

Step 4: Comparison: $\underline{u} \leq u \leq \bar{u}$.

We complete the proof of the theorem by showing that any solution u of (P_0) satisfies $\underline{u} \leq u \leq \bar{u}$, because then we have $Tu = u$, and $T_\alpha u = u$, which shows that $A_T u = Au$, $v(x) \in \partial j(x, u(x))$ for a.e. $x \in \Omega$ and $(F \circ T)u = Fu$, and $Bu = 0$, and therefore, any solution u of (P_0) must be a solution of the original inclusion problem (4.81) satisfying $\underline{u} \leq u \leq \bar{u}$. Let us first prove $u \leq \bar{u}$. To this end, recall that $\bar{u} \in V$ is a supersolution of (4.81); i.e., $\bar{u} \geq 0$ on $\partial\Omega$ and there is a $\bar{v} \in L^q(\Omega)$ such that $\bar{v}(x) \in \partial j(x, \bar{u}(x))$ for a.e. $x \in \Omega$, and

$$
\langle A\bar{u} + F\bar{u} + \bar{v}, \varphi \rangle \geq \langle h, \varphi \rangle \quad \text{for all} \quad \varphi \in V_0 \cap L_+^p(\Omega).
\tag{4.109}
$$

Subtracting (4.109) from (P_0) and taking the special nonnegative test function $\varphi = (u - \bar{u})^+ \in V_0$, we obtain

$$
\langle A_T u - A\bar{u} + (F \circ T)u - F\bar{u} + v - \bar{v} + \lambda B u, (u - \bar{u})^+ \rangle \leq 0.
\tag{4.110}
$$

The definition of A_T given in (4.93) in conjunction with (A2) yields

$$
\begin{aligned}
&\langle A_T u - A\bar{u}, (u - \bar{u})^+ \rangle \\
&= \int_\Omega \sum_{i=1}^N \Big(a_i(x, Tu, \nabla u) - a_i(x, \bar{u}, \nabla\bar{u}) \Big) \frac{\partial(u - \bar{u})^+}{\partial x_i}\, dx \\
&= \int_{\{u > \bar{u}\}} \sum_{i=1}^N \Big(a_i(x, \bar{u}, \nabla u) - a_i(x, \bar{u}, \nabla\bar{u}) \Big) \frac{\partial(u - \bar{u})}{\partial x_i}\, dx \geq 0.
\end{aligned}
\tag{4.111}
$$

The second difference in (4.110) results in

$$
\begin{aligned}
&\langle (F \circ T)u - F\bar{u}, (u - \bar{u})^+ \rangle \\
&= \int_\Omega \Big(f(\cdot, Tu, \nabla Tu) - f(\cdot, \bar{u}, \nabla\bar{u}) \Big)(u - \bar{u})^+\, dx
\end{aligned}
$$

$$= \int_{\{u>\bar{u}\}} \Big(f(\cdot, \bar{u}, \nabla\bar{u}) - f(\cdot, \bar{u}, \nabla\bar{u})\Big)(u - \bar{u})\,dx = 0. \qquad (4.112)$$

If $x \in \{u > \bar{u}\}$, we see from the definition of T_α that $\bar{u}(x) < (T_\alpha u)(x) \le \bar{u}(x) + \alpha$, from which we infer by using hypothesis (H1)(ii)

$$\bar{v}(x) - v(x) \le c_1 \left((T_\alpha u)(x) - \bar{u}(x)\right)^{p-1}.$$

As $(T_\alpha u)(x) \le u(x)$ for $x \in \{u > \bar{u}\}$, we get from the last inequality

$$\langle \bar{v} - v, (u - \bar{u})^+ \rangle = \int_{\{u>\bar{u}\}} (\bar{v}(x) - v(x))(u(x) - \bar{u}(x))\,dx$$

$$\le c_1 \int_{\{u>\bar{u}\}} (u(x) - \bar{u}(x))^p\,dx. \qquad (4.113)$$

The last term on the left-hand side of (4.110) becomes

$$\lambda \langle Bu, u - \bar{u})^+ \rangle = \lambda \int_{\{u>\bar{u}\}} (u(x) - \bar{u}(x))^p\,dx. \qquad (4.114)$$

Taking into account the estimates (4.111)–(4.114), we finally get from (4.110) the inequality

$$(\lambda - c_1) \int_{\{u>\bar{u}\}} (u(x) - \bar{u}(x))^p\,dx \le 0. \qquad (4.115)$$

According to Step 1, the parameter λ was chosen to satisfy $\lambda > C(\eta)/c_4$. Thus, λ can be selected such that

$$\lambda > \max\left\{\frac{C(\eta)}{c_4}, c_1\right\}$$

is satisfied, which by (4.115) implies that

$$\int_{\{u>\bar{u}\}} (u(x) - \bar{u}(x))^p\,dx = \int_\Omega [(u(x) - \bar{u}(x))^+]^p\,dx = 0,$$

and hence, it follows that $(u(x) - \bar{u}(x))^+ = 0$ for a.e. $x \in \Omega$; i.e., $u \le \bar{u}$ a.e. in Ω. In a similar way, we can prove $\underline{u} \le u$, which completes the proof of the theorem.   □

4.3.2 Compactness and Extremality Results

Let $\mathcal{S}$ denote the set of all solutions of the inclusion problem (4.81) within the ordered interval $[\underline{u}, \bar{u}]$ of sub- and supersolutions. In view of Theorem 4.31, $\mathcal{S}$ is nonempty. In this section, we are going to show that $\mathcal{S}$ is compact and directed, which in turn immediately implies the existence of extremal elements of $\mathcal{S}$.

Theorem 4.32. *Let hypotheses (A1)–(A3) and (H1), (H2) be satisfied. Then the solution set S of all solutions of (4.81) within $[\underline{u}, \bar{u}]$ is a compact subset of V_0.*

Proof: Let (u_n) be any sequence in S. We have to show that there is a subsequence of (u_n) that is strongly convergent in V_0 to some $u \in S$. By definition of S, we have $u_n \in [\underline{u}, \bar{u}]$, and

$$u_n \in V_0: \quad \langle Au_n + F(u_n) + v_n, \varphi \rangle = \langle h, \varphi \rangle, \quad \text{for all } \varphi \in V_0, \qquad (4.116)$$

where $v_n \in L^q(\Omega)$ satisfies $v_n(x) \in \partial j(x, u_n(x))$ for a.e. $x \in \Omega$. In view of hypothesis (A3), we get an estimate

$$\langle Au_n, u_n \rangle \geq \nu \|\nabla u_n\|_p^p - \|k_1\|_1, \qquad (4.117)$$

and by (H2) and Young's inequality, the following estimate is readily verified:

$$
\begin{aligned}
|\langle F(u_n), u_n \rangle| &\leq \int_\Omega |f(\cdot, u_n, \nabla u_n) u_n| \, dx \\
&\leq \left(\|k_4\|_q + \|\nabla u_n\|_p^{p-1} \right) \|u_n\|_p \\
&\leq C(\varepsilon) + \varepsilon \|\nabla u_n\|_p^p, \qquad (4.118)
\end{aligned}
$$

for any $\varepsilon > 0$. For (4.118), we have used that (u_n) is bounded in $L^p(\Omega)$ from $u_n \in [\underline{u}, \bar{u}]$. From (H1)(iii), we see that (v_n) is bounded in $L^q(\Omega)$, which shows that

$$|\langle v_n, u_n \rangle| \leq \|v_n\|_q \|u_n\|_p \leq c. \qquad (4.119)$$

Again by Young's inequality, the following inequality holds for any $\delta > 0$:

$$|\langle h, u_n \rangle| \leq \|h\|_{V_0^*} \|u_n\|_{V_0} \leq C(\delta) + \delta \|\nabla u_n\|_p^p, \qquad (4.120)$$

where we have used that $\|u\| = \|\nabla u\|_p$ defines an equivalent norm in V_0. Thus, from (4.116), with the special test function $\varphi = u_n$, and using estimates (4.117)–(4.120), we obtain the inequality

$$(\nu - \varepsilon - \delta) \|\nabla u_n\|_p^p \leq C(\delta, \varepsilon),$$

where $C(\delta, \varepsilon)$ is some constant depending only on δ and ε. Selecting ε and δ sufficiently small such that $\varepsilon + \delta < \nu$, we see from the last inequality that (u_n) is bounded in V_0. Thus, subsequences of (u_n) and (v_n) exist, denoted by (u_k) and (v_k), respectively, such that

$$u_k \rightharpoonup u \text{ in } V_0, \quad v_k \rightharpoonup v \text{ in } L^q(\Omega). \qquad (4.121)$$

The weak convergence of (u_k) in V_0 implies the strong convergence in $L^p(\Omega)$ by the compact embedding $V_0 \subset L^p(\Omega)$. Taking, in addition, the boundedness of (u_n) in V_0 into account, we get from (4.116)

$$\limsup_k \langle Au_k, u_k - u \rangle \leq 0, \tag{4.122}$$

which because of the (S_+)-property of the operator A implies the strong convergence $u_k \to u$ in V_0. Passing to the limit in (4.116) as $k \to \infty$ results in the following equation for the strong limit u:

$$u \in V_0 : \quad \langle Au + F(u) + v, \varphi \rangle = \langle h, \varphi \rangle, \quad \text{for all } \varphi \in V_0.$$

The proof of the compactness is complete provided v satisfies $v(x) \in \partial j(x, u(x))$ for a.e. $x \in \Omega$. To this end, recall that $v_k \rightharpoonup v$ in $L^q(\Omega)$, and for a.e. $x \in \Omega$, we have

$$v_k(x) \in \partial j(x, u_k(x)) = [\underline{g}(x, u_k(x)), \bar{g}(x, u_k(x))].$$

Note that the multivalued mapping $s \mapsto \partial j(x, s) = [\underline{g}(x, s), \bar{g}(x, s)]$ is upper semicontinuous (see Proposition 2.171), and thus $s \mapsto \bar{g}(x, s)$ is an upper semicontinuous function and $s \mapsto \underline{g}(x, s)$ is a lower semicontinuous function. With $\varphi \in L^p_+(\Omega)$ arbitrarily given, we get

$$\int_\Omega \underline{g}(x, u_k(x))\, \varphi(x)\, dx \leq \int_\Omega v_k(x)\, \varphi(x)\, dx \leq \int_\Omega \bar{g}(x, u_k(x))\, \varphi(x)\, dx.$$

The strong convergence of (u_k) in $L^p(\Omega)$ implies the a.e. pointwise convergence of some subsequence, which is again denoted by (u_k). Using $v_k \rightharpoonup v$ in $L^q(\Omega)$, $u_k \to u$ in $L^p(\Omega)$, and a.e. pointwise in Ω as well as the upper semicontinuity of $\bar{g}(\cdot, s)$ and the lower semicontinuity of $\underline{g}(\cdot, s)$ with respect to s, we deduce by applying Fatou's lemma that

$$\int_\Omega \underline{g}(x, u(x))\, \varphi(x)\, dx \leq \int_\Omega v(x)\, \varphi(x)\, dx \leq \int_\Omega \bar{g}(x, u(x))\, \varphi(x)\, dx,$$

which shows that $\underline{g}(x, u(x)) \leq v(x) \leq \bar{g}(x, u(x))$ for a.e. $x \in \Omega$. This process completes the proof. $\qquad \square$

Lemma 4.33. *Let hypotheses (A1)–(A4) and (H1), (H2) be satisfied. If u_1, $u_2 \in S$, then $\max\{u_1, u_2\}$ is a subsolution and $\min\{u_1, u_2\}$ is a supersolution of the inclusion problem (4.81).*

Proof. Denote $u = \max\{u_1, u_2\}$, and define a function $v : \Omega \to \mathbb{R}$ by

$$v(x) = \begin{cases} v_1(x) & \text{if } x \in \{u_1 \geq u_2\}, \\ v_2(x) & \text{if } x \in \{u_2 > u_1\}, \end{cases}$$

where $v_k \in L^q(\Omega)$ satisfy $v_k(x) \in \partial j(x, u_k(x))$ for a.e. $x \in \Omega$, $k = 1, 2$, and

$$\langle Au_k + F(u_k) + v_k, \varphi \rangle = \langle h, \varphi \rangle \quad \text{for all } \varphi \in V_0. \tag{4.123}$$

Obviously, $u \in V_0$ and $v \in L^q(\Omega)$ satisfies $v(x) \in \partial j(x, u(x))$ for a.e. $x \in \Omega$. Thus, $u = \max\{u_1, u_2\}$ is a subsolution of (4.81) provided we can verify the following inequality:

$$\langle Au + F(u) + v, \varphi \rangle \leq \langle h, \varphi \rangle \quad \text{for all} \quad \varphi \in V_0 \cap L_+^p(\Omega). \qquad (4.124)$$

To prove inequality (4.124), we employ the ideas used in the proof of Lemma 4.23 of the previous section. However, from the dependence of the coefficients a_i on both ∇u and u, a more involved special test function technique is needed that is related with the modulus of continuity of $a_i(\cdot, u, \nabla u)$ with respect to u [see (A4)]. As this technique has already been used in the proof of Theorem 3.20 of Chap. 3, we may confine ourself to a sketch of the proof of inequality (4.124). As in Sect. 4.2 for $k = 1$, we use in (4.123) the test function

$$\varphi = \psi \left(1 - \theta_\varepsilon(u_2 - u_1)\right) \in V_0 \cap L_+^p(\Omega),$$

and for $k = 2$, we use the test function

$$\varphi = \psi \, \theta_\varepsilon(u_2 - u_1) \in V_0 \cap L_+^p(\Omega),$$

where $\psi \in \mathcal{D}_+$ with

$$\mathcal{D}_+ = \{\psi \in C_0^\infty(\Omega) : \psi \geq 0 \text{ in } \Omega\}.$$

However, unlike in the previous section, the function $\theta_\varepsilon : \mathbb{R} \to \mathbb{R}$ is defined as follows. In view of hypothesis (A4), for any fixed $\varepsilon > 0$, $\delta(\varepsilon) \in (0, \varepsilon)$ exists such that

$$\int_{\delta(\varepsilon)}^{\varepsilon} \frac{1}{\omega(r)} \, dr = 1.$$

Using this property, we define

$$\theta_\varepsilon(s) = \begin{cases} 0 & \text{if } s < \delta(\varepsilon), \\ \displaystyle\int_{\delta(\varepsilon)}^{s} \frac{1}{\omega(r)} \, dr & \text{if } \delta(\varepsilon) \leq s \leq \varepsilon, \\ 1 & \text{if } s > \varepsilon. \end{cases} \qquad (4.125)$$

Obviously, for each $\varepsilon > 0$, θ_ε is continuous, piecewise differentiable and the derivative is nonnegative and bounded. Thus, the function θ_ε is Lipschitz continuous and nondecreasing and moreover, it satisfies

$$\theta_\varepsilon \to \chi_{\{s>0\}} \quad \text{a.e. as } \varepsilon \to 0,$$

where $\chi_{\{s>0\}}$ is the characteristic function of the set $\{s > 0\} = \{s \in \mathbb{R} : s > 0\}$. In addition, we have

$$\theta_\varepsilon'(s) = \begin{cases} \dfrac{1}{\omega(s)} & \text{if } \delta(\varepsilon) < s < \varepsilon, \\ 0 & \text{if } s \notin [\delta(\varepsilon), \varepsilon]. \end{cases} \qquad (4.126)$$

Taking in (4.123) for $k = 1, 2$ the test functions specified above and adding the resulted equations, we get

$$\langle Au_1 + F(u_1) + v_1, \psi \rangle$$
$$+\langle Au_2 - Au_1 + F(u_2) - F(u_1) + v_2 - v_1, \theta_\varepsilon(u_2 - u_1)\psi \rangle$$
$$= \langle h, \psi \rangle. \tag{4.127}$$

Applying hypothesis (A4) and passing to the limit in (4.127) as $\varepsilon \to 0$, we finally arrive at

$$\langle Au + F(u) + v, \psi \rangle \leq \langle h, \psi \rangle \quad \text{for all} \quad \psi \in \mathcal{D}_+. \tag{4.128}$$

As $\mathcal{D}_+$ is dense in $V_0 \cap L^p_+(\Omega)$, from (4.128), we obtain (4.124). The proof for $\min\{u_1, u_2\}$ being a supersolution can be done in a similar way. $\square$

As an immediate consequence of Theorem 4.32 and Lemma 4.33, we obtain the following extremality result.

Theorem 4.34. *Under hypotheses (A1)–(A4) and (H1), (H2), the solution set $\mathcal{S}$ of all solutions of (4.81) within $[\underline{u}, \bar{u}]$ possesses extremal elements.*

Proof: Lemma 4.33 implies that $\mathcal{S}$ is a directed set, which due to Theorem 4.32, is also compact. With these two properties of $\mathcal{S}$, the proof of the existence of extremal elements follows now the same arguments as in the proof of Theorem 3.22 in Chap. 3. $\square$

4.4 Application: Difference of Multifunctions

In this section, we consider the Dirichlet problem of an elliptic inclusion whose governing multivalued term is given by the difference of Clarke's generalized gradient of some locally Lipschitz function $s \mapsto j(\cdot, s)$ and the subdifferential of some convex function $s \mapsto \beta(\cdot, s)$. More precisely, the following problem will be considered:

$$Au + \partial j(\cdot, u) - \partial\beta(\cdot, u) \ni f, \quad \text{in} \quad \Omega, \quad u = 0, \quad \text{on} \quad \partial\Omega, \tag{4.129}$$

where $\Omega \subset \mathbb{R}^N$ is a bounded domain with Lipschitz boundary $\partial\Omega$ and A is a second-order quasilinear differential operator in divergence form of Leray–Lions type of the form

$$Au(x) = -\sum_{i=1}^{N} \frac{\partial}{\partial x_i} a_i(x, \nabla u(x)). \tag{4.130}$$

Even though the function $s \mapsto j(\cdot, s) - \beta(\cdot, s)$ is locally Lipschitz, its generalized Clarke's gradient, in general, only satisfies

$$\partial(j(\cdot, s) - \beta(\cdot, s)) \subset \partial j(\cdot, s) - \partial\beta(\cdot, s).$$

Therefore, any solution of

$$Au + \partial(j(\cdot, u) - \beta(\cdot, u)) \ni f, \quad \text{in } \Omega, \quad u = 0, \quad \text{on } \partial\Omega,$$

is also a solution of problem (4.129). However, the reverse may be not true. Moreover, the multifunction $s \mapsto \partial j(\cdot, s) - \partial\beta(\cdot, s)$ does not satisfy the one-sided growth condition (H1)(ii), and thus, (4.129) cannot be reduced to those considered in the previous section, and it requires a new approach. However, we will see that methods and results of the previous sections can effectively be used in the treatment of (4.129). Moreover, without assuming sub-supersolutions for (4.129), we are going to prove the existence of extremal solutions and the compactness of its solution set provided certain growth conditions are satisfied. In this sense, the latter can be regarded as sufficient conditions for the existence of sub-supersolutions. The purpose of this section is on the one hand to enlarge the class of the multifunctions and on the other hand to provide a substitute for the existence of sub-supersolutions.

4.4.1 Hypotheses and Main Result

We basically use the notation of Sect. 4.3, and we assume hypotheses (A1)–(A3) for the coefficients $a_i : \Omega \times \mathbb{R}^N \to \mathbb{R}$ of the operator A given by (4.130). Only for the sake of simplicity, the coefficients $a_i(\cdot, \nabla u)$ are supposed to be independent of u. Note that in this case hypothesis (A4) is empty. As in Sect. 4.3, the function $j : \Omega \times \mathbb{R} \to \mathbb{R}$ is assumed to be the primitive of some locally bounded and Borel-measurable function $g : \Omega \times \mathbb{R} \to \mathbb{R}$; i.e.,

$$j(x, s) = \int_0^s g(x, \tau)\, d\tau.$$

The function $\beta : \Omega \times \mathbb{R} \to \mathbb{R}$ is assumed to be the primitive of some Borel-measurable function $h : \Omega \times \mathbb{R} \to \mathbb{R}$ that is monotone nondecreasing in its second variable; i.e.,

$$\beta(x, s) = \int_0^s h(x, \tau)\, d\tau. \tag{4.131}$$

Thus, $\beta(x, \cdot) : \mathbb{R} \to \mathbb{R}$ is convex with $\partial\beta(x, \cdot) : \mathbb{R} \to 2^{\mathbb{R}} \setminus \{\emptyset\}$ denoting the usual subdifferential of β with respect to its second argument, and we have

$$\partial\beta(x, s) = [\underline{h}(x, s), \bar{h}(x, s)], \tag{4.132}$$

where $\underline{h}$ and $\bar{h}$ denote the left-sided and right-sided limits of h, respectively, with respect to the second argument.

As for the function g related with j and the function h related with β, we assume the following hypotheses.

(B1) The function $g : \Omega \times \mathbb{R} \to \mathbb{R}$ satisfies
   (i) g is Borel-measurable in $\Omega \times \mathbb{R}$, and $g(x, \cdot) : \mathbb{R} \to \mathbb{R}$ is locally bounded.

(ii) A constant $c_1 \geq 0$ exists such that

$$g(x, s_1) \leq g(x, s_2) + c_1 (s_2 - s_1)^{p-1},$$

for a.e. $x \in \Omega$, and for all s_1, s_2 with $s_1 < s_2$.

(iii) A function $k_2 \in L_+^q(\Omega)$ and a constant $\mu_1 \geq 0$ exist such that

$$|g(x, s)| \leq k_2(x) + \mu_1 |s|^{p-1},$$

for a.e. $x \in \Omega$, and for all $s \in \mathbb{R}$.

(B2) The function $h : \Omega \times \mathbb{R} \to \mathbb{R}$ is Borel-measurable, monotone nondecreasing in its second argument, and satisfies with some function $k_3 \in L_+^q(\Omega)$ and with some constant $\mu_2 \geq 0$ the growth condition

$$|h(x, s)| \leq k_3(x) + \mu_2 |s|^{p-1}$$

for a.e. $x \in \Omega$ and for all $s \in \mathbb{R}$.

(B3) Let $c_F > 0$ denote the best constant in Poincaré–Friedrichs inequality and denote $\mu = \mu_1 + \mu_2$, where μ_1 and μ_2 are the nonnegative constants of (B1) and (B2), respectively. Then the positive constant ν of (A3) is supposed to satisfy

$$c_F^p \, \mu < \nu.$$

Definition 4.35. *A function $u \in V_0$ is a solution of the inclusion problem (4.129) if there are functions $\eta \in L^q(\Omega)$ and $\kappa \in L^q(\Omega)$ such that the following holds:*

(i) $\eta(x) \in \partial j(x, u(x))$ *and* $\kappa(x) \in \partial \beta(x, u(x))$ *for a.e.* $x \in \Omega$.

(ii) $\langle Au, \varphi \rangle + \int_\Omega (\eta(x) - \kappa(x)) \varphi(x) \, dx = \langle f, \varphi \rangle$, $\forall \varphi \in V_0$, *where* $f \in V_0^*$.

The main result of this section is given by the following theorem.

Theorem 4.36. *Under hypotheses (A1)–(A3) and (B1)–(B3), problem (4.129) possesses extremal solutions and the solution set is compact in V_0.*

4.4.2 A Priori Bounds

We shall prove the existence of a priori bounds for the solutions of (4.129), which are crucial in the proof of our main result. To this end, consider the following auxiliary problems:

$$u \in V_0 : \quad Au = f + k + \mu |u|^{p-1} \quad \text{in } V_0^*, \tag{4.133}$$

$$u \in V_0 : \quad Au = f - k - \mu |u|^{p-1} \quad \text{in } V_0^*, \tag{4.134}$$

where $k \in L_+^q(\Omega)$ is given by $k(x) = k_2(x) + k_3(x)$ and $\mu = \mu_1 + \mu_2$.

Lemma 4.37. *Problems (4.133) and (4.134) possess solutions, and their respective solution sets are bounded in V_0.*

Proof. We prove the existence of solutions and the boundedness of the solution set for the problem (4.133) only, because the same arguments can be applied for (4.134). Let P denote the Nemytskij operator related to the function $s \mapsto \mu |s|^{p-1}$, then $P : L^p(\Omega) \to L^q(\Omega) \subset V_0^*$ is continuous and bounded, and from the compact embedding $V_0 \subset L^p(\Omega)$, it follows that $P : V_0 \to V_0^*$ is completely continuous. As $A : V_0 \to V_0^*$ is bounded, continuous, and monotone, the operator $A - P : V_0 \to V_0^*$ is bounded, continuous, and pseudomonotone. Rewriting the BVP (4.133) in the form

$$u \in V_0 : \quad (A - P)u = f + k \quad \text{in } V_0^*, \tag{4.135}$$

and noting that $f + k \in V_0^*$, solutions of (4.135) exist provided $A - P : V_0 \to V_0^*$ is coercive; i.e., the following holds:

$$\frac{\langle (A - P)u, u \rangle}{\|u\|_{V_0}} \to +\infty \quad \text{as } \|u\|_{V_0} \to \infty. \tag{4.136}$$

By means of (A3) and (B3), we obtain

$$\begin{aligned}
\langle (A - P)u, u \rangle &\geq \nu \|\nabla u\|_{L^p(\Omega)}^p - \|k_1\|_{L^1(\Omega)} - \mu \|u\|_{L^p(\Omega)}^p \\
&\geq (\nu - c_F^p \mu) \|\nabla u\|_{L^p(\Omega)}^p - \|k_1\|_{L^1(\Omega)},
\end{aligned} \tag{4.137}$$

which proves the coercivity from (B3) and the fact that $\|u\| = \|\nabla u\|_{L^p(\Omega)}$ defines an equivalent norm in V_0. The coercivity argument applies also to get the boundedness of the solution set of (4.133). To this end, let u be any solution of the BVP (4.133); then from (4.135) and (4.137), we get

$$\begin{aligned}
(\nu - c_F^p \mu) \|\nabla u\|_{L^p(\Omega)}^p - \|k_1\|_{L^1(\Omega)} &\leq \langle (A - P)u, u \rangle \\
&\leq (\|f\|_{V_0^*} + \|k\|_{L^q(\Omega)}) \|u\|_{V_0},
\end{aligned}$$

which proves the assertion. □

Lemma 4.38. *The solution sets of the BVP (4.133) and (4.134), respectively, are directed and compact sets.*

Proof. We are going to prove the assertion for the BVP (4.133) only, because analogous arguments apply for the BVP (4.134).

Step 1: Directedness of the Solution Set.

Let us denote by $\mathcal{S}$ the solution set of the BVP (4.133). Then $\mathcal{S} \neq \emptyset$ in view of Lemma 4.37. If $u_1, u_2 \in \mathcal{S}$ and $\underline{u}$ is defined by $\underline{u} = \max\{u_1, u_2\} \in V_0$, then $\underline{u}$ is a subsolution of the BVP (4.133) (see Theorem 3.20). Let T denote the following truncation operator:

$$(Tu)(x) = \begin{cases} u(x) & \text{if } \underline{u}(x) \leq u(x), \\ \underline{u}(x) & \text{if } u(x, t) < \underline{u}(x), \end{cases}$$

and consider the auxiliary BVP

$$u \in V_0 : \quad Au = f + k + \mu |Tu|^{p-1} \quad \text{in } V_0^*. \tag{4.138}$$

The same arguments as in the proof of Lemma 4.37 apply to ensure the existence of solutions of the BVP (4.138). We shall show first that S is upward directed. To this end, we only need to verify that any solution u of the BVP (4.138) satisfies $u \geq \underline{u}$, because then $Tu = u$, and thus, it follows that u is a solution of the BVP (4.133), which exceeds the given solutions u_1 and u_2. This result proves S is upward directed. Let u be any solution of (4.138), and recall that $\underline{u}$ is a subsolution of (4.133); i.e., we have

$$\langle A\underline{u}, \varphi \rangle \leq \langle f, \varphi \rangle + \int_\Omega (k + \mu |\underline{u}|^{p-1}) \varphi \, dx, \quad \forall \, \varphi \in V_0 \cap L_+^p(\Omega), \tag{4.139}$$

and u is any solution of (4.138); i.e.,

$$\langle Au, \varphi \rangle = \langle f, \varphi \rangle + \int_\Omega (k + \mu |Tu|^{p-1}) \varphi \, dx, \quad \forall \, \varphi \in V_0. \tag{4.140}$$

Taking as special nonnegative test function $\varphi = (\underline{u} - u)^+ \in V_0 \cap L_+^p(\Omega)$, we obtain by subtracting (4.140) from (4.139) the inequality

$$\int_\Omega \sum_{i=1}^N (a_i(x, \nabla \underline{u}) - a_i(x, \nabla u)) \frac{\partial (\underline{u} - u)^+}{\partial x_i} \, dx$$

$$\leq \mu \int_\Omega (|\underline{u}|^{p-1} - |Tu|^{p-1}) (\underline{u} - u)^+ \, dx$$

$$= \mu \int_{\{\underline{u} > u\}} (|\underline{u}|^{p-1} - |\underline{u}|^{p-1}) (\underline{u} - u) \, dx = 0, \tag{4.141}$$

where $\{\underline{u} > u\} = \{x \in \Omega : \underline{u}(x) > u(x)\}$. By means of (A2), we deduce from (4.141) that $\nabla (\underline{u} - u)^+ = 0$, and thus, $(\underline{u} - u)^+ = 0$, which yields $\underline{u} \leq u$. This process completes the proof for S being upward directed. Noting that for any solutions $u_1, u_2 \in S$, the function $\bar{u} = \min\{u_1, u_2\}$ is a supersolution of the BVP (4.133) (see Theorem 3.20). We can show in a similar way that S is also downward directed and thus the directedness of S.

Step 2: Compactness of the Solution Set.

Let $(u_n) \subset S$, where S denotes the solution set of (4.133). In view of Lemma 4.37, the sequence (u_n) is bounded in V_0 and thus, a subsequence (u_k) exists with

$$u_k \rightharpoonup u \text{ in } V_0 \quad \text{and} \quad u_k \to u \text{ in } L^p(\Omega). \tag{4.142}$$

From (4.133), we get

$$\langle Au_k, u_k - u \rangle = \langle f, u_k - u \rangle + \int_\Omega (k + \mu |u_k|^{p-1}) (u_k - u) \, dx,$$

which in view of (4.142) yields

$$\lim_{k \to \infty} \langle A u_k, u_k - u \rangle = 0,$$

and thus, by the (S_+)-property of the operator A (see Theorem 2.109), we see that the sequence (u_k) is strongly convergent in V_0 to u. Passing to the limit in

$$A u_k = f + k + \mu |u_k|^{p-1} \quad \text{in } V_0^*$$

proves that $u \in \mathcal{S}$.   □

As a consequence of Lemma 4.38, we obtain the following corollary.

Corollary 4.39. *The BVP (4.133) and (4.134) have extremal solutions.*

Proof. Again the proof will be given for the BVP (4.133) only, because for the BVP (4.134), it can be done similarly. Moreover, we will concentrate on the existence of the greatest solution of (4.133), because the existence of the smallest solution follows by obvious dual reasoning. We also note that here the proof of the extremal solutions is not a straightforward application of the arguments used in previous sections, because here the solution sets are not contained in some interval of sub-supersolutions. Let $\mathcal{S}$ be the solution set of the BVP (4.133). First we shall show the existence of a maximal element of $\mathcal{S}$ by means of Zorn's lemma. To this end, let $\mathcal{C} \subset \mathcal{S}$ be any well-ordered chain that is bounded in V_0 by Lemma 4.37 and, thus, in particular, also bounded in $L^p(\Omega)$. Then, an increasing sequence (u_n) of $\mathcal{C}$ exists, which converges strongly in $L^p(\Omega)$ and weakly in V_0 to $w = \sup \mathcal{C}$. This result occurs because the order cone $L_+^p(\Omega)$ is fully regular (see [111, Proposition 5.8.7]). In just the same way as in Step 2 of Lemma 4.38, we see that w belongs to $\mathcal{S}$. Thus, $\mathcal{C}$ possesses an upper bound in $\mathcal{S}$, so that Zorn's lemma can be applied, which ensures the existence of a maximal element w^*. Because $\mathcal{S}$ is, in particular, upward directed, the maximal element is unique and must be the greatest one. Thus, w^* is the greatest solution of (4.133).   □

By means of Corollary 4.39, we can now derive a priori bounds of the original problem (4.129).

Lemma 4.40. *Let w^* be the greatest solution of the BVP (4.133) and w_* be the smallest solution of the BVP (4.134) that exist due to Corollary 4.39. Then any solution u of the inclusion problem (4.129) is contained in $[w_*, w^*]$.*

Proof. Let u be any solution of (4.129); i.e., we have by Definition 4.35,

$$\langle A u, \varphi \rangle + \int_{\Omega} (\eta(x) - \kappa(x)) \, \varphi(x) \, dx = \langle f, \varphi \rangle, \quad \forall \, \varphi \in V_0, \qquad (4.143)$$

where $\eta(x) \in \partial j(x, u(x))$ and $\kappa(x) \in \partial \beta(x, u(x))$ for a.e. $x \in \Omega$. In view of the growth conditions in (B1) and (B2), we have

$$|\eta(x)| \leq k_2(x) + \mu_1 |u(x)|^{p-1}, \quad |\kappa(x)| \leq k_3(x) + \mu_2 |u(x)|^{p-1}. \quad (4.144)$$

From (4.143) and (4.144), we see that u is a subsolution of the BVP (4.133). Now the same arguments as in Step 1 of the proof of Lemma 4.38 apply, which shows that solutions of the BVP (4.133) exist that are greater than u. However, w^* is the greatest of all solutions of (4.133), and thus, it exceeds also u, which proves that w^* is an upper bound of any solution of the original problem (4.129). The proof for w_* being a lower bound is carried out analogously. $\square$

4.4.3 Proof of Theorem 4.36

In this section, we are going to prove our main result. The proof is inspired by an idea used in [36, 39] to treat boundary hemivariational inequalities of d.c.-type and will be given in two steps.

Step 1: Existence of Extremal Solutions of (4.129).

Lemma 4.40 provides a priori bounds w^* and w_* of solutions of (4.129) where w^* is the greatest solution of the BVP (4.133) and w_* is the smallest solution of the BVP (4.134). We are going to prove that (4.129) possesses extremal solutions within the ordered interval $[w_*, w^*]$, which justifies the existence of extremal solutions of (4.129). Let us concentrate on the existence of the greatest solution, because the existence of the smallest solution can be shown similarly.

We recall that the subdifferential $\partial\beta(x, s)$ is generated by the function $h : \Omega \times \mathbb{R} \to \mathbb{R}$, which is monotone nondecreasing in its second argument via

$$\partial\beta(x, s) = [\underline{h}(x, s), \bar{h}(x, s)],$$

where $s \mapsto \underline{h}(x, s)$ and $s \mapsto \bar{h}(x, s)$ are the left- and right-sided limits, respectively, of $s \mapsto h(x, s)$. Denote by $\underline{H}$ and $\bar{H}$ the Nemytskij operator associated with $\underline{h}$ and $\bar{h}$, respectively. By hypothesis (B2), the operators $\underline{H}, \bar{H} : L^p(\Omega) \to L^q(\Omega)$ are well defined, monotone nondecreasing, but not necessarily continuous. Consider the following inclusion of Clarke's gradient type involving a discontinuous nonlinearity:

$$u \in V_0 : \quad Au + \partial j(\cdot, u) \ni f + \bar{H}(u) \quad \text{in } V_0^*. \quad (4.145)$$

Our goal is to show that (4.145) has the greatest solution u^* within $[w_*, w^*]$, and that u^* is the greatest solution of the original problem (4.129). To this end, let us consider first the following inclusion with given right-hand side:

$$u \in V_0 : \quad Au + \partial j(\cdot, u) \ni f + \bar{H}(w^*) \quad \text{in } V_0^*. \quad (4.146)$$

By (B1)(iii) and (B2), and taking into account that w^* is the greatest solution of (4.133), we get for any $\eta^* \in \partial j(\cdot, w^*)$, the estimate

$$Aw^* + \eta^* = f + k + \mu |w^*|^{p-1} + \eta^* \geq f + k_3 + \mu_2 |w^*|^{p-1} \geq f + \bar{H}(w^*),$$

which proves that w^* is a supersolution of (4.146). Analogously we show that w_* is a subsolution of (4.146). Thus, by applying Theorem 4.34 of the previous section with the right-hand side $f + \bar{H}(w^*) \in V_0^*$, extremal solutions of (4.146) exist within the interval $[w_*, w^*]$. Let u_1 denote the greatest solution of (4.146) within $[w_*, w^*]$, and consider next the inclusion

$$u \in V_0: \quad Au + \partial j(\cdot, u) \ni f + \bar{H}(u_1) \quad \text{in } V_0^*. \tag{4.147}$$

By the monotonicity of $\bar{H}$, we have $\bar{H}(u_1) \leq \bar{H}(w^*)$, and thus, u_1 is a supersolution for (4.147). One readily verifies that w_* is a subsolution for (4.147) as well. Again by applying Theorem 4.34, extremal solutions of (4.147) exist within $[w_*, u_1]$. In this way, we can define by induction the following iteration process: Let $u_0 := w^*$ and define by $u_{n+1} \in V_0$ the greatest solution of

$$u \in V_0: \quad Au + \partial j(\cdot, u) \ni f + \bar{H}(u_n) \quad \text{in } V_0^* \tag{4.148}$$

within $[w_*, u_n]$. From $u_{n+1} \in [w_*, u_n]$, this iteration yields a monotone nonincreasing sequence (u_n) that satisfies

$$w_* \leq \cdots \leq u_{n+1} \leq u_n \leq \cdots \leq u_1 \leq u_0 = w^* \tag{4.149}$$

and

$$Au_{n+1} + \eta_{n+1} = f + \bar{H}(u_n) \quad \text{in } V_0^*, \tag{4.150}$$

where $\eta_{n+1} \in \partial j(\cdot, u_{n+1})$, and $\eta_{n+1} \in L^q(\Omega)$. As the sequence (u_n) can easily be seen to be bounded in V_0, and because $(\eta_n) \subset L^q(\Omega)$ is bounded as well, we obtain the following convergence properties:

(i) $u_n \rightharpoonup u^*$ in V_0.
(ii) $u_n \to u^*$ in $L^p(\Omega)$.
(iii) $\eta_n \rightharpoonup \eta^*$ in $L^q(\Omega)$ [for some subsequence that is again denoted by (η_n)].

In (iii), we have $\eta^* \in \partial j(\cdot, u^*)$. The boundedness of $(\bar{H}(u_n))$ in $L^q(\Omega)$ and the convergence properties (i)–(iii) imply that

$$\limsup_{n \to \infty} \langle Au_n, u_n - u^* \rangle \leq 0.$$

As the operator $A : V_0 \to V_0^*$ is, in particular, pseudomonotone, it follows that

(iv) $Au_n \rightharpoonup Au^*$ in V_0^*.

The function $s \mapsto \bar{h}(x, s)$ related with the Nemytskij operator $\bar{H}$ is monotone nondecreasing and right-sided continuous, so that by means of Lebesgue's dominated convergence theorem and from the a.e. monotone pointwise convergence of the sequence (u_n) according to (4.149), we get

$$\int_\Omega \bar{H}(u_n)\, \varphi\, dx \to \int_\Omega \bar{H}(u^*)\, \varphi\, dx, \tag{4.151}$$

for all $\varphi \in L^p(\Omega)$. Finally, the convergence properties (i)–(iv) above and (4.151) allow us to pass to the limit in (4.150) as $n \to \infty$, which shows that $u^* \in [w_*, w^*]$ is a solution of the inclusion (4.145). Moreover, u^* is the greatest solution of (4.145) within $[w_*, w^*]$. To verify this result, let $u \in [w_*, w^*]$ be any solution of (4.145). Then u is, in particular, a subsolution of (4.145). Replacing in the iteration above w_* by u, we see that $u \leq u_n \leq w^*$ holds for all n. Thus, we get $u \leq u^*$; i.e., u^* is the greatest solution of (4.145) in $[w_*, w^*]$. Now, defining $\kappa^* := \bar{H}(u^*)$, then obviously we have $\kappa^*(x) \in \partial\beta(x, u^*(x))$ for a.e. $x \in \Omega$, and thus, u^* satisfies

$$Au^* + \eta^* - \kappa^* = f \quad \text{in} \quad V_0^*,$$

which means that u^* is a solution of the original problem (4.129) as well.

It remains to prove that u^* is the greatest solution of (4.129). To this end, take any solution $\tilde{u}$ of (4.129), which satisfies $\tilde{u} \in [w_*, w^*]$ and

$$A\tilde{u} + \tilde{\eta} - \tilde{\kappa} = f,$$

where $\tilde{\eta} \in \partial j(\cdot, \tilde{u})$ and $\tilde{\kappa} \in \partial\beta(\cdot, \tilde{u}) \subset [\underline{H}(\tilde{u}), \bar{H}(\tilde{u})]$. As $\tilde{\kappa} \leq \bar{H}(\tilde{u})$, we see that $\tilde{u}$ is a subsolution of the inclusion (4.145). By the same iteration procedure introduced above with w_* replaced by $\tilde{u}$, we get $\tilde{u} \leq u_n \leq w^*$, which implies $\tilde{u} \leq u^*$, and thus, u^* must be the greatest solution of the original problem (4.129). The existence of the smallest solution u_* can be shown by obvious dual reasoning, which completes the proof of the extremality result.

Step 2: Compactness of the Solution Set of (4.129).

Denote by $\mathcal{T}$ the set of all solutions of (4.129). Then $\mathcal{T} \subset [u_*, u^*]$, where u_* and u^* is the smallest and the greatest solution of (4.129). Let $(u_n) \subset \mathcal{T}$ be any sequence. Then (u_n) is bounded in V_0, and one has the following convergence properties for some subsequences:

(v) $u_k \rightharpoonup u$ in V_0,
(vi) $u_k \to u$ in $L^p(\Omega)$,
(vii) $\eta_k \rightharpoonup \eta$ and $\kappa_k \rightharpoonup \kappa$ in $L^q(\Omega)$,

where $\eta_k \in \partial j(\cdot, u_k)$ and $\kappa_k \in \partial\beta(\cdot, u_k)$, and we have

$$Au_k + \eta_k - \kappa_k = f \quad \text{in} \quad V_0^*. \tag{4.152}$$

The compact embedding $V_0 \subset L^p(\Omega)$ implies the compact embedding $L^q(\Omega) \subset V_0^*$, which yields

$$\eta_k \to \eta, \quad \kappa_k \to \kappa \quad \text{in} \quad V_0^*, \tag{4.153}$$

where $\eta \in \partial j(\cdot, u)$ and $\kappa \in \partial\beta(\cdot, u)$. Because of (4.153) from (4.152), we get

$$\langle Au_k, u_k - u \rangle = \langle f - \eta_k + \kappa_k, u_k - u \rangle \to 0, \tag{4.154}$$

so that in view of the pseudomonotonicity of A, we infer $Au_k \rightharpoonup Au$ in V_0^* as $k \to \infty$. Passing to the limit as $k \to \infty$ in (4.152) yields

$$Au + \eta - \kappa = f \quad \text{in } V_0^*,$$

and thus, $u \in \mathcal{T}$. Finally, the (S_+)-property of A in conjunction with (4.154) yields the strong convergence $u_k \to u$ in V_0, which completes the compactness proof.   □

We remark the interesting fact that κ^* related with the greatest solution u^* whose existence is proved in Step 1 above is given by $\kappa^* = \max\{\partial\beta(\cdot, u^*)\}$.

Special Case. As a special case of (4.129), we consider the inclusion problem

$$u \in V_0: \quad -\Delta_p u + \partial j(\cdot, u) - \partial\beta(\cdot, u) \ni f, \tag{4.155}$$

where Δ_p denotes the p-Laplacian. Obviously, $-\Delta_p$ satisfies (A1)–(A3). The variational characterization of the first Dirichlet eigenvalue λ_1 of $-\Delta_p$, which is positive and given by

$$\lambda_1 = \inf_{0 \neq u \in V_0} \frac{\int_\Omega |\nabla u|^p \, dx}{\int_\Omega |u|^p \, dx}$$

(see [151]), enables us to sharpen the condition (B3) as follows:

(B4) Let $\mu = \mu_1 + \mu_2 < \lambda_1$ be satisfied.

The following result is an immediate consequence of Theorem 4.36.

Corollary 4.41. *If conditions (B1), (B2), and (B4) are satisfied, then the inclusion (4.155) has extremal solutions and the solution set is compact in V_0.*

Remark 4.42. (i) Our main result (Theorem 4.36) can be extended to more general Leray–Lions operators A such as

$$Au(x) = -\sum_{i=1}^{N} \frac{\partial}{\partial x_i} a_i(x, u(x), \nabla u(x)) + a_0(x, u(x), \nabla u(x)).$$

Only for the sake of simplifying our presentation, and to emphasize the main idea, we have taken a nonlinear, monotone operator A.

(ii) When we assume the existence of an ordered pair $\underline{w} \leq \bar{w}$ that satisfies $\underline{w} \leq 0$ and $\bar{w} \geq 0$ on $\partial\Omega$ as well as the inequalities

$$\bar{w} \in V: \quad A\bar{w} + \bar{\eta} \geq f + \bar{H}(\bar{w}), \quad \text{where } \bar{\eta} \in \partial j(\cdot, \bar{w}),$$

and

$$\underline{w} \in V: \quad A\underline{w} + \underline{\eta} \leq f + \underline{H}(\underline{w}), \quad \text{where } \underline{\eta} \in \partial j(\cdot, \underline{w}),$$

we can prove extremality and compactness of the solution set contained within the ordered interval $[\underline{w}, \bar{w}]$. In this case, hypothesis (B3) can be dropped, and only local growth conditions of g and h are required.

4.5 Parabolic Inclusions with Local Growth

The subject of this section is the parabolic version of the quasilinear elliptic
inclusion (4.81) treated in Sect. 4.3.

Let $\Omega \subset \mathbb{R}^N$ be a bounded domain with Lipschitz boundary $\partial\Omega$, $Q = \Omega \times (0, \tau)$, and $\Gamma = \partial\Omega \times (0, \tau)$, with $\tau > 0$. Our goal is to provide a comparison
principle as well as extremality and compactness results for the quasilinear
initial-boundary value problem

$$u_t + Au + \partial j(\cdot, \cdot, u) \ni Fu + h \quad \text{in} \quad Q,$$
$$u = 0 \text{ in } \Omega \times \{0\}, \quad u = 0 \text{ on } \Gamma, \tag{4.156}$$

where A is given by

$$Au(x, t) = -\sum_{i=1}^{N} \frac{\partial}{\partial x_i} a_i(x, t, u(x, t), \nabla u(x, t)),$$

and F is the Nemytskij operator associated with the Carathéodory function
$f : Q \times \mathbb{R} \to \mathbb{R}$. Unlike in the elliptic case of Sect. 4.3, here we do not
suppose $s \mapsto j(\cdot, \cdot, s)$ to be the primitive of some locally bounded measurable
function. Instead, we only assume $s \mapsto j(\cdot, \cdot, s)$ to be locally Lipschitz and
for its generalized Clarke's gradient to satisfy certain local growth conditions.
Under these more general assumptions on j, the methods of proofs as used
in Sect. 4.3 for the elliptic case cannot be applied here in a straightforward
manner and have to be modified appropriately. However, it should be noted
that the more general approach we are going to develop in this section also
works in the elliptic case.

Problem (4.156) will be treated within the framework of evolution equations as in Sect. 3.3. Therefore, we are going to use the same notation as
in Sect. 3.3; i.e., we set $X = L^p(0, \tau; V)$ and $X_0 = L^p(0, \tau; V_0)$, where $V = W^{1,p}(\Omega)$ and $V_0 = W_0^{1,p}(\Omega)$, and introduce

$$W = \{u \in X : u_t \in X^*\}, \quad (\text{respectively,} \quad W_0 = \{u \in X_0 : u_t \in X_0^*\}),$$

where the derivative $u_t = \partial u/\partial t = u'$ is understood in the sense of vector-valued distributions (see Sect. 2.4). Throughout this section, we assume

$$2 \le p < \infty \quad \text{and} \quad q : 1/q + 1/p = 1.$$

If there is no ambiguity, we use the notation $\langle \cdot, \cdot \rangle$ for any of the dual pairings
between X and X^*, X_0 and X_0^*, V and V^*, and V_0 and V_0^*. For example,
with $f \in X_0^*, u \in X_0$,

$$\langle f, u \rangle = \int_0^\tau \langle f(t), u(t) \rangle \, dt.$$

Finally we note that homogeneous initial and boundary values can be assumed
without loss of generality (see Sect. 3.3).

4.5.1 Comparison Principle

We impose the following hypotheses of Leray-Lions type on the coefficient functions a_i, $i = 1, \ldots, N$, of the operator A.

(A1) $a_i : Q \times \mathbb{R} \times \mathbb{R}^N \to \mathbb{R}$ are Carathéodory functions; i.e., $a_i(\cdot, \cdot, s, \xi) : Q \to \mathbb{R}$ is measurable for all $(s, \xi) \in \mathbb{R} \times \mathbb{R}^N$ and $a_i(x, t, \cdot, \cdot) : \mathbb{R} \times \mathbb{R}^N \to \mathbb{R}$ is continuous for a.e. $(x, t) \in Q$, and

$$|a_i(x, t, s, \xi)| \leq k_0(x, t) + c_0 \left(|s|^{p-1} + |\xi|^{p-1} \right)$$

for a.e. $(x, t) \in Q$ and for all $(s, \xi) \in \mathbb{R} \times \mathbb{R}^N$, for some constant $c_0 > 0$ and some function $k_0 \in L^q(Q)$.

(A2) $\displaystyle\sum_{i=1}^{N} (a_i(x, t, s, \xi) - a_i(x, t, s, \xi'))(\xi_i - \xi_i') > 0$ for a.e. $(x, t) \in Q$, for all $s \in \mathbb{R}$ and all $\xi, \xi' \in \mathbb{R}^N$ with $\xi \neq \xi'$.

(A3) $\displaystyle\sum_{i=1}^{N} a_i(x, t, s, \xi)\xi_i \geq \nu|\xi|^p - k_1(x, t)$ for a.e. $(x, t) \in Q$ and for all $(s, \xi) \in \mathbb{R} \times \mathbb{R}^N$, for some constant $\nu > 0$ and some function $k_1 \in L^1(Q)$.

(A4) $|a_i(x, t, s, \xi) - a_i(x, t, s', \xi)| \leq [k_2(x, t) + |s|^{p-1} + |s'|^{p-1} + |\xi|^{p-1}]\omega(|s - s'|)$ for a.e. $(x, t) \in Q$, for all $s, s' \in \mathbb{R}$ and all $\xi \in \mathbb{R}^N$, for some function $k_2 \in L^q(Q)$ and a continuous function $\omega : [0, +\infty) \to [0, +\infty)$ satisfying

$$\int_{0+} \frac{1}{\omega(r)}\, dr = +\infty.$$

From (A1), the semilinear form $a : X \times X \to \mathbb{R}$ given by

$$a(u, v) = \int_Q \sum_{i=1}^{N} a_i(\cdot, \cdot, u, \nabla u) \frac{\partial v}{\partial x_i}\, dx dt$$

is well defined for any $(u, v) \in X \times X$, and the operator $A : X \to X_0^*$ (respectively, $A : X_0 \to X_0^*$) defined by

$$\langle Au, \varphi \rangle = a(u, \varphi) \quad \text{for all} \ \varphi \in X_0$$

is continuous and bounded. For an appropriate functional analytic setting of the inclusion problem (4.156), we introduce the operator $L = \partial/\partial t$ with domain $D(L)$ given by

$$D(L) = \{u \in X_0 : u_t \in X_0^* \text{ and } u(\cdot, 0) = 0 \text{ in } \Omega\},$$

where $L : D(L) \subset X_0 \to X_0^*$ is defined by

$$\langle Lu, \varphi \rangle = \int_0^\tau \langle u_t(t), \varphi(t) \rangle\, dt \quad \text{for all} \ \varphi \in X_0.$$

We already know that $L : D(L) \subset X_0 \to X_0^*$ is a closed, densely defined, and maximal monotone operator (see Lemma 2.149). This way (4.156) can be rewritten in the form of the operator equation

$$u \in D(L): \quad Lu + Au + \partial j(\cdot, \cdot, u) \ni Fu + h \quad \text{in } X_0^*. \tag{4.157}$$

A solution of (4.156) (respectively, (4.157)) is defined as follows.

Definition 4.43. *A function $u \in D(L)$ is called a solution of (4.156) (respectively, (4.157)) if $Fu \in L^q(Q)$ and if there is a function $\eta \in L^q(Q)$ such that*

(i) $\eta(x,t) \in \partial j(x,t,u(x,t))$ *for a.e.* $(x,t) \in Q$.
(ii) $\langle Lu + Au, \varphi \rangle + \int_Q \eta(x,t)\, \varphi(x,t)\, dx dt = \int_Q (Fu)(x,t)\varphi(x,t)\, dx dt + \langle h, \varphi \rangle$
   for all $\varphi \in X_0$.

Our notion of sub-supersolutions of (4.156) reads as follows.

Definition 4.44. *A function $\bar{u} \in W$ is called a supersolution of problem (4.156) if $F\bar{u} \in L^q(Q)$ and if there is a function $\bar{\eta} \in L^q(Q)$ such that*

(i) $\bar{u}(x,0) \geq 0$ *in Ω and $\bar{u} \geq 0$ on Γ.*
(ii) $\bar{\eta}(x,t) \in \partial j(x,t,\bar{u}(x,t))$ *for a.e.* $(x,t) \in Q$.
(iii) $\langle \bar{u}_t + A\bar{u}, \varphi \rangle + \int_Q \bar{\eta}(x,t)\varphi(x,t)\, dx dt \geq \int_Q (F\bar{u})(x,t)\, \varphi(x,t)\, dx dt + \langle h, \varphi \rangle$
   for all $\varphi \in X_0 \cap L_+^p(Q)$.

Similarly, a function $\underline{u} \in W$ is called a subsolution problem (4.156) if the reversed inequalities hold in Definition 4.44 with $\bar{u}, \bar{\eta}$ replaced by $\underline{u}, \underline{\eta}$.

Let $\underline{u}, \bar{u} \in W$ be an ordered pair of sub- and supersolutions of (4.156). We impose the following local growth hypotheses on j and f.

(H1) The function $j : Q \times \mathbb{R} \to \mathbb{R}$ satisfies:
   (i) $j(\cdot, \cdot, s) : Q \to \mathbb{R}$ is measurable for all $s \in \mathbb{R}$.
   (ii) $j(x,t,\cdot) : \mathbb{R} \to \mathbb{R}$ is locally Lipschitz and constants $\alpha > 0$ and $c_1 \geq 0$ exist such that

$$\xi_1 \leq \xi_2 + c_1(s_2 - s_1)^{p-1}$$

   for a.e. $(x,t) \in Q$, for all $\xi_i \in \partial j(x,t,s_i)$, $i = 1,2$, and for all s_1, s_2 with $\underline{u}(x,t) - \alpha \leq s_1 < s_2 \leq \bar{u}(x,t) + \alpha$.
   (iii) There is a function $k_3 \in L_+^q(Q)$ such that

$$|z| \leq k_3(x,t)$$

   for a.e. $(x,t) \in Q$, for all $s \in [\underline{u}(x,t) - 2\alpha, \bar{u}(x,t) + 2\alpha]$ and all $z \in \partial j(x,t,s)$, where α is the one entering (ii).

(H2) The function $f : Q \times \mathbb{R} \to \mathbb{R}$ is a Carathéodory function and $k_4 \in L^q_+(Q)$ exists such that

$$|f(x, t, s)| \le k_4(x, t)$$

for a.e. $(x, t) \in Q$, for all $s \in [\underline{u}(x, t), \overline{u}(x, t)]$.

In preparation for the comparison principle to be proved in this section, we provide first some preliminaries.

With α given in (H1)(ii), we introduce the truncation operator T_α by

$$(T_\alpha u)(x, t) = \begin{cases} \overline{u}(x, t) + \alpha & \text{if } u(x, t) > \overline{u}(x, t) + \alpha \\ u(x, t) & \text{if } \underline{u}(x, t) - \alpha \le u(x, t) \le \overline{u}(x, t) + \alpha \\ \underline{u}(x, t) - \alpha & \text{if } u(x, t) < \underline{u}(x, t) - \alpha. \end{cases}$$

It is well known that the truncation T_α is continuous and bounded from X into X. Let $\rho : \mathbb{R} \to \mathbb{R}$ be a mollifier function; that is, $\rho \in C^\infty_0((-1, 1))$, $\rho \ge 0$ and

$$\int_{-\infty}^{+\infty} \rho(s)\, ds = 1.$$

For any $\varepsilon > 0$, we define the regularization j^ε of j with respect to the third variable by convolution, i.e.,

$$j^\varepsilon(x, t, s) = \frac{1}{\varepsilon} \int_{-\infty}^{+\infty} j(x, t, s - \zeta) \rho\left(\frac{\zeta}{\varepsilon}\right) d\zeta,$$

and we introduce the operator $J^\varepsilon_\alpha : L^p(Q) \to L^q(Q)$ by

$$J^\varepsilon_\alpha u = (j^\varepsilon)'(\cdot, \cdot, (T_\alpha u)(\cdot, \cdot)), \tag{4.158}$$

where $(j^\varepsilon)'(x, t, s)$ stands for the derivative with respect to s. The definition makes sense because, by (H1)(iii), $k_3 \in L^q(Q)$, and we have

$$|(J^\varepsilon_\alpha u)(x, t)| = |(j^\varepsilon)'(x, t, (T_\alpha u)(x, t))| \le k_3(x, t) \tag{4.159}$$

for a.e. $(x, t) \in Q$, for all $u \in L^p(Q)$ and for all ε with $0 < \varepsilon < \alpha$. To show that (4.159) is true, we see from (H1)(iii) that

$$(j^\varepsilon)'(x, t, (T_\alpha u)(x, t)) \in \frac{1}{\varepsilon} \int_{-\infty}^{+\infty} \partial j(x, t, (T_\alpha u)(x, t) - \zeta) \rho\left(\frac{\zeta}{\varepsilon}\right) d\zeta. \tag{4.160}$$

Here we used the Aubin–Clarke Theorem (see Theorem 2.181) whose application is possible from the inequalities

$$\underline{u}(x, t) - 2\alpha \le \underline{u}(x, t) - \alpha - \zeta \le (T_\alpha u)(x, t) - \zeta \le \overline{u}(x, t) + \alpha - \zeta \le \overline{u}(x, t) + 2\alpha.$$

Using again (H1)(iii), it results in

$$|(j^\varepsilon)'(x, t, (T_\alpha u)(x, t))| \le \frac{1}{\varepsilon} \int_{-\infty}^{+\infty} k_3(x, t) \rho\left(\frac{\zeta}{\varepsilon}\right) d\zeta = k_3(x, t);$$

i.e., (4.159) is true.

Lemma 4.45. *Let* $(u_n) \subset W_0$ *and* $(\varepsilon_n) \subset (0, \alpha)$ *such that* $\varepsilon_n \to 0$ *as* $n \to \infty$.
If

(i) $u_n \rightharpoonup u$ *in* W_0 *as* $n \to \infty$,
(ii) $J_\alpha^\varepsilon u_n \rightharpoonup \eta$ *in* $L^q(Q)$ *as* $n \to \infty$,

then $\eta(x, t) \in \partial j(x, t, (T_\alpha u)(x, t))$ *for a.e.* $(x, t) \in Q$.

Proof. Let us first establish the following inequality:

$$\int_Q \limsup_{n \to \infty} \left(\frac{1}{\varepsilon_n} \int_{-\infty}^{+\infty} j^0(x, t, (T_\alpha u_n)(x, t) - \zeta; w(x, t)) \rho\left(\frac{\zeta}{\varepsilon_n}\right) d\zeta \right) dx\, dt$$
$$\geq \langle \eta, w \rangle_{L^q(Q), L^p(Q)}, \quad \forall\, w \in L^p(Q), \tag{4.161}$$

where the notation j^0 stands for the generalized directional derivative in the sense of Clarke of j with respect to the third variable. For any $w \in L^p(Q)$, using (4.158), (4.160), and [68, Proposition 2.1.2], we have

$$\langle J_\alpha^{\varepsilon_n} u_n, w \rangle_{L^q(Q), L^p(Q)} = \langle (j^{\varepsilon_n})'(T_\alpha u_n), w \rangle_{L^q(Q), L^p(Q)}$$
$$= \int_Q (j^{\varepsilon_n})'(x, t, (T_\alpha u_n)(x, t)) w(x, t)\, dx\, dt$$
$$= \int_Q \left(\frac{1}{\varepsilon_n} \int_{-\infty}^{+\infty} z_n(x, t, \zeta) \rho\left(\frac{\zeta}{\varepsilon_n}\right) d\zeta \right) w(x, t)\, dx\, dt$$
$$\leq \int_Q \left(\frac{1}{\varepsilon_n} \int_{-\infty}^{+\infty} j^0(x, t, (T_\alpha u_n)(x, t) - \zeta; w(x, t)) \rho\left(\frac{\zeta}{\varepsilon_n}\right) d\zeta \right) dx\, dt,$$
$$\tag{4.162}$$

with $z_n(x, t, \zeta) \in \partial j(x, t, (T_\alpha u_n)(x, t) - \zeta)$. Passing to the upper limit in the previous inequality (4.162) and using $J_\alpha^{\varepsilon_n} u_n \rightharpoonup \eta$ in $L^q(Q)$ as well as Fatou's lemma (see Theorem 2.65), we obtain

$$\langle \eta, w \rangle_{L^q(Q), L^p(Q)} = \lim_{n \to \infty} \langle J_\alpha^{\varepsilon_n} u_n, w \rangle_{L^q(Q), L^p(Q)}$$
$$\leq \limsup_{n \to \infty} \int_Q \left(\frac{1}{\varepsilon_n} \int_{-\infty}^{+\infty} j^0(x, t, (T_\alpha u_n)(x, t) - \zeta; w(x, t)) \rho\left(\frac{\zeta}{\varepsilon_n}\right) d\zeta \right) dx\, dt$$
$$\leq \int_Q \limsup_{n \to \infty} \left(\frac{1}{\varepsilon_n} \int_{-\infty}^{+\infty} j^0(x, t, (T_\alpha u_n)(x, t) - \zeta; w(x, t)) \rho\left(\frac{\zeta}{\varepsilon_n}\right) d\zeta \right) dx\, dt,$$

which is (4.161). The application of Fatou's lemma was possible because of the inequalities

$$\frac{1}{\varepsilon_n} \int_{-\infty}^{+\infty} j^0(x, t, (T_\alpha u_n)(x, t) - \zeta; w(x, t)) \rho\left(\frac{\zeta}{\varepsilon_n}\right) d\zeta$$
$$= \frac{1}{\varepsilon_n} \int_{-\infty}^{+\infty} z_n(x, t, \zeta) w(x, t) \rho\left(\frac{\zeta}{\varepsilon_n}\right) d\zeta$$

$$\leq \frac{1}{\varepsilon_n} \int_{-\infty}^{+\infty} k_3(x,t)|w(x,t)|\rho\left(\frac{\zeta}{\varepsilon_n}\right) d\zeta = k_3(x,t)|w(x,t)|,$$

with $k_3 w \in L^1(Q)$, and

$$\int_Q \left(\frac{1}{\varepsilon_n} \int_{-\infty}^{+\infty} j^0(x,t,(T_\alpha u_n)(x,t) - \zeta; w(x,t))\rho\left(\frac{\zeta}{\varepsilon_n}\right) d\zeta \right) dx\,dt$$

$$\geq -\int_Q \left(\frac{1}{\varepsilon_n} \int_{-\infty}^{+\infty} |z_n(x,t,\zeta)|\,|w(x,t)|\rho\left(\frac{\zeta}{\varepsilon_n}\right) d\zeta \right) dx\,dt$$

$$\geq -\int_Q k_3(x,t)|w(x,t)|\,dx\,dt,$$

where $z_n(x,t,\zeta) \in \partial j(x,t,(T_\alpha u_n)(x,t) - \zeta)$ is fixed such that

$$j^0(x,t,(T_\alpha u_n)(x,t) - \zeta; w(x,t)) = z_n(x,t,\zeta)w(x,t).$$

Next we are going to show that

$$\limsup_{n\to\infty} \left(\frac{1}{\varepsilon_n} \int_{-\infty}^{+\infty} j^0(x,t,(T_\alpha u_n)(x,t) - \zeta; w(x,t))\rho\left(\frac{\zeta}{\varepsilon_n}\right) d\zeta \right)$$

$$\leq j^0(x,t,(T_\alpha u)(x,t); w(x,t)), \tag{4.163}$$

for a.e. $(x,t) \in Q$, and for all $w \in L^p(Q)$. From the compact embedding $W_0 \subset L^p(Q)$ and the continuity of T_α, we get $T_\alpha u_n \to T_\alpha u$ in $L^p(Q)$ as $n \to \infty$, which by passing eventually to a subsequence (again denoted by $T_\alpha u_n$) results in

$$(T_\alpha u_n)(x,t) \to (T_\alpha u)(x,t) \text{ for a.e. } (x,t) \in Q \text{ as } n \to \infty. \tag{4.164}$$

Thus, to prove (4.163), it is sufficient to show that (4.163) holds for every $w \in L^p(Q)$ and every point $(x,t) \in Q$ satisfying (4.164) (because (4.164) is valid for a.e. $(x,t) \in Q$). Fix $w \in L^p(Q)$ and any point $(x,t) \in Q$ satisfying (4.164). Let $\varepsilon > 0$ be an arbitrary number. The upper semicontinuity of $j^0(x,t,\cdot; w(x,t))$ yields a number $\delta > 0$ such that for all ξ with $|\xi - (T_\alpha u)(x,t)| < \delta$, we have

$$j^0(x,t,\xi; w(x,t)) < j^0(x,t,(T_\alpha u)(x,t); w(x,t)) + \varepsilon. \tag{4.165}$$

On the other hand, the convergence in (4.164) gives a positive integer n_ε [depending on (x,t)] such that

$$|(T_\alpha u_n)(x,t) - \zeta - (T_\alpha u)(x,t)| \leq |(T_\alpha u_n)(x,t) - (T_\alpha u)(x,t)| + |\zeta|$$
$$\leq |(T_\alpha u_n)(x,t) - (T_\alpha u)(x,t)| + \varepsilon_n < \delta$$

for all $n \geq n_\varepsilon$, and for all $\zeta \in (-\varepsilon_n, \varepsilon_n)$. This result allows us to apply (4.165) with $\xi = (T_\alpha u_n)(x,t) - \zeta$ to get

$$j^0(x,t,(T_\alpha u_n)(x,t) - \zeta; w(x,t)) < j^0(x,t,(T_\alpha u)(x,t); w(x,t)) + \varepsilon$$

for all $n \geq n_\varepsilon$ and all $\zeta \in (-\varepsilon_n, \varepsilon_n)$. Consequently, we may write

$$\frac{1}{\varepsilon_n} \int_{-\infty}^{+\infty} j^0(x, t, (T_\alpha u_n)(x, t) - \zeta; w(x, t)) \rho\left(\frac{\zeta}{\varepsilon_n}\right) d\zeta$$

$$= \frac{1}{\varepsilon_n} \int_{-\varepsilon_n}^{\varepsilon_n} j^0(x, t, (T_\alpha u_n)(x, t) - \zeta; w(x, t)) \rho\left(\frac{\zeta}{\varepsilon_n}\right) d\zeta$$

$$< j^0(x, t, (T_\alpha u)(x, t); w(x, t)) + \varepsilon.$$

Passing to the upper limit in the last inequality as $n \to \infty$, we derive that

$$\limsup_{n \to \infty} \left(\frac{1}{\varepsilon_n} \int_{-\infty}^{+\infty} j^0(x, t, (T_\alpha u_n)(x, t) - \zeta; w(x, t)) \rho\left(\frac{\zeta}{\varepsilon_n}\right) d\zeta\right)$$

$$\leq j^0(x, t, (T_\alpha u)(x, t); w(x, t)) + \varepsilon.$$

As $\varepsilon > 0$ was arbitrary, we conclude that (4.163) holds true. Combining (4.161) and (4.163), we obtain

$$\int_Q \eta(x, t) w(x, t) \, dx \, dt \leq \int_Q j^0(x, t, (T_\alpha u)(x, t); w(x, t)) \, dx \, dt \qquad (4.166)$$

for all $w \in L^p(Q)$. Next we apply a Lebesgue's point argument to (4.166). Let $r \in \mathbb{R}$ be an arbitrarily fixed real, and let $B((\bar{x}, \bar{t}), R)$ be the open ball in Q with radius $R > 0$ centered at some fixed point $(\bar{x}, \bar{t})$. Denote by $\chi_{B((\bar{x}, \bar{t}), R)}$ the characteristic function of $B((\bar{x}, \bar{t}), R)$. Setting $w = \chi_{B((\bar{x}, \bar{t}), R)} r$ in (4.166), we have

$$\int_Q \eta \, \chi_{B((\bar{x}, \bar{t}), R)} \, r \, dx \, dt \leq \int_Q j^0(\cdot, \cdot, (T_\alpha u); \chi_{B((\bar{x}, \bar{t}), R)} \, r) \, dx \, dt.$$

This inequality can be equivalently written as

$$\frac{1}{|B((\bar{x}, \bar{t}), R)|} \int_{B((\bar{x}, \bar{t}), R)} \eta(x, t) r \, dx \, dt$$

$$\leq \frac{1}{|B((\bar{x}, \bar{t}), R)|} \int_{B((\bar{x}, \bar{t}), R)} j^0(x, t, (T_\alpha u)(x, t); r) \, dx \, dt,$$

where $|B((\bar{x}, \bar{t}), R)|$ denotes the Lebesgue-measure of $B((\bar{x}, \bar{t}), R)$. As the functions η and $j^0(\cdot, \cdot, (T_\alpha u)(\cdot, \cdot); r)$ belong to $L^q(Q)$, letting $R \to 0$ in the previous inequality, we arrive at

$$\eta(\bar{x}, \bar{t}) r \leq j^0(\bar{x}, \bar{t}, (T_\alpha u)(\bar{x}, \bar{t}); r), \quad \forall \, r \in \mathbb{R}.$$

The definition of the generalized gradient of Clarke gives

$$\eta(\bar{x}, \bar{t}) \in \partial j(\bar{x}, \bar{t}, (T^\alpha u)(\bar{x}, \bar{t})),$$

which completes the proof of the lemma.   □

The comparison principle for the parabolic inclusion (4.156) now reads as follows.

Theorem 4.46. *Let $\underline{u}$, $\bar{u}$ be sub-supersolutions of (4.156) satisfying $\underline{u} \leq \bar{u}$. If hypotheses (A1)–(A3) and (H1), (H2) are fulfilled, then problem (4.156) admits at least one solution u within the ordered interval $[\underline{u}, \bar{u}]$.*

Proof. The proof will be done in three steps.

Step 1: Auxiliary Problem.

We consider first the following regularized truncated initial-boundary value problem:

$$(P_\varepsilon) \qquad u \in D(L): \quad Lu + A_T u + J_\alpha^\varepsilon u + \lambda B(u) = (F \circ T)u + h \quad \text{in } X_0^*,$$

where J_α^ε is given by (4.158) with $0 < \varepsilon < \alpha$, and λ is some constant sufficiently large. Similarly as in (3.110), the operator A_T is defined by

$$A_T u(x,t) = -\sum_{i=1}^{N} \frac{\partial}{\partial x_i} a_i(x,t,Tu(x,t),\nabla u(x,t)), \qquad (4.167)$$

where T is the truncation between the given sub- and supersolution $\underline{u}$ and $\bar{u}$, respectively. The operator B denotes the Nemytskij operator generated by the cutoff function $b : Q \times \mathbb{R}$, which is given by

$$b(x,t,s) = \begin{cases} (s - \bar{u}(x,t))^{p-1} & \text{if } s > \bar{u}(x,t) \\ 0 & \text{if } \underline{u}(x,t) \leq s \leq \bar{u}(x,t) \\ -(\underline{u}(x,t) - s)^{p-1} & \text{if } s < \underline{u}(x,t). \end{cases}$$

In view of the boundedness of the regularization according to (4.159), the existence of solutions of problem (P_ε) can be proved in the same way as Lemma 3.36 in Sect. 3.3.2. Therefore, for each ε with $0 < \varepsilon < \alpha$, solutions of (P_ε) exist. Let (ε_n) be a sequence satisfying $\varepsilon_n \in (0,\alpha)$ and $\varepsilon_n \to 0$ as $n \to \infty$. For each n, let u_n be a solution of problem (P_{ε_n}). By Remark 2.145, we have for any $u \in D(L)$

$$\langle Lu, u \rangle = \frac{1}{2} \|u(\tau)\|_{L^2(\Omega)}^2, \qquad (4.168)$$

which in conjunction with (A3) and (4.159) yields the boundedness of (u_n) in X_0. This result combined with (P_{ε_n}) implies that (u_n') is bounded in X_0^*. Hence, the sequence (u_n) is bounded in W_0. Thus, a subsequence of (u_n) exists [again denoted by (u_n)] satisfying the hypotheses of Lemma 4.45.

Step 2: Passage to the Limit as $n \to \infty$.

According to the previous step ,there is a sequence $(u_n) \subset W_0$ such that

(i) $u_n \rightharpoonup u$ in W_0 as $n \to \infty$.
(ii) $J_\alpha^\varepsilon u_n \rightharpoonup \eta$ in $L^q(Q)$ as $n \to \infty$ with $\eta(x,t) \in \partial j(x,t,(T_\alpha u)(x,t))$ for a.e. $(x,t) \in Q$.

On the basis of (P_{ε_n}) and (4.168), we have

$$\langle u_t, u_n - u \rangle + \langle A_T u_n, u_n - u \rangle + \langle J^\varepsilon_\alpha u_n, u_n - u \rangle_{L^q(Q), L^p(Q)}$$
$$+ \langle \lambda B(u_n) - (F \circ T)(u_n), u_n - u \rangle$$
$$\leq \langle h, u_n - u \rangle.$$

Passing to the lim sup in the last inequality and using properties (i) and (ii) as well as the fact that $\lambda B - F \circ T : D(L) \subset X_0 \to X_0^*$ is completely continuous with respect to $D(L)$, we obtain

$$\limsup_{n \to \infty} \langle A_T u_n, u_n - u \rangle \leq 0.$$

Taking into account that $u_n \rightharpoonup u$ in W_0, the pseudo-monotonicity of $A_T : X_0 \to X_0^*$ with respect to the graph norm of $D(L)$ yields

$$A_T u_n \rightharpoonup A_T u \quad \text{in } X_0^* \text{ as } n \to \infty.$$

Letting now $n \to \infty$ in problem (P_{ε_n}) and making use of the above convergence properties, we conclude that $u \in W_0$ is a solution of the problem

$$(P_0) \qquad u \in D(L): \quad Lu + A_T u + \eta + \lambda B(u) = (F \circ T)u + h \quad \text{in } X_0^*,$$

where $\eta(x,t) \in \partial j(x,t,(T_\alpha u)(x,t))$ for a.e. $(x,t) \in Q$.

Step 3: Comparison $\underline{u} \leq u \leq \bar{u}$.

We complete the proof of the theorem by showing that any solution u of (P_0) satisfies $\underline{u} \leq u \leq \bar{u}$, because then we have $Tu = u$, and $T_\alpha u = u$, which shows that $A_T u = Au$, $\eta(x,t) \in \partial j(x,t,u(x,t))$ for a.e. $(x,t) \in Q$ and $F \circ Tu = Fu$, and $Bu = 0$, and therefore, any solution u of (P_0) must be a solution of the original parabolic inclusion (4.156) satisfying $\underline{u} \leq u \leq \bar{u}$. We first prove that $u \leq \bar{u}$. To this end, recall that $\bar{u} \in W$ is a supersolution of (4.156), i.e., $\bar{u} \geq 0$ on $\Omega \times \{0\}$, $\bar{u} \geq 0$ on Γ, and there is a $\bar{\eta} \in L^q(\Omega)$ such that $\bar{\eta}(x,t) \in \partial j(x,t,\bar{u}(x,t))$ for a.e. $(x,t) \in Q$, and

$$\langle \bar{u}_t + A\bar{u} + \bar{\eta}, \varphi \rangle \geq \langle F\bar{u} + h, \varphi \rangle \quad \text{for all } \varphi \in X_0 \cap L^p_+(Q). \qquad (4.169)$$

Subtracting (4.169) from (P_0) and taking the special nonnegative test function $\varphi = (u - \bar{u})^+ \in X_0$, we obtain

$$\langle u_t - \bar{u}_t + A_T u - A\bar{u} + \eta - \bar{\eta} + \lambda B(u), (u - \bar{u})^+ \rangle$$
$$\leq \langle (F \circ T)u - F\bar{u}, (u - \bar{u})^+ \rangle. \qquad (4.170)$$

As a consequence of Lemma 2.146 (see Example 2.148), we have

$$\langle (u - \bar{u})_t, (u - \bar{u})^+ \rangle = \frac{1}{2} \| (u - \bar{u})^+ (\cdot, \tau) \|^2_{L^2(\Omega)}. \qquad (4.171)$$

Taking (4.171) into account, we can proceed in a similar way as in Step 4 of the proof of Theorem 4.31 to finally get the following inequality:

$$(\lambda - c_1) \int_{\{u > \bar{u}\}} (u(x,t) - \bar{u}(x,t))^p \, dxdt \leq 0. \tag{4.172}$$

Now we may select λ large enough such that $\lambda - c_1 > 0$, which in view of (4.172) implies that

$$\int_{\{u > \bar{u}\}} (u(x,t) - \bar{u}(x,t))^p \, dxdt = \int_Q [(u(x,t) - \bar{u}(x,t))^+]^p \, dxdt = 0,$$

and hence, it follows that $(u(x,t) - \bar{u}(x,t))^+ = 0$ for a.e. $(x,t) \in Q$; i.e., $u \leq \bar{u}$ a.e. in Q. In a similar way, we can prove $\underline{u} \leq u$, which completes the proof of the theorem. □

Remark 4.47. One possibility to determine sub- and supersolutions of the multivalued problem (4.156) is to replace the problem by the following single-valued one:

$$u \in D(L): \quad u_t + Au + \hat{j}(\cdot, \cdot, u) = Fu + h \quad \text{in } X_0^*, \tag{4.173}$$

where $\hat{j} : Q \times \mathbb{R} \to \mathbb{R}$ may be any single-valued measurable selection of ∂j. Then obviously any subsolution (supersolution) $\underline{u}$ ($\bar{u}$) of the single-valued problem (4.173) is a subsolution (supersolution) of the multivalued one with $\underline{\eta} = \hat{j}(\cdot, \cdot, \underline{u})$ ($\bar{\eta} = \hat{j}(\cdot, \cdot, \bar{u})$).

Example 4.48. Let $p = q = 2$ and $h \in X_0^*$. Consider the initial-Dirichlet boundary value problem

$$(E) \quad u \in D(L): Lu - \sum_{i=1}^N \frac{\partial}{\partial x_i} a_i(\cdot, \cdot, \nabla u) + \partial j(\cdot, \cdot, u) \ni Fu + h \quad \text{in } X_0^*,$$

where $j : Q \times \mathbb{R} \to \mathbb{R}$ verifies condition (H1)(i), and the generalized gradient ∂j satisfies the following global growth conditions:

(ii) $\xi_1 \leq \xi_2 + c_1(s_2 - s_1)$ for a.e. $(x,t) \in Q$ and for all $\xi_i \in \partial j(x, t, s_i)$, $i = 1, 2$, with $s_1 < s_2$, and c_1 some positive constant.

(iii) There is some function $k_5 \in L_+^2(Q)$ such that $|\xi| \leq k_5(x,t) + c_2|s|$ for a.e. $(x,t) \in Q$, for all $s \in \mathbb{R}$ and $\xi \in \partial j(x,t,s)$.

Furthermore, we assume conditions (A1)–(A3) for a_i (note that here (A4) is trivially satisfied) and suppose $f : Q \times \mathbb{R} \to \mathbb{R}$ to be a Carathéodory function having the following global growth:

(iv) $|f(x,t,s)| \leq k_6(x,t) + c_2|s|$, for a.e. $(x,t) \in Q$, for all $s \in \mathbb{R}$, and with some function $k_6 \in L_+^2(Q)$ and a positive constant c_2.

Now we consider the following uniquely solvable single-valued problems: Find $u \in D(L)$ such that

(U) $$Lu - \sum_{i=1}^{N} \frac{\partial}{\partial x_i} a_i(\cdot, \cdot, \nabla u) - (k_5 + c_2|u|) = k_6 + c_2|u| + h \quad \text{in } X_0^*$$

and

(S) $$Lu - \sum_{i=1}^{N} \frac{\partial}{\partial x_i} a_i(\cdot, \cdot, \nabla u) + (k_5 + c_2|u|) = -(k_6 + c_2|u|) + h \quad \text{in } X_0^*.$$

Denote the unique solutions of (U) and (S) by $\bar{u}$ and $\underline{u}$, respectively. Then by comparison, we get $\underline{u} \leq \bar{u}$. Furthermore, $\underline{u}$ and $\bar{u}$ are sub- and supersolutions for problem (E). To verify this result for the case of the supersolution, let $\hat{j}$ be any single-valued measurable selection of ∂j; then the definition is satisfied with $\bar{\eta} = \hat{j}(\cdot, \cdot, \bar{u})$. Similarly, we verify that $\underline{u}$ is a subsolution. We easily see also that all the hypotheses (H1)–(H2) are fulfilled. For instance, the function k_3 in (H1)(iii) is

$$k_3(x,t) = k_5(x,t) + c_2 \max\{|\bar{u}(x,t) + c_0|, |\underline{u}(x,t) - c_0|\},$$

with a constant $c_0(= 2\alpha) > 0$, whereas k_4 required in (H2) is

$$k_4(x,t) = k_6(x,t) + c_2 \max\{|\bar{u}(x,t)|, |\underline{u}(x,t)|\}.$$

Thus, Theorem 4.46 can be applied.

Example 4.49. We give an example where Theorem 4.46 provides nonnegative bounded solutions of an initial-Dirichlet boundary value problem of the form (4.156) in case that the following hypotheses are satisfied:

(i) $a_i(x,t,0,0) = a_i(x,t,1,0) = 0$ for a.e. $(x,t) \in Q$, $i = 1, \ldots, N$.
(ii) $h = 0$.
(iii) $j : Q \times \mathbb{R} \to \mathbb{R}$ is a Carathéodory function and $j(x,t,\cdot)$ is locally Lipschitz for a.e. $(x,t) \in Q$.
(iv)   Constants $\alpha > 0$ and $c_1 \geq 0$ exist such that

$$\xi_1 \leq \xi_2 + c_1(s_2 - s_1)^{p-1}$$

for a.e. $(x,t) \in Q$, for all $\xi_i \in \partial j(x,t,s_i)$, $i = 1, 2$, and for all s_1, s_2 with $-\alpha \leq s_1 < s_2 \leq 1 + \alpha$.
(v)   There is some function $k_3 \in L_+^q(Q)$ such that $|z| \leq k_3(x,t)$ for a.e. $(x,t) \in Q$, for all $s \in [-2\alpha, 1+2\alpha]$ and $z \in \partial j(x,t,s)$.
(vi) $f : Q \times \mathbb{R} \to \mathbb{R}$ is a Carathéodory function for which $k_4 \in L_+^q(Q)$ exists such that

$$|f(x,t,s)| \leq k_4(x,t) \quad \text{for a.e. } (x,t) \in Q, \quad \text{for all } s \in [0,1].$$

(vii) For a.e. $(x, t) \in Q$, the following is supposed:

$$\max\{z : z \in \partial j(x, t, 1)\} \geq f(x, t, 1)$$
$$\min\{z : z \in \partial j(x, t, 0)\} \leq f(x, t, 0).$$

By taking $\bar{\eta}(x, t) = \max\{z : z \in \partial j(x, t, 1)\}$, we readily verify that under hypotheses (i)–(vii), the constant function $\bar{u} = 1$ is a supersolution of (4.156). Similarly, setting $\underline{\eta}(x, t) = \min\{z : z \in \partial j(x, t, 0)\}$, we find that $\underline{u} = 0$ is a subsolution of (4.156). As assumptions (H1) and (H2) are easily seen to be satisfied, Theorem 4.46 can be applied that yields the existence of solutions of problem (4.156) within the ordered interval $[0, 1]$.

4.5.2 Extremality and Compactness Results

Denote by S the set of all solutions of (4.156) within the ordered interval $[\underline{u}, \bar{u}]$ of the given sub- and supersolutions $\underline{u}$ and $\bar{u}$, respectively. By Theorem 4.46, we have that $S \neq \emptyset$. Moreover, under the additional assumption (A4), the following lemma holds true.

Lemma 4.50. *Under hypotheses (A1)–(A4) and (H1–(H2), the solution set S is directed.*

Proof: We are going to show that S is upward directed, because the proof for S being downward directed can be done analogously. To this end, we show the following:

If u_1, $u_2 \in S$, then a $u \in S$ exists satisfying

$$\max\{u_1, u_2\} \leq u. \tag{4.174}$$

The proof of (4.174) will be given in three steps.

Step 1: Regularized Truncated Problem.

We set $u_0 = \max\{u_1, u_2\}$ and assume $0 < \varepsilon < \alpha$. For $k = 0, 1, 2$, we define the truncation mappings T_k related with u_k as follows:

$$(T_k u)(x, t) = \begin{cases} \bar{u}(x, t) & \text{if } u(x, t) > \bar{u}(x, t), \\ u(x, t) & \text{if } u_k(x, t) \leq u(x, t) \leq \bar{u}(x, t), \\ u_k(x, t) & \text{if } u(x, t) < u_k(x, t), \end{cases}$$

and introduce the cutoff function $b_0 : Q \times \mathbb{R} \to \mathbb{R}$ given by

$$b_0(x, t, s) = \begin{cases} (s - \bar{u}(x, t))^{p-1} & \text{if } s > \bar{u}(x, t), \\ 0 & \text{if } u_0(x, t) \leq s \leq \bar{u}(x, t), \\ -(u_0(x, t) - s)^{p-1} & \text{if } s < u_0(x, t). \end{cases}$$

Let us consider the following auxiliary regularized truncated problem:

$$(\hat{P}_\varepsilon) \qquad u \in D(L) : \quad Lu + Au + J_\alpha^\varepsilon u + \lambda B_0(u) = E(u) + h \quad \text{in } X_0^*,$$

where J_α^ε is given by (4.158), $B_0 : L^p(Q) \to L^q(Q)$ is the Nemytskij operator related to b_0, and the operator E is defined by

$$E(u) = (F \circ T_0)u + \sum_{i=1}^{2} |(F \circ T_i)u - (F \circ T_0)u|.$$

The existence of solutions of problem $(\hat{P}_\varepsilon)$ can basically be obtained in the same way as for problem (P_ε), because the operators A, B_0, E of $(\hat{P}_\varepsilon)$ possess the same structure and mapping properties as the corresponding operators A_T, B, $F \circ T$ of (P_ε), which ensures the existence of solutions of $(\hat{P}_\varepsilon)$.

Step 2: Passage to the Limit as $\varepsilon_n \to 0$.

Let $\varepsilon_n \in (0, \alpha)$ with $\varepsilon_n \to 0$ as $n \to \infty$, and let u_n be a solution of $(\hat{P}_{\varepsilon_n})$. In the same way as in the proof of Theorem 4.46, we can show that (u_n) is bounded in W_0, and thus, there is a a subsequence of (u_n) [again denoted by (u_n)] satisfying

(i) $u_n \rightharpoonup u$ in W_0 as $n \to \infty$.
(ii) $J_\alpha^\varepsilon u_n \rightharpoonup \eta$ in $L^q(Q)$ as $n \to \infty$ with $\eta(x, t) \in \partial j(x, t, (T_\alpha u)(x, t))$ for a.e. $(x, t) \in Q$.

As in Step 2 of the proof of Theorem 4.46, we may pass to the limit in $(\hat{P}_{\varepsilon_n})$, which results in the following equation $(\hat{P}_0)$ satisfied by the limit u and $\eta \in \partial j(\cdot, \cdot, T_\alpha u)$:

$$(\hat{P}_0) \qquad u \in D(L) : \quad Lu + Au + \eta + \lambda B_0(u) = E(u) + h \quad \text{in } X_0^*.$$

Step 3: Comparison $u_0 \le u \le \bar{u}$.

To complete the proof of the lemma, we only need to show that any solution u of $(\hat{P}_0)$ satisfies $u_0 \le u \le \bar{u}$, because then $B_0 u = 0$, $T_\alpha u = u$, $T_i u = u$ for $i = 0, 1, 2$, and thus $\eta \in \partial j(\cdot, \cdot, u)$ and $E(u) = Fu$, which shows that $u \in \mathcal{S}$ and $u \ge u_0 = \max\{u_1, u_2\}$.

To prove $u_0 \le u$, we show that $u_k \le u$, $k = 1, 2$. As $u_k \in \mathcal{S}$, it follows that for $k = 1, 2$, $u_k \in W_0$ is a solution of (4.156), i.e.,

$$u_k \in D(L) : \quad Lu_k + Au_k + \eta_k = Fu_k + h \quad \text{in } X_0^*, \tag{4.175}$$

where $\eta_k \in \partial j(\cdot, \cdot, u_k)$. Substracting $(\hat{P}_0)$ from (4.175) results in

$$(u_k - u)' + Au_k - Au + \eta_k - \eta - \lambda B_0 u$$
$$= Fu_k - E(u) \quad \text{in } X_0^*. \tag{4.176}$$

By (A4), for any fixed $\varepsilon > 0$, $\delta(\varepsilon) \in (0, \varepsilon)$ exists such that

$$\int_{\delta(\varepsilon)}^{\varepsilon} \frac{1}{\omega(r)} \, dr = 1.$$

Now we make use of the comparison technique developed in the proof of Theorem 3.41. To this end, we define the function

$$\theta_\varepsilon(s) = \begin{cases} 0 & \text{if } s < \delta(\varepsilon), \\ \displaystyle\int_{\delta(\varepsilon)}^s \frac{1}{\omega(r)}\, dr & \text{if } \delta(\varepsilon) \le s \le \varepsilon, \\ 1 & \text{if } s > \varepsilon. \end{cases}$$

It is clear that, for each $\varepsilon > 0$, the function θ_ε is continuous, piecewise differentiable and the derivative is nonnegative and bounded. Therefore, the function θ_ε is Lipschitz continuous and nondecreasing. In addition, it satisfies

$$\theta_\varepsilon \to \chi_{\{s>0\}} \quad \text{a.e. as } \varepsilon \to 0, \tag{4.177}$$

where $\chi_{\{s>0\}}$ is the characteristic function of the set $\{s > 0\}$. Moreover, we have

$$\theta'_\varepsilon(s) = \begin{cases} \dfrac{1}{\omega(s)} & \text{if } \delta(\varepsilon) < s < \varepsilon \\ 0 & \text{if } s \notin [\delta(\varepsilon), \varepsilon]. \end{cases}$$

Taking in the weak formulation of (4.176) the test function $\theta_\varepsilon(u_k - u) \in X_0 \cap L^p_+(Q)$, it follows that

$$\langle (u_k - u)', \theta_\varepsilon(u_k - u) \rangle + \langle Au_k - Au, \theta_\varepsilon(u_k - u) \rangle$$
$$+ \int_Q (\eta_k - \eta)\theta_\varepsilon(u_k - u)\, dx\, dt - \lambda \int_Q (B_0 u)\theta_\varepsilon(u_k - u)\, dx\, dt$$
$$= \int_Q (Fu_k - E(u))\theta_\varepsilon(u_k - u)\, dx\, dt. \tag{4.178}$$

Let Θ_ε be the primitive of the function θ_ε defined by

$$\Theta_\varepsilon(s) = \int_0^s \theta_\varepsilon(r)\, dr,$$

then for the first term on the left-hand side of (4.178), we obtain

$$\langle (u_k - u)', \theta_\varepsilon(u_k - u) \rangle = \int_\Omega \Theta_\varepsilon(u_k - u)(x, \tau)\, dx \ge 0. \tag{4.179}$$

By using (A2) and (A4), the second term on the left-hand side of (4.178) can be estimated as follows:

$$\langle Au_k - Au, \theta_\varepsilon(u_k - u) \rangle \ge -N \int_{\{\delta(\varepsilon)<u_k-u<\varepsilon\}} k\, |\nabla(u_k - u)|\, dx\, dt, \tag{4.180}$$

where $k = k_2 + |u_k|^{p-1} + |u|^{p-1} + |\nabla u|^{p-1} \in L^q(Q)$. The term on the right-hand side of (4.180) tends to zero as $\varepsilon \to 0$.

By (4.177), the application of Lebesgue's dominated convergence theorem implies that

$$\lim_{\varepsilon \to 0} \int_Q (\eta_k - \eta - \lambda B_0 u - F u_k + E(u)) \theta_\varepsilon (u_k - u) \, dx \, dt$$

$$= \int_Q (\eta_k - \eta - \lambda B_0 u - F u_k + E(u)) \chi_{\{u_k > u\}} \, dx \, dt. \qquad (4.181)$$

Taking (4.179)–(4.181) into account, we get from (4.178) by passing $\varepsilon \to 0$ the inequality

$$-\lambda \int_Q (B_0 u) \chi_{\{u_k > u\}} \, dx \, dt \le \int_Q (\eta - \eta_k + F u_k - E(u)) \chi_{\{u_k > u\}} \, dx \, dt,$$

which yields by applying the definitions of the truncations T_i entering in E as well as of the characteristic function χ the following inequality:

$$-\lambda \int_{\{u_k > u\}} (B_0 u) \, dx \, dt \le \int_{\{u_k > u\}} (\eta - \eta_k) \, dx \, dt. \qquad (4.182)$$

If (x, t) is such that $u(x, t) < u_k(x, t)$, from the definition of T_α, we see that $\underline{u}(x, t) - \alpha \le (T_\alpha u)(x, t) < u_k(x, t) \le \bar{u}(x, t) + \alpha$. Applying (H1)(ii), we derive

$$\eta(x, t) - \eta_k(x, t) \le c_1 (u_k(x, t) - (T_\alpha u)(x, t))^{p-1}.$$

Combining the previous inequality with (4.182) and making use of the definition of b_0 (respectively, B_0), we obtain

$$\lambda \int_{\{u_k > u\}} (u_0 - u)^{p-1} \, dx \, dt = -\lambda \int_{\{u_k > u\}} B_0 u \, dx \, dt$$

$$\le c_1 \int_{\{u_k > u\}} (u_k - T_\alpha u)^{p-1} \, dx \, dt.$$

If (x, t) is such that $u(x, t) < u_k(x, t)$, then by the definition of T_α, we have $(u_k - T_\alpha u)(x, t) \le (u_0 - u)(x, t)$, which yields

$$(\lambda - c_1) \int_{\{u_k > u\}} (u_0 - u)^{p-1} \, dx \, dt \le 0.$$

As $(u_0 - u)(x, t) > 0$ whenever $(u_k - u)(x, t) > 0$, we infer from the previous inequality that the Lebesgue measure of the set $\{u_k > u\}$ is equal to 0 when λ is chosen large enough such that $\lambda > c_1$. This result implies that $u_k \le u$ a.e. in Q, for $k = 1, 2$, and thus, $u_0 \le u$.

To prove $u \le \bar{u}$, we use Definition 4.44 for the supersolution $\bar{u}$, and problem $(\hat{P}_0)$, as well as the test function $\theta_\varepsilon(u - \bar{u}) \in X_0 \cap L_+^p(Q)$ to deduce

$$\langle (u - \bar{u})_t, \theta_\varepsilon(u - \bar{u}) \rangle + \langle Au - A\bar{u}, \theta_\varepsilon(u - \bar{u}) \rangle$$

$$+ \int_Q (\eta - \bar{\eta}) \theta_\varepsilon(u - \bar{u}) \, dx \, dt + \lambda \int_Q (B_0 u) \theta_\varepsilon(u - \bar{u}) \, dx \, dt$$

$$\leq \int_Q (E(u) - F\bar{u})\theta_\varepsilon(u - \bar{u}) \, dx \, dt,$$

where $\eta \in \partial j(\cdot, \cdot, T_\alpha u)$ and $\bar{\eta} \in \partial j(\cdot, \cdot, \bar{u})$. Using similar arguments as in proving (4.182), on the basis of (4.177), we obtain

$$\lambda \int_Q (B_0 u)\chi_{\{u > \bar{u}\}} \, dx \, dt \leq \int_{\{u > \bar{u}\}} (\bar{\eta} - \eta) \, dx \, dt.$$

If (x, t) is such that $u(x, t) > \bar{u}(x, t)$, we have that $\underline{u}(x, t) - \alpha \leq \bar{u}(x, t) < T_\alpha u(x, t) \leq \bar{u}(x, t) + \alpha$. Applying (H1)(ii), we get

$$\bar{\eta}(x, t) - \eta(x, t) \leq c_1(T_\alpha u(x, t) - \bar{u}(x, t))^{p-1}.$$

Consequently, in view of the definition of b_0, we deduce that

$$\lambda \int_{\{u > \bar{u}\}} (u - \bar{u})^{p-1} \, dx \, dt \leq c_1 \int_{\{u > \bar{u}\}} (T_\alpha u - \bar{u})^{p-1} \, dx \, dt.$$

As $T_\alpha u(x, t) \leq u(x, t)$ whenever $u(x, t) > \bar{u}(x, t)$, it follows that

$$(\lambda - c_1) \int_{\{u > \bar{u}\}} (u - \bar{u})^{p-1} \, dx \, dt \leq 0,$$

which by choosing $\lambda > c_1$ implies $u \leq \bar{u}$. $\square$

Next, we prove compactness results for the solution set $\mathcal{S}$ of all solutions of (4.156) within the ordered interval of the given sub- and supersolutions $\underline{u}$ and $\bar{u}$, respectively.

Lemma 4.51. *The solution set $\mathcal{S}$ is weakly sequentially compact in W_0 and compact in X_0.*

Proof: As $\mathcal{S} \subset [\underline{u}, \bar{u}]$, the boundedness of $\mathcal{S}$ in W_0 follows from hypothesis (A3) and the growth conditions (H1)(iii) and (H2). Let $(u_n) \subset \mathcal{S}$ be any sequence. By the reflexivity of W_0, we find a subsequence of (u_n), denoted again by (u_n), such that

$$u_n \rightharpoonup u \text{ in } W_0, \quad u_n \to u \text{ in } L^p(Q) \text{ and a.e. in } Q \text{ as } n \to \infty, \quad (4.183)$$

for some $u \in W_0$, where the compactness of the embedding $W_0 \subset L^p(Q)$ has been used. As L is a closed linear operator, its graph is weakly closed, so $u_n \rightharpoonup u$ in W_0 implies $u \in D(L)$. From the fact that $(u_n) \subset \mathcal{S}$, it follows that

$$u_n \in D(L): \quad Lu_n + Au_n + \eta_n = Fu_n + h \quad \text{in } X_0^* \quad (4.184)$$

with $\eta_n \in \partial j(\cdot, \cdot, u_n)$. Hypothesis (H1)(iii) ensures that (η_n) is bounded in $L^q(Q)$. Thus, a subsequence of (η_n) exists, denoted again by (η_n), such that

$$\eta_n \rightharpoonup \eta \text{ in } L^q(Q) \text{ as } n \to \infty \tag{4.185}$$

for some $\eta \in L^q(Q)$. Next we are going to verify that

$$\eta \in \partial j(\cdot, \cdot, u). \tag{4.186}$$

Using $\eta_n \rightharpoonup \eta$ in $L^q(Q)$, $\eta_n \in \partial j(\cdot, \cdot, u_n)$, the convergence properties (4.183), Fatou's lemma, as well as the upper semicontinuity of $j^0(x, t, \cdot; w(x,t)) : \mathbb{R} \to \mathbb{R}$, we deduce that

$$
\begin{aligned}
\int_Q \eta(x,t) w(x,t)\, dx\, dt &= \lim_{n \to \infty} \int_Q \eta_n(x,t) w(x,t)\, dx\, dt \\
&\leq \limsup_{n \to \infty} \int_Q j^0(x, t, u_n(x,t); w(x,t))\, dx\, dt \\
&\leq \int_Q \limsup_{n \to \infty} j^0(x, t, u_n(x,t); w(x,t))\, dx\, dt \\
&\leq \int_Q j^0(x, t, u(x,t); w(x,t))\, dx\, dt.
\end{aligned}
$$

To use Lebesgue's point argument, fix $r \in \mathbb{R}$, $(\bar{x}, \bar{t}) \in Q$, and $R > 0$ and take $w = \chi_{B((\bar{x},\bar{t}),R)} r$ in the previous inequality, with $\chi_{B((\bar{x},\bar{t}),R)}$ the characteristic function of the open ball $B((\bar{x}, \bar{t}), R)$. We then obtain

$$
\begin{aligned}
&\frac{1}{|B((\bar{x},\bar{t}),R)|} \int_{B((\bar{x},\bar{t}),R)} \eta(x,t) r\, dx\, dt \\
&\leq \frac{1}{|B((\bar{x},\bar{t}),R)|} \int_{B((\bar{x},\bar{t}),R)} j^0(x, t, u(x,t); r)\, dx\, dt. \tag{4.187}
\end{aligned}
$$

Letting $R \to 0$ in inequality (4.187), we infer that

$$\eta(\bar{x}, \bar{t}) r \leq j^0(\bar{x}, \bar{t}, u(\bar{x}, \bar{t}); r) \quad \text{for all } r \in \mathbb{R},$$

which by definition of Clark's generalized gradient shows that (4.186) is satisfied. Testing (4.184) with $u_n - u \in D(L)$ and taking into account the convergence properties (4.183), (4.185) as well as the inequality

$$\langle L u_n, u_n - u \rangle \geq \langle L u, u_n - u \rangle,$$

we arrive at

$$\limsup_{n \to \infty} \langle A u_n, u_n - u \rangle \leq 0. \tag{4.188}$$

By the pseudomonotonicity of A with respect to the graph norm topology of $D(L)$ (w.r.t. to $D(L)$), this inequality and the limit $u_n \rightharpoonup u$ in W_0 imply that $A u_n \rightharpoonup A u$ in X_0^*. This result allows us to pass to the limit as $n \to \infty$ in (4.184), obtaining

$$Lu + Au + \eta = Fu + h \quad \text{in } X_0^*.$$

As η satisfies (4.186), it follows that $u \in \mathcal{S}$, which proves that $\mathcal{S}$ is weakly sequentially compact in W_0. Furthermore, from hypotheses (A1)–(A3), the operator A enjoys the (S_+)-property with respect to $D(L)$ (see Sect. 2.4.4), which by the weak convergence $u_n \rightharpoonup u$ in W_0 in conjunction with (4.188) implies that $u_n \to u$ strongly in X_0, and hence, it follows that $\mathcal{S}$ is compact in X_0. □

The comparison principle given by Theorem 4.46 as well as the directedness and compactness results from Lemma 4.50 and Lemma 4.51 provide the tools to prove the following theorem.

Theorem 4.52. *Under the hypotheses of Lemma 4.50, the solution set $\mathcal{S}$ possesses extremal elements.*

Proof: Let us show the existence of the greatest element of $\mathcal{S}$; i.e., the existence of the greatest solution of (4.156) within the ordered interval $[\underline{u}, \bar{u}]$. As W_0 is separable, we have that $\mathcal{S} \subset W_0$ is separable, so a countable, dense subset $Z = \{z_n : n \in \mathbb{N}\}$ of $\mathcal{S}$ exists. By Lemma 4.50, $\mathcal{S}$ is upward directed, so we can construct an increasing sequence $(u_n) \subset \mathcal{S}$ as follows. Let $u_1 = z_1$. Select $u_{n+1} \in \mathcal{S}$ such that

$$\max\{z_n, u_n\} \leq u_{n+1} \leq \bar{u}.$$

The existence of u_{n+1} follows from Lemma 4.50. By Lemma 4.51, we find a subsequence of (u_n), denoted again (u_n), and an element $u \in \mathcal{S}$ such that $u_n \rightharpoonup u$ in W_0, $u_n \to u$ in $L^p(Q)$ and $u_n(x,t) \to u(x,t)$ a.e. $(x,t) \in Q$. This last property of (u_n) combined with its increasing monotonicity implies that $u = \sup_n u_n$. By construction, we see that

$$\max\{z_1, z_2, \ldots, z_n\} \leq u_{n+1} \leq u \quad \text{for all } n \in \mathbb{N},$$

and thus, $Z \subset [\underline{u}, u]$. As the interval $[\underline{u}, u]$ is closed in W_0, we infer

$$\mathcal{S} \subset \overline{Z} \subset \overline{[\underline{u}, u]} = [\underline{u}, u],$$

which in conjunction with $u \in \mathcal{S}$ ensures that u is the greatest solution of (4.156). The existence of the smallest solution of (4.156) can be proved in a similar way. This process completes the proof. □

Remark 4.53. The results obtained in Sect. 4.5.1 and Sect. 4.5.2 can be used to treat more general multivalued parabolic problems in the form

$$u \in D(L): \quad Lu + Au + \partial j(\cdot, \cdot, u) - \partial \beta(\cdot, \cdot, u) \ni Fu + h \quad \text{in } X_0^*, \quad (4.189)$$

where $s \mapsto \beta(\cdot, \cdot, s)$ is assumed to be a convex function with $s \mapsto \partial \beta(\cdot, \cdot, s)$ denoting the usual subdifferential of β at s. It is well known that $\partial \beta$ has a representation in the form

$$\partial\beta(\cdot,\cdot,s) = [h_1(\cdot,\cdot,s), h_2(\cdot,\cdot,s)],$$

where $s \mapsto h_1(\cdot,\cdot,s)$ is a nondecreasing, left-sided continuous function and $s \mapsto h_2(\cdot,\cdot,s)$ is a nonincreasing, right-sided continuous function. Let H_i, $i = 1, 2$, denote the Nemytskij operators associated with h_i; we define sub-supersolutions for (4.189) as follows.

Definition 4.54. *A function $\bar{u} \in W$ is called a supersolution of problem (4.189) if $F\bar{u} \in L^q(Q)$ and if there is a function $\bar{\eta} \in L^q(Q)$ such that*

(i) $\bar{u}(x,0) \geq 0$ *in Ω and $\bar{u} \geq 0$ on Γ.*
(ii) $\bar{\eta}(x,t) \in \partial j(x,t,\bar{u}(x,t))$ *for a.e. $(x,t) \in Q$.*
(iii) $\langle \bar{u}_t + A\bar{u}, \varphi \rangle + \int_Q \bar{\eta}\varphi\,dxdt \geq \int_Q (F(\bar{u}) + H_2(\bar{u}))\,\varphi\,dxdt + \langle h, \varphi \rangle$ *for all*
      $\varphi \in V_0 \cap L^p_+(Q)$.

Similarly, a function $\underline{u} \in W$ is called a subsolution of problem (4.189) if the reversed inequalities hold with $\bar{u}$, $\bar{\eta}$ replaced by $\underline{u}$, $\underline{\eta}$, respectively, and $H_2(\bar{u})$ replaced by $H_1(\underline{u})$. Assuming an ordered pair of sub-supersolutions in the sense of Definition 4.54, a comparison principle as well as extremality and compactness results for (4.189) can be proved. As for a detailed treatment and proofs, we refer to [60].

Remark 4.55. The case of global growth conditions on Clarke's generalized gradient can effectively be treated within the framework of evolutionary hemivariational inequalities, which will be considered in Chap. 6. We readily can see that any solution of the inclusion (4.156) is also a solution of the hemivariational inequality

$$u \in D(L): \quad \langle Lu + Au - Fu - h, \varphi - u \rangle + \int_Q j^\circ(\cdot,\cdot,u;\varphi-u)\,dxdt \geq 0$$

for all $\varphi \in X_0$. However, the reverse is not true in general.

4.6 An Alternative Concept of Sub-Supersolutions

An alternative notion of sub-supersolution for elliptic and parabolic inclusions with multifunctions of Clarke's generalized gradient can be found, e.g., in [43] or [97]. With the help of the elliptic inclusion (4.81) considered in Sect. 4.3, we are going to point out the differences between this alternative concept of sub-supersolutions and the one introduced in Sect. 4.3. Consider the elliptic inclusion (4.81) written in the form

$$Au - h + Fu \in -\partial j(\cdot, u) \quad \text{in } \Omega, \quad u = 0 \quad \text{on } \partial\Omega, \tag{4.190}$$

and assume that $j : \Omega \times \mathbb{R} \to \mathbb{R}$ is given by (4.82), which results in the following representation:

$$\partial j(\cdot, s) = [\underline{g}(\cdot, s), \bar{g}(\cdot, s)], \quad s \in \mathbb{R}, \tag{4.191}$$

with $\underline{g}$ and $\bar{g}$ as given in Remark 4.30. Noting that

$$-\partial j(\cdot, s) = [-\bar{g}(\cdot, s), -\underline{g}(\cdot, s)], \quad s \in \mathbb{R},$$

the alternative notion of sub-supersolutions is as follows.

Definition 4.56. *A function $\underline{u} \in V$ is a subsolution of (4.190) if $F\underline{u} \in L^q(\Omega)$ and $\bar{g}(\cdot, \underline{u}) \in L^q(\Omega)$ such that*

(i) $\underline{u} \leq 0$ on $\partial\Omega$.
(ii) $A\underline{u} - h + F\underline{u} \leq -\bar{g}(\cdot, \underline{u})$ in V_0^*.

Definition 4.57. *A function $\bar{u} \in V$ is a supersolution of (4.190) if $F\bar{u} \in L^q(\Omega)$ and $\underline{g}(\cdot, \bar{u}) \in L^q(\Omega)$ such that*

(i) $\bar{u} \geq 0$ on $\partial\Omega$.
(ii) $A\bar{u} - h + F\bar{u} \geq -\underline{g}(\cdot, \bar{u})$ in V_0^*.

Comparing Definition 4.56 with Definition 4.28, we can easily see that a subsolution in the sense of Definition 4.56 is also a subsolution in the sense of Definition 4.28. This result is because $\underline{v}(x) \in \partial j(x, \underline{u}(x))$ implies $\underline{v}(x) \leq \bar{g}(x, \underline{u}(x))$. Analogously, any supersolution in the sense of Definition 4.57 is a supersolution in the sense defined in Sect. 4.3.1. Thus, the alternative notions provided by Definition 4.56 and Definition 4.57 are more restrictive than the one introduced in Sect. 4.3. On the other hand, using these alternative notions, a comparison principle can be proved without assuming the one-sided growth condition (H1) (ii). However, the order structure of the solution set $\mathcal{S}$; i.e., the directedness and extremality property of $\mathcal{S}$ seems to be violated if condition (H1) (ii) is dropped. This result shows that the concept of sub-supersolution introduced in this monograph is in some sense a natural generalization of the corresponding notions for equations considered in Chap. 3, because all characteristic features of the comparison principles as well as the properties of the solution set $\mathcal{S}$ are preserved. Moreover, our concept fits nicely into the framework of hemivariational inequalities, which is the subject of Chap. 6.

4.7 Notes and Comments

Differential inclusions considered in this chapter arise, e.g., in mechanical problems governed by nonconvex, possibly nonsmooth energy functionals, called superpotentials, which appear if nonmonotone, multivalued constitutive laws are taken into account (see [177, 180]). In our presentation of comparison principles for elliptic and parabolic inclusions, we have concentrated on Dirichlet and initial-Dirichlet problems, respectively. However, the theory developed here is not restricted to such kind of problems and can straight-forwardly be extended to include boundary conditions of mixed type such as problems in the form

$$Au + Fu = h \quad \text{in } \Omega, \quad u = 0 \text{ on } \partial\Omega \setminus \Gamma, \quad -\frac{\partial u}{\partial \nu} \in \partial j(u) \text{ on } \Gamma, \quad (4.192)$$

where $\Gamma \subset \partial\Omega$ is a relatively open portion of $\partial\Omega$ having a positive surface measure. As for problem (4.192), we refer to [36], and for a related parabolic version of (4.192), see [39]. In the spirit of this chapter, inclusion problems with multifunctions in the form of state-dependent subdifferentials have been treated in bounded as well as in unbounded domains, e.g., in [43] and [34]. As for systems of elliptic and parabolic inclusion problems, we refer to [47] and [45]. In the system case, the corresponding notion of sub-supersolution is replaced by the notion of trapping region, which is defined via outward pointing vector fields. In Chap. 6, we will clarify the interrelation between inclusions of Clarke's gradient type and hemivariational inequalities.

5

Variational Inequalities

The goal of this chapter is starting a systematic study of the sub-supersolution method in variational inequalities. By using subsolutions or supersolutions, we can show the solvability and the existence of extremal solutions of noncoercive inequalities. Despite the nonsymmetric structure of variational inequalities, we show that both supersolutions and subsolutions can be defined in an appropriate manner, which naturally extends the corresponding concepts in equations.

We first consider the case of variational inequalities defined on closed convex sets. After that, in Sect. 5.2, we extend the concepts and results to the more general case of inequalities with convex functionals. We next present a sub-supersolution theory for parabolic variational inequalities on convex sets (Sect. 5.3) and one for systems of elliptic inequalities (Sect. 5.5). When sub- and supersolutions have some additional properties, an approximation scheme is elaborated that yields a monotone approximation for the solutions between sub- and supersolutions while no monotonicity assumption is imposed on the lower order term (Sect. 5.4).

For a simple motivation of the concepts of sub- and supersolutions for variational inequalities, let us consider the equilibrium problem of an elastic string with fixed end points at 0 and l ($l > 0$). The energy of the system is given by $E(u) = \frac{1}{2} \int_0^l [u'(x)]^2 dx$, where $u(x)$ is the displacement at x. In the case without obstacle, at the equilibrium position, u satisfies

$$u \in V_0 : E(u) = \inf_{v \in V_0} E(v),$$

where V_0 can be chosen as $V_0 = W_0^{1,2}(0, l)$. The Euler–Lagrange equation of this minimization problem could be written in the weak form as the following variational equation:

$$u \in V_0 : \int_0^l u' \phi' dx = 0, \ \forall \, \phi \in V_0, \tag{5.1}$$

(which has a unique solution $u \equiv 0$, as expected). The equilibrium problem with an obstacle, represented by $\psi \in C[0, l]$ ($\psi(0) > 0$, $\psi(l) > 0$), is formulated as the minimization problem

$$u \in K : E(u) = \inf_{v \in K} E(v),$$

with $K = \{v \in V_0 : v \leq \psi\}$, which has as the Euler–Lagrange "equation" the following variational inequality:

$$u \in K : \int_0^l u'(\phi - u)' dx \geq 0, \ \forall \, \phi \in K. \tag{5.2}$$

According to Chap. 3, a supersolution of (5.1) is a function $\bar{u} \in V = W^{1,2}(0, l)$ satisfying $\bar{u}(0) \geq 0$, $\bar{u}(l) \geq 0$, and

$$\int_0^l \bar{u}' v' dx \geq 0, \ \forall \, v \in V_0 \cap L_+^2(0, l). \tag{5.3}$$

Similarly a subsolution of (5.1) is a function $\underline{u} \in V$ satisfying $\underline{u}(0) \leq 0$, $\underline{u}(l) \leq 0$, and

$$\int_0^l \underline{u}' v' dx \leq 0, \ \forall \, v \in V_0 \cap L_+^2(0, l). \tag{5.4}$$

To get an idea for the notion of sub-supersolutions for the variational inequality (5.2) with K being a proper closed, convex subset, we reformulate the corresponding inequalities for sub-supersolutions in the variational equation case as follows. Let us consider the inequality (5.4) of the subsolution. Then (5.4) is equivalent to

$$\int_0^l \underline{u}'(-v^+)' dx \geq 0, \ \forall \, v \in V_0. \tag{5.5}$$

Substitution $-v^+ = w - \underline{u}$ in (5.5) yields

$$\int_0^l \underline{u}'(w - \underline{u})' dx \geq 0, \ \forall \, w \in Y, \tag{5.6}$$

where $Y = \{w = \underline{u} - v^+ : v \in V_0\}$. A subset of Y is the following set Z defined by

$$Z = \{w = \underline{u} - (\underline{u} - v)^+ : v \in V_0\} = \{w = \underline{u} \wedge v : v \in V_0\} =: \underline{u} \wedge V_0.$$

We will show later that the set Z is dense in Y, or equivalently that the set M given by

$$M = \{(\underline{u} + v)^+ : v \in V_0\}$$

is dense in $V_0 \cap L_+^2(0, l)$, (see Lemma 5.4). Therefore, we have the following equivalent notion of a subsolution of the variational equation (5.1): The function $\underline{u} \in V$ is a subsolution if $\underline{u}(0) \leq 0$, $\underline{u}(l) \leq 0$, and

$$\int_0^l \underline{u}'(w - \underline{u})' dx \geq 0, \ \forall \ w \in \underline{u} \wedge V_0. \tag{5.7}$$

By similar arguments, $\bar{u} \in V$ is supersolution of (5.1) if $\bar{u}(0) \geq 0$, $\bar{u}(l) \geq 0$, and

$$\int_0^l \bar{u}'(w - \bar{u})' dx \geq 0, \ \forall \ w \in \bar{u} \vee V_0. \tag{5.8}$$

The inequalities (5.7) and (5.8) provide the motivation for the notion of sub- and supersolution of the variational inequality (5.2) in the general case, replacing V_0 by K, which suggests the following definitions.

Subsolution: $\underline{u} \in V$ is a subsolution of (5.2) if $\underline{u}(0) \leq 0$, $\underline{u}(l) \leq 0$, and

$$\int_0^l \underline{u}'(w - \underline{u})' dx \geq 0, \ \forall \ w \in \underline{u} \wedge K. \tag{5.9}$$

Supersolution: $\bar{u} \in V$ is a supersolution of (5.2) if $\bar{u}(0) \geq 0$, $\bar{u}(l) \geq 0$, and

$$\int_0^1 \bar{u}'(w - \bar{u})' dx \geq 0, \ \forall \ w \in \bar{u} \vee K. \tag{5.10}$$

Inequalities (5.9) and (5.10) play a crucial role in generalizing the concept of sub-supersolutions to variational inequalities. Based on these concepts, we can develop existence and comparison principles for inequalities as were done for equations. More details are given in the following sections and chapters.

5.1 Variational Inequalities on Closed Convex Sets

To convey the main ideas, we first consider the following variational inequality:

$$\begin{cases} \int_\Omega A_0(x, \nabla u) \cdot (\nabla v - \nabla u) dx \geq \int_\Omega F(x, u)(v - u) dx, \ \forall \ v \in K \\ u \in K. \end{cases} \tag{5.11}$$

As mentioned in the previous chapters, Ω is a bounded domain in $\mathbb{R}^N$ ($N \geq 1$) with Lipschitz boundary, and $W^{1,p}(\Omega)$ and $W_0^{1,p}(\Omega)$ are the usual Sobolev spaces with $1 < p < \infty$. Moreover, here K is a closed, convex subset of $W_0^{1,p}(\Omega)$. $A_0 : \Omega \times \mathbb{R}^N \to \mathbb{R}^N$ is a Carathéodory, monotone function, and $F : \Omega \times \mathbb{R} \to \mathbb{R}$ is a lower order, perturbing term. Because of the rate of growth of F, the variational inequality (5.11) is, in general, noncoercive.

5.1.1 Solutions and Extremal Solutions above Subsolutions

For simplicity, we use here the notation $V = W^{1,p}(\Omega)$, $V_0 = W_0^{1,p}(\Omega)$,

$$\|u\| = \|u\|_{W^{1,p}(\Omega)}$$

and

$$\|u\|_0 = \|u\|_{W_0^{1,p}(\Omega)} = \left(\int_\Omega |\nabla u|^p \right)^{\frac{1}{p}} dx.$$

It is well known from Poincaré's inequality that $\| \cdot \|$ is equivalent to $\| \cdot \|_0$ on V_0. We also denote by V^* the dual of V and by $\langle \cdot, \cdot \rangle$ the dual pairing between V and V^* and between V_0 and V_0^*.

Let $A_0 : \Omega \times \mathbb{R}^N \to \mathbb{R}^N$ be a Carathéodory function such that

$$\begin{cases} A_0(x,v)v \geq \alpha|v|^p, \\ |A_0(x,v)| \leq \nu|v|^{p-1} + \gamma(x), \end{cases} \tag{5.12}$$

for a.e. $x \in \Omega$, all $v \in \mathbb{R}^N$, where $\alpha > 0$, $\nu \in \mathbb{R}$, and $\gamma \in L^{p'}(\Omega)$ (p' is the Hölder conjugate exponent of p). We assume that $A_0(x, \cdot)$ is monotone; i.e.,

$$[A_0(x,v_1) - A_0(x,v_2)] \cdot (v_1 - v_2) \geq 0, \tag{5.13}$$

for a.e. $x \in \Omega$, for all $v_1, v_2 \in \mathbb{R}^N$. Consider the operator $A : V \to V^*$ defined by

$$\langle A(u), v \rangle = \int_\Omega A_0(x, \nabla u) \cdot \nabla v dx, \ \forall \, u, v \in V.$$

It can be checked that A given above is well defined, continuous, bounded, and coercive in V_0 in the sense that

$$\lim_{\|u\| \to \infty, u \in V_0} \frac{\langle A(u), u - \phi \rangle}{\|u - \phi\|_0} = +\infty, \tag{5.14}$$

for $\phi \in V_0$ fixed. This property of A is a direct consequence of (5.12) and Hölder's inequality. Assume that $F : \Omega \times \mathbb{R} \to \mathbb{R}$ is a Carathéodory function that satisfies a certain growth condition to be specified later (cf. Sect. 5.1.1).

We are concerned here with the existence of solutions and extremal solutions of the variational inequality (5.11). As usual, an element $u \in K$ is called a solution of (5.11) if $F(\cdot, u) \in L^{p'}(\Omega)$ and (5.11) is satisfied.

Note that although A is coercive, the operator $u \mapsto A(u) - \int_\Omega F(\cdot, u)dx$, and thus the variational inequality (5.11), are not coercive in general. In the sections that follow, we use the sub- and supersolution method to show the existence of solutions of (5.11) either above a subsolution (or under a supersolution) or between a subsolution and a supersolution. We show furthermore the existence of smallest and/or greatest solutions.

Definitions of Sub- and Supersolutions in Variational Inequalities

In this section, we define subsolutions and supersolutions of the variational inequality (5.11). Assume that v and w are functions defined on Ω and L and M are sets of functions defined on Ω. We use the notation $w \wedge v = \min\{w, v\}$, $L \wedge M = \{w \wedge v : w \in L, v \in M\}$, and $w \wedge M = \{w\} \wedge M$. Similarly, $w \vee v = \max\{w, v\}$, $L \vee M = \{w \vee v : w \in L, v \in M\}$, and $w \vee M = \{w\} \vee M$.

Definition 5.1. *A function $\underline{u} \in V$ is called a subsolution of (5.11) if*

$$\underline{u} \leq 0 \quad on \ \partial\Omega, \tag{5.15}$$

$$F(\cdot, \underline{u}) \in L^{p'}(\Omega), \tag{5.16}$$

and

$$\langle A(\underline{u}), w - \underline{u} \rangle \geq \int_\Omega F(\cdot, \underline{u})(w - \underline{u})dx, \ \forall \ w \in \underline{u} \wedge K. \tag{5.17}$$

We also consider maxima of a finite number of subsolutions; i.e., functions $\underline{u}$ of the form

$$\underline{u} = \max\{\underline{u}_1, \ldots, \underline{u}_k\}, \tag{5.18}$$

where $\underline{u}_1, \ldots, \underline{u}_k$ are subsolutions of (5.11).

We have similar definitions for supersolutions of (5.11):

Definition 5.2. *A function $\bar{u} \in V$ is called a supersolution of (5.11) if*

$$\bar{u} \geq 0 \quad on \ \partial\Omega, \tag{5.19}$$

$$F(\cdot, \bar{u}) \in L^{p'}(\Omega), \tag{5.20}$$

and

$$\langle A(\bar{u}), w - \bar{u} \rangle \geq \int_\Omega F(\cdot, \bar{u})(w - \bar{u})dx, \ \forall \ w \in \bar{u} \vee K. \tag{5.21}$$

Before showing some existence results for (5.11) based on the new concept of sub- and supersolutions introduced above, we make some remarks about relationships between Definition 5.1 and Definition 5.2 and the definitions of sub- and supersolutions in variational equations (see Sect. 3.2 and cf. [83], [133], or [29]).

Remark 5.3. (a) If K satisfies the condition $K \wedge K \subset K$ (respectively, $K \vee K \subset K$), i.e.,

$$w, v \in K \Rightarrow w \wedge v \in K, \tag{5.22}$$

(respectively,

$$w, v \in K \Rightarrow w \vee v \in K), \tag{5.23}$$

then any solution of (5.11) is also a subsolution (respectively, supersolution).

In fact, because $u \in K$, $u = 0$ on $\partial \Omega$. Moreover, it follows from (5.22) that $u \wedge K \subset K$. Hence, (5.11) implies (5.17); i.e., u is a subsolution of (5.11).

(b) We note that both (5.22) and (5.23) are satisfied in several types of convex sets usually occur in applications. Some examples are the following.

$K = W_0^{1,p}(\Omega)$. In this case, (5.11) is an equation and (5.22) and (5.23) are immediate consequences of the lattice structure of $W_0^{1,p}(\Omega)$ and $W^{1,p}(\Omega)$ (cf. [99] or Chap. 2).

$K = \{u \in V_0 : u(x) \geq \psi(x)$, for a.e. $x \in \Omega_0\}$; i.e., (5.11) is an obstacle problem. Here, Ω_0 is a subset of Ω, and ψ is a given function on Ω_0, representing the obstacle.

$K = \{u \in V_0 : \psi_1(x) \leq u(x) \leq \psi_2(x)$ for a.e. $x \in \Omega_0\}$; i.e., (5.11) is a biobstacle problem. Here, ψ_1, ψ_2 are given functions on Ω_0.

K can be given by certain gradient condition, as in the elasto-plastic torsion problem (cf. [197], [95], or [145]):

$$K = \{u \in V_0 : |\nabla u(x)| \leq c, \text{ for a.e. } x \in \Omega_0\},$$

$c \geq 0$ is a given number. This example can be generalized as:

$K = \{u \in V_0 : \partial_i u(x) \leq (\geq) \psi_i(x)$, for a.e. $x \in \Omega_0$ $(i \in I)\}$, where $I \subset \{1, \ldots, N\}$ and the ψ_i's $(i \in I)$ are given functions.

It can be easily checked that both (5.22) and (5.23) are satisfied in all of these examples.

Coherence with Sub-Supersolution Concepts in Equations

Before stating some existence theorems resulted from the above concepts of sub-supersolutions, we note that the definitions above extend in a natural way the classic concepts of sub- and supersolutions for (smooth) variational equations. In fact, let us consider the case where $K = W_0^{1,p}(\Omega)$; i.e., (5.11) is an equation on $W_0^{1,p}(\Omega)$.

Let us show that when $K = W_0^{1,p}(\Omega)$ is the entire space, the definition above reduces to the usual definition of sub- and supersolutions for variational equations [see Sect. 3.2 (cf. [83] or [133])]. For the proof, the following density result is crucial.

Lemma 5.4. *Let* $\underline{u} \in V$ *with* $\underline{u}|_{\partial \Omega} \leq 0$. *If* M *is the set defined by*

$$M = \{(\underline{u} + v)^+ : v \in V_0\} = \{v^+ : v \in V, v|_{\partial \Omega} = \underline{u}|_{\partial \Omega}\},$$

then the closure of M *in* V_0 *results in*

$$\overline{M}^{V_0} = V_0 \cap L_+^p(\Omega). \tag{5.24}$$

Similarly, if $\bar{u} \in V$ with $\bar{u}|_{\partial\Omega} \geq 0$, then the set N given by

$$N = \{(v - \bar{u})^+ : v \in V_0\} = \{v^+ : v \in V, \ v|_{\partial\Omega} = -\bar{u}|_{\partial\Omega}\}$$

satisfies

$$\overline{N}^{V_0} = V_0 \cap L^p_+(\Omega).$$

Proof: Let us prove (5.24) only, because the density for N follows the same arguments. Obviously, $M \subset V_0 \cap L^p_+(\Omega)$. First we note that if $v \in V_0 \cap L^p_+(\Omega)$ and v has compact support in Ω, then $v \in M$. To prove this, we assume $\kappa = \operatorname{supp} v$ is a compact subset of Ω. Then, a function $\varphi \in C^\infty_0(\Omega)$ exists such that

$$0 \leq \varphi(x) \leq 1, \ \forall \, x \in \Omega \ \text{ and } \ \varphi(x) = 1, \ \forall \, x \in \kappa,$$

(cf. [115]). Put

$$\tilde{v} = v + (1 - \varphi)\min\{\underline{u}, 0\}.$$

As φ is smooth, $\tilde{v} \in V$. Also, because $0 \leq 1 - \varphi \leq 1$ and $1 - \varphi = 0$ on κ, we immediately have

$$\tilde{v}^+ = v. \tag{5.25}$$

Furthermore, $\tilde{v} = \min\{\underline{u}, 0\}$ on $\Omega \setminus \operatorname{supp}\varphi$, because $\kappa \subset \operatorname{supp}\varphi$. Thus, in view of $\underline{u}|_{\partial\Omega} \leq 0$, we get

$$\tilde{v}|_{\partial\Omega} = \min\{\underline{u}, 0\}|_{\partial\Omega} = \underline{u}|_{\partial\Omega}. \tag{5.26}$$

(5.25) and (5.26) show that $v \in M$.

To prove (5.24), we just note that any function v in $V_0 \cap L^p_+(\Omega)$ can always be approximated in V_0 by functions in $V_0 \cap L^p_+(\Omega)$ with compact support. This fact follows from the density of $C^\infty_0(\Omega)$ in V_0 and the continuity of the truncation operator: $V_0 \rightarrow V_0$, $v \mapsto \max\{v, 0\}$. $\qquad\square$

In view of (5.24), we see that (5.17) is equivalent to

$$\langle A(\underline{u}), v \rangle \leq \int_\Omega F(\cdot, \underline{u})v\,dx, \ \forall \, v \in V_0 \cap L^p_+(\Omega).$$

Hence, $\underline{u}$ is a subsolution in the usual sense of Sect. 3.2 (cf. [133] or [83]). We have a similar observation for supersolutions. Hence, Definition 5.1 and Definition 5.2 are extensions to variational inequalities of the usual sub- and supersolution concepts in equations.

Existence of Solutions above Subsolutions

We show in this section the existence of solutions of (5.11) that lie above several subsolutions. Assume that the variational inequality (5.11) has subsolutions $\underline{u}_1, \ldots, \underline{u}_k$. Let us denote

$$u = \max\{\underline{u}_1, \ldots, \underline{u}_k\} \tag{5.27}$$

and $\underline{u}_0 = \min\{\underline{u}_1, \ldots, \underline{u}_k\} \in V$. We assume that F has the following growth condition:

$$|F(x, u)| \leq a(x) + b|u|^{\sigma}, \tag{5.28}$$

for a.e. $x \in \Omega$, all $u \in \mathbb{R}$, $u \geq \underline{u}_0(x)$. Here, $a \in L^{p'}(\Omega)$, b is a positive constant, and $0 \leq \sigma < p - 1$. Under those conditions, we have the following existence theorem of solutions above subsolutions.

Theorem 5.5. *Assume (5.27), (5.28), and*

$$\underline{u}_j \vee K \subset K, \ 1 \leq j \leq k. \tag{5.29}$$

Then, a solution u of (5.11) exists such that

$$u \geq \underline{u}. \tag{5.30}$$

Proof: The proof is motivated by the truncation method already used in the previous chapters (see also in [83], [113], [133], and [29]). We first define some auxiliary mappings. Put

$$b(x, t) = -\{[\underline{u}(x) - t]^+\}^{p-1} = \begin{cases} 0 & \text{if } t \geq \underline{u}(x) \\ -[\underline{u}(x) - t]^{p-1} & \text{if } t < \underline{u}(x), \end{cases} \tag{5.31}$$

for $x \in \Omega, t \in \mathbb{R}$. b is clearly a Carathéodory function and, for some $c_0 > 0$ depending only on p,

$$\begin{aligned} |b(x, t)| &\leq (|t| + |\underline{u}(x)|)^{p-1} \\ &\leq c_0(|t|^{p-1} + |\underline{u}(x)|^{p-1}) \\ &= a_1(x) + c_0|t|^{p-1}, \end{aligned} \tag{5.32}$$

with $a_1 = c_0|\underline{u}|^{p-1} \in L^{p'}(\Omega)$. Hence, $u \mapsto b(\cdot, u)$ is a continuous mapping from $L^p(\Omega)$ to $L^{p'}(\Omega)$. In the sequel, we use the set notation

$$\{g < h\} = \{x \in \Omega : g(x) < h(x)\} \text{ and } \{g \leq h\} = \{x \in \Omega : g(x) \leq h(x)\}.$$

Elementary calculations show that positive constants c_1 and c_2 exist depending only on p such that $-(u - t)^{p-1}t \geq c_1|t|^p - c_2|u|^{p-1}|t|$, for all $u, t \in \mathbb{R}$, $u \geq t$. Hence, for $u \in L^p(\Omega)$,

$$\begin{aligned} \int_{\Omega} b(\cdot, u)u \, dx &= \int_{\{u < \underline{u}\}} -[\underline{u} - u]^{p-1}u \, dx \\ &\geq \int_{\{u < \underline{u}\}} (c_1|u|^p - c_2|\underline{u}|^{p-1}|u|) \, dx \\ &\geq c_3 \int_{\{u < \underline{u}\}} |u|^p dx - c_4 \int_{\Omega} |\underline{u}|^p dx \\ &= c_3 \int_{\{u < \underline{u}\}} |u|^p dx - c_5, \end{aligned} \tag{5.33}$$

where c_3, c_4, c_5 are positive constants independent of u. For $j \in \{1, \ldots, k\}$, we define

$$T_j(u)(x) = (u \vee u_j)(x) = \begin{cases} u_j(x) & \text{if } u(x) < u_j(x) \\ u(x) & \text{if } u(x) \geq u_j(x), \end{cases} \qquad (5.34)$$

$$T(u)(x) = (u \vee \underline{u})(x) = \begin{cases} \underline{u}(x) & \text{if } u(x) < \underline{u}(x) \\ u(x) & \text{if } u(x) \geq \underline{u}(x). \end{cases} \qquad (5.35)$$

It is known that T_j, T are continuous mappings from $L^p(\Omega)$ to itself, and from V into itself; see Chap. 2 or [43]. Moreover, for all $u \in L^p(\Omega)$, $T_j(u), T(u) \geq \underline{u}_0$ a.e. on Ω. Together with the growth condition (5.28), this implies that the mappings $u \mapsto F(\cdot, T_j(u))$ and $u \mapsto F(\cdot, T(u))$ are bounded and continuous from $L^p(\Omega)$ to $L^{p'}(\Omega)$ ($j \in \{1, \ldots, k\}$). Let us consider the following variational inequality:

$$\begin{cases} \langle A(u) + \beta B(u) - H(u), v - u \rangle \geq 0, \ \forall \, v \in K \\ u \in K, \end{cases} \qquad (5.36)$$

where β is a positive constant,

$$\langle B(u), \phi \rangle = \int_\Omega b(\cdot, u)\phi dx, \qquad (5.37)$$

and

$$\langle H(u), \phi \rangle = \int_\Omega \left[F(\cdot, T(u)) + \sum_{j=1}^k |F(\cdot, T_j(u)) - F(\cdot, T(u))| \right] \phi \, dx, \qquad (5.38)$$

for all $\phi \in V$. As the mappings $u \mapsto b(\cdot, u)$, $u \mapsto F(\cdot, T_j(u))$, and $u \mapsto F(\cdot, T(u))$ are bounded and continuous from $L^p(\Omega)$ to $L^{p'}(\Omega)$, by using the compact embedding $W^{1,p}(\Omega) \subset L^p(\Omega)$ and the continuous embedding $L^{p'}(\Omega) \subset [W^{1,p}(\Omega)]^*$, we immediately obtained that $\beta B - H$ is a completely continuous mapping from V_0 into V_0^*.

As A is monotone and bounded, $A + \beta B - H$ is pseudomonotone and bounded. On the other hand, for $u, v \in V$,

$$|\langle H(u), v \rangle| \leq \int_\Omega \left[|F(\cdot, T(u))|(k+1) + \sum_{j=1}^k |F(\cdot, T_j(u))| \right] |v| \, dx. \qquad (5.39)$$

Now, for $j \in \{1, \ldots, k\}$,

$$\int_{\Omega} |F(\cdot, T_j(u))|\,|v|\,dx$$

$$\leq \int_{\{u \geq \underline{u}_0\}} |F(\cdot, u)|\,|v|\,dx + \int_{\{u < \underline{u}\}} |F(\cdot, \underline{u}_j)|\,|v|\,dx \ \text{(since } \underline{u}_0 \leq \underline{u}_j \leq \underline{u})$$

$$\leq \int_{\{u \geq \underline{u}_0\}} (a + b|u|^\sigma)\,|v|\,dx + \int_{\{u < \underline{u}\}} |F(\cdot, \underline{u}_j)|\,|v|\,dx \ \text{(by (5.28))}$$

$$\leq \|a\|_{L^{p'}(\Omega)} \|v\|_{L^p(\Omega)} + b|\Omega|^{1 - \frac{\sigma+1}{p}} \|u\|_{L^p(\Omega)}^\sigma \|v\|_{L^p(\Omega)}$$

$$+ \|F(\cdot, \underline{u}_j)\|_{L^{p'}(\Omega)} \|v\|_{L^p(\Omega)}.$$

$$(5.40)$$

A similar estimate holds for the integral with T_j replaced by T. Hence, from (5.39) and (5.40), we get

$$|\langle H(u), v \rangle| \leq (2k + 1) \Big[\|a\|_{L^{p'}(\Omega)} \|v\|_{L^p(\Omega)}$$

$$+ b|\Omega|^{1 - \frac{\sigma+1}{p}} \|u\|_{L^p(\Omega)}^\sigma \|v\|_{L^p(\Omega)} + \|F(\cdot, \underline{u}_j)\|_{L^{p'}(\Omega)} \|v\|_{L^p(\Omega)} \Big]$$

$$= c_6 \|v\|_{L^p(\Omega)} + c_7 \|u\|_{L^p(\Omega)}^\sigma \|v\|_{L^p(\Omega)}, \ \text{for all } u, v \in V,$$

$$(5.41)$$

where c_6, c_7 are positive constants independent of u, v. Now, fix $\phi \in K$. From (5.12), (5.32), (5.33), (5.41), and the continuous embedding $W_0^{1,p}(\Omega) \subset L^p(\Omega)$, we get

$$\langle A(u) + \beta B(u) - H(u), u - \phi \rangle$$

$$\geq \langle A(u), u \rangle - |\langle A(u), \phi \rangle| + \beta \langle B(u), u \rangle - \beta |\langle B(u), \phi \rangle| - |\langle H(u), u - \phi \rangle|$$

$$\geq \alpha \int_\Omega |\nabla u|^p dx - \int_\Omega \left(\nu |\nabla u|^{p-1} + \gamma \right) |\nabla \phi| dx + \beta c_3 \int_{\{u < \underline{u}\}} |u|^p dx - \beta c_5$$

$$- \beta \int_\Omega (a_1 + c_0 |u|^{p-1}) |\phi| dx - c_6 \|u - \phi\|_{L^p(\Omega)} - c_7 \|u\|_{L^p(\Omega)}^\sigma \|u - \phi\|_{L^p(\Omega)}$$

$$\geq \alpha \|u\|_0^p - c_8 \left(\|u\|_0^{p-1} + \|\gamma\|_{L^{p'}(\Omega)} \right) \|\phi\|_0 - \beta c_5 - \beta \|a_1\|_{L^{p'}(\Omega)} \|\phi\|_{L^p(\Omega)}$$

$$- \beta c_0 \|u\|_{L^p(\Omega)}^{p-1} \|\phi\|_{L^p(\Omega)} - c_6 \left(\|u\|_{L^p(\Omega)} + \|\phi\|_{L^p(\Omega)} \right) - c_7 \|u\|_{L^p(\Omega)}^{\sigma+1}$$

$$- c_7 \|u\|_{L^p(\Omega)}^\sigma \|\phi\|_{L^p(\Omega)}$$

$$\geq \alpha \|u\|_0^p - c_9 \|u\|_0^{p-1} - c_{10} \|u\|_0 - c_{11} \|u\|_0^{\sigma+1} - c_{12} \|u\|_0^\sigma - c_{13},$$

$$(5.42)$$

for all $u \in V_0$, where the c_i's $(0 \leq i \leq 13)$ are positive constants that do not depend on u. As $\sigma + 1 < p$, the right-hand side of (5.42) is bounded from below by $\frac{\alpha}{2} \|u\|_0^p$, for all $u \in K$ with $\|u\|_0$ sufficiently large. Hence,

$$\lim_{\|u\|_0 \to \infty, \, u \in K} \frac{\langle A(u) + \beta B(u) - H(u), u - \phi \rangle}{\|u - \phi\|_0} = +\infty;$$

i.e., $A + \beta B - H$ is coercive on K. By classic existence results for variational inequalities (cf. [152], [124]), (5.36) has a solution $u \in K$.

We now check that $u \geq \underline{u}_j$, for all $j \in \{1, \ldots, k\}$, which implies that $u \geq \underline{u}$. Let $q \in \{1, \ldots, k\}$. As $u \in K$, we have $\underline{u}_q \wedge u \in \underline{u}_q \wedge K$, and thus, (5.17) (with $\underline{u}$ replaced by $\underline{u}_q$) gives

$$\langle A(\underline{u}_q), \underline{u}_q \wedge u - \underline{u}_q \rangle \geq \int_\Omega F(\cdot, \underline{u}_q)(\underline{u}_q \wedge u - \underline{u}_q)dx.$$

Because $\underline{u}_q \wedge u = \underline{u}_q - (\underline{u}_q - u)^+$, this inequality is the same as

$$-\langle A(\underline{u}_q), (\underline{u}_q - u)^+ \rangle \geq - \int_\Omega F(\cdot, \underline{u}_q)(\underline{u}_q - u)^+ dx. \tag{5.43}$$

On the other hand, it follows from (5.29) that $\underline{u}_q \vee u \in K$. Letting $v = \underline{u}_q \vee u$ in (5.36), we get

$$\langle A(u) + \beta B(u) - H(u), \underline{u}_q \vee u - u \rangle \geq 0.$$

As $\underline{u}_q \vee u = u + (\underline{u}_q - u)^+$, this gives

$$\langle A(u) + \beta B(u) - H(u), (\underline{u}_q - u)^+ \rangle \geq 0. \tag{5.44}$$

Adding (5.43) and (5.44) yields

$$\langle A(u) - A(\underline{u}_q), (\underline{u}_q - u)^+ \rangle + \beta \int_\Omega b(\cdot, u)(\underline{u}_q - u)^+ dx$$
$$+ \int_\Omega \left[F(\cdot, \underline{u}_q) - F(\cdot, T(u)) - \sum_{j=1}^k |F(\cdot, T_j(u)) - F(\cdot, T(u))| \right] (\underline{u}_q - u)^+ dx$$
$$\geq 0. \tag{5.45}$$

Now, from (5.13),

$$\langle A(u) - A(\underline{u}_q), (\underline{u}_q - u)^+ \rangle$$
$$= - \int_{\{\underline{u}_q - u > 0\}} \left[A_0(\cdot, \nabla \underline{u}_q) - A_0(\cdot, \nabla u) \right] \cdot \nabla(\underline{u}_q - u)dx \tag{5.46}$$
$$\leq 0.$$

On the other hand,

$$\int_\Omega \left[F(\cdot, \underline{u}_q) - F(\cdot, T(u)) - \sum_{j=1}^k |F(\cdot, T_j(u)) - F(\cdot, T(u))| \right] (\underline{u}_q - u)^+ dx$$
$$= \int_{\{\underline{u}_q > u\}} \left[F(\cdot, \underline{u}_q) - F(\cdot, T(u)) - \sum_{j=1}^k |F(\cdot, T_j(u)) - F(\cdot, T(u))| \right] (\underline{u}_q - u)dx$$
$$\leq 0. \tag{5.47}$$

In fact, for x such that $\underline{u}_q(x) > u(x)$, we also have $\underline{u}(x) \geq u(x)$. Hence, in view of (5.34) and (5.35),

$$T_q(u) = \underline{u}_q \quad \text{and} \quad T(u) = \underline{u}.$$

Thus,

$$F(\cdot, \underline{u}_q) - F(\cdot, T(u)) - \sum_{j=1}^{k} |F(\cdot, T_j(u)) - F(\cdot, T(u))|$$
$$\leq F(\cdot, \underline{u}_q) - F(\cdot, T(u)) - |F(\cdot, T_q(u)) - F(\cdot, T(u))|$$
$$= F(\cdot, \underline{u}_q) - F(\cdot, \underline{u}) - |F(\cdot, \underline{u}_q) - F(\cdot, \underline{u})|$$
$$\leq 0,$$

implying (5.47). Now, (5.45), (5.46), and (5.47) imply that

$$0 \leq \int_{\Omega} b(\cdot, u)(\underline{u}_q - u)^+ dx$$
$$= \int_{\{\underline{u}_q > u\}} b(\cdot, u)(\underline{u}_q - u) dx$$
$$= -\int_{\{\underline{u}_q > u\}} (\underline{u} - u)^{p-1}(\underline{u}_q - u) dx \quad (\text{because } \underline{u} \geq \underline{u}_q)$$
$$\leq 0.$$

Consequently,

$$0 = \int_{\{\underline{u}_q > u\}} (\underline{u} - u)^{p-1}(\underline{u}_q - u) dx \geq \int_{\{\underline{u}_q > u\}} (\underline{u}_q - u)^p dx \geq 0.$$

This result shows that $\underline{u}_q - u = 0$ a.e. on $\{\underline{u}_q > u\}$; i.e., the measure of $\{\underline{u}_q > u\}$ is 0. We have proved that $u \geq \underline{u}_q$ a.e. in Ω. As this holds for all $q \in \{1, \ldots, k\}$, $u \geq \underline{u}$.

From (5.31), $b(x, u) = 0$ a.e. in Ω; i.e., $B(u) = 0$. Also, $T(u) = T_j(u) = u$, $\forall j \in \{1, \ldots, k\}$ and $F(\cdot, u) \in L^{p'}(\Omega)$ by (5.28). Hence,

$$\langle H(u), \phi \rangle = \int_{\Omega} F(\cdot, u)\phi \, dx.$$

(5.36), therefore, reduces to (5.11); i.e., u is a solution of (5.11). Hence, (5.11) has a solution $u \geq \underline{u}$.   □

Remark 5.6. Under some obvious modifications (for example, reversing the inequality and the min-max signs in (5.27), (5.28), (5.29), and (5.30)), we can prove existence results, similar to Theorem 5.5, for solutions of the inequality (5.11) that are bounded from above by supersolutions of (5.11).

Existence of Extremal Solutions

In this section, we show that if K has a certain lattice structure, then the variational inequality (5.11) has a greatest solution. First, we need an estimate for solutions of (5.11). Let

$$S = \{u \in V_0 : u \geq \underline{u} \text{ and } u \text{ is a solution of (5.11)}\}.$$

By Theorem 5.5, $S \neq \emptyset$. Moreover, we have the following lemma.

Lemma 5.7. *S is bounded in V_0.*

Proof: We fix an element $\phi \in K$. For $u \in S$, we have $u(x) \geq \underline{u}(x)$ a.e. in Ω, and

$$\int_\Omega A_0(\cdot, \nabla u) \cdot \nabla(\phi - u) dx \geq \int_\Omega F(\cdot, u)(\phi - u) dx.$$

It follows from (5.12), the growth condition (5.28), and Hölder's inequality that

$$\begin{aligned}
\alpha \|u\|_0^p &\leq \int_\Omega A_0(\cdot, \nabla u) \cdot \nabla u \, dx \\
&\leq \int_\Omega A_0(\cdot, \nabla u) \cdot \nabla \phi \, dx - \int_\Omega F(\cdot, u)(\phi - u) dx \\
&\leq \int_\Omega |A_0(\cdot, \nabla u)| \, |\nabla \phi| dx + \int_\Omega |F(\cdot, u)| \, (|\phi| + |u|) dx \\
&\leq \nu \|u\|_0^{p-1} \|\phi\|_0 + \|\gamma\|_{L^{p'}(\Omega)} \|\phi\|_0 + \|a\|_{L^{p'}(\Omega)} \left(\|u\|_{L^p(\Omega)} + \|\phi\|_{L^p(\Omega)} \right) \\
&\quad + b|\Omega|^{1 - \frac{\sigma+1}{p}} \|u\|_{L^p(\Omega)}^\sigma \left(\|u\|_{L^p(\Omega)} + \|\phi\|_{L^p(\Omega)} \right) \\
&\leq c_{14} \left(\|u\|_0^{p-1} + \|u\|_0 + \|u\|_0^\sigma + \|u\|_0^{\sigma+1} + 1 \right) \\
&\leq c_{15} \left(\|u\|_0^{p-1} + \|u\|_0^{\sigma+1} + 1 \right),
\end{aligned}$$

where c_{14}, c_{15} do not depend on u. As $\sigma + 1 < p$, it implies that $\|u\|_0$ is uniformly bounded for all $u \in S$. $\qquad \square$

We consider on $S \subset V$ the usual ordering:

$$u \leq v \iff u(x) \leq v(x) \text{ a.e. } x \in \Omega.$$

Now, we show the following improvement of Theorem 5.5, about the existence of greatest solutions of (5.11).

Theorem 5.8. *If K satisfies (5.29), (5.22), and (5.23), then (5.11) has a greatest solution $u^* \geq \underline{u}$; i.e., u^* is a solution of (5.11), and if u is any solution of (5.11) such that $u \geq \underline{u}$, then $u \leq u^*$.*

Proof: The greatest solution of (5.11) is the greatest element of S with respect to the (partial) ordering $\leq$. We shall show the existence of such an element by using Zorn's lemma. To apply this lemma, we first prove that each (nonempty) chain $\mathcal{C}$ in S has an upper bound.

Let $u_0 \in \mathcal{C}$ and $\mathcal{C}_0 = \{u \in \mathcal{C} : u \geq u_0\}$. Then, $u_0 \in \mathcal{C}_0$ and any upper bound of $\mathcal{C}_0$ is also an upper bound of $\mathcal{C}$. As $\mathcal{C}_0 \subset S$, the set $\{\|u\|_{L^p(\Omega)} : u \in \mathcal{C}_0\}$ is bounded by Lemma 5.7. Let

$$\alpha_0 = \sup\{\|u\|_{L^p(\Omega)} : u \in \mathcal{C}_0\}(< \infty).$$

By considering $\mathcal{C}_0 - u_0$ instead of $\mathcal{C}_0$, we can assume, without loss of generality, that $\mathcal{C}_0 \subset L_+^p(\Omega)$. If $u \in \mathcal{C}_0$ exists such that $\|u\|_{L^p(\Omega)} = \alpha_0$, then u is an upper bound of $\mathcal{C}_0$. In fact, $u \in S$ and for any $v \in \mathcal{C}_0$, we either have $v \leq u$ or $v \geq u$. In the second case,

$$\alpha_0 = \int_\Omega u^p dx \leq \int_\Omega v^p dx \leq \alpha_0.$$

Hence, $\int_\Omega u^p dx = \int_\Omega v^p dx$. As $v \geq u \geq 0$ in Ω, this holds only if $u = v$ a.e. on Ω; i.e., $u = v$. Thus, $u \geq v$, for all $v \in \mathcal{C}_0$; i.e., u is an upper bound of $\mathcal{C}_0$ (and then of $\mathcal{C}$).

Now, assume otherwise that $\|u\|_{L^p(\Omega)} < \alpha_0, \forall\, u \in \mathcal{C}_0$. By the definition of α_0, we have $u_1 \in \mathcal{C}_0$ such that

$$\alpha_0 > \|u_1\|_{L^p(\Omega)} > \alpha_0 - 1.$$

Inductively, we can choose, for each $n \in \mathbb{N}$, an element $u_n \in \mathcal{C}_0$ such that

$$\alpha_0 > \|u_n\|_{L^p(\Omega)} > \max\left\{\|u_{n-1}\|_{L^p(\Omega)}, \alpha_0 - \frac{1}{n}\right\}. \tag{5.48}$$

We must have $u_n \geq u_{n-1}$. In fact, if this does not hold, then $u_{n-1} \geq u_n$, because $\mathcal{C}_0$ is a chain. It follows that

$$\|u_{n-1}\|_{L^p(\Omega)}^p \geq \|u_n\|_{L^p(\Omega)}^p,$$

which contradicts the choice of u_n. It means that (u_n) is an increasing sequence in $L_+^p(\Omega)$. Letting $u = \sup\{u_n : n \in \mathbb{N}\}$, we have $u_n \uparrow u$ a.e. in Ω. By the Monotone convergence theorem, we get

$$\int_\Omega u_n^p dx \rightarrow \int_\Omega u^p dx.$$

From (5.48), $\int_\Omega u^p dx = \lim \int_\Omega u_n^p dx = \alpha_0^p$. Therefore, $u \in L^p(\Omega)$ and $\|u\|_{L^p(\Omega)} = \alpha_0$. Also, because

$$0 \leq |u_n - u|^p = (u - u_n)^p \leq u^p,$$

an application of the Dominated convergence theorem gives

$$\int_\Omega |u_n - u|^p dx \rightarrow 0, \ n \rightarrow \infty,$$

i.e.,

$$u_n \rightarrow u \ \text{in} \ L^p(\Omega). \tag{5.49}$$

We now check that u is an upper bound for $\mathcal{C}_0$. Let $v \in \mathcal{C}_0$. If $v \leq u_n$ for some n, then $v \leq u$. Assume that $v \not\leq u_n, \forall\, n$. As $\mathcal{C}_0$ is a chain, we must have $u_n \leq v, \forall\, n$. Using arguments as presented previously, we have $\|u_n\|_{L^p(\Omega)}^p \leq$

$\|v\|^p_{L^p(\Omega)}$, $\forall$ n. Letting $n \to \infty$, we get $\alpha_0 \leq \|v\|^p_{L^p(\Omega)}$, which contradicts our assumption about α_0. Hence, u is an upper bound of C_0.

We now prove that $u \in S$. As $u_n \geq \underline{u}$, for all n, we have $u \geq \underline{u}$. Also, as (u_n) is bounded in V_0, by passing to a subsequence if necessary, we can assume that

$$u_n \rightharpoonup u \text{ in } V_0.$$

As A is monotone, continuous, and bounded, it is pseudomonotone (cf. [152]). On the other hand, as K is closed and convex in V_0, it is weakly closed in this space. Hence, $u \in K$. Now, because $u_n \in S$,

$$\langle A(u_n), v - u_n \rangle \geq \int_\Omega F(\cdot, u_n)(v - u_n)dx, \forall\, v \in K, \forall\, n \in \mathbb{N}. \tag{5.50}$$

Letting $v = u$ in this inequality, we have

$$\langle A(u_n), u - u_n \rangle \geq \int_\Omega F(\cdot, u_n)(u - u_n)dx.$$

As $u_n \to u$ in $L^p(\Omega)$ and $u_n, u \geq \underline{u}$, it follows from (5.28) that $F(\cdot, u_n) \to F(\cdot, u)$ in $L^{p'}(\Omega)$. Hence,

$$\int_\Omega F(\cdot, u_n)(u - u_n)dx \to 0.$$

Thus, $\limsup \langle A(u_n), u - u_n \rangle \leq 0$. As A is pseudomonotone,

$$\liminf \langle A(u_n), u_n - v \rangle \geq \langle A(u), u - v \rangle, \,\forall\, v \in K. \tag{5.51}$$

It follows, from (5.50) and (5.51), that

$$\langle A(u), v - u \rangle \geq \limsup \langle A(u_n), v - u_n \rangle$$
$$\geq \limsup \int_\Omega F(\cdot, u_n)(v - u_n)dx$$
$$= \int_\Omega F(\cdot, u)(v - u)dx,$$

for all $v \in K$. Thus, u is a solution of (5.11); i.e., $u \in S$.

We have shown that every chain $\mathcal{C}$ in S has an upper bound in S. By Zorn's lemma, S has a maximal element u^*. We prove that u^* is, in fact, the greatest element of S. Assume otherwise that an element $v \in S$ exists such that $v \not\leq u^*$. As K satisfies (5.22), both v and u^* satisfy (5.15)–(5.17). Moreover, as $v, u^* \geq \underline{u} \geq \underline{u}_0$, (5.28) holds. Also, as $v, u^* \in K$, we have $v \vee K \subset K$ and $u^* \vee K \subset K$, by (5.23). Hence, by Theorem 5.5, a solution $w \in K$ exists such that $w \geq \tilde{u} = \max\{v, u^*\}(\geq \underline{u})$. We have $w \in S$ and $w \geq u^*$. Hence, $w = u^* \geq v$, contradicting the choice of v. This contradiction shows that u^* is the greatest element of S. $\qquad\square$

Remark 5.9. A part of the proof of Theorem 5.8 was to prove that for a L^p-bounded chain $\mathcal{C}$, an increasing sequence (u_n) exists that converges in $L^p(\Omega)$ to $\sup \mathcal{C} \in L^p(\Omega)$. This fact has already been used also in previous chapters to show the existence of extremal solutions by using Zorn's lemma.

We can prove furthermore that under the above conditions of Theorem 5.8, the variational inequality (5.11) has a smallest solution. In fact, we have the following result.

Theorem 5.10. *If K satisfies (5.29), (5.22), and (5.23), then (5.11) has a greatest solution u^* and a smallest solution u_* such that*

$$\underline{u} \leq u_* \leq u^*,$$

in the sense that for any solution u of (5.11) that satisfies $u \geq \underline{u}$, we have

$$u_* \leq u \leq u^*.$$

The proof of this theorem, which requires some arguments in Sect. 5.1.2, will be presented in that section.

Remark 5.11. (a) Theorem 5.10 establishes the existence of solutions and both the greatest and smallest solutions, from solely the existence of subsolutions of (5.11). We have similar results for solutions and extremal solutions below supersolutions of variational inequalities.

(b) To convey the ideas and somewhat simplify the calculations, we consider here perturbing functions F that depend only on x and u. However, Theorem 5.5, Theorem 5.8, and Theorem 5.10 still hold in the case where $F = F(x, u, \nabla u)$ depends also on the gradient of u. The arguments and calculations used above in the proof of Theorem 5.5 can be simplified; however, we keep the above approach due to its direct way to be extended to this more general situation. More detailed discussions in this case will be presented in the next section.

5.1.2 Comparison Principle and Extremal Solutions

In this section, we prove that if the inequality (5.11) has a pair of subsolution and supersolution, then at least a solution between them exists. Moreover, both greatest and smallest solutions in that interval exist. In this case (where both sub- and supersolutions exist), we can weaken the growth condition imposed on the lower order term F.

We shall keep the notation of the previous section and fix a number q such that

$$1 < q < p^*, \tag{5.52}$$

where p^* is the Sobolev critical exponent (corresponding to p). As $\partial\Omega$ is Lipschitzian, the embedding

$$W^{1,p}(\Omega) \subset L^q(\Omega) \tag{5.53}$$

is compact. In this section, we relax condition (5.16) in the definition of sub-(super)solutions to the weaker condition

$$F(\cdot, \underline{u}) \in L^{q'}(\Omega). \tag{5.54}$$

If we choose $q = p$ in (5.52), then this condition reduces to (5.16).

Now, we assume that (5.11) has subsolutions $\underline{u}_1, \ldots, \underline{u}_k$ and supersolutions $\bar{u}_1, \ldots, \bar{u}_m$. Define $\underline{u}$ as in (5.27) and put

$$\bar{u} = \min\{\bar{u}_1, \ldots, \bar{u}_m\}.$$

Suppose furthermore that $\underline{u} \leq \bar{u}$. Let $u_0 = \min\{\underline{u}_1, \ldots, \underline{u}_k\}$ and $u^0 = \max\{\bar{u}_1, \ldots, \bar{u}_m\}$. It is clear that $u_0, u^0 \in V$. We assume that F has the growth condition

$$|F(x, u)| \leq a(x), \text{ for a.e. } x \in \Omega, \text{ all } u \in [u_0(x), u^0(x)], \tag{5.55}$$

with $a \in L^{q'}(\Omega)$, $b \geq 0$. Note that if F satisfies the seemingly more general growth condition

$$|F(x, u)| \leq a(x) + b|u|^{q-1}, \text{ for a.e. } x \in \Omega, \text{ all } u \in [u_0(x), u^0(x)],$$

then F also satisfies (5.55) (with a different function a). We are now ready to prove the following existence result, which can be considered as the analog to the comparison principle for variational equations proved in Sect. 3.2 of Chap. 3.

Theorem 5.12. *Assume (5.55), (5.27), (5.29), and*

$$\bar{u}_i \wedge K \subset K, \, 1 \leq i \leq m. \tag{5.56}$$

Then, a solution u of (5.11) exists such that

$$\underline{u} \leq u \leq \bar{u}.$$

Proof: The ideas are similar to those in the proof of Theorem 5.5; however, the calculations are somewhat more involved and employ truncation techniques already used in the proof of Theorem 3.6 of Chap. 3. We define the mapping

$$b(x, t) = \begin{cases} [t - \bar{u}(x)]^{q-1} & \text{if } t > \bar{u}(x) \\ 0 & \text{if } \underline{u}(x) \leq t \leq \bar{u}(x) \\ -[\underline{u}(x) - t]^{q-1} & \text{if } t < \underline{u}(x), \end{cases} \tag{5.57}$$

for $x \in \Omega$, $t \in \mathbb{R}$. b is a Carathéodory function and

$$\begin{aligned} |b(x,t)| &\leq (|t| + |\bar{u}(x)| + |\underline{u}(x)|)^{q-1} \\ &\leq c_{16} \left\{ \left[|u_0(x)| + |u^0(x)| \right]^{q-1} + |t|^{q-1} \right\} \\ &= a_2(x) + c_{17}|t|^{q-1}, \end{aligned} \tag{5.58}$$

with $a_2 \in L^{q'}(\Omega)$, $c_{16} > 0$ being a constant. Hence, the mapping $u \mapsto b(\cdot, u)$ is continuous from $L^q(\Omega)(\supset W^{1,p}(\Omega))$ to $L^{q'}(\Omega)(\subset [W^{1,p}(\Omega)]^*)$. We have, as in (5.33),

$$
\begin{aligned}
&\int_\Omega b(\cdot, u) u \, dx \\
&= \int_{\{u > \bar{u}\}} (u - \bar{u})^{q-1} u \, dx - \int_{\{u < \underline{u}\}} (\underline{u} - u)^{q-1} u \, dx \\
&\geq \int_{\{u > \bar{u}\}} (c_{17}|u|^q - c_{18}|\bar{u}|^{q-1}|u|) dx + \int_{\{u < \underline{u}\}} (c_{19}|u|^q - c_{20}|\underline{u}|^{q-1}|u|) dx \\
&\geq c_{21} \int_\Omega |u|^q dx - c_{21} \int_{\{\underline{u} \leq u \leq \bar{u}\}} |u|^q dx - c_{22} \int_\Omega (|\bar{u}|^{q-1} + |\underline{u}|^{q-1})|u| dx - c_{23} \\
&\quad (c_{21} = \min\{c_{17}, c_{19}\} > 0) \\
&\geq c_{21} \int_\Omega |u|^q dx - c_{21} \int_\Omega (|\bar{u}| + |\underline{u}|)^q dx - c_{22} \int_\Omega (|\bar{u}|^{q-1} + |\underline{u}|^{q-1})|u| dx - c_{23} \\
&\geq c_{21} \|u\|_{L^q(\Omega)}^q - c_{24},
\end{aligned}
$$

$$\tag{5.59}$$

for all $u \in L^q(\Omega)$, where the c_i's ($17 \leq i \leq 24$) are positive constants independent of u. Let $1 \leq i \leq k$, $1 \leq j \leq m$. For each $u \in V$, we define

$$
T_{i0}(u)(x) = \begin{cases} \underline{u}_i(x) & \text{if } u(x) < \underline{u}_i(x) \\ u(x) & \text{if } \underline{u}_i(x) \leq u(x) \leq \bar{u}(x) \\ \bar{u}(x) & \text{if } u(x) > \bar{u}(x), \end{cases}
\tag{5.60}
$$

$$
T_{0j}(u)(x) = \begin{cases} \underline{u}(x) & \text{if } u(x) < \underline{u}(x) \\ u(x) & \text{if } \underline{u}(x) \leq u(x) \leq \bar{u}_j(x) \\ \bar{u}_j(x) & \text{if } u(x) > \bar{u}_j(x), \end{cases}
\tag{5.61}
$$

and

$$
T(u)(x) = \begin{cases} \underline{u}(x) & \text{if } u(x) < \underline{u}(x) \\ u(x) & \text{if } \underline{u}(x) \leq u(x) \leq \bar{u}(x) \\ \bar{u}(x) & \text{if } u(x) > \bar{u}(x), \end{cases}
\tag{5.62}
$$

($x \in \Omega$). It is easy to check that T_{i0}, T_{0j}, and T are bounded, continuous mappings from $W^{1,p}(\Omega)$ into itself and from $L^q(\Omega)$ into itself. Now, as $u_0 \leq \underline{u}_i, \underline{u}, \bar{u}, \bar{u}_j \leq u^0$, the growth condition (5.55) implies that the mappings

$$u \mapsto F(\cdot, T_{i0}(u)), \quad u \mapsto F(\cdot, T_{0j}(u)), \quad \text{and} \quad u \mapsto F(\cdot, T(u)),$$

are bounded and continuous from $L^q(\Omega)$ into $L^{q'}(\Omega)$. Consider the following variational inequality:

$$
\begin{cases} \langle A(u) + \beta B(u) - H(u), v - u \rangle \geq 0, \; \forall \, v \in K \\ u \in K. \end{cases}
\tag{5.63}
$$

Here, β is a fixed positive number,

$$\langle B(u), \phi \rangle = \int_\Omega b(\cdot, u)\phi dx, \tag{5.64}$$

and

$$\langle H(u), \phi \rangle = \int_\Omega \left[F(\cdot, T(u)) + \sum_{i=1}^k |F(\cdot, T_{i0}(u)) - F(\cdot, T(u))| \right.$$
$$\left. - \sum_{j=1}^m |F(\cdot, T_{0j}(u)) - F(\cdot, T(u))| \right] \phi dx, \tag{5.65}$$

for all $u, \phi \in V_0$. From the continuity and boundedness of the mappings

$$u \mapsto b(\cdot, u), u \mapsto F(\cdot, T_{i0}(u)), u \mapsto F(\cdot, T_{0j}(u)), \text{ and } u \mapsto F(\cdot, T(u)),$$

from $L^q(\Omega)$ to $L^{q'}(\Omega)$ the compactness of the embedding $W_0^{1,p}(\Omega) \subset L^q(\Omega)$, and the continuity of the embedding $L^{q'}(\Omega) \subset [W_0^{1,p}(\Omega)]^*$, we have that the mapping $\beta B - H$ is bounded and completely continuous from $W_0^{1,p}(\Omega)$ to $[W_0^{1,p}(\Omega)]^*$. It follows that $A + \beta B - H$ is pseudomonotone and bounded. Also, it follows from (5.55) that

$$|F(x, T_{i0}(u)(x))|, |F(x, T_{0j}(u)(x))| \leq a(x),$$

and thus,

$$|\langle H(u), v \rangle| \leq c_{25}\|v\|_{L^q(\Omega)}, \tag{5.66}$$

for some constant $c_{25} > 0$, fixed. As in the proof of Theorem 5.5, it follows from (5.12), (5.59), and (5.66) that for $\phi \in K$ fixed, for all $u \in K$,

$$\langle A(u) + \beta B(u) - H(u), u - \phi \rangle$$
$$\geq \langle A(u), u \rangle + \beta \langle B(u), u \rangle - |\langle A(u), \phi \rangle| - \beta|\langle B(u), \phi \rangle| - |\langle H(u), u \rangle|$$
$$- |\langle H(u), \phi \rangle|$$
$$\geq \alpha\|u\|_0^p + \beta c_{21}\|u\|_{L^q(\Omega)}^q - \beta c_{24} - \nu\|u\|_0^{p-1}\|\phi\|_0 - \|\gamma\|_{L^q(\Omega)}\|\phi\|_0$$
$$- \beta\left(\|a_2\|_{L^{q'}(\Omega)} + c_{17}\|u\|_{L^q(\Omega)}^{q-1}\right)\|\phi\|_{L^q(\Omega)} - c_{25}\left(\|u\|_{L^q(\Omega)} + \|\phi\|_{L^q(\Omega)}\right)$$
$$\geq \alpha\|u\|_0^p + \beta c_{21}\|u\|_{L^q(\Omega)}^q - c_{26}\|u\|_0^{p-1} - c_{27}\|u\|_{L^q(\Omega)}^{q-1} - c_{28}\|u\|_{L^q(\Omega)} - c_{29}. \tag{5.67}$$

Here, the c_i's are again positive constants depending on p, q, Ω, ϕ, and other fixed functions, but not on u. Thus, for some constants $c_{30}, c_{31}, c_{32} > 0$, we have

$$\langle A(u) + \beta B(u) - H(u), u - \phi \rangle \geq c_{30}\|u\|_0^p + c_{31}\|u\|_{L^q(\Omega)}^q - c_{32},$$

for all u with $\|u\|_0$ sufficiently large. Hence,

$$\lim_{\|u\|_0 \to \infty, u \in K} \frac{\langle A(u) + \beta B(u) - H(u), u - \phi \rangle}{\|u - \phi\|_0} = +\infty,$$

which proves the coercivity of $A + \beta B - H$ on K. We have the existence of solutions of (5.63) by well-known existence results for variational inequalities, as in Theorem 5.5.

In the next step, we show that any solution u of (5.63) must satisfy

$$\underline{u}_q \leq u \leq \bar{u}_r, \ \forall \, q \in \{1, \ldots, k\}, \ r \in \{1, \ldots, m\}. \tag{5.68}$$

Let us prove the first inequality, the proof of the second being similar. Because $\underline{u}_q$ satisfies (5.15)–(5.17), we also have (5.43) as in the proof of Theorem 5.5. On the other hand, arguing as in that theorem, using (5.29), we also have (5.44), with B, H defined by (5.64) and (5.65). Adding (5.43) and (5.44), we get

$$
\begin{aligned}
\langle A(u) - A(\underline{u}_q), (\underline{u}_q - u)^+ \rangle + \beta \int_\Omega b(\cdot, u)(\underline{u}_q - u)^+ dx \\
+ \int_\Omega \Big[F(\cdot, \underline{u}_q) - F(\cdot, T(u)) - \sum_i |F(\cdot, T_{i0}(u)) - F(\cdot, T(u))| \\
+ \sum_j |F(\cdot, T_{0j}(u)) - F(\cdot, T(u))| \Big] (\underline{u}_q - u)^+ dx \\
\geq 0.
\end{aligned}
\tag{5.69}
$$

For $x \in \Omega$ such that $\underline{u}_q(x) > u(x)$, we have $\underline{u}(x) > u(x)$, and thus,

$$T_{q0}(u)(x) = \underline{u}_q(x), \ T_{0j}(u)(x) = T(u)(x) = \underline{u}(x), \ \forall \, j \in \{1, \ldots, m\}.$$

Therefore,

$$
\begin{aligned}
\int_\Omega \Big[F(\cdot, \underline{u}_q) - F(\cdot, T(u)) - \sum_i |F(\cdot, T_{i0}(u)) - F(\cdot, T(u))| \\
+ \sum_j |F(\cdot, T_{0j}(u)) - F(\cdot, T(u))| \Big] (\underline{u}_q - u)^+ dx \\
= \int_{\{\underline{u}_q > u\}} \Big[F(\cdot, \underline{u}_q) - F(\cdot, T(u)) - \sum_i |F(\cdot, T_{i0}(u)) - F(\cdot, T(u))| \\
+ \sum_j |F(\cdot, T_{0j}(u)) - F(\cdot, T(u))| \Big] (\underline{u}_q - u)^+ dx \\
\leq 0,
\end{aligned}
\tag{5.70}
$$

because, for x in the set $\{\underline{u}_q > u\}$,

$$F(x, \underline{u}_q(x)) - F(x, T(u)(x)) - \sum_i |F(x, T_{i0}(u)(x)) - F(x, T(u)(x))|$$

$$+ \sum_j |F(x, T_{0j}(u)(x)) - F(x, T(u)(x))|$$

$$= F(x, \underline{u}_q(x)) - F(x, T(u)(x)) - \sum_i |F(x, T_{i0}(u)(x)) - F(x, T(u)(x))|$$

$$\leq F(x, \underline{u}_q(x)) - F(x, T(u)(x)) - |F(x, T_{q0}(u)(x)) - F(x, T(u)(x))|$$

$$= F(x, \underline{u}_q(x)) - F(x, \underline{u}(x)) - |F(x, \underline{u}_q(x)) - F(x, \underline{u}(x))|$$

$$\leq 0.$$

$$(5.71)$$

Now, it follows from (5.69), (5.70), and (5.46) that

$$0 \leq \int_\Omega b(\cdot, u)(\underline{u}_q - u)^+ dx$$

$$= \int_{\{\underline{u}_q > u\}} b(\cdot, u)(\underline{u}_q - u) dx$$

$$= - \int_{\{\underline{u}_q > u\}} (\underline{u} - u)^{q-1}(\underline{u}_q - u) dx \quad (\text{since } \underline{u}_q \leq \underline{u})$$

$$\leq 0.$$

$$(5.72)$$

This result means that

$$0 = \int_{\{\underline{u}_q > u\}} (\underline{u} - u)^{q-1}(\underline{u}_q - u) dx$$

$$\geq \int_{\{\underline{u}_q > u\}} (\underline{u}_q - u)^q dx$$

$$= \int_\Omega \left[(\underline{u}_q - u)^+ \right]^q dx.$$

Hence, $(\underline{u}_q - u)^+ = 0$ in Ω; i.e., $u \geq \underline{u}_q$ a.e. in Ω. Similarly, we can show by using the same mappings T_{i0}, T_{0j}, and T that

$$u \leq \bar{u}_r, \ \forall \, r \in \{1, \ldots, m\}.$$

Consequently,

$$\max\{\underline{u}_q : q \in \{1, \ldots, k\}\} = \underline{u} \leq u \leq \bar{u} = \min\{\bar{u}_r : r \in \{1, \ldots, m\}\}.$$

These inequalities imply that $b(\cdot, u) = 0$ in Ω and $T_{i0}(u) = T_{0j}(u) = T(u) = u$, $\forall \, i, j$. It thus follows from (5.55) that $F(\cdot, u) \in L^{q'}(\Omega)$. Also,

$$\langle H(u), \phi \rangle = \int_\Omega F(\cdot, u)\phi dx.$$

(5.63) becomes the variational inequality (5.11); i.e., u is a solution of (5.11). $\square$

Existence of Extremal Solutions

We show in this section the existence of extremal (smallest and greatest) solutions of (5.11), between sub- and supersolutions. Namely, we have the following result.

Theorem 5.13. *Assume (5.55), (5.56), (5.29), and (5.27) are satisfied, and that K has the lattice structure (5.22)–(5.23). Then, the variational inequality (5.11) has a greatest solution u^* and a smallest solution u_* such that*

$$\underline{u} \leq u_* \leq u^* \leq \bar{u};$$

i.e., if u is a solution of (5.11) such that $\underline{u} \leq u \leq \bar{u}$, then $u_* \leq u \leq u^*$.

Proof: The proof follows the same line as that of Theorem 5.8; thus, we present only the outline here. Let

$$S = \{u \in K : u \text{ is a solution of (5.11) and } \underline{u} \leq u \leq \bar{u}\}.$$

From Theorem 5.12, $S \neq \emptyset$. We show that S has the greatest and the smallest elements with respect to the ordering $\leq$. We apply again Zorn's lemma.

Let $C \subset S$ be a nonempty chain. Fix an element $u_0 \in C$, and by considering $C_0 = \{u \in C : u \geq u_0\}$, we can assume that C has a least element u_0. Again, by replacing C by $C - u_0$, we can also assume that $C \subset L_+^q(\Omega)$. Now, let $u \in S$. As $\underline{u} \leq u \leq \bar{u}$, we have

$$|u(x)| \leq |\underline{u}(x)| + |\bar{u}(x)|, \quad \text{a.e. in } \Omega,$$

and thus,

$$\|u\|_{L^q(\Omega)} \leq c_{33} \left(\|\underline{u}\|_{L^q(\Omega)} + \|\bar{u}\|_{L^q(\Omega)} \right).$$

This result means that S is bounded in $L^q(\Omega)$. Using the same arguments as in Theorem 5.8, with $L^q(\Omega)$, $L^{q'}(\Omega)$, and $\| \cdot \|_{L^q(\Omega)}$ instead of $L^p(\Omega)$, $L^q(\Omega)$, and $\| \cdot \|_{L^p(\Omega)}$, we come to the same conclusion as in that theorem; i.e., C has an upper bound in S.

We have shown that each nonempty chain in S has an upper bound (also in S). By Zorn's lemma, S has a maximal element. Using arguments as in Theorem 5.8 (with Theorem 5.12 instead of Theorem 5.5), we see that this maximal element is, in fact, the greatest element of S. By reversing the inequality signs, we can prove, in the same way, the existence of the smallest element of S. □

Now, we are ready to complete the proof of Theorem 5.10, which is based on Theorem 5.13.

Proof of Theorem 5.10: It follows from Theorem 5.8 that a greatest solution $u^* \geq \underline{u}$ exists. As K satisfies (5.23), u^* is also a supersolution of (5.11). For $u_0(x) \leq u(x) \leq u^*(x)$, we have from (5.28) that

$$|F(x, u(x))| \le a(x) + b|u(x)|^\sigma \le a(x) + b(|u_0(x)| + |u^*(x)|)^\sigma. \qquad (5.73)$$

As $u_0, u^* \in L^p(\Omega)$ and $0 \le \sigma < p - 1$, $(|u_0| + |u^*|)^\sigma \in L^{p/\sigma}(\Omega) \subset L^q(\Omega)$. Hence, the right-hand side of (5.73) is in $L^q(\Omega)$. Therefore, we have (5.55) with $q = p\,(< p^*)$.

On the other hand, as $u^* \in K$, (5.56) follows from (5.22). We see that all conditions of Theorem 5.13 are satisfied. Consequently, (5.11) has a greatest solution u_0^* and a smallest solution u_*^0 between $\underline{u}$ and u^*; i.e.,

$$\underline{u} \le u_*^0 \le u_0^* \le u^*,$$

and u_0^* and u_*^0 are, respectively, the greatest and smallest solutions (i.e., the greatest and smallest elements) of the set

$$S_1 = \{u \in V : \underline{u} \le u \le u^*,\ u \text{ is a solution of } (5.11)\}.$$

Now, recalling that $S = \{u \in V : \underline{u} \le u,\ u \text{ is a solution of } (5.11)\}$, we have that $S = S_1$ by Theorem 5.8 and the definition of u^*. Therefore, u_0^* and u_*^0 are also the greatest and smallest solutions (i.e., the greatest and smallest elements) in S. Hence, $u_0^* = u^*$ and the proof of Theorem 5.10 is complete.

$\square$

We conclude this section with some remarks.

Remark 5.14. (a) To convey the main ideas, we assume here that F depends only on x and u. However, the arguments given above for $F = F(x, u)$ can be easily extended to the more general case where $F = F(x, u, \nabla u)$ also depends on the gradient of the unknown function u (see the following sections).

(b) In the definitions of sub- and supersolutions presented above, we do not require that sub- and supersolutions are elements of K, which extends the scope of application of the subsolution and supersolution method. For example, consider the variational equation

$$u \in W_0^{1,p}(\Omega) : \quad \langle A(u), v \rangle = \int_\Omega F(x, u, \nabla u)v\,dx,\ \forall\, v \in W_0^{1,p}(\Omega), \qquad (5.74)$$

which correspondinds to the variational inequality (5.11) when $K = W_0^{1,p}(\Omega)$. Then, if $\underline{u}$ (respectively, $\bar{u}$) is a subsolution (respectively, supersolution) of (5.74) in the usual sense of sub- and supersolutions of variational equations of Chap. 3 (cf. [113], [83], and [133]), then $\underline{u}$ (respectively, $\bar{u}$) is also a subsolution (respectively, supersolution) of (5.11) in the the sense of Definition 5.1.

In fact, assume that $\underline{u}$ is a subsolution of (5.74). Then, we have (5.15) and (5.16). Moreover,

$$\langle A(u), v \rangle \le \int_\Omega F(x, u, \nabla u)v\,dx,\ \forall\, v \in V_0 \cap L_+^p(\Omega), \qquad (5.75)$$

(cf. Chap. 3 and [113], [83], and [133]). Let $w \in \underline{u} \wedge K$, $w = \underline{u} \wedge w_1$ for some $w_1 \in K$. It follows that $w - \underline{u} \le 0$ on Ω. As $\underline{u}|_{\partial\Omega} \le 0$ on $\partial\Omega$, $w_1|_{\partial\Omega} = 0$ on

$\partial\Omega$, we have $w|_{\partial\Omega} = \underline{u}|_{\partial\Omega}$. Hence, $w - \underline{u} \in V_0$. Using (5.75) with $v = -w + \underline{u}$, we have $v \in V_0$, $v \geq 0$. Thus, in this case, (5.75) reduces to (5.17). A similar proof holds for supersolutions.

(c) The applications of the above general results to some specific, but interesting, noncoercive variational inequalities and equations will be given in a following section. In the next section, we present an extension of the above concepts and arguments to a more general variational inequality, containing convex functionals not necessarily indicators of convex sets. We also study a sub-supersolution approach for evolutionary variational inequalities.

5.2 Variational Inequalities with Convex Functionals

In this section, we extend the above discussions to more general variational inequalities that contain convex functionals. We are concerned here with the existence of solutions and extremal solutions of noncoercive variational inequalities of the form:

$$\begin{cases} \langle \mathcal{A}(u), v - u \rangle - \langle G(u), v - u \rangle + j(v) - j(u) \geq 0, \ \forall \, v \in V_0 \\ u \in V_0. \end{cases} \tag{5.76}$$

Here, $\mathcal{A}$ is (the weak form of) the second-order quasi-linear elliptic operator

$$-\sum_{i=1}^{N} \frac{\partial}{\partial x_i}[a_i(x, u, \nabla u)] + a_0(x, u, \nabla u) \tag{5.77}$$

G is the lower order term [cf. (5.78) and (5.83)], and j is a convex functional, representing obstacles or unilateral conditions imposed on the solutions. Depending on the choice of j, the variational inequality (5.76) is the weak form of an equation or a complementarity problem that contains the operator (5.77) with various types of free boundaries or constraints (cf. [124, 11, 95]).

This section is the next step of our study plan proposed in Section 5.1 on sub-supersolution methods applied to variational inequalities on closed convex sets; that is, the particular case of (5.76) where j is the indicator function of a closed convex set K:

$$j(u) = \begin{cases} 0 & \text{if } u \in K \\ +\infty & \text{if } u \notin K. \end{cases}$$

However, many interesting problems in mechanics and applied mathematics lead to other types of convex functionals; for example,

$$j(u) = \int_\Omega \psi(x, u(x)) dx \ \text{ or } \ j(u) = \int_{\partial\Omega} \psi(x, u|_{\partial\Omega}(x)) dS,$$

(cf. [95, 89]). Because of the nonsymmetric nature of the problem, sub-supersolution methods for smooth equations (cf. [113], [75], [41], or [133]) and the arguments in [141] for inequalities on convex sets are not directly applicable to (5.76). The goal of this section is to extend some results in the previous section to the inequality (5.76) with more generality on the convex functional j.

5.2.1 General Settings—Sub- and Supersolutions

In this section, we consider some assumptions on the inequality (5.76) and next define sub- and supersolutions for it. Although similar conditions are considered before, the main assumptions are presented here for completeness. In (5.76), $\mathcal{A}$ is a mapping from V to V^*, defined by

$$\langle \mathcal{A}(u), v \rangle = \int_\Omega \Big[\sum_{i=1}^N a_i(x, u, \nabla u)\partial_i v + a_0(x, u, \nabla u)v \Big] dx, \ \forall \ u, v \in V, \quad (5.78)$$

where, for each $i \in \{0, 1, \ldots, N\}$, a_i is a Carathéodory function from $\Omega \times \mathbb{R}^{N+1}$ to $\mathbb{R}$. For $i \in \{1, \ldots, N\}$,

$$|a_i(x, u, \xi)| \le a_0(x) + b_0(|u|^{p-1} + |\xi|^{p-1}) \quad (5.79)$$

and

$$|a_0(x, u, \xi)| \le a_1(x) + b_1(|u|^{q-1} + |\xi|^{\frac{p}{q'}}), \quad (5.80)$$

for almost all $x \in \Omega$, all $u \in \mathbb{R}$, $\xi \in \mathbb{R}^N$ with $b_0, b_1 > 0$, $a_0 \in L^{p'}(\Omega)$, $a_1 \in L^{p'}(\Omega)$, $1 < q < p^*$. (As usual, p' is the Hölder conjugate of p and p^* is its Sobolev conjugate.) Moreover,

$$\sum_{i=1}^N [a_i(x, u, \xi) - a_i(x, u', \xi')](\xi_i - \xi_i') + [a_0(x, u, \xi) - a_0(x, u', \xi')](u - u') > 0,$$
$$(5.81)$$

if $(u, \xi) \ne (u', \xi')$, and

$$\sum_{i=1}^N a_i(x, u, \xi)\xi_i + a_0(x, u, \xi)u \ge \alpha(|\xi|^p + |u|^p) - \beta(x), \quad (5.82)$$

for a.e. $x \in \Omega$, all $u \in \mathbb{R}$, $\xi \in \mathbb{R}^N$, where $\alpha > 0$ and $\beta \in L^1(\Omega)$. The lower order operator G is defined by

$$\langle G(u), v \rangle = \int_\Omega F(x, u, \nabla u)v dx, \quad (5.83)$$

where $F : \Omega \times \mathbb{R}^{N+1} \to \mathbb{R}$ is a Carathéodory function with a certain growth condition to be specified later. We also assume that j is a mapping from V to

$\mathbb{R} \cup \{+\infty\}$ such that the restriction $j|_{V_0}$ is convex and lower semicontinuous on V_0 with nonempty effective domain. Before stating our theorem about existence of solutions, we need to define subsolutions and supersolutions for inequalities with convex functionals. These definitions extend those definitions presented in Sect. 5.1 for inequalities on closed convex sets.

Definition 5.15. *A function $\underline{u} \in V$ is called a subsolution of (5.76) if a functional J (depending on $\underline{u}$) exists:*

$$J = J_{\underline{u}} : V \to \mathbb{R} \cup \{+\infty\},$$

such that

$$
\begin{aligned}
&(i) \ \underline{u} \leq 0 \ on \ \partial\Omega \\
&(ii) \ F(\cdot, \underline{u}, \nabla\underline{u}) \in L^{q'}(\Omega) \\
&(iii) \ J(\underline{u}) < \infty,
\end{aligned}
\tag{5.84}
$$

and

$$j(v \vee \underline{u}) + J(v \wedge \underline{u}) \leq j(v) + J(\underline{u}), \ \forall \ v \in V_0 \cap D(j) \tag{5.85}$$

and

$$(iv)\langle \mathcal{A}(\underline{u}), v - \underline{u}\rangle - \langle G(\underline{u}), v - \underline{u}\rangle + J(v) - J(\underline{u}) \geq 0, \tag{5.86}$$

for all $v \in \underline{u} \wedge [V_0 \cap D(j)]$ $(D(j) = \{v \in V : j(v) < \infty\}$ is the effective domain of j). We have a similar definition for supersolutions: $\bar{u}$ is a supersolution of (5.76) if $J = J_{\bar{u}} : V \to \mathbb{R} \cup \{+\infty\}$ exists such that:

$$
\begin{aligned}
&(i) \ \bar{u} \geq 0 \ on \ \partial\Omega \\
&(ii) \ F(\cdot, \bar{u}, \nabla\bar{u}) \in L^{q'}(\Omega) \\
&(iii) \ J(\bar{u}) < \infty,
\end{aligned}
\tag{5.87}
$$

and

$$j(v \wedge \bar{u}) + J(v \vee \bar{u}) \leq j(v) + J(\bar{u}), \ \forall \ v \in V_0 \cap D(j) \tag{5.88}$$

and

$$(iv)\langle \mathcal{A}(\bar{u}), v - \bar{u}\rangle - \langle G(\bar{u}), v - \bar{u}\rangle + J(v) - J(\bar{u}) \geq 0 \tag{5.89}$$

for all $v \in \bar{u} \vee [V_0 \cap D(j)]$.

Suppose subsolutions $\underline{u}_1, \ldots, \underline{u}_k$ and supersolutions $\bar{u}_1, \ldots, \bar{u}_m$ of (5.76) exist. As above, we put

$$\underline{u} = \max\{\underline{u}_i : 1 \leq i \leq k\}, \quad \bar{u} = \min\{\bar{u}_l : 1 \leq l \leq m\},$$

$$\underline{u}_0 = \min\{\underline{u}_i : 1 \leq i \leq k\}, \quad \bar{u}_0 = \max\{\bar{u}_l : 1 \leq l \leq m\},$$

and assume that F has the following growth condition:

$$|F(x, u, \xi)| \leq a_2(x) + b_2|\xi|^{p/q'} \tag{5.90}$$

for a.e. $x \in \Omega$, all $\xi \in \mathbb{R}^N$, all u such that $\underline{u}_0(x) \leq u \leq \bar{u}_0(x))$, where $a_2 \in L^{q'}(\Omega)$, $b_2 \geq 0$, $q < p^*$.

We conclude this section with some remarks.

Remark 5.16. (i) If u is a solution of (5.76), then u is a subsolution of (5.76), provided j satisfies the following condition:

$$j(v \vee u) + j(v \wedge u) \leq j(v) + j(u), \tag{5.91}$$

for all $u, v \in W^{1,p}(\Omega)$. In fact, if u is a solution of (5.76), then it satisfies (i)–(ii). By choosing $J = j$, we see that (5.85) follows from (5.91). If $v = u \wedge w$, $w \in V_0$, then $v = 0$ on $\partial\Omega$; i.e., $v \in V_0$. Hence, (5.86) is a consequence of (5.76). Similarly, if (5.91) holds, then any solution is a supersolution.

(ii) (5.91) is satisfied for several usual convex functionals j. For example, if j is given by

$$j(u) = \int_E \psi(x, u)dx, \tag{5.92}$$

where E is a subset of Ω or $\partial\Omega$, $\psi : \Omega \times \mathbb{R} \to \mathbb{R} \cup \{+\infty\}$, is a Carathéodory function such that

$$\psi(x, u) \geq a_3(x) + b_3|u|^s, \ x \in \Omega, \ u \in \mathbb{R}, \tag{5.93}$$

where $a_3 \in L^1(\Omega)$ and $0 \leq s < p^*$. The functional j is well defined from $W^{1,p}(\Omega)$ to $\mathbb{R} \cup \{+\infty\}$ and j is convex if $\psi(x, \cdot)$ is convex for a.e. $x \in \Omega$. Also, by Fatou's lemma, j is weakly lower semicontinuous. Let $u, v \in V$, and denote

$$\Omega_1 = \{x \in \Omega : v(x) < u(x)\}, \ \Omega_2 = \{x \in \Omega : v(x) \geq u(x)\}.$$

Then,

$$\begin{aligned} j(v \wedge u) + j(v \vee u) &= \left(\int_{\Omega_1} + \int_{\Omega_2}\right)\psi(v \wedge u)dx + \left(\int_{\Omega_1} + \int_{\Omega_2}\right)\psi(v \vee u)dx \\ &= \int_\Omega \psi(x, u)\, dx + \int_\Omega \psi(x, v)dx \\ &= j(u) + j(v). \end{aligned} \tag{5.94}$$

Hence, (5.91) is satisfied. Note that from (5.93), $\psi(x, u)$ is bounded from below by a function in $L^1(\Omega)$. Thus, the integrals in (5.94) are in $\mathbb{R} \cup \{+\infty\}$, and we can split and combine them as done.

(iii) If $j = I_K$, K is a closed convex set in V_0, then we recover the cases considered in [141] (see also Sect. 5.1). Moreover, (5.91) holds provided K satisfies the condition

$$u, v \in K \Longrightarrow u \wedge v, \ u \vee v \in K. \tag{5.95}$$

As noted in [141], (5.95) is satisfied whenever K is defined by obstacles or by certain conditions on the gradients. We can also check that by using (5.92).

(iv) If $j = 0$, we have an equation in (5.76). By choosing $J = 0$ also, we see that (5.85)–(5.88) obviously hold and (5.86)–(5.89) reduce to the usual definitions of sub- and supersolutions of equations. If $j = I_K$ as in (iii), then by choosing $J = 0$, we see that the definition of subsolutions in [141] is equivalent to the definition in (i)-(iv) here. Thus, Definition 5.15 is an extension of that in [141].

(v) By choosing $J = 0$ in (5.85) and (5.86), we see that if $\underline{u}$ is a subsolution of the equation

$$\langle \mathcal{A}(u), v \rangle - \langle G(u), v \rangle = 0, \ \forall \, v \in V_0$$

and $j(v \vee \underline{u}) \leq j(v), \ \forall \, v \in W_0^{1,p}(\Omega) \cap D(j)$, then $\underline{u}$ is a subsolution of (5.76). Similar observations hold for supersolutions.

(vi) Compared with the definitions in Chap. 3 or [113, 75, 41, 133, 141], the new ingredient here is the introduction of the functional J in Definition 5.15, which permits more flexibility in constructing sub- and supersolutions (by choosing different J).

5.2.2 Existence and Comparison Results

In this section, we state and prove our existence results for solutions and extremal solutions of (5.76), based on the concepts of sub- and supersolutions in Sect. 5.2.1.

Theorem 5.17. *Assume (5.76) has a subsolution $\underline{u}$ and a supersolution $\bar{u}$ such that $\underline{u} \leq \bar{u}$ and that (5.90) holds. Then, (5.76) has a solution u such that $\underline{u} \leq u \leq \bar{u}$.*

Proof: We follow the usual truncation–penalization technique as in the previous sections (see [113, 41, 133, 141]). Therefore, we just outline the main arguments and present only the different points and modifications needed for our situation here. Let b be defined as in (5.57). We have the estimates (5.58) and (5.59).

We also define T_{i0}, T_{0j}, and T as in (5.60), (5.61), and (5.62). Let us consider the variational inequality

$$\begin{cases} \langle \mathcal{A}(u) + \beta B(u) - H(u), v - u \rangle + j(v) - j(u) \geq 0, \ \forall \, v \in V_0 \\ u \in V_0, \end{cases} \tag{5.96}$$

with $\beta > 0$ sufficiently large, B given by (5.64), and H by

$$\langle H(u), \phi \rangle$$

$$= \int_{\Omega} \Big[F(\cdot, T(u), \nabla T(u)) + \sum_{i=1}^{k} |F(\cdot, T_{i0}(u), \nabla T_{i0}(u)) - F(\cdot, T(u), \nabla T(u))| $$

$$- \sum_{j=1}^{m} |F(\cdot, T_{0j}(u), \nabla T_{0j}(u)) - F(\cdot, T(u), \nabla T(u))| \Big] \phi \, dx$$

$$(5.97)$$

for all $u, \phi \in V_0$. Let us prove that $\mathcal{H} = \mathcal{A} + \beta B - H$ is pseudomonotone on V. In fact, assume $w_n \rightharpoonup w$ in V and

$$\limsup_{n \to \infty} \langle \mathcal{H}(w_n), w_n - w \rangle \leq 0. \qquad (5.98)$$

We show that

$$\lim_{n \to \infty} \langle \mathcal{H}(w_n), w_n - v \rangle \geq \langle \mathcal{H}(w), w - v \rangle, \ \forall \, v \in V. \qquad (5.99)$$

As the embedding $W^{1,p}(\Omega) \subset L^q(\Omega)$ is compact, we have $w_n \to w$ in $L^q(\Omega)$. As the sequence (w_n) is bounded in V, the sequences

$$(F(\cdot, T_{i0}(w_n), \nabla T_{i0}(w_n))), \quad (F(\cdot, T_{0j}(w_n), \nabla T_{0j}(w_n))),$$

and $(F(\cdot, T(w_n), \nabla T(w_n)))$ are uniformly bounded in $L^{q'}(\Omega)$. From (5.97), it follows that the sequence $((\beta B - H)(w_n))$ is bounded in $L^{q'}(\Omega)$, and thus, the strong convergence of (u_n) in $L^q(\Omega)$ implies

$$\langle (\beta B - H)(w_n), w_n - w \rangle \to 0.$$

Hence, from (5.98), we get

$$\limsup \langle \mathcal{A}(w_n), w_n - w \rangle \leq 0. \qquad (5.100)$$

As $\{a_i\}$ $(i = 0, 1, \ldots, n)$ satisfy the Leray–Lions conditions (5.79)–(5.81), the operator $\mathcal{A} : W_0^{1,p}(\Omega) \to (W_0^{1,p}(\Omega))^*$ is pseudomonotone, and thus by Definition 2.97, we have

$$\mathcal{A} w_n \rightharpoonup \mathcal{A} w \quad \text{and} \quad \langle \mathcal{A} w_n, w_n \rangle \to \langle \mathcal{A} w, w \rangle,$$

which shows (5.99) (even equality holds), and thus, $\mathcal{A} + \beta B - H$ is pseudomonotone. Using arguments similar to those in [141], we can prove that $\mathcal{A} + \beta B - H$ is coercive on V_0. Moreover, this mapping is obviously continuous and bounded. Classic existence results for variational inequalities (cf. [152, 124]) give the existence of at least one solution $u \in V_0$ of (5.96). Also, it is clear that $u \in D(j)$. We prove that $\underline{u} \leq u$. Let $\underline{u}_q$ $(1 \leq q \leq k)$ be a subsolution. As $u \in V_0 \cap D(j)$, (5.86) with $\underline{u} = \underline{u}_q$ and $v = \underline{u} \wedge u$ gives

$$\langle \mathcal{A}(\underline{u}_q), \underline{u}_q \wedge u - \underline{u}_q \rangle - \langle G(\underline{u}_q), \underline{u}_q \wedge u - \underline{u}_q \rangle + J(\underline{u}_q \wedge u) - J(\underline{u}_q) \geq 0.$$

As $\underline{u}_q \wedge u = \underline{u}_q - (\underline{u}_q - u)^+$, the above inequality becomes

$$-\langle \mathcal{A}(\underline{u}_q), (\underline{u}_q - u)^+ \rangle - \langle G(\underline{u}_q), (\underline{u}_q - u)^+ \rangle + J(\underline{u}_q \wedge u) - J(\underline{u}_q) \geq 0. \quad (5.101)$$

On the other hand, as $\underline{u}_q \vee u = 0$ on $\partial\Omega$, $v = \underline{u}_q \vee u \in V_0$. Letting v into (5.96) and noting that $\underline{u}_q \vee u = u + (\underline{u}_q - u)^+$, we get

$$\langle \mathcal{A}(u) + \beta B(u) - H(u), (\underline{u}_q - u)^+ \rangle + j(\underline{u}_q \vee u) - j(u) \geq 0. \quad (5.102)$$

Adding (5.101) and (5.102), we get

$$\langle \mathcal{A}(u) - \mathcal{A}(\underline{u}_q), (\underline{u}_q - u)^+ \rangle + \langle G(\underline{u}_q) + \beta B(u) - H(u), (\underline{u}_q - u)^+ \rangle$$
$$+ j(\underline{u}_q \vee u) - j(u) + J(\underline{u}_q \wedge u) - J(\underline{u}_q) \geq 0.$$

From (5.85), we get

$$j(\underline{u}_q \vee u) - j(u) + J(\underline{u}_q \wedge u) - J(\underline{u}_q) \leq 0.$$

Using the integral formulation of B, H, and G, we get

$$\langle \mathcal{A}(u) - \mathcal{A}(\underline{u}_q), (\underline{u}_q - u)^+ \rangle + \int_\Omega F(x, \underline{u}_q, \nabla \underline{u}_q)(\underline{u}_q - u)^+ \, dx$$
$$+ \beta \int_\Omega b(x, u)(\underline{u}_q - u)^+ dx$$
$$- \int_\Omega \left[F(x, T(u), \nabla T(u)) + \sum_i |F(x, T_{i0}(u), \nabla T_{i0}(u)) - F(x, T(u), \nabla T(u))| \right.$$
$$\left. - \sum_j |F(x, T_{0j}(u), \nabla T_{0j}(u)) - F(x, T(u), \nabla T(u))| \right] (\underline{u}_q - u)^+ dx$$
$$\geq 0.$$
$$(5.103)$$

We have also the following estimate:

$$\langle \mathcal{A}(u) - \mathcal{A}(\underline{u}_q), (\underline{u}_q - u)^+ \rangle + \beta \int_\Omega b(\cdot, u)(\underline{u}_q - u)^+ dx$$
$$+ \int_\Omega \left[F(\cdot, \underline{u}_q, \nabla \underline{u}_q) - F(\cdot, T(u), \nabla T(u)) \right.$$
$$- \sum_i |F(\cdot, T_{i0}(u), \nabla T_{i0}(u)) - F(\cdot, T(u), \nabla T(u))|$$
$$\left. + \sum_j |F(\cdot, T_{0j}(u), \nabla T_{0j}(u)) - F(\cdot, T(u), \nabla T(u))| \right] (\underline{u}_q - u)^+ dx$$
$$\geq 0.$$

Using calculations as in (5.70) in the proof of Theorem 5.12, we finally obtain the following estimate:

$$0 \leq \int_\Omega b(\cdot, u)(\underline{u}_q - u)^+ dx$$
$$= - \int_{\{\underline{u}_q > u\}} (u - u)^{q-1}(\underline{u}_q - u)dx \quad (\text{because } \underline{u}_q \leq u)$$
$$\leq 0.$$

Thus, $0 = \int_\Omega [(\underline{u}_q - u)^+]^q dx$ and $(\underline{u}_q - u)^+ = 0$ a.e. in Ω; i.e., $u \geq \underline{u}_q$ a.e. in Ω. Using these arguments for all $q \in \{1,\ldots,k\}$, we see that $u \geq \underline{u}$. We can show in the same way that $u \leq \bar{u}$. Now, from (5.57), we have $b(x,u(x)) = 0$ for almost all $x \in \Omega$; i.e., $B = 0$. Also, $T_{il}(u) = T(u) = u$, for all i, l, and thus,

$$\langle H(u), \phi \rangle = \int_\Omega F(\cdot, u, \nabla u)\phi dx = \langle G(u), \phi \rangle.$$

Hence, as u satisfies (5.96), it also satisfies (5.76); i.e., u is a solution of (5.76) and $\underline{u} \leq u \leq \bar{u}$. □

We now note that (5.76) has a greatest and a smallest solution within the interval between $\underline{u}$ and $\bar{u}$.

Theorem 5.18. *Assume (5.76) has a subsolution $\underline{u}$ and a supersolution $\bar{u}$ such that $\underline{u} \leq \bar{u}$. Moreover, (5.90) and (5.91) hold. Then, (5.76) has a greatest solution u^* and a smallest solution u_* such that*

$$\underline{u} \leq u_* \leq u^* \leq \bar{u}; \tag{5.104}$$

that is, u_* and u^* are solutions of (5.76) that satisfy (5.104), and if u is a solution of (5.76) such that $\underline{u} \leq u \leq \bar{u}$, then $u_* \leq u \leq u^*$ on Ω.

The proof is similar to that of the particular case $j = I_K$, which was already presented in Theorem 5.13, Sect. 5.1 (see also [141]). Therefore, it is omitted.

As in the case of variational inequalities on convex sets, we still have the existence of solutions and extremal solutions provided only subsolutions (or supersolutions) exist together with certain one-sided growth conditions. We have in fact the following result.

Theorem 5.19. *Assume (5.76) has subsolutions $\underline{u}_1,\ldots,\underline{u}_k$ and F has the growth condition*

$$|F(x, u, \xi)| \leq a_3(x) + b_3(|u|^\sigma + |\xi|^\sigma) \tag{5.105}$$

for a.e. $x \in \Omega$, all u such that $\underline{u}_0(x) \leq u$, all $\xi \in \mathbb{R}^N$, where $0 \leq \sigma < p-1$, $a \in L^{p'}(\Omega)$, and

$$\underline{u}_0 = \min\{\underline{u}_i : 1 \leq i \leq k\}.$$

Hence, (5.76) has a solution u such that $u \geq \underline{u} = \max\{\underline{u}_i : 1 \leq i \leq k\}$.

The idea of the proof of this result is a combination of Theorem 5.17 stated above and an extension of Theorem 1 in [141] (cf. Theorem 5.5). We omit the proof and refer to [141] and Sect. 5.1 above for more details. By looking closely at the set of solutions of (5.76), we can improve Theorem 5.19 and get the following stronger result.

Theorem 5.20. *Under the assumptions of Theorem 5.19, (5.76) has a greatest solution* u^* *and a smallest solution* u_* *such that*

$$\underline{u} \leq u_* \leq u^*; \tag{5.106}$$

that is, u_* *and* u^* *are solutions of (5.76) that satisfy (5.106), and if* u *is a solution of (5.76) such that* $\underline{u} \leq u$, *then* $u_* \leq u \leq u^*$ *a.e. on* Ω.

The proof follows the same line as that in Theorem 2, [141], and it is therefore omitted. For more details, we refer to [141] and [142].

Remark 5.21. Note that if $a_i = a_i(x, \xi)$ $(i = 1, \ldots, N)$ do not depend on u, then we can choose $a_0 = 0$ and all results stated above still hold.

5.2.3 Some Examples

We now apply these general results to establish the existence of solutions and extremal solutions in some particular variational inequalities.

Example 5.22. In this example, we study a quasi-linear elliptic variational inequality that contains a "unilateral" term given by an integral. Assume that for $i = 0, 1, \ldots, N$, a_i satisfies

$$a_i(x, u, 0) = 0 \tag{5.107}$$

for a.e. $x \in \Omega$, all $u \in \mathbb{R}$, and consider the variational inequality

$$\begin{cases} \langle A(u), v - u \rangle - \lambda \int_\Omega F(x, u, \nabla u)(v - u) dx + j(v) - j(u) \geq 0, \ \forall \, v \in V_0 \\ u \in V_0. \end{cases}$$

$$\tag{5.108}$$

Here, A and F are defined as in (5.78), (5.79), and (5.80) of Section 5.2.1. λ is a real parameter and

$$j(u) = \int_\Omega \psi(x, u(x)) dx, \tag{5.109}$$

where $\psi : \Omega \times \mathbb{R} \to \mathbb{R} \cup \{+\infty\}$ is a Carathéodory function such that

$$\psi(x, u) \geq -a(x) - b|u|^p, \tag{5.110}$$

where $a \in L^1(\Omega), b \geq 0$. It follows from this inequality that for $u \in V$, $x \mapsto \psi(x, u(x))$ is measurable and because $-a - b|u|^p \in L^1(\Omega)$, j is well defined and $j(u) \in \mathbb{R} \cup \{+\infty\}$. Assume also that for almost all $x \in \Omega$, $\psi(x, \cdot)$ is convex. Hence, j is convex on V. It follows from Fatou's lemma that j is lower semicontinuous on that space. The following lemma shows the existence of constant sub- and supersolutions of (5.108).

Lemma 5.23. (a) *Assume* $B \in \mathbb{R}$, $B \leq 0$ *is such that*

$$\text{(i)} \ \ F(x, B, 0) \geq 0 \ \ \text{for a.e. } x \in \Omega,$$
$$\text{(ii)} \ \ F(\cdot, B, 0) \in L^{q'}(\Omega), \tag{5.111}$$

and

$$\text{(iii)} \ \ \psi(x, B) \leq \psi(x, v), \ \forall \, v \leq B; \tag{5.112}$$

then B is a subsolution of (5.108).

(b) *Similarly, if $A \in \mathbb{R}$, $A \geq 0$ and*

$$\text{(i)} \ \ F(x, A, 0) \leq 0 \ \ \text{for a.e. } x \in \Omega,$$
$$\text{(ii)} \ \ F(\cdot, A, 0) \in L^{q'}(\Omega), \tag{5.113}$$

and

$$\text{(iii)} \ \ \psi(x, A) \geq \psi(x, v), \ \forall \, v \geq A; \tag{5.114}$$

then A is a supersolution of (5.108).

Proof: (a) Choosing $J = 0$, we see that $\underline{u} = B$ satisfies conditions (i)–(iii) of Definition 5.15. Moreover, (5.85) becomes, in this case,

$$j(v \vee B) \leq j(v), \ v \in V_0 \cap D(j); \tag{5.115}$$

i.e.,

$$\int_\Omega \psi(x, v(x) \vee B) dx \leq \int_\Omega \psi(x, v(x)) dx.$$

In view of (5.109) and (5.110), this is equivalent to

$$\begin{aligned}
&\int_{\{x \in \Omega : v(x) > B\}} \psi(x, v) dx + \int_{\{x \in \Omega : v(x) \leq B\}} \psi(x, B) dx \\
&\leq \int_{\{x \in \Omega : v(x) > B\}} \psi(x, v) dx + \int_{\{x \in \Omega : v(x) \leq B\}} \psi(x, v) dx.
\end{aligned} \tag{5.116}$$

Now, from (5.112), we have

$$\psi(x, B) \leq \psi(x, v(x)) \ \text{ on } \{x \in \Omega : v(x) \leq B\},$$

and thus,

$$\int_{\{x \in \Omega : v(x) \leq B\}} \psi(x, B) dx \leq \int_{\{x \in \Omega : v(x) \leq B\}} \psi(x, v) dx,$$

which implies (5.116) and thus (5.115).

To check (5.86), we assume that $v = B \wedge w$ with some $w \in V_0 \cap D(j)$. From (5.107) and the definition of $\mathcal{A}$, $\mathcal{A}(B) = 0$. As $v - B \leq 0$, we have from (5.111)(i) that

$$\langle G(B), v - B \rangle = \int_\Omega F(x, B, 0)(v - B)dx \leq 0.$$

This result implies (5.86), which completes the proof of (a). The proof of (b) is similar.   □

By using Theorem 5.18, Theorem 5.20, and Lemma 5.23, we have the following existence result for (5.108).

Theorem 5.24. (a) *Assume $B \in \mathbb{R}$ satisfies (5.111), ψ satisfies (5.112), and that*

$$|F(x, u, \xi)| \leq a(x) + b(|u|^\sigma + |\xi|^\sigma)$$

for a.e. $x \in \Omega$, $u \geq B$, $\xi \in \mathbb{R}^N$, with $0 \leq \sigma < p - 1$, $a \in L^{p'}(\Omega)$. Then, (5.108) has a smallest solution u_* and a greatest solution u^* such that $B \leq u_* \leq u^*$.
(b) *Assume $A, B \in \mathbb{R}$ $(A \geq B)$ satisfy (5.111)–(5.114) and that F has the growth condition*

$$|F(x, u, \xi)| \leq a(x) + b|\xi|^{p/q'}$$

for a.e. $x \in \Omega$, $\xi \in \mathbb{R}^N$, $u \in [A, B]$ with $q < p^*$, $a \in L^{p'}(\Omega)$. Then, (5.108) has a smallest solution u_* and a greatest solution u^* such that $B \leq u_* \leq u^* \leq A$.

Example 5.25. We consider in this example a variational inequality that contains the p-Laplacian, that is, the inequality (5.76) with

$$\langle \mathcal{A}(u), v \rangle = \int_\Omega |\nabla u|^{p-2} \nabla u \cdot \nabla v \, dx.$$

In this case, $a_i = |\nabla u|^{p-2}\partial_i u$, $(1 \leq i \leq N)$, and $a_0 = 0$. The coefficients a_i $(i = 0, 1, \ldots, N)$ clearly satisfy (5.79) and (5.80). For each $K > 0$, suppose that the function

$$x \mapsto \sup\{|F(x, u, \xi)| : 0 \leq u \leq K, |\xi| \leq K\} \tag{5.117}$$

belongs to $L^{q'}(\Omega)$. We also assume the following behavior of $F(x, u, \xi)$ when $u(> 0)$ is very small or very large:

$$\liminf_{u \to 0^+, |\xi| \to 0} \frac{F(x, u, \xi)}{u^{p-1}} > \frac{\lambda_0}{\lambda} > \limsup_{u \to \infty, \xi \in \mathbb{R}^N} \frac{F(x, u, \xi)}{u^{p-1}}, \tag{5.118}$$

where λ_0 is the principal eigenvalue of the p-Laplacian,

$$\lambda_0 = \inf\left\{ \left(\int_\Omega |u|^p dx \right)^{-1} \int_\Omega |\nabla u|^p dx : u \in V_0 \setminus \{0\} \right\}.$$

Let ϕ_0 be the (unique) eigenfunction corresponding to λ_0 such that $\phi_0(x) > 0$ for all $x \in \Omega$. (It is known, see [178], that $\phi_0 \in C^{1,\alpha}(\bar\Omega)$ for some $\alpha \in (0,1)$.) By choosing $J = 0$ and using the arguments in [140] (Lemma 1), we can show that the function $\underline{u} = \varepsilon\phi_0$ satisfies (5.86) for all $\varepsilon > 0$ sufficiently small. On the other hand, let $\tilde{\Omega}$ be a bounded open region that contains $\bar{\Omega}$ and let $\tilde{\lambda}$ be the principal eigenvalue of the p-Laplacian on $\tilde{\Omega}$ and $\tilde{\phi}$ the corresponding eigenfunction on $\tilde{\Omega}$ such that $\tilde{\phi} > 0$ on $\tilde{\Omega}$. Then, we can prove that $\bar{u} = R\tilde{\phi}|_{\bar{\Omega}}$ satisfies (5.89) (with $J = 0$) for $R > 0$, sufficiently large. The proofs of these statements are somewhat lengthy; we refer to [140] for more details. The following lemma is about the construction of sub- and supersolutions of (5.76) based on the eigenfunctions ϕ_0 and $\tilde{\phi}$ of the p-Laplacian.

Lemma 5.26. (a) *If $C_1 > 0$ exists such that ψ is nonincreasing on $(-\infty, C_1)$, i.e.,*

$$\psi(x, u) \leq \psi(x, v), \text{ for a.e. } x \in \Omega, \text{ for all } u, v \text{ such that } v \leq u < C_1,$$
(5.119)

then, for $\varepsilon > 0$ sufficiently small, $\underline{u} = \varepsilon\phi_0$ is a subsolution of (5.108).

(b) *Similarly, if $C_2 > 0$ exists such that ψ is nondecreasing on (C_2, ∞), i.e.,*

$$\psi(x, u) \geq \psi(x, v), \text{ for a.e. } x \in \Omega, \text{ for all } u, v \text{ such that } u \geq v > C_2,$$
(5.120)

then for $R > 0$ sufficiently large, $\bar{u} = R\tilde{\phi}|_{\Omega}$ is a supersolution of (5.108).

Proof: (a) We need only to check (5.85); i.e.,

$$j(v \vee \varepsilon\phi_0) \leq j(v), \ \forall \ v \in V_0 \cap D(j).$$

This result is equivalent to

$$\int_{\{x \in \Omega : v < \varepsilon\phi_0\}} \psi(x, \varepsilon\phi_0)dx + \int_{\{x \in \Omega : v \geq \varepsilon\phi_0\}} \psi(x, v)dx$$
$$\leq \left(\int_{\{x \in \Omega : v < \varepsilon\phi_0\}} + \int_{\{x \in \Omega : v \geq \varepsilon\phi_0\}} \right) \psi(x, v)dx;$$

that is,

$$\int_{\{x \in \Omega : v < \varepsilon\phi_0\}} \psi(x, \varepsilon\phi_0)dx \leq \int_{\{x \in \Omega : v < \varepsilon\phi_0\}} \psi(x, v)dx.$$
(5.121)

Now, as $\phi_0 \in L^\infty(\Omega)$, $\varepsilon\phi_0(x) < C_1$, for a.e. $x \in \Omega$ for $\varepsilon > 0$ small. Hence, for $v < \varepsilon\phi_0 < C_1$, (5.119) implies $\psi(x, v(x)) \geq \psi(x, \varepsilon\phi_0(x))$ for a.e. $x \in \Omega$. This result implies (5.121). Hence, $\varepsilon\phi_0$ is a subsolution of (5.108). The proof of (b) is similar. $\square$

As a consequence of Lemma 5.26 and Theorem 5.18, we have the following result.

Theorem 5.27. *Under the conditions (5.118) and (5.117), a smallest solution u_* and a greatest solution u^* of (5.108) exist such that*

$$(0 <)\varepsilon\phi_0 \leq u_* \leq u^* \leq R\tilde{\phi}|_\Omega,$$

where $\varepsilon > 0$ sufficiently small and $R > 0$ sufficiently large. In particular, if F has the growth condition (5.117) and λ satisfies (5.118), then, (5.108) has a positive solution.

Remark 5.28. (5.108) can be seen as an eigenvalue problem for a variational inequality. We have proved that for λ in certain appropriate interval [given by (5.118)]; then (5.108) has a positive eigenfunction.

5.3 Evolutionary Variational Inequalities

Let $\Omega \subset \mathbb{R}^N$ be as above, $Q = \Omega \times (0, \tau)$ and $\Gamma = \partial\Omega \times (0, \tau)$, $\tau > 0$. This section is concerned with existence and comparison results of the following parabolic variational inequality:

$$u \in W_0 \cap K, \ u(\cdot, 0) = 0: \quad \langle u_t + A(u) + F(u) - h, v - u \rangle \geq 0, \quad \forall \, v \in K, \tag{5.122}$$

where K is a closed, convex subset of $X_0 := L^p(0, \tau; W_0^{1,p}(\Omega))$, $W_0 = \{u \in X_0 : u_t \in X_0^*\}$, $\langle \cdot, \cdot \rangle$ denotes the duality pairing between X_0^* and X_0, and $p \in [2, \infty)$. The operator $A : X_0 \to X_0^*$ is related with a nonlinear elliptic operator of Leray–Lions type in divergence form given by

$$A(u)(x, t) = -\sum_{i=1}^{N} \frac{\partial}{\partial x_i} a_i(x, t, \nabla u(x, t)),$$

and F is the Nemytskij operator associated with the Carathéodory function $f : Q \times \mathbb{R} \times \mathbb{R}^N \to \mathbb{R}$ by

$$F(u)(x, t) = f(x, t, u(x, t), \nabla u(x, t)).$$

We assume that $h \in L^{p'}(Q) \subset X_0^*$ (as above, p' is the Hölder conjugate of p).

Solutions of the variational inequality (5.122) are usually referred to as *strong* solutions (cf. [152]). Many papers deal with parabolic inequalities under different structure and regularity hypotheses of the data such as [65, 67, 81, 84, 89, 95, 152, 175, 184, 188, 214, 218] and the recent survey paper [198]. The aim of this section is to develop the method of sub-supersolutions for the parabolic variational inequality (5.122).

5.3.1 General Settings

Only for completeness here we recall the evolutionary framework that has already been introduced in Chap. 3 and Chap. 4, in which the parabolic variational inequality is embedded. Throughout this section, we set $X = L^p(0, \tau; W^{1,p}(\Omega))$ and its dual space $X^* = L^{p'}(0, \tau; (W^{1,p}(\Omega))^*)$. We also denote

$$W = \{u \in X : u_t \in X^*\},$$

where the derivative $\partial/\partial t$ is understood in the sense of vector-valued distributions (cf. [222]), which is characterized by

$$\int_0^\tau u'(t)\phi(t)\, dt = -\int_0^\tau u(t)\phi'(t)\, dt, \quad \forall\, \phi \in C_0^\infty(0, \tau).$$

The space W endowed with the graph norm

$$\|u\|_W = \|u\|_X + \|u_t\|_{X^*}$$

is a Banach space that is separable and reflexive because of the separability and reflexivity of X and X^*, respectively.

Let $\|\cdot\|_X$ and $\|\cdot\|_{X_0}$ be the usual norms defined on X and X_0 (and similarly on X^* and X_0^*) :

$$\|u\|_X = \left(\int_0^\tau \|u(t)\|_{W^{1,p}(\Omega)}^p\, dt\right)^{1/p}, \quad \|u\|_{X_0} = \left(\int_0^\tau \|u(t)\|_{W_0^{1,p}(\Omega)}^p\, dt\right)^{1/p}.$$

We use the notation $\langle\cdot, \cdot\rangle$ for any of the dual pairings between X and X^*, X_0 and X_0^*, $V = W^{1,p}(\Omega)$ and V^*, and $V_0 = W_0^{1,p}(\Omega)$ and $V_0^* = W^{-1,p'}(\Omega)$. For example, with $f \in X^*, u \in X$,

$$\langle f, u\rangle = \int_0^\tau \langle f(t), u(t)\rangle\, dt.$$

Let $L = \partial/\partial t$ and its domain of definition $D(L)$ given by

$$D(L) = \{u \in X_0 : u_t \in X_0^* \text{ and } u(0) = 0\}.$$

The linear operator $L : D(L) \to X_0^*$ is closed, densely defined, and maximal monotone. We assume that $a_i : Q \times \mathbb{R}^N \to \mathbb{R}$ and $f : Q \times \mathbb{R} \times \mathbb{R}^N \to \mathbb{R}$ are Carathéodory functions, where f has certain growth conditions to be specified later and a_i satisfies:

$$|a_i(x, t, \xi)| \le c_1|\xi|^{p-1} + c_2(x, t), \tag{5.123}$$

$$\sum_{i=1}^N [a_i(x, t, \xi) - a_i(x, t, \xi')](\xi_i - \xi_i') > 0, \tag{5.124}$$

$$\sum_{i=1}^{N} a_i(x,t,\xi)\xi_i \geq c_3|\xi|^p - c_4(x,t), \tag{5.125}$$

for almost all $(x,t) \in Q$, all $\xi, \xi' \in \mathbb{R}^N$ with $\xi' \neq \xi$, where $c_1, c_3 \in (0, \infty)$, $c_2 \in L^{p'}(Q)$, and $c_4 \in L^1(Q)$.

The operators $A : X \to X^* \subset X_0^*$ related with the quasilinear elliptic operator, and $F : X \to X^* \subset X_0^*$, as well as $h \in L^{p'}(Q) \subset X_0^*$, are defined as follows:

$$
\begin{aligned}
\langle A(u), v \rangle &= \sum_{i=1}^{N} \int_Q a_i(x,t,\nabla u)v_{x_i}\,dxdt, \\
\langle F(u), v \rangle &= \int_Q f(\cdot,\cdot,u,\nabla u)v\,dxdt, \\
\langle h, v \rangle &= \int_Q h(x,t)v(x,t)\,dxdt,
\end{aligned}
\tag{5.126}
$$

for all $v, u \in X$. Thus, the variational inequality (5.122) may be rewritten as:

$$u \in D(L) \cap K : \langle Lu + A(u) - F(u) - h, v - u \rangle \geq 0, \ \forall v \in K. \tag{5.127}$$

Furthermore, as above, for $u, v \in X$, $U, V \subset X$, we use the notation $u \wedge v = \min\{u,v\}$, $u \vee v = \max\{u,v\}$, $U * V = \{u * v : u \in U, v \in V\}$, and $u * U = \{u\} * U$ with $* \in \{\wedge, \vee\}$.

Our basic notion of sub-and supersolution of (5.122) is defined as follows.

Definition 5.29. *A function $\underline{u} \in W$ is called a subsolution of (5.122) if*

    (i) $F\underline{u} \in L^{p'}(Q)$,
    (ii) $\underline{u}(\cdot,0) \leq 0$ *a.e. in* Ω, $\underline{u} \leq 0$ *on* Γ, *and*
    (iii) $\langle \underline{u}_t, v - \underline{u} \rangle + \langle A(\underline{u}), v - \underline{u} \rangle + \langle F(\underline{u}), v - \underline{u} \rangle \geq \langle h, v - \underline{u} \rangle, \ \forall v \in \underline{u} \wedge K.$

$$\tag{5.128}$$

We have a similar definition for supersolutions of (5.122).

Definition 5.30. *A function $\bar{u} \in W$ is called a supersolution of (5.122) if*

    (i) $F\bar{u} \in L^{p'}(Q)$,
    (ii) $\bar{u}(\cdot,0) \geq 0$ *a.e. in* Ω, $\bar{u} \geq 0$ *on* Γ, *and*
    (iii) $\langle \bar{u}_t, v - \bar{u} \rangle + \langle A(\bar{u}), v - \bar{u} \rangle + \langle F(\bar{u}), v - \bar{u} \rangle \geq \langle h, v - \bar{u} \rangle, \ \forall v \in \bar{u} \vee K.$

$$\tag{5.129}$$

Definition 5.31. *Let $C \neq \emptyset$ be a closed and convex subset of a reflexive Banach space X. A bounded, hemicontinuous and monotone operator $P : X \to X^*$ is called a penalty operator associated with $C \subset X$ if*

$$P(u) = 0 \iff u \in C.$$

We assume that a pair of sub-supersolutions $\underline{u}$ and $\bar{u}$ of (5.122) exists such that $\underline{u} \le \bar{u}$ a.e. in Q and that f has the following growth between $\underline{u}$ and $\bar{u}$:

(H1)
$$|f(x,t,u,\xi)| \le c_5(x,t) + c_6|\xi|^{p-1}, \qquad (5.130)$$

for some $c_5 \in L^{p'}(Q)$, for a.e. $(x,t) \in Q$, all $\xi \in \mathbb{R}^N$, and all $u \in [\underline{u}(x,t), \bar{u}(x,t)]$. Moreover, suppose that a penalty operator $P : X_0 \to X_0^*$ exists associated with $K \subset X_0$ with the following properties.

(H2) For each $u \in D(L)$, $w = w(u) \in X_0$ exists such that

$$\begin{aligned} &\text{(i) } \langle u_t + Au, w \rangle \ge 0, \ \text{ and} \\ &\text{(ii) } \langle Pu, w \rangle \ge D\|Pu\|_{X_0^*}\|w\|_{L^p(Q)}, \end{aligned} \qquad (5.131)$$

for some constant $D > 0$ independent of u and w.

5.3.2 Comparison Principle

Now, let us prove the following existence and comparison result for the inequality (5.122).

Theorem 5.32. *Assume (5.122) has an ordered pair of sub- and supersolutions $\underline{u}$ and $\bar{u}$ and that (5.123)–(5.125) and (H1)–(H2) are satisfied. Suppose furthermore that $D(L) \cap K \ne \emptyset$ and*

$$\underline{u} \vee K \subset K, \bar{u} \wedge K \subset K. \qquad (5.132)$$

Then, (5.122) has a solution u such that $\underline{u} \le u \le \bar{u}$ a.e. in Q.

Proof: The proof is a combination of arguments for parabolic variational equations [29] (see Chap. 3) with those for elliptic variational inequalities in [141] (see Sect. 5.1 and Sect. 5.2). As in previous sections, we define the cutoff function b and truncation operator T:

$$b(x,t,u) = \begin{cases} [u - \bar{u}(x,t)]^{p-1} & \text{if } u > \bar{u}(x,t) \\ 0 & \text{if } \underline{u}(x,t) \le u \le \bar{u}(x,t) \\ -[\underline{u}(x,t) - u]^{p-1} & \text{if } u < \underline{u}(x,t), \end{cases}$$

for $(x,t,u) \in \Omega \times (0,\tau) \times \mathbb{R}$ and

$$(Tu)(x,t) = \begin{cases} \bar{u}(x,t) & \text{if } u(x,t) > \bar{u}(x,t) \\ u(x,t) & \text{if } \underline{u}(x,t) \le u(x,t) \le \bar{u}(x,t) \\ \underline{u}(x,t) & \text{if } u(x,t) < \underline{u}(x,t), \end{cases}$$

for $(x,t) \in Q$, $u \in X$. It is easy to check that b is a Carathéodory function with the growth condition

$$|b(x,t,u)| \le c_7(x,t) + c_8|u|^{p-1}, \quad \text{for a.e. } (x,t) \in Q, \text{ all } u \in \mathbb{R}, \qquad (5.133)$$

with $c_7 \in L^{p'}(Q)$, $c_8 > 0$. Hence, the operator $B : X_0 \to X_0^*$ given by

$$\langle Bu, v \rangle = \int_Q b(\cdot, \cdot, u)\, v \, dxdt \quad (u, v \in X), \tag{5.134}$$

is well defined. Moreover, there are $c_9, c_{10} > 0$ such that

$$\int_Q b(\cdot, \cdot, u)u \, dxdt \geq c_9 \|u\|^p_{L^p(Q)} - c_{10}, \ \forall \, u \in X_0. \tag{5.135}$$

We define the operator $\mathcal{C}$ from X_0 to X_0^* by

$$\mathcal{C}(u) = \gamma Bu + F \circ T(u), \ u \in X_0 \tag{5.136}$$

(γ is a positive constant to be determined later and $F \circ T$ denotes the composition of F and T) and consider the following auxiliary variational inequality in X_0:

$$u \in D(L) \cap K : \langle Lu + A(u) + \mathcal{C}(u) - h, v - u \rangle \geq 0, \ \forall \, v \in K. \tag{5.137}$$

Using usual arguments, we readily verify that $A + \mathcal{C}$ is pseudomonotone with respect to $D(L)$. Let us check that $A + \mathcal{C}$ is coercive on X_0 in the following sense:

$$\lim_{\|u\|_{X_0} \to \infty} \frac{\langle (A + \mathcal{C})(u), u - \varphi \rangle}{\|u\|_{X_0}} = +\infty, \tag{5.138}$$

for any $\varphi \in X_0$. In fact, from (5.125), we have

$$\langle Au, u \rangle \geq c_3 \|\|\nabla u\|\|^p_{L^p(Q)} - c_{11}, \ \forall \, u \in X_0, \tag{5.139}$$

with some constant $c_{11} > 0$. Using Stampacchia's theorem (cf. [124, 99]) and Young's inequality together with (5.130), we have for each $\varepsilon > 0$, constants $c_{12} = c_{12}(\varepsilon), c_{13} > 0$ such that for all $u \in X_0$,

$$\begin{aligned}
|\langle F \circ T(u), u \rangle| &= \left| \int_Q (F \circ T)(u)u \, dxdt \right| \\
&\leq \|c_5\|_{L^{p'}(Q)} \|u\|_{L^p(Q)} + c_6 \|\|\nabla u\|\|^{p-1}_{L^p(Q)} \|u\|_{L^p(Q)} \\
&\leq \varepsilon \|\|\nabla u\|\|^p_{L^p(Q)} + c_{12} \|u\|^p_{L^p(Q)} + c_{13}.
\end{aligned} \tag{5.140}$$

Combining (5.135) with (5.139) and (5.140), we get

$$\langle (A+\mathcal{C})(u), u \rangle \geq (c_3 - \varepsilon) \|\|\nabla u\|\|^p_{L^p(Q)} + (\gamma c_9 - c_{12}) \|u\|^p_{L^p(Q)} - (c_{11} + \gamma c_{10} + c_{13}),$$

for all $u \in X_0$. Choosing $\varepsilon = c_3/2$ and $\gamma = c_3 c_{12} c_9^{-1}$, we have $c_{14}, c_{15} > 0$ such that

$$\langle (A + \mathcal{C})(u), u \rangle \geq c_{14} \|u\|^p_{X_0} - c_{15}, \ \forall \, u \in X_0. \tag{5.141}$$

For any $\varphi \in X_0$ fixed, it is inferred from (5.123), (5.133), and (5.130) that

$$|\langle (A + \mathcal{C})(u), \varphi \rangle| \leq c_{16}(\|u\|_{X_0}^{p-1} + 1)\|\varphi\|_{X_0}, \ \forall \ u \in X_0, \tag{5.142}$$

for some constant $c_{16} = c_{16}(\varphi) > 0$. From (5.141) and (5.142), we obtain (5.138).

It follows from the pseudo-monotonicity and coercivity of $A + \mathcal{C}$ with respect to $D(L)$ that the variational inequality (5.137) has a solution u. The proof of this claim is given in Lemma 5.33 below. Now, let us show that any solution u of (5.137) satisfies: $\underline{u} \leq u \leq \bar{u}$ a.e. in Q. We verify that $\underline{u} \leq u$, the second inequality is proved in the same way. Because $u \in K$, it follows from (5.132) that

$$u + (\underline{u} - u)^+ = \underline{u} \vee u \in K.$$

Letting $v = u + (\underline{u} - u)^+$ into (5.137), we get

$$\langle u_t, (\underline{u} - u)^+ \rangle + \langle Au + \gamma Bu + F(Tu), (\underline{u} - u)^+ \rangle \geq \langle h, (\underline{u} - u)^+ \rangle. \tag{5.143}$$

On the other hand, as $\underline{u}$ is a subsolution, it follows from (5.128)(iii), with

$$v = \underline{u} - (\underline{u} - u)^+ = \underline{u} \wedge u \in \underline{u} \wedge K,$$

that

$$-\langle \underline{u}_t, (\underline{u} - u)^+ \rangle - \langle A\underline{u}, (\underline{u} - u)^+ \rangle - \langle F(\underline{u}), (\underline{u} - u)^+ \rangle \geq -\langle h, (\underline{u} - u)^+ \rangle. \tag{5.144}$$

Adding (5.143) and (5.144), we get

$$\begin{aligned}\langle (u - \underline{u})_t, (\underline{u} - u)^+ \rangle &+ \langle Au - A\underline{u} + \gamma Bu, (\underline{u} - u)^+ \rangle \\ &+ \langle F(Tu) - F(\underline{u}), (\underline{u} - u)^+ \rangle \geq 0.\end{aligned} \tag{5.145}$$

We have $\underline{u} - u \in W$ and $(\underline{u} - u)^+(\cdot, 0) = 0$, and thus,

$$\langle (\underline{u} - u)_t, (\underline{u} - u)^+ \rangle = \frac{1}{2}\|(\underline{u} - u)^+(\cdot, \tau)\|_{L^2(\Omega)}^2 \geq 0. \tag{5.146}$$

On the other hand, it is easy to check from (5.124) that

$$\langle A\underline{u} - Au, (\underline{u} - u)^+ \rangle \geq 0. \tag{5.147}$$

Moreover,

$$\langle F(Tu) - F(\underline{u}), (\underline{u} - u)^+ \rangle = \int_{Q^+} [f(\cdot, \cdot, Tu, \nabla(Tu)) - f(\cdot, \cdot, \underline{u}, \nabla\underline{u})](\underline{u} - u)\, dx dt,$$

where $Q^+ = \{(x, t) \in Q : \underline{u}(x, t) \geq u(x, t)\}$. But because of

$$Tu = \underline{u} \text{ and } \nabla(Tu) = \nabla\underline{u} \text{ a.e. on } Q^+,$$

we have

$$\langle F(Tu) - F(\underline{u}), (\underline{u} - u)^+ \rangle = 0. \tag{5.148}$$

Combining (5.146)–(5.148) with (5.145), we obtain

$$0 \le \gamma \langle Bu, (\underline{u} - u)^+ \rangle$$
$$= -\int_{Q^+} (\underline{u} - u)^p \, dx dt$$
$$\le 0.$$

This result proves that $\underline{u} - u = 0$ a.e. on Q^+, and thus, $\underline{u} \le u$ a.e. on Q. A similar proof shows that $u \le \bar{u}$. From $\underline{u} \le u \le \bar{u}$, we have $Bu = 0$ and $Tu = u$. Consequently, u is also a solution of (5.127).   □

To complete the proof of the Theorem 5.32, we need to show the solvability of the inequality (5.137), which is given in the following lemma.

Lemma 5.33. *Under the assumptions of Theorem 5.32, the variational inequality (5.137) has solutions.*

Proof: The penalty arguments we use here are motivated by Deuel and Hess' paper [84]. For $\varepsilon > 0$, let us consider the following penalized equation:

$$u \in D(L) : \langle u_t, v \rangle + \langle (A + \mathcal{C})(u), v \rangle + \frac{1}{\varepsilon} \langle Pu, v \rangle = \langle h, v \rangle, \ \forall \, v \in X_0, \tag{5.149}$$

where P is a penalty operator (associated to K) that satisfies (5.131).

Because $A + \mathcal{C}$ is pseudomonotone with respect to $D(L)$ and $\varepsilon^{-1}P$ is monotone, $A + \mathcal{C} + \varepsilon^{-1}P$ is also pseudomonotone with respect to $D(L)$. Moreover, it is bounded and hemicontinuous on X_0. From the coercivity of $A + \mathcal{C}$, see (5.138), and the monotonicity of $\varepsilon^{-1}P$, it is easy to see that $A + \mathcal{C} + \varepsilon^{-1}P$ is coercive on X_0:

$$\lim_{\|u\|_{X_0} \to \infty} \frac{\langle (A + \mathcal{C} + \varepsilon^{-1}P)(u), u - \varphi \rangle}{\|u\|_{X_0}} = +\infty, \tag{5.150}$$

for any $\varphi \in X_0$ (fixed). According to existence results for solutions of parabolic variational equalities (see Chap. 3, and cf. [152, 19, 20]), for each $\varepsilon > 0$, (5.149) has solutions. Let u_ε be a solution of (5.149). We show that the family $\{u_\varepsilon : \varepsilon > 0, \text{ small}\}$ is bounded with respect to the graph norm of $D(L)$. In fact, let u_0 be a (fixed) element of $D(L) \cap K$. Putting $v = u_\varepsilon - u_0$ into (5.149) (with u_ε) and noting the monotonicity of L and that $Pu_0 = 0$, we get

$$\langle h - u_{0t}, u_\varepsilon - u_0 \rangle$$
$$= \langle u_{\varepsilon t} - u_{0t}, u_\varepsilon - u_0 \rangle + \langle (A + \mathcal{C})(u_\varepsilon), u_\varepsilon - u_0 \rangle + \frac{1}{\varepsilon} \langle Pu_\varepsilon - Pu_0, u_\varepsilon - u_0 \rangle$$
$$\ge \langle (A + \mathcal{C})(u_\varepsilon), u_\varepsilon - u_0 \rangle.$$

Thus,

$$\frac{\langle (A+C)(u_\varepsilon), u_\varepsilon - u_0 \rangle}{\|u_\varepsilon - u_0\|_{X_0}} \le \|h - u_{0t}\|_{X_0^*},$$

for all $\varepsilon > 0$. From (5.138), we have that $\|u_\varepsilon\|_{X_0}$ is bounded. As a consequence, we see that $\{Au_\varepsilon\}$ and $\{Cu_\varepsilon\}$ are bounded sequences in X_0^*. Moreover, from the growth conditions of b and F and the definition of T, we can also prove that $\{Cu_\varepsilon\}$ is a bounded sequence in $L^{p'}(Q)$.

Next, we check that the sequence $\{\varepsilon^{-1} Pu_\varepsilon\}$ is also bounded in X_0^*. To see this, for each ε, we choose $w = w_\varepsilon$ to be an element satisfying (5.131) with $u = u_\varepsilon$. From (5.149), we have

$$\langle u_{\varepsilon t}, w_\varepsilon \rangle + \langle (A+C)(u_\varepsilon), w_\varepsilon \rangle + \frac{1}{\varepsilon}\langle Pu_\varepsilon, w_\varepsilon \rangle = \langle h, w_\varepsilon \rangle.$$

From (5.131)(i), $\langle u_{\varepsilon t}, w_\varepsilon \rangle + \langle Au_\varepsilon, w_\varepsilon \rangle \ge 0$. Therefore,

$$\frac{1}{\varepsilon}\langle Pu_\varepsilon, w_\varepsilon \rangle \le \langle h - C(u_\varepsilon), w_\varepsilon \rangle. \tag{5.151}$$

As $\{\|Cu_\varepsilon\|_{L^{p'}(Q)}\}$ is bounded, a constant $c > 0$ exists such that

$$|\langle h - Cu_\varepsilon, w_\varepsilon \rangle| \le c\|w_\varepsilon\|_{L^p(Q)}, \ \forall \ \varepsilon.$$

This result and (5.131)(ii) imply that

$$\frac{1}{\varepsilon}\|Pu_\varepsilon\|_{X_0^*} \le \frac{c}{D}, \ \forall \ \varepsilon.$$

On the other hand, as

$$u_{\varepsilon t} = h - (A + C + \varepsilon^{-1}P)(u_\varepsilon)$$

in X_0^*, the above estimate implies that $\{u_{\varepsilon t}\}$ is also bounded in X_0^*. We have shown that $\{u_\varepsilon\}$ is bounded with respect the graph norm of $D(L)$. As a consequence, $u \in X_0$ (with $u_t \in X_0^*$) and a subsequence of $\{u_\varepsilon\}$ exist, still denoted by $\{u_\varepsilon\}$, such that

$$u_\varepsilon \rightharpoonup u \text{ in } X_0, \ u_{\varepsilon t} \rightharpoonup u_t \text{ in } X_0^* \ (\varepsilon \to 0^+). \tag{5.152}$$

As $D(L)$ is closed in W and convex, it is weakly closed in W, and thus, $u \in D(L)$.

Now, we prove that u is a solution of the variational inequality (5.137). First, note that $Pu = 0$. In fact, we have $Pu_\varepsilon \to 0$ in X_0^*. It follows from the monotonicity of P that

$$\langle Pv, v - u \rangle \ge 0, \ \forall \ v \in X_0.$$

As in the proof of Minty's lemma (cf. [124]), we obtain from this inequality that

$$\langle Pu, v \rangle \ge 0, \ \forall \ v \in X_0.$$

Hence, $Pu = 0$ in X_0^*, that is, $u \in K$. On the other hand, (5.152) and Aubin's lemma (see [152]) imply that

$$u_\varepsilon \to u \text{ in } L^p(Q).$$ (5.153)

As a consequence, we get

$$\langle Cu_\varepsilon, u_\varepsilon - u \rangle \to 0 \text{ as } \varepsilon \to 0^+.$$ (5.154)

For $w \in K$, letting $v = w - u_\varepsilon$ in (5.149) (with $u = u_\varepsilon$), we get

$$\langle u_{\varepsilon t}, w - u_\varepsilon \rangle + \langle (A+C)(u_\varepsilon), w - u_\varepsilon \rangle - \langle h, w - u_\varepsilon \rangle$$
$$= \frac{1}{\varepsilon} \langle -Pu_\varepsilon, w - u_\varepsilon \rangle \geq 0.$$ (5.155)

By choosing $w = u$ in (5.155), we have

$$\langle Au_\varepsilon, u - u_\varepsilon \rangle$$
$$\geq \langle h, u - u_\varepsilon \rangle - \langle Cu_\varepsilon, u - u_\varepsilon \rangle - \langle u_t, u - u_\varepsilon \rangle + \langle u_t - u_{\varepsilon t}, u - u_\varepsilon \rangle$$
$$\geq \langle h, u - u_\varepsilon \rangle - \langle Cu_\varepsilon, u - u_\varepsilon \rangle - \langle u_t, u - u_\varepsilon \rangle.$$

As a consequence, we get

$$\liminf_{\varepsilon \to 0^+} \langle Au_\varepsilon, u - u_\varepsilon \rangle \geq 0.$$

Because A is of class (S_+) with respect to $D(L)$ (see Chap. 2, and cf. [19, 20] or [43]), we infer from (5.152) and this limit that

$$u_\varepsilon \to u \text{ in } X_0.$$ (5.156)

Letting $\varepsilon \to 0$ in (5.155) and taking (5.152) and (5.156) into account, we obtain

$$\langle u_t, w - u \rangle + \langle (A+C)(u), w - u \rangle - \langle h, w - u \rangle \geq 0.$$

This result holds for all $w \in K$, proving that u is in fact a solution of (5.137).
□

Remark 5.34. (a) Theorem 5.32 can be extended to the case in which $\underline{u}$ is the maximum of some subsolutions and $\bar{u}$ is the minimum of some supersolutions. In fact, assume that

$$\underline{u} := \max\{\underline{u}_1, \ldots, \underline{u}_k\} \leq \bar{u} := \min\{\bar{u}_1, \ldots, \bar{u}_m\},$$

where $\underline{u}_1, \ldots, \underline{u}_k$ (respectively, $\bar{u}_1, \ldots, \bar{u}_m$) are subsolutions (respectively, supersolutions) of (5.122). If f has the growth condition (5.130) for a.e. $(x, t) \in Q$, all $\xi \in \mathbb{R}^N$, all u in the interval

$$[\min\{\underline{u}_1, \ldots, \underline{u}_k\}(x, t), \max\{\bar{u}_1, \ldots, \bar{u}_m\}(x, t)],$$

then (5.122) has a solution within the interval $[\underline{u}, \bar{u}]$.

(b) If K satisfies $K \wedge K \subset K$ (respectively, $K \vee K \subset K$), then any solution of (5.122) is also a subsolution (respectively, supersolution).

(c) The following result, whose proof is given in [51], is about the existence of extremal (i.e., greatest and smallest) solutions of (5.122).

Corollary 5.35. *Let the hypotheses of Theorem 5.32 be satisfied, and assume, in addition, that K satisfies $K \wedge K \subset K$ (respectively, $K \vee K \subset K$). Let S denote the set of all solutions of (5.122) in $[\underline{u}, \bar{u}]$. If S is bounded in W_0, then the variational inequality (5.122) possesses extremal solutions within $[\underline{u}, \bar{u}]$; i.e., the greatest solution u^* and the smallest solution u_* of (5.122) in $[\underline{u}, \bar{u}]$ exists such that for any other solution u of (5.122) in $[\underline{u}, \bar{u}]$, we have $u_* \leq u \leq u^*$.*

5.3.3 Obstacle Problem

As an example of the applicability of the general results of the preceding sections, we consider an obstacle problem, where the convex set K is given by

$$K = \{u \in X_0 : u \leq \psi \text{ a.e. on } Q\},$$

with ψ a function in W such that $\psi(\cdot, 0) \geq 0$ on Ω, $\psi \geq 0$ on Γ, and $\psi_t + A\psi \geq 0$ in X_0^*; i.e.,

$$\langle \psi_t + A\psi, v \rangle \geq 0, \ \forall \, v \in X_0 \cap L_+^p(Q).$$

The penalty function P can be chosen as

$$\langle Pu, v \rangle = \int_Q [(u - \psi)^+]^{p-1} \, v \, dx dt, \tag{5.157}$$

for all $u, v \in X_0$. It is easy to verify that P satisfies (5.131). To check (5.131), for each $u \in D(L)$, we choose $w = (u - \psi)^+$. Then, $w \in X_0$ and (5.131)(i) is satisfied. In fact, because $(u - \psi)^+(\cdot, 0) = 0$, we have

$$\langle u_t - \psi_t, (u - \psi)^+ \rangle = \frac{1}{2} \|(u - \psi)^+(\cdot, \tau)\|_{L^2(\Omega)}^2 \geq 0.$$

On the other hand, as above, we infer easily from (5.124) that

$$\langle Au - A\psi, (u - \psi)^+ \rangle \geq 0.$$

These inequalities imply that

$$\langle u_t + Au, (u - \psi)^+ \rangle \geq \langle \psi_t + A\psi, (u - \psi)^+ \rangle \geq 0,$$

because $(u - \psi)^+ \in X_0 \cap L_+^p(Q)$. We have checked (i) of (5.131). To verify (5.131)(ii), we note that

$$\langle Pu, w \rangle = \int_Q [(u - \psi)^+]^p \, dx = \|(u - \psi)^+\|_{L^p(Q)}^p. \tag{5.158}$$

From (5.157) and Hölder's inequality, we have

$$|\langle Pu, v \rangle| \leq \|(u - \psi)^+\|_{L^p(Q)}^{p-1} \|v\|_{L^p(Q)},$$

for all $v \in X_0$. Hence,

$$\|Pu\|_{X_0^*} \le c\|(u - \psi)^+\|_{L^p(Q)}^{p-1}, \ \forall \, u \in X_0,$$

for some constant $c > 0$. This result, together with (5.158), implies (5.131)(ii).

For our example of K, $\bar{u} \wedge K \subset K$ for every $\bar{u} \in W$ and $\underline{u} \vee K \subset K$ if $\underline{u} \le \psi$ on Q. Moreover, the conditions $K \wedge K \subset K$ (respectively, $K \vee K \subset K$) are satisfied, which allows us to apply Theorem 5.32. As far as the existence of extremal solutions is concerned, Corollary 5.35 cannot be applied directly. However, if we assume the existence of special sub- and supersolutions, we can prove the existence of extremal solutions by a penalty approach (see Section 5.4).

Finally, let us conclude this section by verifying that the notion of sub-supersolutions of the parabolic variational inequality (5.122) introduced here is consistent with the usual notion of (weak) sub-supersolutions of the corresponding nonlinear parabolic boundary value problem; i.e., the case of variational equalities, that is when $K = X_0$, which has been considered in Chap. 3. We show that in this case, the definitions given above agree with those in Chap. 3, Sect. 3.3 (see also [29]) for sub- and supersolutions of equations. It is enough to show that if $\underline{u}$ satisfies (5.128) (with $K = X_0$), then it satisfies the inequality

$$\langle \underline{u}_t, v \rangle + \langle A\underline{u}, v \rangle + \langle F\underline{u}, v \rangle \le \langle h, v \rangle, \tag{5.159}$$

for all $v \in X_0 \cap L_+^p(Q)$. Note that, because $\underline{u} \wedge w = \underline{u} - (\underline{u} - w)^+$, the inequality in (5.128)(iii) is equivalent to that in (5.159) for all $v \in M$, where

$$\begin{aligned} M &= \{(\underline{u} + w)^+ : w \in X_0\} \\ &= \{v^+ : v \in X \text{ and } v(t)|_{\partial\Omega} = \underline{u}(t)|_{\partial\Omega} \text{ for a.e. } t \in (0, \tau)\}. \end{aligned} \tag{5.160}$$

As $\underline{u}(t) \le 0$ for a.e. $t \in (0, \tau)$, we have $M \subset X_0 \cap L_+^p(Q)$. To show that (5.128)(iii) is equivalent to (5.159), we only need to verify the following density result, which is the analog to Lemma 5.4.

Lemma 5.36. *If M is the set given by (5.160), then M is dense in $X_0 \cap L_+^p(Q)$; i.e.,*

$$\overline{M}^{X_0} = X_0 \cap L_+^p(Q). \tag{5.161}$$

Proof: First, we observe that if $v \in X_0 \cap L_+^p(Q)$ and there is a compact subset κ of Ω (independent of t) such that

$$\operatorname{supp} v(t) \subset \kappa \quad \text{for a.e. } t \in (0, \tau), \tag{5.162}$$

then $v \in M$. In fact, we can choose $\varphi \in C_0^\infty(\Omega)$ such that $\varphi(x) \in [0, 1]$, $\forall \, x \in \Omega$ and $\varphi(x) = 1$, $\forall \, x \in \kappa$ (cf. [115]). We define

$$\tilde{v}(x, t) = v(x, t) + [1 - \varphi(x)] \min\{\underline{u}(x, t), 0\} \ ((x, t) \in Q).$$

As $\min\{\underline{u}, 0\} \in X$ and $1 - \varphi$ is smooth on $\overline{\Omega}$, we have $\tilde{v} \in X$. Moreover, because $1 - \varphi(x) \in [0, 1]$, $\forall\, x$, and $1 - \varphi(x) = 0$, $\forall\, x \in \kappa$, and

$$v(x, t) = 0, \quad \text{for a.e. } x \in \Omega \setminus \kappa, \text{ a.e. } t \in (0, \tau),$$

we have

$$\tilde{v}^+(x, t) = v(x, t) \text{ for a.e. } (x, t) \in Q.$$

On the other hand, for almost all $t \in (0, \tau)$, because

$$\tilde{v}(t) = (1 - \varphi) \min\{\underline{u}(t), 0\} = \min\{\underline{u}(t), 0\},$$

a.e. on $\Omega \setminus \operatorname{supp} \varphi$, we have

$$\tilde{v}(t)|_{\partial\Omega} = \min\{\underline{u}(t), 0\}|_{\partial\Omega} = \underline{u}(t)|_{\partial\Omega}.$$

This result shows that $v = \tilde{v}^+ \in M$.

Now, let $v \in X_0 \cap L^p_+(Q)$. Then v can be approximated (in X_0) by polynomials of the form

$$v_n(x, t) = \sum_{i=0}^{m_n} a_{in}(x) t^i, \tag{5.163}$$

where $a_{in} \in W^{1,p}_0(\Omega)$ (cf. Chapter 23, [222]). As $C^\infty_0(\Omega)$ is dense in $W^{1,p}_0(\Omega)$ (with respect to the norm topology), we can choose the functions a_{in} above to be in $C^\infty_c(\Omega)$. Because the truncation operator

$$v \mapsto \max\{v, 0\}$$

is continuous from V_0 to V_0 and thus from X_0 to X_0 (cf. [43]), we have

$$w_n := \max\{v_n, 0\} \to \max\{v, 0\} = v \text{ in } X_0. \tag{5.164}$$

It is clear that $w_n \in X_0 \cap L^p_+(\Omega)$. Moreover, for almost all $t \in (0, \tau)$,

$$\operatorname{supp} w_n(t) \subset \operatorname{supp} v_n(t) \subset \bigcup_{i=0}^{m_n} \operatorname{supp} a_{in},$$

where $\bigcup_{i=0}^{m_n} \operatorname{supp} a_{in}$ is a compact subset of Ω. This result means that w_n satisfies (5.162). From the above arguments, $w_n \in M$. This finding and (5.164) show (5.161). □

5.4 Sub-Supersolutions and Monotone Penalty Approximations

In this section, we consider the following quasilinear variational inequality:

$$u \in K : \langle Au + F(u), v - u \rangle \geq 0, \ \forall \, v \in K, \tag{5.165}$$

with

$$K = V_0 \cap \mathcal{K}, \tag{5.166}$$

where $\mathcal{K}$ is a closed convex subset of V ($V = W^{1,p}(\Omega)$ and $V_0 = W_0^{1,p}(\Omega)$ as defined above). The operator $A : V \to V_0^*$ is assumed to be a second-order quasilinear differential operator in divergence form given by

$$Au(x) = -\sum_{i=1}^{N} \frac{\partial}{\partial x_i} a_i(x, \nabla u(x)), \tag{5.167}$$

and F is the Nemytskij operator generated by a Carathéodory function $f : \Omega \times \mathbb{R} \times \mathbb{R}^N \to \mathbb{R}$, and is defined by

$$\langle F(u), \varphi \rangle = \int_{\Omega} f(\cdot, u, \nabla u) \, \varphi \, dx, \ \varphi \in V_0. \tag{5.168}$$

From the structure and growth conditions that will be imposed on f, the operator $A + F : V_0 \to V_0^*$ is neither monotone nor coercive, in general, and thus, standard existence results for (5.165) cannot be applied.

The main goal of this section is to provide a monotone approximation scheme for the extremal solutions by a sequence of penalty problems assuming the existence of specific super- and subsolutions. We show the convergence of the extremal solutions of the penalty equations to the corresponding solutions of the original problem. We note that the sequences of approximated solutions are monotone (increasing or decreasing), but no monotonicity condition is imposed on the lower order term $F(u)$. As a model problem for (5.165), we deal with an obstacle problem given by (5.165), (5.166) with

$$\mathcal{K} = \{ v \in V : v \leq \psi \}. \tag{5.169}$$

5.4.1 Hypotheses and Preliminary Results

The coefficient functions a_i, $i = 1, \ldots, N$, of the operator A are assumed to satisfy the following hypotheses of Leray–Lions type:

(A1) Each $a_i : \Omega \times \mathbb{R}^N \to \mathbb{R}$ satisfies the Carathéodory condition, and a constant $c_0 > 0$ and a function $k_0 \in L^{p'}(\Omega), 1/p + 1/p' = 1$, exist such that

$$|a_i(x, \xi)| \leq k_0(x) + c_0 |\xi|^{p-1},$$

for a.a. $x \in \Omega$ and for all $\xi \in \mathbb{R}^N$.

(A2) $\sum_{i=1}^{N} (a_i(x, \xi) - a_i(x, \xi'))(\xi_i - \xi_i') > 0$ for a.a. $x \in \Omega$, and for all $\xi, \xi' \in \mathbb{R}^N$ with $\xi \neq \xi'$.

(A3) $\sum_{i=1}^{N} a_i(x, \xi)\xi_i \geq \nu |\xi|^p - k_1(x)$ for a.a. $x \in \Omega$, and for all $\xi \in \mathbb{R}^N$ with some constant $\nu > 0$ and some function $k_1 \in L^1(\Omega)$.

As a consequence of (A1), (A2) the semilinear form a associated with the operator A by

$$\langle Au, \varphi \rangle := a(u, \varphi) = \int_\Omega \sum_{i=1}^N a_i(x, \nabla u) \frac{\partial \varphi}{\partial x_i} \, dx, \quad \forall \, \varphi \in V_0$$

is well defined for any $u \in V$, and the operator $A : V_0 \to V_0^*$ is continuous, bounded, and monotone. The notion of super- and subsolutions introduced in Sect. 5.1 (cf. [141]) is equivalent to the following definitions.

Definition 5.37. *A function $\underline{u} \in V$ is called a subsolution of (5.165) if the following holds:*

(i) $\underline{u} \leq 0$ on $\partial\Omega$.
(ii) $F(\underline{u}) \in L^{p'}(\Omega)$.
(iii) $\langle A\underline{u} + F(\underline{u}), (\underline{u} - v)^+ \rangle \leq 0, \quad \forall \, v \in K$.

Definition 5.38. *$\bar{u} \in V$ is a supersolution of (5.165) if the following holds:*

(i) $\bar{u} \geq 0$ on $\partial\Omega$.
(ii) $F(\bar{u}) \in L^{p'}(\Omega)$.
(iii) $\langle A\bar{u} + F(\bar{u}), (v - \bar{u})^+ \rangle \geq 0, \quad \forall \, v \in K$.

For a given pair $\bar{u}, \underline{u}$ of super-subsolutions of (5.165) satisfying $\underline{u} \leq \bar{u}$, we assume the following hypothesis for f:

(H)   The function $f : \Omega \times \mathbb{R} \times \mathbb{R}^N \to \mathbb{R}$ is a Carathéodory function satisfying the growth condition

$$|f(x, s, \xi)| \leq k_2(x) + c_1 |\xi|^{p-1},$$

for a.a. $x \in \Omega$, for all $s \in [\underline{u}(x), \bar{u}(x)]$, and for all $\xi \in \mathbb{R}^N$, with $k_2 \in L_+^{p'}(\Omega)$ and $c_1 > 0$.

In the construction of our approximate scheme, we make use of the following result (see [215, Theorem 4.19] and Definition 5.31).

Lemma 5.39. *Let $C \neq \emptyset$ be a closed and convex subset of a reflexive Banach space V, and let $\mathcal{A} : V \to V^*$ be a pseudomonotone and coercive operator, and $f \in V^*$ be given. If $P : V \to V^*$ is a penalty operator associated with C, then a sequence (u_n) exists, where each u_n satisfies*

$$u_n \in V, \quad \mathcal{A}u_n + \frac{1}{\varepsilon_n} P(u_n) = f \quad \text{in} \ \ V^*, \tag{5.170}$$

with $\varepsilon_n \to 0^+$ as $n \to \infty$, which converges weakly in V toward a solution of the variational inequality

$$u \in C : \ \langle \mathcal{A}u - f, v - u \rangle \geq 0, \ \forall \, v \in C. \tag{5.171}$$

5.4.2 Obstacle Problem

In this section, we consider the variational inequality (5.165) with the convex set K given by

$$K = \{v \in V_0 : v \leq \psi\}, \tag{5.172}$$

which is of the form (5.166) with $\mathcal{K}$ given by (5.169). Let us assume $\psi \in L^p(\Omega)$. Then it can easily be seen that a penalty opeartor $P : V_0 \to V_0^*$ associated with K is given by

$$\langle P(u), \varphi \rangle = \int_\Omega \{(u - \psi)^+\}^{p-1} \varphi \, dx, \ \varphi \in V_0. \tag{5.173}$$

Furthermore, we assume the existence of functions $\bar{u}$, $\underline{u} \in V$ with the properties:

(O1) $\underline{u}$, $\bar{u} \in V$ and $\underline{u} \leq \bar{u}$.
(O2) (i) $\underline{u} \leq 0$ on $\partial\Omega$, and $\underline{u} \leq \psi$.
    (ii) $F(\underline{u}) \in L^{p'}(\Omega)$.
    (iii) $A\underline{u} + F(\underline{u}) \leq 0$ in V_0^*.
(O3) (i) $\bar{u} \geq 0$ on $\partial\Omega$.
    (ii) $F(\bar{u}) \in L^{p'}(\Omega)$.
    (iii) $A\bar{u} + F(\bar{u}) \geq 0$ in V_0^*.

Hypotheses (O2) and (O3) imply, in particular, that $\bar{u}$ and $\underline{u}$ are supersolution and subsolution, respectively, of (5.165), (5.172) according to Definition 5.37 and Definition 5.38. Moreover, $\bar{u}$ and $\underline{u}$ satisfy the lattice conditions $\underline{u} \vee K \subset K$, $\bar{u} \wedge K \subset K$, $K \vee K \subset K$, $K \wedge K \subset K$.

For later purposes, we introduce the cutoff function $b : \Omega \times \mathbb{R} \to \mathbb{R}$ and truncation operator T related with the functions $\underline{u}$, $\bar{u}$, and given by [cf. (5.57) and (5.62)]

$$b(x, s) = \begin{cases} (s - \bar{u}(x))^{p-1} & \text{if } s > \bar{u}(x), \\ 0 & \text{if } \underline{u}(x) \leq s \leq \bar{u}(x), \\ -(\underline{u}(x) - s)^{p-1} & \text{if } s < \underline{u}(x), \end{cases} \tag{5.174}$$

and

$$(Tu)(x) = \begin{cases} \bar{u}(x) & \text{if } u(x) > \bar{u}(x), \\ u(x) & \text{if } \underline{u}(x) \leq u(x) \leq \bar{u}(x), \\ \underline{u}(x) & \text{if } u(x) < \underline{u}(x). \end{cases} \tag{5.175}$$

It is known that the truncation operator T is continuous and bounded from V into V (see [43, Chap. C.4]). We readily verify that b is a Carathéodory function satisfying the growth condition [cf. (5.58)]

$$|b(x, s)| \leq k_3(x) + c_2 |s|^{p-1} \tag{5.176}$$

for a.a. $x \in \Omega$, for all $s \in \mathbb{R}$, with some function $k_3 \in L_+^{p'}(\Omega)$, and some positive constant c_2. Moreover, we have the following estimate [cf. (5.59)]:

$$\int_\Omega b(x, u(x)) \, u(x) \, dx \geq c_3 \, \|u\|_{L^p(\Omega)}^p - c_4, \quad \forall \, u \in L^p(\Omega), \tag{5.177}$$

where c_3 and c_4 are some positive constants. In view of (5.176), the Nemytskij operator $B : L^p(\Omega) \to L^{p'}(\Omega)$ defined by

$$Bu(x) = b(x, u(x))$$

is continuous and bounded.

Now we associate with the obstacle problem (5.165), (5.172), the following penalty problems (P_n):

$$u \in V_0 : \quad Au + F(u) + \frac{1}{\varepsilon_n} P(u) = 0 \quad \text{in } V_0^*, \tag{5.178}$$

where (ε_n) is a sequence of positive penalty parameters tending to zero; e.g., let us take $\varepsilon_n = 1/(n+1)$. The main result of this section is given by the following theorem.

Theorem 5.40. *Assume hypotheses (A1)–(A3), and let (H) be satisfied with respect to the order interval given by the pair of functions $\underline{u}$, $\bar{u}$, which satisfies the above hypotheses (O1)–(O3). Then the greatest solution of the obstacle problem (5.165), (5.172) within $[\underline{u}, \bar{u}]$ can be obtained as the limit of a monotone decreasing sequence of solutions of the penalty problems (5.178).*

Proof: Note first that $u \in [\underline{u}, \bar{u}]$ is a solution of the variational inequality (5.165), (5.172) if and only if it is a solution of the following auxiliary variational inequality:

$$u \in K : \quad \langle Au + (F \circ T)(u) + \lambda B(u), v - u \rangle \geq 0, \ \forall \, v \in K. \tag{5.179}$$

The term $\lambda B(u)$ has been introduced to apply Lemma 5.39 where $\lambda \geq 0$ is some constant, which from (5.177) can be specified in such a way that $\lambda B(u)$ is a coercivity generating term for the operator $\mathcal{A} = A + F \circ T + \lambda B$. The corresponding penalty problems are then given by

$$u \in V_0 : \quad Au + (F \circ T)(u) + \lambda B(u) + \frac{1}{\varepsilon_n} P(u) = 0 \quad \text{in } V_0^*. \tag{5.180}$$

In view of the definition of B and T, a function u within the interval $[\underline{u}, \bar{u}]$ is a solution of (5.178) if and only if it is a solution of (5.180). Moreover, we readily verify that the given $\bar{u}$ is a supersolution and $\underline{u}$ is a subsolution of (5.180) [and of (5.178)] for any penalty parameter ε_n in the usual sense of variational equations. Let $\varepsilon_n = 1/(n+1)$, and consider (5.180) for $n = 0$. As $\underline{u}$ and $\bar{u}$ are sub- and supersolutions, respectively, of (5.180), extremal solutions of (5.180) exist within the interval $[\underline{u}, \bar{u}]$, i.e., the greatest and smallest solutions (see Chap. 3). Denote the greatest solution of (5.180) within $[\underline{u}, \bar{u}]$ by u_0. Then we observe that u_0 is a supersolution of the problem (5.180) for $n = 1$, because $\varepsilon_1 < \varepsilon_0$ and $Pu \geq 0$ for any u. Thus, applying the extremality result for (5.180) with $n = 1$, extremal solutions of (5.180) exist within $[\underline{u}, u_0]$. Denote the greatest one by u_1. By induction, we obtain a monotone sequence of extremal solutions (u_n) of (5.180) satisfying $u_n \in [\underline{u}, u_{n-1}]$. One can show that (u_n) is bounded in V_0 and thus converges weakly in V_0 and strongly in $L^p(\Omega)$ due to its monotonicity and the compact embedding of V_0 into $L^p(\Omega)$. Let $\lim_{n \to \infty} u_n = u$; then we can verify in a similar way as in the proof of Lemma 5.39 that this limit is a solution of the variational inequality (5.179), and trivially u satisfies $\underline{u} \leq u \leq \bar{u}$. Because of the latter, we have $Bu = 0$ and $Tu = u$, and thus, the limit u must be a solution of the original obstacle problem (5.165), (5.172). Finally, we show that the limit u is the greatest solution u^* of (5.165), (5.172) within $[\underline{u}, \bar{u}]$. To this end, let $\hat{u} \in [\underline{u}, \bar{u}]$ be any solution of (5.165), (5.172). Then we readily verify that $\hat{u}$ satisfies, in particular, hypothesis (O2) with $\underline{u}$ replaced by $\hat{u}$. Obviously (O2)(i) and (O2)(ii) are fulfilled by $\hat{u}$. To see that (O2) (iii) is valid too, take as a special function $v \in K$ in (5.165) $v = \hat{u} - \varphi$, where $\varphi \in V_0 \cap L_+^p(\Omega)$. Thus, $\hat{u}$ is a subsolution for the penalty problems (5.180) for any parameter ε_n, and for the sequence constructed above, we have $\hat{u} \leq u_n$ for all n, which shows $\hat{u} \leq u$; i.e., $u = u^*$. $\square$

Remark 5.41. Note that Theorem 5.40 provides a monotone scheme only for the greatest solution of the obstacle problem (5.172). A similar monotone scheme can be established for the smallest solution of the obstacle problem

$$K = \{v \in V_0 : v \geq \psi\}, \tag{5.181}$$

within the interval $[\underline{u}, \bar{u}]$, where the condition (O2)(i) has to be replaced by

$$\underline{u} \leq 0 \quad \text{on} \quad \partial\Omega,$$

and condition (O3)(i) has to be replaced by

$$\bar{u} \geq 0 \quad \text{on} \quad \partial\Omega, \text{ and } \bar{u} \geq \psi.$$

5.4.3 Generalized Obstacle Problem

In this section, we extend our arguments above to general types of obstacle problems. We consider the variational inequality (5.165) with the convex set K given by

$$K = V_0 \cap \mathcal{K}, \tag{5.182}$$

where $\mathcal{K}$ is a closed convex subset of V. Assume that $\mathcal{K}$ has a penalty operator P. Moreover, suppose P has the following properties:

$$\langle P(u) - P(v), (u - v)^+ \rangle \geq 0, \ \forall \, u, v \in V \qquad (5.183)$$

and

$$\langle P(u), v \rangle \geq 0, \ \forall \, u \in V, v \in V_0^+ := V_0 \cap L_+^p(\Omega). \qquad (5.184)$$

Note that P is not necessarily given by an integral over Ω. Also, (5.183) is a monotonicity condition on the positive parts and (5.184) means that P is a positive operator with respect to the positive cone V_0^+ in V_0. Assumption (O2)(i) now becomes:

(O2) (i') $\underline{u} \leq 0$ on $\partial\Omega$, $\underline{u} \in \mathcal{K}$.

Assume that a pair of functions $\underline{u}$ and $\bar{u}$ exists satisfying (O1)–(O3) with condition (O2)(i) now replaced by (O2)(i'). The inequality (5.165) is approximated by a sequence of equations (P_n) [cf. (5.178)], with a general penalty operator P having the above properties (5.183) and (5.184). The definitions of sub- and supersolutions for (5.178) are similar to those for variational equations (see Chap. 3, and [29, 43, 75, 113, 133]). For example, a function $\underline{w} \in V$ is a subsolution of (5.178) if $\underline{w} \leq 0$ on $\partial\Omega$, $F(\underline{w}) \in L^{p'}(\Omega)$, and

$$A\underline{w} + F(\underline{w}) + \frac{1}{\varepsilon_n} P(\underline{w}) \leq 0 \ \text{ in } V_0^*. \qquad (5.185)$$

The existence of the least and greatest solutions of (5.178) in this more general setting is given in the following result.

Theorem 5.42. *Assume $\underline{w}, \bar{w} \in [\underline{u}, \bar{u}]$ are sub- and supersolutions of (5.178). Then, the least and greatest solutions of (5.178) exist within the interval $[\underline{w}, \bar{w}]$.*

Proof: The proof is similar to that for variational equations as given in Chap. 3 and is only outlined with the necessary differences (see also [43, 75, 133]). Let b and T be the cutoff function and truncation operator, respectively, associated with $\underline{w}$ and $\bar{w}$. As $A + F \circ T + \lambda B : V_0 \to V_0^*$ is pseudomonotone and $P : V_0 \to V_0^*$ is monotone, the operator $A + F \circ T + \lambda B + \varepsilon_n^{-1} P$ is pseudomonotone on V_0. Also, it is coercive with λ chosen sufficiently large. Hence the equation

$$Au + (F \circ T)(u) + \lambda B(u) + \frac{1}{\varepsilon_n} P(u) = 0 \qquad (5.186)$$

has a solution $u \in V_0$. Let us show that $u \geq \underline{w}$. By (5.185) and (5.186), we obtain

$$\left\langle A\underline{w} - Au + F(\underline{w}) - F(Tu) - \lambda B(u) + \frac{1}{\varepsilon_n} P(\underline{w}) - \frac{1}{\varepsilon_n} P(u), v \right\rangle$$

$$\leq 0, \quad \forall \, v \in V_0^+.$$

Choosing $v = (\underline{w} - u)^+ \in V_0^+$ in this inequality, we get

$$\langle A\underline{w} - Au, (\underline{w} - u)^+ \rangle + \frac{1}{\varepsilon_n} \langle P(\underline{w}) - P(u), (\underline{w} - u)^+ \rangle - \lambda \langle B(u), (\underline{w} - u)^+ \rangle$$
$$\leq \langle F(Tu) - F(\underline{w}), (\underline{w} - u)^+ \rangle = 0.$$

From the assumption on A and (5.183), we obtain $\langle B(u), (\underline{w} - u)^+ \rangle \geq 0$. This result implies that $(\underline{w} - u)^+ = 0$ on Ω, and therefore, $\underline{w} \leq u$. A similar proof shows that $u \leq \bar{w}$.

By modifying the terms $F \circ T$ and B appropriately, we can extend the above arguments to the case $\underline{w}$ is the maximum of a finite number of subsolutions and $\bar{w}$ is the minimum of a finite number of supersolutions of (5.178). Using this fact, we can proceed as in the case of usual equations to show the existence of extremal solutions of (5.178) (again, we refer to Chap. 3 for more details). □

The following simple lemma gives us some relations between solutions of (5.165) and (5.178) in terms of the so-called recession cone, cf. [196].

Definition 5.43. *Let $K \neq \emptyset$ be a closed and convex subset of a reflexive Banach space V. The recession cone* rc K *of K is defined by* rc $K := \bigcap_{t>0} t\,(K - x_0)$, *where x_0 is any (fixed) element of K, and $K - x_0 = \{x - x_0 : x \in K\}$.*

The above definition does not depend on $x_0 \in K$. As K is a closed and convex set, we can prove that the following holds:

$$v \in \text{rc}\,K \iff x_0 + tv \in K, \ \forall\, t > 0 \iff x + tv \in K, \ \forall\, x \in K, \ \forall\, t > 0. \tag{5.187}$$

It follows from Definition 5.43 and (5.187) that 0 is always in rc K and rc K is a closed and convex cone (not necessarily contained in K in general). As for these properties, we refer to the book by Rockafellar [196]. Even though in [196] the finite-dimensional case has been treated only, the above properties hold for the infinite dimensional case as well.

Lemma 5.44. (a) *If u is a solution of (P_n) within $[\underline{u}, \bar{u}]$, then it is a supersolution of (P_{n+1}).*

(b) *Assume K has the following property:*

$$V_0^- := V_0 \cap L_-^p(\Omega) \subset \text{rc}\,K. \tag{5.188}$$

Then all solutions of (5.165) are subsolutions of (P_n) for any $n \in \mathbb{N}$.

Proof: (a) Assume u is a solution of (P_n) within the interval $[\underline{u}, \bar{u}]$. For $v \in V_0^+$, in view of (5.184),

$$\left\langle Au + F(u) + \frac{1}{\varepsilon_{n+1}} P(u), v \right\rangle = \left(\frac{1}{\varepsilon_{n+1}} - \frac{1}{\varepsilon_n} \right) \langle P(u), v \rangle \geq 0.$$

This result shows that u is a supersolution of (P_{n+1}).

(b) Assume u is a solution of (5.165). Let $w \in V_0^+$. We have $-w \in \mathrm{rc}\,K$. Properties of recession cones (cf. [196]) imply that $u-w \in K$. Letting $v = u-w$ in (5.165) and noting that $P(u) = 0$, we get

$$0 \geq \langle Au + F(u), w \rangle = \left\langle Au + F(u) + \frac{1}{\varepsilon_n}P(u), w \right\rangle.$$

$\square$

Theorem 5.45. *Assume (5.165) has a pair of sub- and supersolution $\underline{u}$ and $\bar{u}$ satisfying (O1)–(O3) with (O2)(i) replaced by (O2)(i'), and the convex set K satisfies (5.182), (5.183), (5.184), and (5.188). Then the greatest solution of (5.165) within the interval $[\underline{u}, \bar{u}]$ can be obtained as the limit of a monotone decreasing sequence of solutions $u_n \in [\underline{u}, \bar{u}]$ of (5.178).*

Proof: The proof follows the same lines as that of Theorem 5.40 and is just outlined here. As $P(\underline{u}) = 0$, we see that $\underline{u}$ is a subsolution of (5.178) for any n. Also, from (5.184), $\bar{u}$ is a supersolution of (5.178)(for any n). For $n = 0$, from Theorem 5.42, (P_0) has a greatest solution within $[\underline{u}, \bar{u}]$, which is denoted by u_0. From Lemma 5.44, u_0 is a supersolution of (P_1). Assume $u_n \in [\underline{u}, u_{n-1}]$ is the greatest solution of (P_n). From Lemma 5.44 and Theorem 5.42, u_n is a supersolution of (P_{n+1}), and thus, (P_{n+1}) has a greatest solution u_{n+1} within the interval $[\underline{u}, u_n] \subset [\underline{u}, \bar{u}]$. By induction, we have a decreasing sequence (u_n) of solutions of (5.178) in the interval $[\underline{u}, \bar{u}]$. Note that any solution of (5.178) within $[\underline{u}, \bar{u}]$ is also a solution of the associated auxiliary problem (5.180) and vice versa.

As (u_n) is bounded in $L^p(\Omega)$, the coercivity of $\mathcal{A} = A + F \circ T + \lambda B$ and monotonicity of P imply that (u_n) is also bounded in V_0. Because (u_n) is a monotone sequence, there is $u^* \in V_0$ such that $u_n \rightharpoonup u^*$ in V_0. It can be checked that $u^* \in K$ and is a solution of (5.165), and moreover, $\underline{u} \leq u \leq \bar{u}$. As $u_n - u^* \geq 0$, we get by (5.184) that $\langle P(u_n), u_n - u^* \rangle \geq 0$, and thus,

$$\limsup_{n \to \infty} \langle \mathcal{A}u_n, u_n - u^* \rangle \leq 0. \tag{5.189}$$

Because $u_n \rightharpoonup u^*$ in V_0 and $\mathcal{A}$ possesses the (S_+)–property (cf. Theorem 2.109 in Chap. 2), we infer that the convergence of u_n to u^* is in fact a strong convergence in V.

To show that u^* is the greatest solution of (5.165) within $[\underline{u}, \bar{u}]$, we assume that $\hat{u}$ is any solution of (5.165) such that $\underline{u} \leq \hat{u} \leq \bar{u}$. From Lemma 5.44, $\hat{u}$ is a subsolution of (P_0). Then there is a solution u of (P_0) such that $\underline{u} \leq \hat{u} \leq u \leq \bar{u}$. We have $u \leq u_0$ and thus $\hat{u} \leq u_0$. Using induction again, we have $\hat{u} \leq u_n$ for all n. Thus, $\hat{u} \leq u^*$. $\square$

Some Examples

Example 5.46. In the obstacle problem, we considered in Sect. 5.4.2, the penalty operator P, given by

$$\langle P(u), v \rangle = \int_{\Omega} \{[u - \psi]^+\}^{p-1} v dx, \ \forall \ u, v \in V,$$

satisfies (5.183)–(5.184). The closed and convex set K is of the form (5.182) with $\mathcal{K} = \{u \in V : u \leq \psi$ a.e. on $\Omega\}$. Moreover, rc $K = \{u \in V_0 : u \leq 0$ a.e. on $\Omega\} = V_0^-$ and so (5.188) also holds. To see that rc $K = V_0^-$, let $u \in K$ and v be given such that $u + tv \in K$ for all $t > 0$. Then first $v \in V_0$ from $u + tv \in V_0$ and $u \in V_0$, and second, in view of $u \leq \psi$ and $u + tv \leq \psi$ for all $t > 0$, it follows that (letting $t \to \infty$) $v \leq 0$ a.e. on Ω, and thus by (5.187), we get $v \in$ rc $K = V_0^-$.

We can extend this result to a local obstacle ψ defined on a measurable set M in Ω. Assuming that $\psi \in L^p(M)$, the penalty operator P is now given by

$$\langle P(u), v \rangle = \int_{M} \{[u - \psi]^+\}^{p-1} v dx, \ \forall \ u, v \in V.$$

Example 5.47. Assume S is an $(N - 1)$-dimensional surface and $S \subset \Omega$. We consider a thin obstacle problem with an obstacle ψ on S. In this case, the set K in (5.165) is given by

$$K = \{u \in V_0 : u \leq \psi \ \mu\text{-a.e. on } S\},$$

where $\psi \in L^p(S)$ and $d\mu = dS$ is the surface measure on S (we refer to [123] and the references therein for more discussions on low-dimensional obstacle problems).

The penalty operator P is now given by

$$\langle P(u), v \rangle = \int_{S} \{[u - \psi]^+\}^{p-1} v d\mu \ (\forall \ u, v \in V).$$

K and P also satisfy the assumptions we mentioned above.

Example 5.48. Assume $p > N$. As $V \subset C(\overline{\Omega})$, our penalization approach above is applicable to obstacle problems with even lower dimensions. The set S in (b) is now an m-dimensional manifold in Ω $(m \leq N)$.

Remark 5.49. If instead of (5.188), we have $V_0^+ \subset$ rc K and P satisfies

$$\langle P(u), v \rangle \leq 0, \ \forall \ v \in V_0^+,$$

instead of (5.184), then similar arguments to those we had before show that the least solution of (5.165) can be approximated by a monotone sequence of solutions of the penalized problem (5.178).

An example for this case is the lower obstacle problem, in which the convex set K in (5.165) is given by

$$K = \{v \in V_0 : v \geq \psi\}.$$

Remark 5.50. If monotone approximation is wanted for both smallest and greatest solutions of (5.165) by our approach here, then in view of Lemma 5.44, we need

$$V_0^+, V_0^- \subset \mathrm{rc}\, K.$$

As $\mathrm{rc}\, K$ is a convex cone, this implies that $\mathrm{rc}\, K = {}^-K = V_0$; i.e., (5.165) is a variational equation. This result also shows that we cannot use our arguments here directly to convex sets defined by constraints on ∇u, such as $K = \{u \in V_0 : |\nabla u| \leq 1 \text{ a.e. on } \Omega\}$.

Remark 5.51. When f does not depend on the gradient and $s \mapsto f(x, s)$ is Lipschitz continuous with sufficiently small Lipschitz constant, then the penalty problem (5.178) has a unique solution u_n and the greatest solution of (5.165) within $[\underline{u}, \bar{u}]$ is the limit of the iterates u_n. Numerical realizations of monotone iteration schemes for coercive semilinear obstacle problems have been proved, e.g., in [14, 127].

Remark 5.52. Our approach can be extended to variational inequalities (5.165) involving general (nonmonotone) Leray–Lions operators A in the form

$$Au(x) = -\sum_{i=1}^{N} \frac{\partial}{\partial x_i} a_i(x, u(x), \nabla u(x)).$$

5.5 Systems of Variational Inequalities

Let $V = W^{1,p}(\Omega)$ and $V_0 = W_0^{1,p}(\Omega)$ be as in the previous sections. We consider here the following system of variational inequalities: For $k = 1, \ldots, m$

$$u_k \in K_k : \langle A_k u_k + F_k(u, \nabla u), v_k - u_k \rangle \geq 0, \ \forall \, v_k \in K_k, \tag{5.190}$$

where $u = (u_1, \ldots, u_m)$, $\nabla u = (\nabla u_1, \ldots, \nabla u_m)$, and K_k is a closed and convex set of V_0, and $\langle \cdot, \cdot \rangle$ denotes the duality pairing between V_0^* and V_0. The operator A_k is assumed to be a second -order quasilinear differential operator in divergence form of Leray–Lions type given by

$$A_k v(x) = -\sum_{i=1}^{N} \frac{\partial}{\partial x_i} a_i^{(k)}(x, \nabla v(x)), \tag{5.191}$$

and F_k is the Nemytskij operator generated by some Carathéodory function $f_k : \Omega \times \mathbb{R}^m \times \mathbb{R}^{mN} \to \mathbb{R}$, and defined by

$$\langle F_k(u, \nabla u), \varphi \rangle = \int_\Omega f_k(\cdot, u, \nabla u) \, \varphi \, dx, \ \varphi \in V_0. \tag{5.192}$$

Our main goal in this section is to prove existence and enclosure of solutions for (5.190) in terms of an appropriately defined rectangle $\mathcal{R}$ formed by an ordered pair of vectors $\underline{u}$, $\bar{u} \in W^{1,p}(\Omega; \mathbb{R}^m)$, which generalizes the notion of super- and subsolutions in the scalar case. More precisely, we are going to prove the existence of minimal and maximal (in the set theoretical sense) solutions within some ordered interval of an appropriately defined pair of sub- and supersolutions. Furthermore, for weakly coupled quasimonotone systems of variational inequalities, the existence of smallest and greatest solutions, i.e., extremal solutions, is proved.

5.5.1 Notations and Assumptions

We impose the following hypotheses of Leray–Lions type on the coefficient functions $a_i^{(k)}$, $i = 1, \ldots, N$, of the operators A_k.

(A1) Each $a_i^{(k)} : \Omega \times \mathbb{R}^N \to \mathbb{R}$ satisfies Carathéodory conditions; i.e., $a_i^{(k)}(x, \xi)$ is measurable in $x \in \Omega$ for all $\xi \in \mathbb{R}^N$ and continuous in ξ for almost all $x \in \Omega$. Constants $c_0^{(k)} > 0$ and functions $\kappa_0^{(k)} \in L^q(\Omega)$, $1/p + 1/q = 1$ exist, such that

$$|a_i^{(k)}(x, \xi)| \leq \kappa_0^{(k)}(x) + c_0^{(k)} |\xi|^{p-1},$$

for a.e. $x \in \Omega$ and for all $\xi \in \mathbb{R}^N$.

(A2) $\sum_{i=1}^N (a_i^{(k)}(x, \xi) - a_i(x, \xi'))(\xi_i - \xi_i') > 0$ for a.e. $x \in \Omega$, and for all $\xi, \xi' \in \mathbb{R}^N$ with $\xi \neq \xi'$.

(A3) $\sum_{i=1}^N a_i^{(k)}(x, \xi)\xi_i \geq \nu_k |\xi|^p$ for a.e. $x \in \Omega$, and for all $\xi \in \mathbb{R}^N$ with some constants $\nu_k > 0$.

As a consequence of (A1) and (A2), the operators $A_k : V \to V^*$ defined by

$$\langle A_k u, \varphi \rangle := a_k(u, \varphi) = \sum_{i=1}^N \int_\Omega a_i^{(k)}(x, \nabla u) \frac{\partial \varphi}{\partial x_i} \, dx$$

are continuous, bounded, and monotone and, hence, in particular, pseudomonotone. If u, $w \in L^p(\Omega; \mathbb{R}^m)$, then a partial ordering is given by $u \leq w$ if and only if $u_k \leq w_k$, for $k \in \{1, \ldots, m\}$; i.e., $L^p(\Omega; \mathbb{R}^m)$ is equipped with the componentwise partial ordering, which induces a corresponding partial ordering in the spaces $X := W^{1,p}(\Omega; \mathbb{R}^m)$ and $X_0 := W_0^{1,p}(\Omega; \mathbb{R}^m)$. Furthermore, if $s \in \mathbb{R}^m$, then we denote

$$[s]_k := (s_1, \ldots, s_{k-1}, s_{k+1}, \ldots, s_m) \in \mathbb{R}^{m-1},$$
$$(t, [s]_k) := (s_1, \ldots, s_{k-1}, t, s_{k+1}, \ldots, s_m) \in \mathbb{R}^m,$$

and for $\eta = (\eta_1, \ldots, \eta_m) \in \mathbb{R}^{mN}$ with $\eta_k \in \mathbb{R}^N$ and $\tau \in \mathbb{R}^N$, we denote

$$[[\eta]]_k := (\eta_1, \ldots, \eta_{k-1}, \eta_{k+1}, \ldots, \eta_m) \in \mathbb{R}^{(m-1)N},$$
$$(\tau, [[\eta]]_k) := (\eta_1, \ldots, \eta_{k-1}, \tau, \eta_{k+1}, \ldots, \eta_m) \in \mathbb{R}^{mN}.$$

Thus, we may write, e.g., $f_k(\cdot, s, \eta) = f_k(\cdot, s_k, \eta_k, [s]_k, [[\eta]]_k)$.

In the following definition, we introduce our basic notion of a pair of super- and subsolutions.

Definition 5.53. *The two vectors $\bar{u}$, $\underline{u} \in X$ are said to form a pair of super- and subsolutions of (5.190) if the following conditions hold:*

(i) $\underline{u} \leq \bar{u}$ *in Ω and $\underline{u} \leq 0 \leq \bar{u}$ on $\partial\Omega$.*

(ii) $\langle A_k \underline{u}_k + F_k(\underline{u}_k, \nabla \underline{u}_k, [w]_k, [[\nabla w]]_k), (\underline{u}_k - v_k)^+\rangle \leq 0, \ \forall \ v_k \in K_k \ and \ \forall \ w:$ $[\underline{u}]_k \leq [w]_k \leq [\bar{u}]_k,$

(iii) $\langle A_k \bar{u}_k + F_k(\bar{u}_k, \nabla \bar{u}_k, [w]_k, [[\nabla w]]_k), (v_k - \bar{u}_k)^+\rangle \geq 0, \ \forall \ v_k \in K_k \ and \ \forall \ w:$ $[\underline{u}]_k \leq [w]_k \leq [\bar{u}]_k.$

For a given pair of super- and subsolutions, we assume the following hypothesis on the f_k:

(H) For $k \in \{1, \ldots, m\}$, the functions $f_k : \Omega \times \mathbb{R}^m \times \mathbb{R}^{mN} \to \mathbb{R}$ are Carathéodory and satisfy the growth conditions

$$|f_k(x, s, \eta)| \leq \varrho_k(x) + c_1^{(k)}(|\eta_1|^{p-1} + \cdots + |\eta_m|^{p-1}),$$

for a.e. $x \in \Omega$, for all $s \in [\underline{u}(x), \bar{u}(x)]$, and for all $\eta \in \mathbb{R}^{mN}$, where $\varrho_k \in L_+^q(\Omega)$.

Let $K = K_1 \times \cdots \times K_m$, and $Au = (A_1 u_1, \ldots, A_m u_m)$, as well as

$$F(u, \nabla u) = (F_1(u, \nabla u), \ldots, F_m(u, \nabla u)).$$

We denote

$$\langle Au + F(u, \nabla u), v \rangle = \sum_{k=1}^{m} \langle A_k u_k + F_k(u, \nabla u), v_k \rangle.$$

Then problem (5.190) is equivalent to

$$u \in K : \ \langle Au + F(u, \nabla u), v - u \rangle \geq 0, \ \forall \ v \in K. \tag{5.193}$$

5.5.2 Preliminaries

Let $\bar{u}$, $\underline{u}$ be an ordered pair of super- and subsolutions, and let hypotheses (A1)–(A3) and (H) be satisfied throughout this section. We are going to prove an existence result for some related auxiliary system of variational inequalities that is crucial in the proof of our main result. To this end, we introduce the following truncation operators:

$$(T_k u_k)(x) = \begin{cases} \bar{u}_k(x) & \text{if } u_k(x) > \bar{u}_k(x), \\ u_k(x) & \text{if } \underline{u}_k(x) \leq u_k(x) \leq \bar{u}_k(x), \\ \underline{u}_k(x) & \text{if } u_k(x) < \underline{u}_k(x), \end{cases} \tag{5.194}$$

which are known to be continuous and bounded from V into V (see Chap. 2). The related truncated vector function Tu is given by $Tu = (T_1u_1, \ldots, T_mu_m)$. Next we introduce cutoff functions $b_k : \Omega \times \mathbb{R} \to \mathbb{R}$ given by

$$b_k(x, s) = \begin{cases} (s - \bar{u}_k(x))^{p-1} & \text{if } s > \bar{u}_k(x), \\ 0 & \text{if } \underline{u}_k(x) \leq s \leq \bar{u}_k(x), \\ -(\underline{u}_k(x) - s)^{p-1} & \text{if } s < \underline{u}_k(x). \end{cases} \quad (5.195)$$

As seen in previous sections, b_k is a Carathéodory function satisfying the growth condition

$$|b_k(x, s)| \leq \rho_k(x) + c_2^{(k)}|s|^{p-1} \quad (5.196)$$

for a.e. $x \in \Omega$, for all $s \in \mathbb{R}$, with some function $\rho_k \in L_+^q(\Omega)$. Moreover, we have the following estimate

$$\int_\Omega b_k(x, u_k(x))\, u_k(x)\, dx \geq c_3^{(k)}\|u_k\|_{L^p(\Omega)}^p - c_4^{(k)}, \quad \forall\, u_k \in L^p(\Omega), \quad (5.197)$$

where $c_3^{(k)}$ and $c_4^{(k)}$ are some positive constants. In view of (5.196), the Nemytskij operator $B_k : L^p(\Omega) \to L^q(\Omega)$ defined by

$$B_k u_k(x) = b_k(x, u_k(x))$$

is continuous and bounded. Let $Bu = (B_1u_1, \ldots, B_mu_m)$; then

$$B : L^p(\Omega; \mathbb{R}^m) \to L^q(\Omega; \mathbb{R}^m)$$

is continuous and bounded. For $\lambda \in \mathbb{R}_+^m$, we define $\lambda \cdot Bu$ as

$$\lambda \cdot Bu = (\lambda_1 B_1u_1, \ldots, \lambda_m B_m u_m),$$

and we consider the following auxiliary, truncated variational inequality:

$$u \in K : \ \langle Au + F(Tu, \nabla Tu) + \lambda \cdot Bu, v - u \rangle \geq 0, \ \forall\, v \in K, \quad (5.198)$$

where λ will be specified later. The existence result for (5.198) given by the next lemma plays an important role in the proof of our main result.

Lemma 5.54. *For $\lambda \in \mathbb{R}_+^m$ suitably chosen, the system of variational inequalities (5.198) possesses solutions.*

Proof: By the compact embedding $X_0 \subset L^p(\Omega; \mathbb{R}^m)$ the operator $\lambda \cdot B : X_0 \to X_0^*$ is completely continuous and bounded. As $T : X_0 \to X_0$ is continuous and bounded, the composed operator $F \circ T : X_0 \to X_0^*$ defined by $F \circ T(u) = F(Tu, \nabla Tu)$ is continuous and bounded as well due to (H). Moreover, the Leray–Lions conditions (A1)–(A3) together with (H) finally imply that $A + F \circ T + \lambda \cdot B : X_0 \to X_0^*$ is a bounded, continuous, and pseudomonotone operator. Thus, the existence of solutions for (5.198) follows provided the operator $A + F \circ T + \lambda \cdot B : X_0 \to X_0^*$ can be shown to be coercive relative to K, i.e., provided the following holds: An $v_0 \in K$ exists such that

$$\frac{\langle (A + F \circ T + \lambda \cdot B)u, u - v_0 \rangle}{\|u\|_{X_0}} \to 0 \text{ as } \|u\|_{X_0} \to \infty. \tag{5.199}$$

In view of (A3), we get the estimate

$$\langle Au, u \rangle \geq \sum_{k=1}^{m} \nu_k \|\nabla u_k\|_{L^p(\Omega)}^p, \tag{5.200}$$

and by (A1), we obtain

$$
\begin{aligned}
&|\langle Au, v_0 \rangle| \\
&\leq \sum_{k=1}^{m} \left(\|\kappa_0^{(k)}\|_{L^q(\Omega)} \|\nabla v_{0,k}\|_{L^p(\Omega)} + c_0^{(k)} \|\nabla u_k\|_{L^p(\Omega)}^{p-1} \|\nabla v_{0,k}\|_{L^p(\Omega)} \right) \\
&\leq C + \sum_{k=1}^{m} \left(\delta_k \|\nabla u_k\|_{L^p(\Omega)}^p + c_k(\delta_k) \|\nabla v_{0,k}\|_{L^p(\Omega)}^p \right) \\
&= C(\delta, v_0) + \sum_{k=1}^{m} \delta_k \|\nabla u_k\|_{L^p(\Omega)}^p,
\end{aligned}
\tag{5.201}
$$

where $\delta = (\delta_1, \ldots, \delta_m)$ may be any strictly positive vector of $\mathbb{R}^m$, and $C(\delta, v_0)$ is some constant only depending on δ and v_0. For the estimate (5.201), we have made use of Young's inequality. Next we provide an estimate for the term $\langle F \circ Tu, u - v_0 \rangle$. To this end, we first prove an estimate for the components by using (H).

$$
\begin{aligned}
&|\langle F_k \circ Tu, u_k - v_{0,k} \rangle| \\
&\leq \int_\Omega |f_k(\cdot, Tu, \nabla Tu)(u_k - v_{0,k})| \, dx \\
&\leq \int_\Omega \Big(\varrho_k(x) + c_1^{(k)}(|\nabla u_1|^{p-1} + \cdots + |\nabla u_m|^{p-1} + |\nabla \bar{u}_1|^{p-1} + \cdots + |\nabla \bar{u}_m|^{p-1} \\
&\quad + |\nabla \underline{u}_1|^{p-1} + \cdots + |\nabla \underline{u}_m|^{p-1} \Big)(|u_k| + |v_{0,k}|) \, dx \\
&\leq c_k(\|u_k\|_{L^p(\Omega)} + \|v_{0,k}\|_{L^p(\Omega)}) \\
&\quad + \varepsilon_k \sum_{j=1}^{m} \|\nabla u_j\|_{L^p(\Omega)}^p + c_k(\varepsilon_k)\|u_k\|_{L^p(\Omega)}^p + c_k(\varepsilon_k)\|v_{0,k}\|_{L^p(\Omega)}^p \\
&= C(\varepsilon_k, v_0) + C(\varepsilon_k, v_0)\|u_k\|_{L^p(\Omega)}^p + \varepsilon_k \sum_{j=1}^{m} \|\nabla u_j\|_{L^p(\Omega)}^p,
\end{aligned}
\tag{5.202}
$$

where $\varepsilon_k \in \mathbb{R}$ may be any positive constant and $C(\varepsilon_k, v_0)$ is some constant only depending on ε_k and v_0. By estimate (5.202), we get

$$|\langle F \circ Tu, u - v_0 \rangle|$$
$$\leq \sum_{k=1}^{m} \left(C(\varepsilon_k, v_0) + C(\varepsilon_k, v_0) \|u_k\|_{L^p(\Omega)}^{p} + \varepsilon_k \sum_{j=1}^{m} \|\nabla u_j\|_{L^p(\Omega)}^{p} \right)$$
$$\leq C(\varepsilon, v_0) + C(\varepsilon, v_0) \sum_{k=1}^{m} \|u_k\|_{L^p(\Omega)}^{p} + (\varepsilon_1 + \cdots + \varepsilon_m) \sum_{k=1}^{m} \|\nabla u_k\|_{L^p(\Omega)}^{p}.$$

$$(5.203)$$

By using (5.197), we obtain

$$\langle \lambda_k B_k u_k, u_k \rangle \geq \lambda_k \, c_3^{(k)} \|u_k\|_{L^p(\Omega)}^{p} - c_4^{(k)}, \tag{5.204}$$

and in view of (5.196), we have

$$|\langle \lambda_k B_k u_k, v_{0,k} \rangle| \leq c_k(v_{0,k}) + \lambda_k \, c_2^{(k)} \|u_k\|_{L^p(\Omega)}^{p-1} \|v_{0,k}\|_{L^p(\Omega)}$$
$$\leq C_k(\lambda_k, v_{0,k}, \theta_k) + \theta_k \, \lambda_k \, \|u_k\|_{L^p(\Omega)}^{p}, \tag{5.205}$$

where $\theta_k \in \mathbb{R}$ may be any positive constant and $C_k(\lambda_k, v_{0,k}, \theta_k)$ is some constant only depending on θ_k, $v_{0,k}$, and λ_k. Thus, (5.204) and (5.205) yields

$$\langle \lambda \cdot Bu, u - v_0 \rangle \geq \sum_{k=1}^{m} \lambda_k \, (c_3^{(k)} - \theta_k) \|u_k\|_{L^p(\Omega)}^{p} - C(\lambda, v_0, \theta). \tag{5.206}$$

From (5.200), (5.203), and (5.206), we obtain

$$\langle (A + F \circ T + \lambda \cdot B)u, u - v_0 \rangle$$
$$\geq \sum_{k=1}^{m} \left(\nu_k - \delta_k - \sum_{j=1}^{m} \varepsilon_j \right) \|\nabla u_k\|_{L^p(\Omega)}^{p}$$
$$+ \sum_{k=1}^{m} \left(\lambda_k(c_3^{(k)} - \theta_k) - C(\varepsilon, v_0) \right) \|u_k\|_{L^p(\Omega)}^{p}$$
$$- C(\delta, v_0) - C(\varepsilon, v_0) - C(\lambda, v_0, \theta). \tag{5.207}$$

From (5.207), we see that by chosing ε_k, δ_k, and θ_k sufficiently small and λ_k large enough, we arrive at

$$\langle (A + F \circ T + \lambda \cdot B)u, u - v_0 \rangle \geq \mu \|u\|_{X_0}^{p} - C(\varepsilon, \delta, \theta, v_0), \tag{5.208}$$

for some positive constant μ, which proves the coercivity of the operator $A + F \circ T + \lambda \cdot B$. $\qquad\square$

5.5.3 Comparison Principle for Systems

The main result of this section is the following theorem.

Theorem 5.55. *Let $\bar{u}, \underline{u} \in X$ be a pair of super- and subsolutions, and let hypotheses (A1)–(A3) and (H) be satisfied. Assume $\underline{u}_k \vee K_k \subset K_k$ and $\bar{u}_k \wedge K_k \subset K_k$; then the system of variational inequalities (5.190) has solutions within $[\underline{u}, \bar{u}]$.*

Proof: By Lemma 5.54, solutions of the auxiliary variational inequality (5.198) exist. We are going to prove that any solution of (5.198) belongs to the rectangle $[\underline{u}, \bar{u}]$, which verifies that (5.190) has solutions within $[\underline{u}, \bar{u}]$, because then we have $Tu = u$ and $Bu = 0$. Let u be a solution of (5.198). We show first $\underline{u} \leq u$.

By Definition 5.53 $\underline{u}$ satisfies $\underline{u} \leq 0$ on $\partial\Omega$ and

$$\langle A_k\underline{u}_k + F_k(\underline{u}_k, \nabla\underline{u}_k, [w]_k, [[\nabla w]]_k), (\underline{u}_k - v_k)^+\rangle \leq 0, \tag{5.209}$$

for all $v_k \in K_k$ and for all $w : [\underline{u}]_k \leq [w]_k \leq [\bar{u}]_k$, $k = 1, \ldots, m$. The solution u of (5.198) satisfies

$$u_k \in K_k : \langle A_k u_k + F_k(Tu, \nabla Tu) + \lambda_k B_k u_k, v_k - u_k\rangle \geq 0, \ \forall \ v_k \in K_k, \tag{5.210}$$

for $k = 1, \ldots, m$. In (5.209), we may take, in particular, $w = Tu$ and $v_k = u_k$, and in (5.210), we may take $v_k = \underline{u}_k \vee u_k = u_k + (\underline{u}_k - u_k)^+$, which yields

$$\langle A_k\underline{u}_k + F_k(\underline{u}_k, \nabla\underline{u}_k, [Tu]_k, [[\nabla Tu]]_k), (\underline{u}_k - u_k)^+\rangle \leq 0 \tag{5.211}$$

and

$$\langle A_k u_k + F_k(Tu, \nabla Tu) + \lambda_k B_k u_k, (\underline{u}_k - u_k)^+\rangle \geq 0. \tag{5.212}$$

Subtracting (5.212) from (5.211), we obtain

$$\begin{aligned}&\langle A_k\underline{u}_k - A_k u_k - \lambda_k B_k u_k, (\underline{u}_k - u_k)^+\rangle \\ &\leq -\langle F_k(\underline{u}_k, \nabla\underline{u}_k, [Tu]_k, [[\nabla Tu]]_k) - F_k(Tu, \nabla Tu), (\underline{u}_k - u_k)^+\rangle\end{aligned} \tag{5.213}$$

The right-hand side can easily be seen to become zero, and because of

$$\langle A_k\underline{u}_k - A_k u_k, (\underline{u}_k - u_k)^+\rangle \geq 0,$$

we get from (5.213)

$$\langle -\lambda_k B_k u_k, (\underline{u}_k - u_k)^+\rangle \leq 0, \tag{5.214}$$

which by definition of the cutoff function yields

$$-\lambda_k \int_{\{\underline{u}_k > u_k\}} -(\underline{u}_k - u_k)^p \, dx \leq 0, \tag{5.215}$$

and therefore,

$$0 \leq \lambda_k \int_\Omega \left((\underline{u}_k - u_k)^+\right)^p dx \leq 0. \tag{5.216}$$

From (5.216), we infer $(\underline{u}_k - u_k)^+ = 0$; i.e., $\underline{u}_k \leq u_k$. Similarly, we show $u_k \leq \bar{u}_k$, which completes the proof of the theorem.   □

Only for simplicity we have assumed operators A_k that are monotone. Theorem 5.55 can be extended to general Leray–Lions operators A_k of the form

$$A_k v(x) = -\sum_{i=1}^N \frac{\partial}{\partial x_i} a_i^{(k)}(x, v(x), \nabla v(x)).$$

5.5.4 Generalization, Minimal and Maximal Solutions

Our existence theorem above can be extended to solutions between a finite number of subsolutions and supersolutions. Assume $u^{(1)}, \ldots, u^{(l)} \in L^p(\Omega, \mathbb{R}^m)$, we use the notation

$$\vee_{r=1}^l u^{(r)} = \max\{u^{(1)}, \ldots, u^{(l)}\}$$
$$= (\max\{u_1^{(1)}, \ldots, u_1^{(l)}\}, \ldots, \max\{u_m^{(1)}, \ldots, u_m^{(l)}\})$$

and

$$\wedge_{r=1}^l u^{(r)} = \min\{u^{(1)}, \ldots, u^{(l)}\} = (\min\{u_1^{(1)}, \ldots, u_1^{(l)}\}, \ldots, \min\{u_m^{(1)}, \ldots, u_m^{(l)}\}).$$

It is clear that if $u^{(1)}, \ldots, u^{(l)} \in X$ (respectively, X_0), then $\vee_{r=1}^l u^{(r)}, \wedge_{r=1}^l u^{(r)} \in X$ (respectively, X_0). Also, for $u \in X$, we denote

$$u \wedge K = \prod_{k=1}^m (u_k \wedge K_k), \ u \vee K = \prod_{k=1}^m (u_k \vee K_k)$$

We have the following theorem.

Theorem 5.56. *Let $\underline{u}^{(r)}$ $(r = 1, \ldots, R)$ and $\bar{u}^{(s)}$ $(s = 1, \ldots, S)$ are vectors in X and*

$$\underline{u} = \vee_{r=1}^R \underline{u}^{(r)}, \ \bar{u} = \wedge_{s=1}^S \bar{u}^{(s)}.$$

Suppose that for each $r = 1, \ldots, R$, each $s = 1, \ldots, S$, the vectors $\underline{u}^{(r)}, \bar{u}$ and $\underline{u}, \bar{u}^{(s)}$ form pairs of sub- and supersolutions of (5.190). Assume furthermore hypotheses (A1)–(A3) and the following growth condition:

(H') *For any $k \in \{1, \ldots, m\}$,*

$$|f_k(x, s, \eta)| \le \varrho_k(x) + c_1^{(k)} \sum_{k=1}^m |\eta_k|^{p-1},$$

for a.e. $x \in \Omega$, for all $s \in [\wedge_{r=1}^R \underline{u}^{(r)}(x), \vee_{s=1}^S \bar{u}^{(s)}(x)]$, and all $\eta \in \mathbb{R}^{mN}$, where $\varrho_k \in L_+^q(\Omega)$.

If $\underline{u}^{(r)}, \bar{u}^{(s)}$ and K satisfy the lattice assumption

$$\underline{u}^{(r)} \vee K \subset K, \ \bar{u}^{(s)} \wedge K \subset K, \ r = 1, \ldots, R, \ s = 1, \ldots, S, \tag{5.217}$$

(i.e., $\underline{u}_k^{(r)} \vee K_k \subset K_k$ and $\bar{u}_k^{(s)} \wedge K_k \subset K_k$, for $k = 1, \ldots, m$, $r = 1, \ldots, R$, $s = 1, \ldots, S$), then the system (5.190) has solutions within $[\underline{u}, \bar{u}]$.

The proof follows the same line as that of Theorem 5.55 and is omitted. We refer to [50] for more details.

We now consider some properties of the set of solutions within the interval $[\underline{u}, \bar{u}]$, whose verifications could be done as in Theorem 5.8 and Theorem 5.13.

Theorem 5.57. *Under the conditions of Theorem 5.55 or Theorem 5.56, within the interval $[\underline{u}, \bar{u}]$ a maximal solution u^* and a minimal solution u_* exist in a set theoretical sense; that is, u^* and u_* are solutions of (5.190),*

$$\underline{u} \leq u_*, u^* \leq \bar{u},$$

and if u is a solution of (5.190) within $[\underline{u}, \bar{u}]$ and $u \geq u^*$ (respectively, $u \leq u_*$), then $u = u^*$ (respectively, $u = u_*$).

5.5.5 Weakly Coupled Systems and Extremal Solutions

Note that different from the scalar case, solutions of systems of equations or inequalities are not, in general, sub- or supersolutions. However, this property holds if certain monotonicity conditions are imposed. We assume in the sequel that the lower order term F does not depend on ∇u. As above, we assume that $\underline{u} = \vee_{r=1}^R \underline{u}^{(r)}$ and $\bar{u} = \wedge_{s=1}^S \bar{u}^{(s)}$ are as in the assumptions of Theorem 5.56 and that f_k ($k = 1, \ldots, m$) are quasi-monotone in the following sense:

$$f_k(x, u) \leq f_k(x, u_k, [w]_k), \tag{5.218}$$

for a.e. $x \in \Omega$, all $u, w \in \mathbb{R}^m$ such that $[u]_k \geq [w]_k$, or rather,

$$f_k(x, u(x)) \leq f_k(x, u_k(x), [w(x)]_k), \tag{5.219}$$

for a.e. $x \in \Omega$, all $u, w \in X$ such that $\underline{u} \leq u, w \leq \bar{u}$ and $[u]_k \geq [w]_k$. Under condition (5.218) (or (5.219)), we have the following result.

Lemma 5.58. *If u is a solution of (5.190) within the interval $[\underline{u}, \bar{u}]$ such that*

$$u \wedge K \subset K \text{ (respectively, } u \vee K \subset K\text{),} \tag{5.220}$$

then $u, \bar{u}$ (respectively, $\underline{u}, u$) form a pair of sub-supersolution for (5.190).

Proof: Let $s = 1, \ldots, S$. It is clear that u and $\bar{u}^{(s)}$ satisfy conditions (i) and (iii) in Definition 5.53. Let us check condition (ii) in that definition. Because u is a solution of (5.190),

$$\langle A_k u_k + F_k(u), \tilde{v}_k - u_k \rangle \geq 0, \ \forall \tilde{v}_k \in K_k. \tag{5.221}$$

For any $v_k \in K_k$, by choosing

$$\tilde{v}_k = u_k - (u_k - v_k)^+ = u_k \wedge v_k \in K_k,$$

in (5.221), we get

$$\langle A_k u_k + F_k(u), (u_k - v_k)^+ \rangle \leq 0.$$

Also, according to (5.219), if $[w]_k \geq [u]_k$, then

$$\langle F_k(u), (u_k - v_k)^+ \rangle \geq \langle F_k(u_k, [w]_k), (u_k - v_k)^+ \rangle.$$

Thus,

$$\langle A_k u_k + F_k(u_k, [w]_k), (u_k - v_k)^+ \rangle \leq \langle A_k u_k + F_k(u), (u_k - v_k)^+ \rangle \leq 0.$$

We have condition (ii) in Definition 5.53. Similarly, we can prove that $\underline{u}^{(r)}, u$ form a pair of sub-supersolution of (5.190) by verifying condition (iii) in Definition 5.53. □

Now we can prove an extremality result of the solution set S enclosed by $\underline{u}$ and $\bar{u}$.

Theorem 5.59. *Under assumption (5.218) [or (5.219)] and the following lattice condition*

$$K \wedge K \subset K, \ K \vee K \subset K \tag{5.222}$$

(that is, $K_k \wedge K_k \subset K_k$, $K_k \vee K_k \subset K_k$, for $k = 1, \ldots, m$), or rather

$$\{v \in K : v \geq \underline{u}\} \wedge K \subset K, \ \{v \in K : v \leq \bar{u}\} \vee K \subset K,$$

the set S of solutions of (5.190) between $\underline{u}$ and $\bar{u}$ has greatest and smallest elements u^* and u_*, respectively, with respect to the ordering $\leq$; i.e., for any $u \in S$, we have $u_* \leq u \leq u^*$.

Proof: Let u^* be a maximal element of S, whose existence is proved previously in section 5.5.4. We show that u^* is in fact the greatest element of S; that is, $u \leq u^*$ for all $u \in S$. Assume otherwise that there is $v \in S$ such that $v \not\leq u^*$. From the above lemma, $u^*, \bar{u}$ and $v, \bar{u}$ are pairs of sub-supersolutions of (5.190). Condition (5.222) implies that $u^* \vee K \subset K$, $v \vee K \subset K$. The growth condition (H') of Theorem 5.56 also holds for $u^*, \bar{u}$ and $v, \bar{u}$ because it is satisfied for $\underline{u}, \bar{u}$. According to Theorem 5.56, a solution $\tilde{u}$ of (5.190) exists such that

$$(\underline{u} \leq) u^* \vee v \leq \tilde{u} \leq \bar{u}.$$

Hence, $\tilde{u} \in S$. As $\tilde{u} \geq u^*$, we must have $u^* = \tilde{u} \geq v$. This result contradicts the choice of v and proves that u^* is the greatest element of S. The existence of the smallest element u_* of S is established in the same way. □

Note the difference between the notion of minimal and maximal solutions (see Theorem 5.57) on the one hand and of smallest and greatest solutions (see Theorem 5.59) on the other hand. As for the existence of extremal solutions, i.e., greatest and smallest solutions, for general quasilinear (scalar) elliptic variational equalities and inequalities, we refer to Chap. 3 and Chap. 4, and e.g., [35, 36, 43, 141, 146, 142].

Let us conclude this section with some remarks on particular cases of the above definitions and results and a simple example. When $m = 1$, the system (5.190) becomes a single (scalar) variational inequality and the definitions above reduce to those in [141] (see previous sections).

In the case where $K_k = V_0$ ($k = 1, \ldots, m$), (5.190) becomes a system of variational equations. In this case, Definition 5.53 reduces to the usual definition of sub- and supersolutions in equations (see Chap. 3 and cf. [83] for the scalar case and [40, 111, 134] for systems). The proof of this definition coherence follows the same lines as in the case of a single variational inequality in Section 5.1.

Example 5.60. To illustrate the above concepts and assumptions, we consider a system of obtacle problems. Let us consider problem (5.190) with K_k given by

$$K_k = \{v_k \in V_0 : v_k \leq g_k \text{ a.e. on } G_k, \ v_k \geq h_k \text{ a.e. on } H_k\},$$

where G_k, H_k are measurable subsets of Ω and $g_k : G_k \to [-\infty, +\infty]$, $h_k : H_k \to [-\infty, +\infty]$ are given measurable functions. Assume that $K_k \neq \emptyset$ for $k = 1, \ldots, m$. It is clear that K_k are closed, convex subsets of V_0. Also, condition (5.222) always holds for K. Let $u_k \in V_0$. Then $u_k \vee K_k \subset K_k$ if $u_k \leq g_k$ on G_k and $u_k \wedge K_k \subset K_k$ if $u_k \geq h_k$ on H_k.

5.6 Notes and Comments

The solvability of noncoercive variational inequalities has been studied extensively recently by various methods, such as bifurcation, recession, variational, and topological/fixed point approaches (cf. [131, 190, 207, 137] (bifurcation methods), [12, 10, 7] (recession arguments), [162, 211, 143] (variational approaches), [209, 210] (topological/fixed point methods), and the extensive references therein). Chapter 5 is about another way to study the solvability of noncoercive variational inequalities, that of sub- and supersolutions. As discussed, compared with the other methods, this approach when applicable (i.e., when sub- and supersolutions exist) usually permits more flexible requirements on the growth rate of the perturbing term $F(x, u, \nabla u)$. Moreover, based on the lattice structure of the spaces $W^{1,p}(\Omega)$ and $W_0^{1,p}(\Omega)$, the sub- and supersolution method could also give insight into properties of the solution set between the sub- and supersolutions such as its compactness or directedness and, especially, the existence of *extremal* solutions.

The sub-supersolution method for classic (or strong) solutions of equations, motivated by the well-known Perron arguments on sub- and superharmonic functions, were used by Nagumo, Akô, Sattinger ([176, 4, 202]) to study the solvability of quasilinear and semilinear equations and systems. The method was extended later by Bebernes and Schmitt ([15]) to parabolic equations with perturbing terms also depending on the gradient of the unknown function. The existence of extremal solutions was also established in [4, 202, 15].

The sub-supersolution argument was later employed by Deuel and Hess ([83, 84]) to study the existence of weak solutions of equations in variational forms. The existence of weak extremal solutions of nonlinear equations was considered recently by Dancer/Sweers, Kura, Carl, Heikkilä, Lakshmikan-tham, Papageorgiou, Le/Schmitt, and others (cf. [75, 133, 29, 41, 46, 110, 183, 146, 138] and the references therein; see also Chap. 3 and Chap. 4). All of these cited papers are concerned with solutions of various equations. From the symmetric structure of equations (i.e., the equality of left- and right-hand sides) in both classic and weak formulations, subsolutions and supersolutions for equations are defined in a natural and straightforward manner, by replacing the equality by the corresponding inequalities.

The situation is, however, different for variational inequalities. Because of the intrinsic asymmetry of variational inequalities (where the problems are stated as inequalities rather than equalities), it is more difficult to define sub- and supersolutions for variational inequalities. An attempt in this direction has been made in [184], where a class of parabolic variational inequalities is studied. However, because of the presence of unilateral constraints (and/or obstacles) and the asymmetry of the problem, the authors only defined supersolutions for these inequalities. As a consequence, the existence of minimal solutions (but not maximal solutions) of those inequalities is established in [184], in cases where supersolutions exists. Another restriction in [184] is that the supersolutions are assumed to be in the convex sets $K(t)$; hence, they must have a zero boundary condition, which is somewhat restrictive, compared with the usual requirements on supersolutions in equations (in equations with zero Dirichlet boundary condition, supersolutions are only assumed to be nonnegative on the boundary).

In this chapter, we proposed a systematic investigation of the subsolution-supersolution method in variational inequalities. Despite the nonsymmetric structure of variational inequalities, we showed that both supersolutions and subsolutions could be defined in an appropriate manner, which naturally extends the corresponding concepts in equations. Moreover, the presence of unilateral constraints and obstacles in variational inequalities does not, in many interesting situations, preclude the existence of both sub- and supersolutions. Consequently, this finding permits us to establish the existence of both smallest and greatest solutions in inequalities and other interesting properties of solution sets such as their directedness or compactness. Moreover, under some growth condition from above (respectively, from below) on the perturbing functions $F(x, u, \nabla u)$, we can show that the existence of subsolutions (respectively, supersolutions) alone would imply the solvability and then the existence of both smallest and greatest solutions. However, we note that the sub-supersolution methods are presented here only in function spaces with some lattice structure [such as $W^{1,p}(\Omega)$]. That is the reason why we apply the method here mainly to problems with second-order operators.

Variational inequalities similar to (5.122) [or (5.137)] were studied in [175] by Rothe's method (see also [121]). In addition to coercivity, smoothness conditions are usually required for the coefficients of the principal operators and lower order terms. These smoothness conditions are relaxed here, and if sub-supersolutions exist, we also have existence in noncoercive cases.

6

Hemivariational Inequalities

Hemivariational inequalities have been introduced by P. D. Panagiotopoulos (see [179, 180]) to describe, e.g., problems in mechanics and engineering governed by nonconvex, possibly nonsmooth energy functionals (so-called superpotentials). This kind of energy functionals appear if nonmonotone, possibly multivalued constitutive laws are taken into account. Hemivariational inequalities are of the following abstract setting:

Let $A : B \to B^*$ be a pseudomonotone and coercive operator from a reflexive Banach space B into its dual B^*, and let $f \in B^*$ be some given element. Find $u \in B$ such that

$$\langle Au - f, v \rangle + J^o(u; v) \geq 0 \quad \text{for all } v \in B, \tag{6.1}$$

where $J^o(u; v)$ denotes the generalized directional derivative in the sense of Clarke of a locally Lipschitz functional $J : B \to \mathbb{R}$ (see Chap. 2, Sect. 2.5). An equivalent multivalued formulation of (6.1) is given by

$$u \in B : \quad -Au + f \in \partial J(u) \quad \text{in } B^*, \tag{6.2}$$

where $\partial J(u) : B \to 2^{B^*} \setminus \{\emptyset\}$ denotes Clarke's generalized gradient (see Chap. 2, Sect. 2.5). Abstract existence results for (6.1) [respectively, (6.2)] can be found in [177].

In particular, if $J : B \to \mathbb{R}$ is convex (note that $\text{dom}(J) = B$), then ∂J coincides with the usual subdifferential in convex analysis, and therefore, problem (6.2) reduces to the variational inequality:

$$u \in B : \quad \langle Au - f, v - u \rangle + J(v) - J(u) \geq 0, \quad \text{for all } v \in B. \tag{6.3}$$

In this sense, hemivariational inequalities are a generalization of variational inequalities. To indicate this finding and for a later treatment of hemivariational inequalities under constraints (see Chap. 7), we will write (6.1) in the following equivalent form:

$$u \in B : \quad \langle Au - f, v - u \rangle + J^o(u; v - u) \geq 0 \quad \text{for all } v \in B. \tag{6.4}$$

In this chapter, we deal with concrete realizations of (6.1) and its corresponding dynamic counterpart in the form

$$\frac{\partial u}{\partial t} + Au - f \in \partial J(u), \tag{6.5}$$

where A is assumed to be a quasilinear (in the case of (6.5) also time-dependent) elliptic differential operator of Leray–Lions type, J is some integral functional, and B is a subspace of the Sobolev space $W^{1,p}(\Omega)$. More precisely, we are going to consider integral functionals J of the form

$$J(u) = \int_\Omega j(x, u(x)) \, dx, \tag{6.6}$$

where the function $s \mapsto j(x, s)$ is assumed to be locally Lipschitz, and $x \mapsto j(x, s)$ is measurable in Ω with Ω being some bounded Lipschitz domain in $\mathbb{R}^N$. Other integral functionals defined on some portion of the boundary $\partial \Omega$ whose Clarke's gradient representing certain multivalued flux boundary conditions will also be considered in Chap. 7. Assuming a suitable growth condition on $s \mapsto \partial j(x, s)$ to be specified later, we can show that $J : L^p(\Omega) \to \mathbb{R}$ is locally Lipschitz and that the following relation holds:

$$J^o(u; v) \le \int_\Omega j^o(x, u(x); v(x)) \, dx, \tag{6.7}$$

where $j^o(\cdot, s; r)$ denotes the generalized directional derivative of j at the point s in the direction r. Motivated by the weak solution of elliptic (parabolic) boundary value problems, and taking into account (6.7), we further relax the problem (6.1) in that we are going to treat the problem

$$u \in B : \quad \langle Au - f, v \rangle + \int_\Omega j^o(\cdot, u; v) \, dx \ge 0 \quad \text{for all } v \in B. \tag{6.8}$$

Obviously, any solution of (6.1) is also a solution of (6.8) (in the above specified situation of integral functionals). The reverse, in general, is not true, and it only holds under additional assumptions on j such as to require j to be regular in the sense of Clarke (see Sect. 2.5). Hemivariational inequalities of the form (6.8) are closely related to and in fact generalize problems for differential inclusions considered in Chap. 4, because any solution of

$$u \in B : \quad Au - f + \partial j(\cdot, u) \ni 0 \quad \text{in } B^* \tag{6.9}$$

is also a solution of (6.8). To prove the latter, we only need to apply the definition of Clarke's generalized gradient. In the special case that j is given by the primitive of a locally bounded function $g : \mathbb{R} \to \mathbb{R}$ satisfying some growth condition, problem (6.9) reduces to the inclusion

$$u \in B : \quad Au - f + [\underline{g}(u), \bar{g}(u)] \ni 0 \quad \text{in } B^*. \tag{6.10}$$

In this sense, the hemivariational inequality (6.8) may be considered as a generalization of the inclusion problems in Chap. 4 and, thus, in turn of boundary value problems with discontinuous nonlinearities.

The goal of this chapter is to establish comparison principles for hemivariational inequalities in the form (6.8) and its evolutionary counterpart, which are based on an appropriate notion for sub-solutions. As hemivariational inequalities with integral functionals include differential inclusions considered in Chap. 4 as a special case, the new notion of sub-supersolution should be compatible with the corresponding notion for inclusions. In fact we will see that if Clarke's generalized gradient of j satisfies certain global growth conditions, then hemivariational inequalities and their corresponding inclusion problems are equivalent. The new technique to be developed here will allow us to study hemivariational inequalities under additional restrictions represented, in general, by some convex, lower semicontinuous functionals $\Phi : B \to \mathbb{R} \cup \{+\infty\}$, which leads to the subject of variational–hemivariational inequalities (see Chap. 7).

6.1 Notion of Sub-Supersolution

In this section, we provide a motivation for the new notion of sub-supersolution for hemivariational inequalities in the form (6.8) with the help of a simple example.

Let $\Omega \subset \mathbb{R}^N$ be a bounded domain with Lipschitz boundary $\partial\Omega$, and let $V = W^{1,2}(\Omega)$ and $V_0 = W_0^{1,2}(\Omega)$ denote the usual Sobolev spaces. Consider the hemivariational inequality

$$u \in V_0 : \quad \langle -\Delta u - f, v - u \rangle + \int_\Omega j^\circ(u; v - u)\, dx \geq 0, \quad \forall\, v \in V_0, \quad (6.11)$$

where $f \in V_0^*$ is given, and the function $j : \mathbb{R} \to \mathbb{R}$ is supposed to satisfy the following structure and growth condition.

(H) The function $j : \mathbb{R} \to \mathbb{R}$ is locally Lipschitz and its Clarke's generalized gradient ∂j satisfies the following growth conditions:

(i)   A constant $c_1 \geq 0$ exists such that

$$\xi_1 \leq \xi_2 + c_1(s_2 - s_1)$$

for all $\xi_i \in \partial j(s_i)$, $i = 1, 2$, and for all s_1, s_2 with $s_1 < s_2$.

(ii) A constant $c_2 \geq 0$ exists such that

$$\xi \in \partial j(s) : \quad |\xi| \leq c_2(1 + |s|), \quad \forall\, s \in \mathbb{R}.$$

In addition, consider the inclusion problem

$$-\Delta u + \partial j(u) \ni f \quad \text{in } \Omega, \quad u = 0 \quad \text{on } \partial\Omega. \qquad (6.12)$$

When $j : \mathbb{R} \to \mathbb{R}$ is regular in the sense of Clarke (see Sect. 2.5), problem (6.11) can equivalently be written in the form

$$u \in V_0 : \quad -\Delta u + \partial(J \circ i)(u) \ni f \quad \text{in } V_0^*, \tag{6.13}$$

where $i : V_0 \to L^2(\Omega)$ is the embedding operator and $J : L^2(\Omega) \to \mathbb{R}$ is the integral functional given by

$$J(u) = \int_\Omega j(u(x)) \, dx. \tag{6.14}$$

From hypothesis (H)(ii) and as j has been assumed to be regular in the sense of Clarke, it follows that $J : L^2(\Omega) \to \mathbb{R}$ is locally Lipschitz, and moreover, we have

$$\eta \in \partial J(u) \iff \eta \in L^2(\Omega) \text{ and } \eta(x) \in \partial j(u(x)) \text{ for a.e. } x \in \Omega. \tag{6.15}$$

The characterization of Clarke's gradient (6.15) in conjunction with the chain rule for Clarke's gradient immediately imply that (6.11) and (6.12) are equivalent, where (6.12) has to be understood in the generalized sense; i.e., u is a solution of (6.12) if the following is satisfied:

$$u \in V_0 : \quad -\Delta u + \eta = f \quad \text{in } V_0^*, \tag{6.16}$$

where $\eta \in L^2(\Omega)$ satisfies $\eta(x) \in \partial j(u(x))$.

To develop a proper notion of sub-supersolution for the hemivariational inequality (6.11), we have to take into account that it should be compatible with the corresponding notion for sub-supersolutions for inclusion problems given in Chap. 4. So our point of departure will be the notion of sub-supersolution for the inclusion (6.12). Let $\underline{u} \in V$ be a subsolution of (6.12); i.e.,

(i) $\underline{u} \leq 0$ on $\partial\Omega$.
(ii)   There is a $\underline{\eta} \in L^2(\Omega)$ with $\underline{\eta}(x) \in \partial j(\underline{u}(x))$ for a.e. $x \in \Omega$.
(iii) $\langle -\Delta\underline{u} - f, \varphi \rangle + \int_\Omega \underline{\eta}\varphi \, dx \leq 0, \quad \forall \varphi \in V_0 \cap L_+^2(\Omega)$.

Inequality (iii) is, in particular, satisfied for $\varphi = (\underline{u} - \psi)^+ \in V_0 \cap L_+^2(\Omega)$ with $\psi \in V_0$, which yields

$$\langle -\Delta\underline{u} - f, -(\underline{u} - \psi)^+ \rangle + \int_\Omega \underline{\eta}\left(-(\underline{u} - \psi)^+\right) dx \geq 0, \quad \forall \psi \in V_0. \tag{6.17}$$

As $v = \underline{u} - (\underline{u} - \psi)^+ = \underline{u} \wedge \psi$, we see that (6.17) is equivalent with

$$\langle -\Delta\underline{u} - f, v - \underline{u} \rangle + \int_\Omega \underline{\eta}\,(v - \underline{u}) \, dx \geq 0, \quad \forall v \in \underline{u} \wedge V_0. \tag{6.18}$$

By definition of Clarke's gradient, we have

$$j^\circ(\underline{u}; v - \underline{u}) \geq \underline{\eta}\,(v - \underline{u}),$$

and thus (6.18) results in the following inequality:

$$\langle -\Delta \underline{u} - f, v - \underline{u} \rangle + \int_\Omega j^\circ(\underline{u}; v - \underline{u}) \, dx \geq 0, \quad \forall \, v \in \underline{u} \wedge V_0. \tag{6.19}$$

It is relation (6.19), which leads to our basic notion of subsolution for the hemivariational inequality (6.11). Similar considerations can be done for the supersolution. Our new definition for sub-supersolution for (6.11) reads as follows.

Definition 6.1. *The function $\underline{u} \in V$ is a subsolution of (6.11) if it satisfies*

(i) $\underline{u} \leq 0$ *on* $\partial\Omega$.
(ii) $\langle -\Delta\underline{u} - f, v - \underline{u} \rangle + \int_\Omega j^\circ(\underline{u}; v - \underline{u}) \, dx \geq 0, \quad \forall \, v \in \underline{u} \wedge V_0$.

Definition 6.2. *The function $\bar{u} \in V$ is a supersolution of (6.11) if it satisfies*

(i) $\bar{u} \geq 0$ *on* $\partial\Omega$.
(ii) $\langle -\Delta\bar{u} - f, v - \bar{u} \rangle + \int_\Omega j^\circ(\bar{u}; v - \bar{u}) \, dx \geq 0, \quad \forall \, v \in \bar{u} \vee V_0$.

An immediate consequence of these definitions for sub-supersolutions of the hemivariational inequality (6.11) is the following corollary.

Corollary 6.3. *If u is a solution of (6.11), then it is both a subsolution and a supersolution of (6.11).*

The following result is crucial in the study of the relation between the hemivariational inequality (6.11) and the associated differential inclusion (6.12).

Theorem 6.4. *Assume $j : \mathbb{R} \to \mathbb{R}$ is locally Lipschitz and let the growth condition (H)(ii) be satisfied. Then $\underline{u}$ ($\bar{u}$) $\in V$ is a subsolution (supersolution) of (6.11) if and only if it is a subsolution (supersolution) of (6.12).*

Proof: As the definitions given above for sub- and supersolutions of (6.11) arise from the corresponding notions for the inclusion (6.12), we only need to prove that any subsolution (supersolution) of (6.11) is also a subsolution (supersolution) of (6.12). Let $\underline{u}$ be a subsolution of (6.11) according to Definition 6.1. As $v \in \underline{u} \wedge V_0$, it is of the form $v = \underline{u} \wedge \psi = \underline{u} - (\underline{u} - \psi)^+$ with $\psi \in V_0$, and thus, from the inequality (ii) of Definition 6.1, we obtain

$$\langle -\Delta\underline{u} - f, -(\underline{u} - \psi)^+ \rangle + \int_\Omega j^\circ(\underline{u}; -(\underline{u} - \psi)^+) \, dx \geq 0, \quad \forall \, \psi \in V_0. \tag{6.20}$$

Denoting $\varphi = (\underline{u} - \psi)^+ \in V_0 \cap L_+^2(\Omega)$ and taking into account the density result given by Lemma 5.4, the set $\{\varphi = (\underline{u} - \psi)^+ : \psi \in V_0\}$ is dense in $V_0 \cap L_+^2(\Omega)$, and thus we get

$$\langle -\Delta\underline{u} - f, \varphi \rangle - \int_\Omega j^\circ(\underline{u}; -1) \, \varphi \, dx \leq 0, \quad \forall \, \varphi \in V_0 \cap L_+^2(\Omega). \tag{6.21}$$

By the properties of the generalized Clarke's gradient, a function $\underline{\eta} : \Omega \to \mathbb{R}$ exists such that

$$\underline{\eta}(x) \in \partial j(\underline{u}(x)) \quad \text{and} \quad j^o(\underline{u}(x); -1) = (-1)\underline{\eta}(x) \quad \text{for a.e. } x \in \Omega. \quad (6.22)$$

We are going to show that $\underline{\eta}$ is in fact an element from $L^2(\Omega)$. The function $s \mapsto j^o(s; -1)$ is upper semicontinuous, and by hypothesis (H)(ii), it satisfies the growth condition

$$|j^o(s; -1)| \leq c_2(1 + |s|), \quad \forall \, s \in \mathbb{R}.$$

By applying general approximation results for lower (upper) semicontinuous functions in Hilbert spaces (see [8]), a sequence (j_n) of locally Lipschitz functions $j_n : \mathbb{R} \to \mathbb{R}$ exists which converge pointwise to $j^o(\cdot; -1)$; i.e.,

$$j_n(s) \to j^o(s; -1), \quad \forall \, s \in \mathbb{R}$$

as $n \to \infty$. As the functions j_n are superpositionally measurable, it follows that $s \mapsto j^o(s; -1)$ is superpositionally measurable as well, which means that the function $x \mapsto j^o(u(x); -1)$ is measurable whenever $u : \Omega \to \mathbb{R}$ is a measurable function. Thus, from (6.22), we infer that $\underline{\eta} = -j^o(\underline{u}; -1)$ is a measurable function, which from (H)(ii) satisfies

$$|\underline{\eta}(x)| \leq c_2(1 + |\underline{u}(x)|) \quad \text{for a.e. } x \in \Omega,$$

and therefore, $\underline{\eta} \in L^2(\Omega)$. In view of (6.21), the latter implies

$$\langle -\Delta\underline{u} - f, \varphi \rangle + \int_\Omega \underline{\eta}\,\varphi\,dx \leq 0, \quad \forall \, \varphi \in V_0 \cap L^2_+(\Omega),$$

and thus $\underline{u}$ is a subsolution for the inclusion (6.12). In a similar way, we can prove the result for the supersolution. $\qquad \Box$

Note that for Theorem 6.4 to be valid, j is not required to be regular in the sense of Clarke. We next will see that if in addition to the assumption of Theorem 6.4 j satisfies also (H)(i), then problems (6.11) and (6.12) are in fact equivalent, and comparison principles hold for both problems.

Theorem 6.5. *Assume $j : \mathbb{R} \to \mathbb{R}$ satisfies hypothesis (H). If $\underline{u}$ and $\bar{u}$ are sub- and supersolutions of (6.11) satisfying $\underline{u} \leq \bar{u}$, then solutions of (6.11) within $[\underline{u}, \bar{u}]$ exist, and the solution set S of all solutions of (6.11) in $[\underline{u}, \bar{u}]$ is compact and possesses extremal elements. Moreover, problems (6.11) and (6.12) are equivalent.*

Proof: We already know by just applying the definition of Clarke's gradient ∂j that any solution of the inclusion (6.12) is a solution of (6.11). Now, let u be a solution of (6.11); then u is both a sub- and supersolution of (6.11), and by Theorem 6.4, u is a sub- and supersolution of (6.12) as well. This result means there are pairs $(u, \underline{\eta})$ and $(u, \bar{\eta})$ such that $\underline{\eta} \in \partial j(u)$, $\bar{\eta} \in \partial j(u)$ and

$$-\Delta u + \underline{\eta} \le f, \quad -\Delta u + \bar{\eta} \ge f. \tag{6.23}$$

Note that $\underline{\eta}$ and $\bar{\eta}$ need not be the same, because $\underline{\eta}$ is chosen such that $j^{\circ}(u; -1) = \underline{\eta}(-1)$ and $\bar{\eta}$ is chosen such that $j^{\circ}(u; 1) = \bar{\eta} 1$. But $-j^{\circ}(u; -1 \ne j^{\circ}(u; 1)$, in general, so $\underline{\eta} \ne \bar{\eta}$. Now, from hypothesis (H), we may apply the comparison principle for the inclusion (6.12) (see Theorem 4.11 in Chap. 4), which implies the existence of a solution of (6.12) within the interval $[u, u] = \{u\}$; i.e., there is a $\eta \in L^2(\Omega)$ with $\eta \in \partial j(u)$ such that

$$-\Delta u + \eta = f, \tag{6.24}$$

which shows that the solution u of (6.11) is in fact a solution of (6.12). A further application of Theorem 4.11 in Chap. 4 completes the proof of the theorem. $\qquad \square$

Remark 6.6. An alternative proof of the equivalence of the problems (6.11) and (6.12) is based on Lemma 4.10 of Chap. 4, which states that under hypothesis (H)(i) given above, the function j admits a representation in the form

$$j(s) = \hat{j}(s) - \frac{c_1}{2} s^2, \tag{6.25}$$

where $\hat{j} : \mathbb{R} \to \mathbb{R}$ is a convex function. As convex functions are regular in the sense of Clarke, it follows from (6.25) that j must be regular in the sense of Clarke as well, which implies the equivalence of (6.11) and (6.12). We note that the equivalence of the two problems requires the growth conditions (H) (i) and (ii) on ∂j to be satisfied globally.

6.2 Quasilinear Elliptic Hemivariational Inequalities

Let $\Omega \subset \mathbb{R}^N$ be a bounded domain with Lipschitz boundary $\partial \Omega$, and let $V = W^{1,p}(\Omega)$ and $V_0 = W_0^{1,p}(\Omega)$, $1 < p < \infty$, denote the usual Sobolev spaces with their dual spaces V^* and V_0^*, respectively. In this section, we deal with the following quasilinear hemivariational inequality:

$$u \in V_0 : \quad \langle Au - f, v - u \rangle + \int_{\Omega} j^{\circ}(u; v - u) \, dx \ge 0, \quad \forall \, v \in V_0, \tag{6.26}$$

where $j^{\circ}(s; r)$ denotes the generalized directional derivative of the locally Lipschitz function $j : \mathbb{R} \to \mathbb{R}$ at s in the direction r, and the operator $A : V \to V_0^*$ is assumed to be a second-order quasilinear differential operator in divergence form

$$Au(x) = -\sum_{i=1}^{N} \frac{\partial}{\partial x_i} a_i(x, \nabla u(x)). \tag{6.27}$$

For j we assume the following hypothesis that is the analog to hypothesis (H) in the preceding section and again denoted by (H).

(H) The function $j : \mathbb{R} \to \mathbb{R}$ is locally Lipschitz, and its Clarke's generalized gradient ∂j satisfies the following growth conditions:
   (i) A constant $c_1 \geq 0$ exists such that

$$\xi_1 \leq \xi_2 + c_1(s_2 - s_1)^{p-1}$$

   for all $\xi_i \in \partial j(s_i)$, $i = 1, 2$, and for all s_1, s_2 with $s_1 < s_2$.
   (ii) A constant $c_2 \geq 0$ exists such that

$$\xi \in \partial j(s) : \quad |\xi| \leq c_2 \left(1 + |s|^{p-1}\right), \quad \forall\, s \in \mathbb{R}.$$

Basically, similar arguments as for problems (6.11) and (6.12) apply also here to show that under the hypothesis (H) of this section the hemivariational inequality (6.26) is equivalent with the differential inclusion

$$u \in V_0 : \quad Au + \partial j(u) \ni f \quad \text{in } V_0^*. \tag{6.28}$$

Only for the sake of demonstrating an alternative technique we prove the comparison principle for the hemivariational inequality (6.26) independently without reducing it to the inclusion (6.28). Without difficulties and using the tools developed in the previous chapters, the comparison results for (6.26) can be extended to hemivariational inequalities involving more general quasilinear elliptic operators of Leray–Lions type and functions $j : \Omega \times \mathbb{R} \to \mathbb{R}$ depending, in addition, on the space variable x.

6.2.1 Comparison Principle

We assume $f \in V_0^*$ and impose the following hypotheses of Leray–Lions type on the coefficient functions a_i, $i = 1, \ldots, N$, of the operator A:

(A1) Each $a_i : \Omega \times \mathbb{R}^N \to \mathbb{R}$ satisfies the Carathéodory conditions; i.e., $a_i(x, \xi)$ is measurable in $x \in \Omega$ for all $\xi \in \mathbb{R}^N$ and continuous in ξ for almost all $x \in \Omega$. A constant $c_0 > 0$ and a function $k_0 \in L^q(\Omega), 1/p + 1/q = 1$, exist such that
$$|a_i(x, \xi)| \leq k_0(x) + c_0 |\xi|^{p-1},$$
   for a.e. $x \in \Omega$ and for all $\xi \in \mathbb{R}^N$.
(A2) $\sum_{i=1}^{N}(a_i(x, \xi) - a_i(x, \xi'))(\xi_i - \xi_i') > 0$ for a.e. $x \in \Omega$, and for all $\xi, \xi' \in \mathbb{R}^N$ with $\xi \neq \xi'$.
(A3) $\sum_{i=1}^{N} a_i(x, \xi)\xi_i \geq \nu |\xi|^p - k_1(x)$ for a.e. $x \in \Omega$, and for all $\xi \in \mathbb{R}^N$ with some constant $\nu > 0$ and some function $k_1 \in L^1(\Omega)$.

From (A1), (A2), the semilinear form a associated with the operator A by

$$\langle Au, \varphi \rangle := a(u, \varphi) = \int_\Omega \sum_{i=1}^N a_i(x, \nabla u) \frac{\partial \varphi}{\partial x_i}\, dx, \quad \forall\, \varphi \in V_0$$

is well defined for any $u \in V$, and the operator $A : V_0 \to V_0^*$ is continuous, bounded, and monotone. For functions w, $z : \Omega \to \mathbb{R}$ and sets W and Z of functions defined on Ω, we use, as in Chap. 5, the notations: $w \wedge z = \min\{w, z\}$, $w \vee z = \max\{w, z\}$, $W \wedge Z = \{w \wedge z : w \in W, z \in Z\}$, $W \vee Z = \{w \vee z : w \in W, z \in Z\}$, and $w \wedge Z = \{w\} \wedge Z$, $w \vee Z = \{w\} \vee Z$. From Sect. 6.1, the analogous definitions for sub- and supersolutions for (6.26) read as follows.

Definition 6.7. *A function $\underline{u} \in V$ is called a subsolution of (6.26) if the following holds:*

(i) $\underline{u} \leq 0$ *on* $\partial\Omega$.
(ii) $\langle A\underline{u} - f, v - \underline{u} \rangle + \int_\Omega j^\circ(\underline{u}; v - \underline{u})\, dx \geq 0, \quad \forall\, v \in \underline{u} \wedge V_0.$

Definition 6.8. *$\bar{u} \in V$ is a supersolution of (6.26) if the following holds:*

(i) $\bar{u} \geq 0$ *on* $\partial\Omega$.
(ii) $\langle A\bar{u} - f, v - \bar{u} \rangle + \int_\Omega j^\circ(\bar{u}; v - \bar{u})\, dx \geq 0, \quad \forall\, v \in \bar{u} \vee V_0.$

Remark 6.9. In the same way as in Sect. 6.1, we can show that u is a subsolution (supersolution) of (6.26) if and only if u is a subsolution (supersolution) of (6.28). Under hypothesis (H) of this section, problems (6.26) and (6.28) are equivalent. Therefore, the hemivariational inequality (6.26) can be treated via the differential inclusions (6.28). As mentioned, we provide in the following an independent proof of comparison principles and related properties for (6.26) by using hemivariational formulation.

In the proof of the comparison principle, we make use of the cutoff function $b : \Omega \times \mathbb{R} \to \mathbb{R}$ related with an ordered pair of functions $\underline{u}$, $\bar{u}$, and given by

$$b(x, s) = \begin{cases} (s - \bar{u}(x))^{p-1} & \text{if } s > \bar{u}(x), \\ 0 & \text{if } \underline{u}(x) \leq s \leq \bar{u}(x), \\ -(\underline{u}(x) - s)^{p-1} & \text{if } s < \underline{u}(x). \end{cases} \tag{6.29}$$

We already know from previous chapters that b is a Carathéodory function satisfying the growth condition

$$|b(x, s)| \leq k_2(x) + c_3\, |s|^{p-1} \tag{6.30}$$

for a.e. $x \in \Omega$, for all $s \in \mathbb{R}$, with some function $k_2 \in L_+^q(\Omega)$. Moreover, we have the following estimate:

$$\int_\Omega b(x, u(x))\, u(x)\, dx \geq c_4\, \|u\|_{L^p(\Omega)}^p - c_5, \quad \forall\, u \in L^p(\Omega), \tag{6.31}$$

where c_4 and c_5 are some positive constants. In view of (6.30), the Nemytskij operator $B : L^p(\Omega) \to L^q(\Omega)$ defined by

$$Bu(x) = b(x, u(x))$$

is continuous and bounded, and thus from the compact embedding $V \subset L^p(\Omega)$, it follows that $B : V_0 \to V_0^*$ is completely continuous. Now we prove the following existence and comparison theorem.

Theorem 6.10. *Assume hypotheses (A1)–(A3), (H), and let $\bar{u}$ and $\underline{u}$ be super- and subsolutions of (6.26), respectively, satisfying $\underline{u} \leq \bar{u}$. Then solutions of (6.26) exist within the ordered interval $[\underline{u}, \bar{u}]$.*

Proof: Let us consider the auxiliary hemivariational inequality

$$u \in V_0 : \quad \langle Au - f + \lambda B(u), v - u \rangle + \int_\Omega j^o(u; v - u)\, dx \geq 0, \quad \forall\, v \in V_0, \tag{6.32}$$

where $\lambda \geq 0$ is some constant that is at our disposal and that will be specified later. We note that any solution $u \in [\underline{u}, \bar{u}]$ of (6.26) is a solution of (6.32), and any solution of (6.32) that is in $[\underline{u}, \bar{u}]$ solves (6.26). Let us introduce the function $J : L^p(\Omega) \to \mathbb{R}$ by

$$J(v) = \int_\Omega j(v(x))\, dx, \quad \forall\, v \in L^p(\Omega).$$

Using the growth condition (H)(ii) and Lebourg's mean value theorem, we note that the function J is well defined and Lipschitz continuous on bounded sets in $L^p(\Omega)$, thus locally Lipschitz. Moreover, the Aubin–Clarke theorem (see Theorem 2.181, Chap. 2) ensures that, for each $u \in L^p(\Omega)$, we have

$$\xi \in \partial J(u) \implies \xi \in L^q(\Omega) \text{ with } \xi(x) \in \partial j(u(x)) \text{ for a.e. } x \in \Omega.$$

Consider now the multivalued operator $F : V_0 \to 2^{V_0^*}$ defined by

$$F(v) = Av + \lambda B(v) + \partial(J|_{V_0})(v), \quad \forall\, v \in V_0,$$

where $J|_{V_0}$ denotes the restriction of J to V_0. We readily verify that the operator $A + \lambda B : V_0 \to V_0^*$ is continuous, bounded, strictly monotone, and thus, in particular, pseudomonotone. By Lemma 4.16 of Chap. 4, the multivalued operator $\partial(J|_{V_0}) : V_0 \to 2^{V_0^*}$ is bounded and pseudomonotone in the sense of Definition 2.120. Lemma 4.18 of Chap. 4 holds likewise also for the operator F defined above, which states that $F : V_0 \to 2^{V_0^*}$ is bounded, pseudomonotone, and coercive for $\lambda > 0$ sufficiently large. Thus by Theorem 2.125 of Chap. 2, it follows that F is surjective; i.e., $u \in V_0$ exists such that $f \in F(u)$; i.e., there is an $\xi \in \partial J(u)$ such that $\xi \in L^q(\Omega)$ with $\xi(x) \in \partial j(u(x))$ for a.e. $x \in \Omega$ and

$$Au - f + \lambda B(u) + \xi = 0 \quad \text{in } V_0^*, \tag{6.33}$$

where

$$\langle \xi, \varphi \rangle = \int_{\Omega} \xi(x)\, \varphi(x)\, dx \quad \text{for all } \varphi \in V_0, \tag{6.34}$$

and thus by definition of Clarke's generalized gradient ∂j, from (6.34), we get

$$\langle \xi, \varphi \rangle = \int_{\Omega} \xi(x)\, \varphi(x)\, dx \leq \int_{\Omega} j^{\circ}(u(x); \varphi(x))\, dx \quad \text{for all } \varphi \in V_0. \tag{6.35}$$

From (6.33) and (6.35), we conclude that $u \in V_0$ is a solution of the auxiliary hemivariational inequality (6.32). To complete the proof, we only need to show that there are solutions of (6.32) that belong to the interval $[\underline{u}, \bar{u}]$. In fact, we are going to prove that any solution u of (6.32) belongs to this interval.

Let us show: $u \leq \bar{u}$. By definition $\bar{u}$ satisfies $\bar{u} \geq 0$ on $\partial \Omega$ and

$$\langle A\bar{u} - f, v - \bar{u} \rangle + \int_{\Omega} j^{\circ}(\bar{u}; v - \bar{u})\, dx \geq 0, \quad \forall\, v \in \bar{u} \vee V_0.$$

In view of $v = \bar{u} \vee \varphi = \bar{u} + (\varphi - \bar{u})^{+}$ with $\varphi \in V_0$, this implies the following inequality:

$$\langle A\bar{u} - f, (\varphi - \bar{u})^{+} \rangle + \int_{\Omega} j^{\circ}(\bar{u}; (\varphi - \bar{u})^{+})\, dx \geq 0, \quad \forall\, \varphi \in V_0. \tag{6.36}$$

Taking in (6.32) the special test function $v = u - (u - \bar{u})^{+}$ and $\varphi = u$ in (6.36) and adding the resulting inequalities, we obtain

$$\langle Au - A\bar{u}, (u - \bar{u})^{+} \rangle + \lambda \langle B(u), (u - \bar{u})^{+} \rangle$$
$$\leq \int_{\Omega} \Big(j^{\circ}(\bar{u}; (u - \bar{u})^{+}) + j^{\circ}(u; -(u - \bar{u})^{+}) \Big)\, dx. \tag{6.37}$$

Next we estimate the right-hand side of (6.37) by using the facts from nonsmooth analysis (cf. Chap. 2).

The function $r \mapsto j^{\circ}(s; r)$ is finite and positively homogeneous; $\partial j(s)$ is a nonempty, convex and compact subset of $\mathbb{R}$; and we have

$$j^{\circ}(s; r) = \max\{\xi\, r : \xi \in \partial j(s)\}.$$

By using (H) and the properties of j° and ∂j, we get for certain $\bar{\xi}(x) \in \partial j(\bar{u}(x))$ and $\xi(x) \in \partial j(u(x))$, the following estimate:

$$\int_{\Omega} \Big(j^{\circ}(\bar{u}; (u - \bar{u})^{+}) + j^{\circ}(u; -(u - \bar{u})^{+}) \Big)\, dx$$
$$= \int_{\{u > \bar{u}\}} \Big(j^{\circ}(\bar{u}; u - \bar{u}) + j^{\circ}(u; -(u - \bar{u})) \Big)\, dx$$
$$= \int_{\{u > \bar{u}\}} \Big(\bar{\xi}(x)(u(x) - \bar{u}(x)) + \xi(x)(-(u(x) - \bar{u}(x))) \Big)\, dx$$

$$= \int\limits_{\{u>\bar{u}\}} (\bar{\xi}(x) - \xi(x))(u(x) - \bar{u}(x)) \, dx$$

$$\leq \int\limits_{\{u>\bar{u}\}} c_1 \left(u(x) - \bar{u}(x)\right)^p \, dx. \tag{6.38}$$

As $\langle Au - A\bar{u}, (u - \bar{u})^+ \rangle \geq 0$ and

$$\langle B(u), (u - \bar{u})^+ \rangle = \int\limits_{\{u>\bar{u}\}} (u - \bar{u})^p \, dx,$$

we get from (6.37) and (6.38) the estimate

$$(\lambda - c_1) \int\limits_{\{u>\bar{u}\}} (u - \bar{u})^p \, dx \leq 0. \tag{6.39}$$

Selecting the free parameter $\lambda \geq 0$ in such a way that $\lambda - c_1 > 0$ then (6.39) yields

$$\int_\Omega ((u - \bar{u})^+)^p \, dx \leq 0,$$

which implies $(u - \bar{u})^+ = 0$ and thus $u \leq \bar{u}$. The proof for the inequality $\underline{u} \leq u$ can be carried out in a similar way that completes the proof of the theorem.   □

6.2.2 Extremal Solutions and Compactness Results

Let $\mathcal{S}$ denote the set of all solutions of (6.26) within the interval $[\underline{u}, \bar{u}]$ of the ordered pair of sub- and supersolutions $\underline{u}$ and $\bar{u}$ of problem (6.26). In this section, we are going to show that $\mathcal{S}$ possesses the smallest and greatest element with respect to the given partial ordering.

Lemma 6.11. *The solution set $\mathcal{S}$ is a directed set.*

Proof: By Theorem 6.10 we have $\mathcal{S} \neq \emptyset$. Given $u_1, u_2 \in \mathcal{S}$ we shall show that there is a $u \in \mathcal{S}$ such that $u_k \leq u$, $k = 1, 2$, which means $\mathcal{S}$ is upward directed. To this end, we consider the following auxiliary hemivariational inequality:

$$u \in V_0: \quad \langle Au - f + \lambda B(u), v - u \rangle + \int_\Omega j^\circ(u; v - u) \, dx \geq 0, \quad \forall \, v \in V_0, \tag{6.40}$$

where $\lambda \geq 0$ is a free parameter to be chosen later, but unlike in the proof of Theorem 6.10, the operator B is now given by the following cutoff function $b : \Omega \times \mathbb{R} \to \mathbb{R}$:

$$b(x,s) = \begin{cases} (s - \bar{u}(x))^{p-1} & \text{if } s > \bar{u}(x), \\ 0 & \text{if } u_0(x) \le s \le \bar{u}(x), \\ -(u_0(x) - s)^{p-1} & \text{if } s < u_0(x), \end{cases} \quad (6.41)$$

where $u_0 = \max\{u_1, u_2\}$. By similar arguments as in the proof of Theorem 6.10, we obtain the existence of solutions of (6.40). The set $\mathcal{S}$ is shown to be upward directed provided that any solution u of (6.40) satisfies $u_k \le u \le \bar{u}$, $k = 1, 2$, because then $Bu = 0$ and thus $u \in \mathcal{S}$ exceeding u_k. By assumption $u_k \in \mathcal{S}$, which means $u_k \in [\underline{u}, \bar{u}]$ satisfies

$$u_k \in V_0 : \quad \langle Au_k - f, v - u_k \rangle + \int_\Omega j^\circ(u_k; v - u_k)\, dx \ge 0, \quad \forall\, v \in V_0. \quad (6.42)$$

Taking the special functions $v = u + (u_k - u)^+$ in (6.40) and $v = u_k - (u_k - u)^+$ in (6.42) and adding the resulting inequalities, we obtain

$$\langle Au_k - Au, (u_k - u)^+ \rangle - \lambda \langle B(u), (u_k - u)^+ \rangle$$
$$\le \int_\Omega \Big(j^\circ(u; (u_k - u)^+) + j^\circ(u_k; -(u_k - u)^+) \Big)\, dx. \quad (6.43)$$

Similarly to (6.38), we get for the right-hand side of (6.43) the estimate

$$\int_\Omega \Big(j^\circ(u; (u_k - u)^+) + j^\circ(u_k; -(u_k - u)^+) \Big)\, dx$$
$$\le \int_{\{u_k > u\}} c_1\, (u_k(x) - u(x))^p\, dx. \quad (6.44)$$

For the terms on the left-hand side of (6.43), we have

$$\langle Au_k - Au, (u_k - u)^+ \rangle \ge 0 \quad (6.45)$$

and (6.41) yields

$$\langle B(u), (u_k - u)^+ \rangle = -\int_{\{u_k > u\}} (u_0(x) - u(x))^{p-1}(u_k(x) - u(x))\, dx$$
$$\le -\int_{\{u_k > u\}} (u_k(x) - u(x))^p\, dx. \quad (6.46)$$

By means of (6.44)–(6.46), we get from (6.43) the inequality

$$(\lambda - c_1) \int_{\{u_k > u\}} (u_k(x) - u(x))^p\, dx \le 0. \quad (6.47)$$

Selecting λ such that $\lambda > c_1$ from (6.47), we obtain $u_k \le u$. The proof for $u \le \bar{u}$ follows similar arguments, and thus, $\mathcal{S}$ is upward directed. By obvious modifications of the auxiliary problem, we can show analogously that $\mathcal{S}$ is also downward directed. $\qquad\square$

Lemma 6.12. *The solution set S is compact in V_0.*

Proof: First we prove that S is bounded in V_0. As any $u \in S$ belongs to the interval $[\underline{u}, \bar{u}]$, it follows that S is bounded in $L^p(\Omega)$. Moreover, any $u \in S$ solves (6.26), i.e., we have

$$u \in V_0 : \quad \langle Au - f, v - u \rangle + \int_\Omega j^o(u; v - u)\, dx \geq 0, \quad \forall\, v \in V_0,$$

and thus by taking $v = 0$, we obtain

$$\langle Au, u \rangle \leq \langle f, u \rangle + \int_\Omega j^o(u; -u)\, dx,$$

which yields by applying (A3), (H)(ii), and Young's inequality

$$\nu \|\nabla u\|^p_{L^p(\Omega)} \leq \|k_1\|_{L^1(\Omega)} + c(\varepsilon) \|f\|^q_{V_0^*} + \varepsilon \|u\|^p_{V_0} + \tilde{\alpha} \left(\|u\|_{L^p(\Omega)} + \|u\|^p_{L^p(\Omega)} \right),$$

with a constant $\tilde{\alpha} > 0$, for any $\varepsilon > 0$, and hence, the boundedness of S in V_0 follows by choosing ε sufficiently small [Note: S is bounded in $L^p(\Omega)$].

Let $(u_n) \subset S$. Then there is a subsequence (u_k) of (u_n) with

$$u_k \rightharpoonup u \text{ in } V_0, \quad u_k \to u \text{ in } \quad L^p(\Omega), \quad \text{and} \quad u_k(x) \to u(x) \text{ a.e. in } \Omega. \quad (6.48)$$

Obviously $u \in [\underline{u}, \bar{u}]$. As u_k solve (6.26) we get with $v = u$ in (6.26),

$$\langle Au_k, u_k - u \rangle \leq \langle f, u_k - u \rangle + \int_\Omega j^o(u_k; u - u_k)\, dx. \quad (6.49)$$

From (6.48) and because $(s, r) \mapsto j^o(s; r)$ is upper semicontinuous, we get by applying Fatou's lemma

$$\limsup_k \int_\Omega j^o(u_k; u - u_k)\, dx \leq \int_\Omega \limsup_k j^o(u_k; u - u_k)\, dx = 0. \quad (6.50)$$

In view of (6.50), we thus obtain from (6.48) and (6.49) the relation

$$\limsup_k \langle Au_k, u_k - u \rangle \leq 0. \quad (6.51)$$

As the operator A enjoys the (S_+)-property, the weak convergence of (u_k) in V_0 along with (6.51) imply the strong convergence $u_k \to u$ in V_0. Moreover, the limit u belongs to S as can be seen by passing to the $\limsup$ on the left-hand side of the following inequality:

$$\langle Au_k - f, v - u_k \rangle + \int_\Omega j^o(u_k; v - u_k)\, dx \geq 0, \quad (6.52)$$

where we have used Fatou's lemma and the strong convergence of (u_k) in V_0. $\qquad \square$

By means of Lemma 6.11 and Lemma 6.12, we can prove the following extremality result.

Theorem 6.13. *The solution set $\mathcal{S}$ possesses extremal elements.*

Proof: We show the existence of the greatest element of $\mathcal{S}$. As V_0 is separable, we have that $\mathcal{S} \subset V_0$ is separable too, so a countable, dense subset $Z = \{z_n : n \in \mathbb{N}\}$ of $\mathcal{S}$ exists. By Lemma 6.11, $\mathcal{S}$ is upward directed, so we can construct an increasing sequence $(u_n) \subset \mathcal{S}$ as follows. Let $u_1 = z_1$. Select $u_{n+1} \in \mathcal{S}$ such that

$$\max\{z_n, u_n\} \leq u_{n+1} \leq \overline{u}.$$

The existence of u_{n+1} is from Lemma 6.11. By Lemma 6.12, we find a subsequence of (u_n), denoted again (u_n), and an element $u \in \mathcal{S}$ such that $u_n \to u$ in V_0, and $u_n(x) \to u(x)$ a.e. in Ω. This last property of (u_n) combined with its increasing monotonicity implies that the entire sequence is convergent in V_0, and moreover, $u = \sup_n u_n$. By construction, we see that

$$\max\{z_1, z_2, \ldots, z_n\} \leq u_{n+1} \leq u, \quad \forall n;$$

thus, $Z \subset [\underline{u}, u]$. As the interval $[\underline{u}, u]$ is closed in V_0, we infer

$$\mathcal{S} \subset \overline{Z} \subset \overline{[\underline{u}, u]} = [\underline{u}, u],$$

which in conjunction with $u \in \mathcal{S}$ ensures that u is the greatest solution. The existence of the least solution of (1.1) can be proved in a similar way.   □

6.2.3 Application

Let us assume throughout this subsection the assumptions (A1)–(A3) for A and (H) for the function $j : \mathbb{R} \to \mathbb{R}$ as before. Here we are going to deal with the following quasilinear hemivariational inequality with a multivalued right-hand side:

$$u \in V_0, \ \eta \in \partial\Psi(u) : \ \langle Au - f, v - u \rangle + \int_\Omega j^\circ(u; v - u)\, dx \geq \langle \eta, v - u \rangle,$$
$$(6.53)$$

for all $v \in V_0$, where $\partial\Psi(u)$ is the subdifferential of the continuous and convex functional $\Psi : L^p(\Omega) \to \mathbb{R}$ given by

$$\Psi(u) = \int_\Omega \left(\int_0^{u(x)} h(\tau)\, d\tau \right) dx, \qquad (6.54)$$

with $h : \mathbb{R} \to \mathbb{R}$ being some monotone nondecreasing (not necessarily continuous) function satisfying a certain growth condition specified later. If ψ denotes the primitive of h given by

$$\psi(s) = \int_0^s h(\tau)\, d\tau, \qquad (6.55)$$

then $\psi : \mathbb{R} \to \mathbb{R}$ is continuous (but not necessarily smooth) and convex from the monotonicity of h, and its subdifferential is given by

$$\partial\psi(s) = [\underline{h}(s), \bar{h}(s)], \qquad (6.56)$$

where $\underline{h}$ and $\bar{h}$ denote the left-sided and right-sided limits of h. Furthermore, the following characterization of the subdifferential $\partial\Psi(u)$ holds:

$$\eta \in \partial\Psi(u) \iff \eta \in L^q(\Omega) \text{ and } \eta(x) \in \partial\psi(x) \text{ for a.e. } x \in \Omega. \qquad (6.57)$$

We denote by $\underline{H}$ and $\bar{H}$ the Nemytskij operators associated with $\underline{h}$ and $\bar{h}$, respectively, and define sub- and supersolutions for problem (6.53) as follows.

Definition 6.14. *The function $\underline{u} \in V$ is called a subsolution of (6.53) if the following holds:*

(i) $\underline{u} \le 0$ *on* $\partial\Omega$.
(ii) $\langle A\underline{u} - f, v - \underline{u} \rangle + \int_\Omega j^o(\underline{u}; v - \underline{u}) \, dx \ge \langle \underline{H}(\underline{u}), v - \underline{u} \rangle, \quad \forall v \in \underline{u} \wedge V_0.$

Definition 6.15. *The function $\bar{u} \in V$ is a supersolution of (6.53) if the following holds:*

(i) $\bar{u} \ge 0$ *on* $\partial\Omega$.
(ii) $\langle A\bar{u} - f, v - \bar{u} \rangle + \int_\Omega j^o(\bar{u}; v - \bar{u}) \, dx \ge \langle \bar{H}(\bar{u}), v - \bar{u} \rangle, \quad \forall v \in \bar{u} \vee V_0.$

Remark 6.16. Under the hypothesis (H) for j, the hemivariational inequality (6.53) can be shown to be equivalent to the following inclusion problem:

$$u \in V_0: \quad Au + \partial j(u) - \partial\psi(u) \ni f \quad \text{in } V_0^*. \qquad (6.58)$$

This equivalence can be seen as follows. Let $u \in V_0$ be a solution of (6.58); i.e., there are functions $\xi \in L^q(\Omega)$ and $\eta \in L^q(\Omega)$ satisfying $\xi(x) \in \partial j(u(x))$ and $\eta(x) \in \partial\psi(u(x))$ for a.e. $x \in \Omega$ and

$$Au + \xi - \eta = f \quad \text{in } V_0^*.$$

By definition of Clarke's generalized gradient, we have

$$\int_\Omega j^o(u; v - u) \, dx \ge \int_\Omega \xi(v - u) \, dx,$$

and thus,

$$\langle Au - f, v - u \rangle + \int_\Omega j^o(u; v - u) \, dx \ge \langle \eta, v - u \rangle,$$

where $\eta \in L^q(\Omega)$ is an element of $\partial\Psi(u)$ from (6.57). This result proves that any solution of (6.58) is also a solution of (6.53). As for the reverse, let $u \in V_0$ be a solution of (6.53) i.e., there is an element $\eta \in \partial\Psi(u)$ such that inequality (6.53) holds. Again by (6.57), it follows that $\eta \in L^q(\Omega)$ and $\eta(x) \in \partial\psi(u(x))$

for a.e. $x \in \Omega$. If we set $\tilde{f} = f + \eta$, then u satisfies the hemivariational inequality

$$u \in V_0 : \quad \langle Au - \tilde{f}, v - u \rangle + \int_\Omega j^\circ(u; v - u) \, dx \geq 0,$$

for all $v \in V_0$. Now, under hypothesis (H), we conclude in just the same way as in the proof of Theorem 6.5 that u satisfies the inclusion

$$Au + \partial j(u) \ni \tilde{f},$$

which means that there is a function $\xi \in L^q(\Omega)$ such that $\xi(x) \in \partial j(u(x))$ and

$$Au + \xi = \tilde{f} \quad \text{in } V_0^*.$$

In view of the definition of $\tilde{f}$, the last equation shows that u is a solution of the inclusion (6.58). Elliptic problems governed by the difference of multifunctions have been treated already in Sect. 4.4, where j has been assumed to be the primitive of some locally bounded function without assuming sub-supersolutions. Here we continue the study of this kind of problems by using alternative techniques within the framework of hemivariational inequalities.

We make the following assumption on h:

(H-h) The function $h : \mathbb{R} \to \mathbb{R}$ is monotone nondecreasing and satisfies

$$|h(s)| \leq c_3 \left(1 + |s|^{p-1}\right), \quad \forall s \in \mathbb{R}.$$

The main result for problem (6.53) reads as follows.

Theorem 6.17. *Let hypotheses (A1)–(A3) and (H)–(H-h) be satisfied, and let $\underline{u}$ and $\bar{u}$ be sub- and supersolutions of (6.53) with $\underline{u} \leq \bar{u}$. Then the hemivariational inequality (6.53) possesses extremal solutions within the order interval $[\underline{u}, \bar{u}]$, and the solution set of all solutions of (6.53) within $[\underline{u}, \bar{u}]$ is a compact subset in V_0.*

Proof: Step 1: Existence of Extremal Solutions.

The hypothesis (H-h) ensures that the functional $\Psi : L^p(\Omega) \to \mathbb{R}$ is well defined, convex, and locally Lipschitz continuous, and so it is the restriction of Ψ to V_0 denoted $\Psi|_{V_0}$. As $V_0 \subset L^p(\Omega)$ is densely embedded, we get in view of (6.57) the following characterization of the subgradients of $\partial(\Psi|_{V_0})(u)$ from [64, Theorem 2.3]:

$$\eta \in \partial(\Psi|_{V_0})(u) \iff \eta \in L^q(\Omega) \text{ with } \eta(x) \in \partial\psi(u(x)) \text{ for a.e. } x \in \Omega. \tag{6.59}$$

In view of (6.56), the inclusion on the right-hand side is equivalent with $\eta \in [\underline{H}(u), \bar{H}(u)]$. Therefore, u is a solution of (6.53) if the following holds: $u \in V_0$, and a $\eta \in L^q(\Omega)$ with $\eta \in [\underline{H}(u), \bar{H}(u)]$ exists such that

$$\langle Au - f, v - u \rangle + \int_\Omega j^\circ(u; v - u) \, dx \geq \langle \eta, v - u \rangle, \quad \forall \, v \in V_0. \tag{6.60}$$

Let us consider the following hemivariational inequality related with (6.60):

$$u \in V_0 : \langle Au - f, v - u \rangle + \int_\Omega j^\circ(u; v - u) \, dx \geq \langle \bar{H}(u), v - u \rangle, \tag{6.61}$$

for all $v \in V_0$. Note that in view of (H-h), the Nemytskij operator $\bar{H} : L^p(\Omega) \to L^q(\Omega)$ is well defined, but not necessarily continuous, which makes the treatment of (6.61) more difficult. We are going to show that (6.61) has the greatest solution u^* within the interval $[\underline{u}, \bar{u}]$, and that u^* is at the same time the greatest solution of the original problem (6.53) within $[\underline{u}, \bar{u}]$. To this end, we consider first the following hemivariational inequality with given right-hand side $\bar{H}(\bar{u}) \in L^q(\Omega) \subset V_0^*$:

$$u \in V_0 : \langle Au - f, v - u \rangle + \int_\Omega j^\circ(u; v - u) \, dx \geq \langle \bar{H}(\bar{u}), v - u \rangle, \tag{6.62}$$

for all $v \in V_0$. By hypothesis $\bar{u}$ is a supersolution of (6.53) and thus, in particular, a supersolution of (6.62). Because of $\underline{H}(\underline{u}) \leq \bar{H}(\underline{u}) \leq \bar{H}(\bar{u})$, we readily can see that the given subsolution of (6.53) is also a subsolution of (6.62). Therefore, we may apply Theorem 6.13 with f replaced by $f + \bar{H}(\bar{u}) \in V_0^*$, which ensures the existence of extremal solutions of (6.62) within $[\underline{u}, \bar{u}]$. We denote by u_1 the greatest solution of (6.62) within $[\underline{u}, \bar{u}]$ and consider next the hemivariational inequality with $\bar{H}(\bar{u})$ replaced by $\bar{H}(u_1)$; i.e.,

$$u \in V_0 : \langle Au - f, v - u \rangle + \int_\Omega j^\circ(u; v - u) \, dx \geq \langle \bar{H}(u_1), v - u \rangle, \tag{6.63}$$

for all $v \in V_0$. As $u_1 \leq \bar{u}$, we get $\bar{H}(u_1) \leq \bar{H}(\bar{u})$, which shows that u_1 is a supersolution of (6.63). Furthermore, in view of $\underline{u} \leq u_1$, we have $\underline{H}(\underline{u}) \leq \bar{H}(\underline{u}) \leq \bar{H}(u_1)$, and this implies that $\underline{u}$ is also a subsolution of (6.63). Again by applying Theorem 6.13, extremal solutions of (6.63) within $[\underline{u}, u_1]$ exist, and we denote the greatest one by u_2. Continuing this process, we get by induction the following iteration: $u_0 = \bar{u}$, and $u_{n+1} \in [\underline{u}, u_n]$ is the greatest solution of

$$u \in V_0 : \langle Au - f, v - u \rangle + \int_\Omega j^\circ(u; v - u) \, dx \geq \langle \bar{H}(u_n), v - u \rangle, \tag{6.64}$$

for all $v \in V_0$, which yields a monotone nonincreasing sequence (u_n) satisfying

$$\underline{u} \leq \cdots \leq u_{n+1} \leq u_n \leq \cdots \leq u_1 \leq u_0 = \bar{u}, \tag{6.65}$$

and (6.64) with u replaced by u_{n+1}; i.e., we have

$$u_{n+1} \in V_0 : \langle Au_{n+1} - f, v - u_{n+1} \rangle + \int_\Omega j^\circ(u_{n+1}; v - u_{n+1}) \, dx$$

$$\geq \langle \bar{H}(u_n), v - u_{n+1} \rangle, \tag{6.66}$$

for all $v \in V_0$. From (6.65), the sequence (u_n) is $L^p(\Omega)$-bounded, which implies that the sequence $(\bar{H}(u_n))$ is $L^q(\Omega)$-bounded, and thus from (6.66), we get by taking $v = 0$ the following estimate:

$$\langle Au_{n+1}, u_{n+1} \rangle \leq \langle \bar{H}(u_n) + f, u_{n+1} \rangle + \int_\Omega j^o(u_{n+1}; -u_{n+1}) \, dx. \tag{6.67}$$

As $j^o(s;r) = \max\{\zeta r : \zeta \in \partial j(s)\}$, from (6.67), we get by using (A3) and (H)(ii), the boundedness of (u_n) in V_0; i.e.,

$$\|u_n\|_{V_0} \leq c, \quad \forall \, n. \tag{6.68}$$

The boundedness (6.68) and the monotonicity of the sequence (u_n) as well as the compact embedding $V_0 \subset L^p(\Omega)$ imply the following convergence properties:

(i) $u_n(x) \to u^*(x)$ a.e. in Ω.
(ii) $u_n \to u^*$ in $L^p(\Omega)$.
(iii) $u_n \rightharpoonup u^*$ in V_0.

Replacing v in (6.66) by u^*, we get

$$\langle Au_{n+1}, u_{n+1} - u^* \rangle \leq \int_\Omega j^o(u_{n+1}; u^* - u_{n+1}) \, dx + \langle f + \bar{H}(u_n), u_{n+1} - u^* \rangle. \tag{6.69}$$

As $(s,r) \mapsto j^o(s;r)$ is upper semicontinuous, Fatou's lemma yields

$$\limsup_n \int_\Omega j^o(u_n; u^* - u_n) \, dx \leq \int_\Omega \limsup_n j^o(u_n; u^* - u_n) \, dx = 0. \tag{6.70}$$

From (6.69), (6.70), the boundedness of $(\bar{H}(u_n))$ in $L^q(\Omega)$ and the convergence properties (i)–(iii) above, we obtain

$$\limsup_n \langle Au_n, u_n - u^* \rangle \leq 0. \tag{6.71}$$

Hypotheses (A1)–(A3) imply that the operator A enjoys the (S_+)-property (see Chap. 2), which in view of (iii) and (6.71) yields the strong convergence

(iv) $u_n \to u^*$ in V_0.

Furthermore, because the function $s \mapsto \bar{h}(s)$ is monotone nondecreasing and right-sided continuous, we get by means of Lebesgue's dominated convergence theorem and the a.e. monotone pointwise convergence of the sequence (u_n)

$$\int_\Omega \bar{H}(u_n) \, v \, dx \to \int_\Omega \bar{H}(u^*) \, v \, dx, \quad \forall \, v \in L^p(\Omega); \tag{6.72}$$

that is, $\bar{H}(u_n) \rightharpoonup \bar{H}(u^*)$ in $L^q(\Omega)$, which from the compact embedding $L^q(\Omega) \subset V_0^*$ results in

(v) $\bar{H}(u_n) \to \bar{H}(u^*)$ in V_0^*.

Thus, taking into account the convergencies (i)–(v) and passing to the lim sup in (6.66), we arrive at

$$u^* \in V_0 : \langle Au^* - f, v - u^* \rangle + \int_\Omega j^o(u^*; v - u^*) \, dx \geq \langle \bar{H}(u^*), v - u^* \rangle,$$

$$(6.73)$$

for all $v \in V_0$, which shows that $u^* \in [\underline{u}, \bar{u}]$ is a solution of (6.61). Moreover, u^* is also the greatest solution of (6.61) within $[\underline{u}, \bar{u}]$. To see this, let $u \in [\underline{u}, \bar{u}]$ be any solution of (6.61). Because (6.61) holds, in particular, for all $v \in u \wedge V_0$; i.e., v is of the form $v = u - (u - w)^+$ with $w \in V_0$, and as $\bar{H}(u) \leq \bar{H}(\bar{u})$, we infer that u is a subsolution of (6.62). Replacing $\underline{u}$ by u in the iteration above and noticing that the iterates u_n are defined as the greatest solutions, the same iterates as before satisfy $u \leq u_n \leq \bar{u}$ for all n, and thus, $u \leq u^*$, which proves that u^* is the greatest solution of (6.61) within $[\underline{u}, \bar{u}]$. If we set $\eta^* = \bar{H}(u^*)$, then, in particular, $\eta^* \in [\underline{H}(u^*), \bar{H}(u^*)]$ and we have

$$u^* \in V_0 : \langle Au^* - f, v - u^* \rangle + \int_\Omega j^o(u^*; v - u^*) \, dx \geq \langle \eta^*, v - u^* \rangle, \quad (6.74)$$

for all $v \in V_0$, which proves that u^* is a solution of (6.53). Finally, we shall show that u^* is the greatest solution of (6.53) in $[\underline{u}, \bar{u}]$ as well. To this end, let $\tilde{u}$ be any solution of (6.53) in $[\underline{u}, \bar{u}]$, which means that there is an $\tilde{\eta} \in \partial\Psi(\tilde{u}) = [\underline{H}(\tilde{u}), \bar{H}(\tilde{u})]$ such that (6.53) holds with u and η replaced by $\tilde{u}$ and $\tilde{\eta}$, respectively. As $\tilde{\eta} \leq \bar{H}(\tilde{u}) \leq \bar{H}(\bar{u})$, similar arguments as above imply that $\tilde{u}$ is a subsolution of (6.62), which by interchanging the role of $\underline{u}$ and $\tilde{u}$ yields the following inequality for the iterates $(u_n) : \tilde{u} \leq u_n \leq \bar{u}$, and thus, we get $\tilde{u} \leq u^*$, which proves that u^* is the greatest solution of (6.53) within $[\underline{u}, \bar{u}]$. By similar reasoning, the existence of the smallest solution u_* can be proved.

Step 2: Compactness of the Solution Set.

Let $\mathcal{S} \subset [\underline{u}, \bar{u}]$ denote the set of all solutions of (6.53) within $[\underline{u}, \bar{u}]$, and let $(u_n) \subset \mathcal{S}$ be any sequence; i.e., we have: There are $\eta_n \in \partial\Psi(u_n) = [\underline{H}(u_n), \bar{H}(u_n)]$ such that

$$u_n \in V_0 : \langle Au_n - f, v - u_n \rangle + \int_\Omega j^o(u_n; v - u_n) \, dx \geq \langle \eta_n, v - u_n \rangle, \quad (6.75)$$

for all $v \in V_0$. The sequence (u_n) is bounded in $L^p(\Omega)$, and hence, (η_n) is bounded in $L^q(\Omega)$ in view of (H), which by similar reasoning as in Step 1 implies the boundedness of (u_n) in V_0. Thus, subsequences (u_k) and (η_k) of (u_n) and (η_n) exist, respectively, satisfying

(1) $u_k(x) \to u(x)$ a.e. in Ω,
(2) $u_k \to u$ in $L^p(\Omega)$,
(3) $u_k \rightharpoonup u$ in V_0,

(4) $\eta_k \rightharpoonup \eta$ in $L^q(\Omega)$,

where $\eta_k \in \partial\Psi(u_k)$, and $\eta \in \partial\Psi(u)$. By means of the convergence properties (1)–(4), we get similarly as in Step 1 the following:

$$\limsup_k \langle Au_k, u_k - u \rangle \leq 0,$$

which by the (S_+)-property of A implies the strong convergence

(5) $u_k \to u$ in V_0.

Passing to the limit in (6.75) with u_n replaced by (u_k) as $k \to \infty$ shows that the limit u belongs to $\mathcal{S}$. $\qquad\square$

In the next section, we shall continue our treatment of hemivariational inequalities with evolutionary hemivariational inequalities.

6.3 Evolutionary Hemivariational Inequalities

In this section, we are going to study the evolutionary counterpart to the elliptic hemivariational inequality treated in Sect. 6.2. The problem we are dealing with is the following evolutionary hemivariational inequality:

$$u \in W_0, \ u(\cdot, 0) = 0 \text{ in } \Omega,$$

$$\langle u' + Au - f, v - u \rangle + \int_Q j^\circ(u; v - u) \, dxdt \geq 0, \quad \forall\, v \in X_0, \quad (6.76)$$

where $Q = \Omega \times (0, \tau)$, $X_0 = L^p(0, \tau; V_0)$, with $V_0 = W_0^{1,p}(\Omega)$, and $W_0 = \{w \in X_0 : w' \in X_0^*\}$. Throughout this section, we make use of the same notations of the evolutionary framework used in the study of parabolic inclusions in Sect. 4.5 of Chap. 4. As in Sect. 4.5, we will assume $2 \leq p < \infty$, and the same assumptions (A1)–(A4) for the operator A, which are given here again only for convenience.

(A1) $a_i : Q \times \mathbb{R} \times \mathbb{R}^N \to \mathbb{R}$ are Carathéodory functions; i.e., $a_i(\cdot, \cdot, s, \xi) : Q \to \mathbb{R}$ is measurable for all $(s, \xi) \in \mathbb{R} \times \mathbb{R}^N$ and $a_i(x, t, \cdot, \cdot) : \mathbb{R} \times \mathbb{R}^N \to \mathbb{R}$ is continuous for a.e. $(x, t) \in Q$. In addition, we have

$$|a_i(x, t, s, \xi)| \leq k_0(x, t) + c_0 \left(|s|^{p-1} + |\xi|^{p-1} \right)$$

for a.e. $(x, t) \in Q$ and for all $(s, \xi) \in \mathbb{R} \times \mathbb{R}^N$, for some constant $c_0 > 0$ and some function $k_0 \in L^q(Q)$.

(A2) $\displaystyle\sum_{i=1}^N (a_i(x, t, s, \xi) - a_i(x, t, s, \xi'))(\xi_i - \xi_i') > 0$ for a.e. $(x, t) \in Q$, for all $s \in \mathbb{R}$ and all $\xi, \xi' \in \mathbb{R}^N$ with $\xi \neq \xi'$.

(A3) $\displaystyle\sum_{i=1}^{N} a_i(x,t,s,\xi)\xi_i \geq \nu|\xi|^p - k_1(x,t)$ for a.e. $(x,t) \in Q$ and for all $(s,\xi) \in$
$\mathbb{R} \times \mathbb{R}^N$, for some constant $\nu > 0$ and some function $k_1 \in L^1(Q)$.

(A4) $|a_i(x,t,s,\xi) - a_i(x,t,s',\xi)| \leq [k_2(x,t)+|s|^{p-1}+|s'|^{p-1}+|\xi|^{p-1}]\omega(|s-s'|)$
for a.e. $(x,t) \in Q$, for all $s, s' \in \mathbb{R}$ and all $\xi \in \mathbb{R}^N$, for some function
$k_2 \in L^q(Q)$ and a continuous function $\omega : [0,+\infty) \to [0,+\infty)$ satisfying

$$\int_{0+} \frac{1}{\omega(r)}\, dr = +\infty.$$

As in Sect. 4.5, let $L := \partial/\partial t$ and its domain of definition $D(L)$ given by

$$D(L) = \{u \in X_0 : u' \in X_0^* \text{ and } u(\cdot,0) = 0 \text{ in } \Omega\}.$$

Thus, the evolutionary hemivariational inequality (6.76) may be rewritten as

$$u \in D(L) : \langle Lu + A(u) - f, v - u\rangle + \int_Q j^\circ(u; v - u)\, dx dt \geq 0, \quad \forall\, v \in X_0.$$
$$(6.77)$$

As mentioned in Remark 4.55, the case of quasilinear parabolic inclusions with global growth conditions on Clarke's generalized gradient of j can effectively be treated within the framework of evolutionary hemivariational inequalities in the form (6.77). We note that only by applying the definition of Clarke's generalized gradient ∂j and without any additional assumtions, we readily see that any solution of the parabolic inclusion

$$u \in D(L): \quad Lu + Au + \partial j(u) \ni f \quad \text{in } X_0^* \qquad (6.78)$$

(as for parabolic inclusions, see Chap. 4) is also a solution of the hemivariational inequality (6.77). The reverse, in general, is not true. However, like in the elliptic case, we will see that under the following (global) assumption (H) on j the inclusion (6.78) and the hemivariational problem (6.77) are in fact equivalent.

(H) The function $j : \mathbb{R} \to \mathbb{R}$ is locally Lipschitz and its Clarke's generalized gradient ∂j satisfies the following growth conditions:
(i) A constant $c_1 \geq 0$ exists such that

$$\xi_1 \leq \xi_2 + c_1(s_2 - s_1)^{p-1}$$

for all $\xi_i \in \partial j(s_i)$, $i = 1, 2$, and for all $s_1, s_2 \in \mathbb{R}$ with $s_1 < s_2$.
(ii) A constant $c_2 \geq 0$ exists such that

$$\xi \in \partial j(s): \quad |\xi| \leq c_2(1 + |s|^{p-1}), \quad \forall\, s \in \mathbb{R}.$$

6.3.1 Sub-Supersolutions and Equivalence of Problems

A motivation for the notion of sub-supersolution for the hemivariational inequality (6.77) can be given in a similar way as in Sect. 6.1 for the elliptic case. For example, let $\bar{u}$ be a supersolution of the inclusion (6.78); i.e., $\bar{u} \in W$, and there is a function $\eta \in L^q(Q)$ such that $\bar{u}(\cdot, 0) \geq 0$ in Ω, $\bar{u} \geq 0$ on $\Gamma = \partial\Omega \times (0, \tau)$, $\eta(x, t) \in \partial j(\bar{u}(x, t))$ and the following inequality holds:

$$\langle \bar{u}' + A\bar{u} - f, \varphi \rangle + \int_Q \eta(x, t)\varphi(x, t)\, dxdt \geq 0, \quad \forall\, \varphi \in X_0 \cap L^p_+(Q). \quad (6.79)$$

(Note: $W = \{u \in X : u' \in X^*\}$ and $X = L^p(0, \tau; V)$ with $V = W^{1,p}(\Omega)$.)
Thus (6.79), in particular, holds for φ in the form $\varphi = (w - \bar{u})^+$, for any $w \in X_0$, which yields by applying the definition of Clarke's generalized gradient the following inequality:

$$\langle \bar{u}' + A\bar{u} - f, (w - \bar{u})^+ \rangle + \int_Q j^\circ(\bar{u}; (w - \bar{u})^+)\, dxdt \geq 0, \quad \forall\, w \in X_0. \quad (6.80)$$

As $\bar{u} \vee w = \bar{u} + (w - \bar{u})^+$, from (6.80), we get

$$\langle \bar{u}' + A\bar{u} - f, v - \bar{u} \rangle + \int_Q j^\circ(\bar{u}; v - \bar{u})\, dxdt \geq 0, \quad \forall\, v \in \bar{u} \vee X_0. \quad (6.81)$$

Similar arguments can be applied for the subsolution. This process leads to the following notion of sub-supersolution for the hemivariational inequality (6.77).

Definition 6.18. *A function $\underline{u} \in W$ is called a subsolution of (6.77) if the following holds:*

(i) $\underline{u}(\cdot, 0) \leq 0$ *in Ω, $\underline{u} \leq 0$ on Γ.*
(ii) $\langle \underline{u}' + A\underline{u} - f, v - \underline{u} \rangle + \int_Q j^\circ(\underline{u}; v - \underline{u})\, dxdt \geq 0, \quad \forall\, v \in \underline{u} \wedge X_0.$

Definition 6.19. *A function $\bar{u} \in W$ is a supersolution of (6.77) if the following holds:*

(i) $\bar{u}(\cdot, 0) \geq 0$ *in Ω, $\bar{u} \geq 0$ on Γ.*
(ii) $\langle \bar{u}_t + A\bar{u} - f, v - \bar{u} \rangle + \int_Q j^\circ(\bar{u}; v - \bar{u})\, dxdt \geq 0, \quad \forall\, v \in \bar{u} \vee X_0.$

An immediate consequence of these definitions is that any solution of (6.77) is both a subsolution and a supersolution of (6.77) in the sense of the definitions above. Next, we will see that the global growth condition (H)(ii) is enough to prove that sub-supersolutions of (6.77) are also sub-supersolutions of (6.78) and, thus, the equivalence of the notions of sub-supersolution for (6.77) and (6.78).

Theorem 6.20. *Let $j : \mathbb{R} \to \mathbb{R}$ be locally Lipschitz and satisfy condition (H) (ii). Then $\underline{u}\,(\bar{u}) \in W$ is a subsolution (supersolution) of (6.77) if and only if it is a subsolution (supersolution) of (6.78).*

Proof: In fact we only need to show that any subsolution (supersolution) of the hemivariational inequality (6.77) is also a subsolution (supersolution) of the inclusion (6.78), because the reverse has been proved above. Let $\underline{u}$ be a subsolution of (6.77) according to Definition 6.18. As $v \in \underline{u} \wedge X_0$ can be written in the form $v = \underline{u} - (\underline{u} - \psi)^+$ with $\psi \in X_0$, we get

$$\langle \underline{u}' + A\underline{u} - f, -(\underline{u} - \psi)^+ \rangle + \int_Q j^\circ(\underline{u}; -(\underline{u} - \psi)^+) \, dx dt \geq 0, \quad \forall \, \psi \in X_0.$$

$$(6.82)$$

By the density result of Lemma 5.36, we get from (6.82)

$$\langle \underline{u}' + A\underline{u} - f, \phi \rangle - \int_Q j^\circ(\underline{u}; -1)\phi \, dx dt \leq 0, \quad \forall \, \phi \in X_0 \cap L^p_+(Q). \quad (6.83)$$

The properties of Clarke's generalized gradient imply the existence of some function $\eta : Q \to \mathbb{R}$ such that

$$\eta(x, t) \in \partial j(\underline{u}(x, t)) \quad \text{and} \quad j^\circ(\underline{u}(x, t); -1) = \eta(x, t)\,(-1).$$

Similar arguments as in the proof of Theorem 6.4 apply to ensure that $s \mapsto j^\circ(s; -1)$ is superpositionally measurable, which implies that

$$\eta(x, t) = -j^\circ(\underline{u}(x, t); -1), \quad (x, t) \in Q,$$

is measurable. Taking the growth condition (H) (ii) of ∂j into account, we infer that $\eta \in L^q(Q)$, which in view of (6.83) shows that $\underline{u}$ is in fact a subsolution of the inclusion (6.78). The proof for the supersolution is analogous and can be omitted. $\qquad \Box$

Now we will show that if j satisfies, in addition, (H)(i), then the two problems (6.77) and (6.78) are in fact equivalent.

Theorem 6.21. *Assume $j : \mathbb{R} \to \mathbb{R}$ fulfills hypothesis (H). Then u is a solution of the hemivariational inequality (6.77) if and only if u is a solution of the inclusion (6.78).*

Proof: We already know that [even without assumption (H)] any solution of the inclusion (6.78) solves also (6.77). Thus, let us prove the reverse, and assume u is a solution of (6.77). As u is trivially both a subsolution and a supersolution of (6.77), it must be also a subsolution and a supersolution of (6.78) from Theorem 6.20. For u to be a subsolution of (6.78) means that there is an $\underline{\eta} \in L^q(Q)$ such that

$$\underline{\eta} \in \partial j(u) : \quad u' + Au + \underline{\eta} \leq f \quad \text{in } X_0^* \qquad (6.84)$$

(note: u satisfies homogeneous initial and boundary conditions). Similarly, for u to be a supersolution of (6.78) means that there is an $\bar{\eta} \in L^q(Q)$ such that

$$\bar{\eta} \in \partial j(u): \quad u' + Au + \bar{\eta} \geq f \quad \text{in } X_0^*. \tag{6.85}$$

Note that η and $\bar{\eta}$ need not be the same. The hypothese (A1)–(A3) and (H) allow us to apply the comparison principle for parabolic inclusions given by Theorem 4.46 of Sect. 4.5 in Chap. 4, which ensures the existence of solutions between an ordered pair of sub- and supersolutions. Therefore, a solution of (6.78) exists within the trivial interval $[u, u] = \{u\}$, which means the existence of an $\eta \in L^q(Q)$ such that

$$\eta \in \partial j(u): \quad u' + Au + \eta = f \quad \text{in } X_0^*, \tag{6.86}$$

where in view of inequalities (6.84) and (6.85), we get, in addition, $\eta \in [\underline{\eta}, \bar{\eta}]$. This process completes the proof.   □

6.3.2 Existence and Comparison Results

In this section we are going to establish existence and comparison results for the hemivariational inequality (6.77) based on the the notion of sub-supersolution. In preparation of our main result, we will first provide some preliminaries used later.

Consider the function $J : L^p(Q) \to \mathbb{R}$ defined by

$$J(v) = \int_Q j(v(x,t)) \, dx dt, \quad \forall \, v \in L^p(Q). \tag{6.87}$$

Using the growth condition (H)(ii) and Lebourg's mean value theorem, we note that the function J is well defined and Lipschitz continuous on bounded sets in $L^p(Q)$, thus locally Lipschitz so that Clarke's generalized gradient $\partial J : L^p(Q) \to 2^{L^q(Q)} \setminus \{\emptyset\}$ is well defined. Moreover, the Aubin–Clarke theorem (see Chap. 2 and [68, p. 83]) ensures that, for each $u \in L^p(Q)$, we have

$$\xi \in \partial J(u) \implies \xi \in L^q(Q) \text{ with } \xi(x,t) \in \partial j(u(x,t)) \text{ for a.e. } (x,t) \in Q. \tag{6.88}$$

We already know that the operator $L = \partial/\partial t : D(L) \subset X_0 \to X_0^*$ is closed, densely defined, and maximal monotone, and under hypotheses (A1)–(A3), the operator $A : X_0 \to X_0^*$ is pseudomonotone w.r.t. $D(L)$. Denote the restriction of J to X_0 by $J|_{X_0}$; then the following result holds.

Lemma 6.22. *Hypothesis (H)(ii) implies that Clarke's generalized gradient* $\partial(J|_{X_0}) : X_0 \to 2^{X_0^*}$ *is pseudomonotone w.r.t.* $D(L)$.

Proof: The growth condition (H)(ii) implies that $\partial(J|_{X_0}) : X_0 \to 2^{X_0^*}$ is bounded. From the calculus of Clarke's generalized gradient (see Chap. 2, Sect. 2.5, or [68, Chap. 2]), we know that $\partial(J|_{X_0})(u)$ is nonempty, closed, and convex. Condition (ii) in Definition 2.154 is also satisfied (see Chap. 2, Proposition 2.171, or [68, p.29]). Therefore, we only need to show that $\partial(J|_{X_0})$ satisfies property (iii) of Definition 2.154. To this end, let $(u_n) \subset D(L)$ with

$u_n \rightharpoonup u$ in X_0, $Lu_n \rightharpoonup Lu$ in X_0^*, $u_n^* \in \partial(J|_{X_0})(u_n)$ with $u_n^* \rightharpoonup u^*$ in X_0^*. We are going to show that already under these assumptions, we get $u^* \in \partial(J|_{X_0})(u)$ and $\langle u_n^*, u_n \rangle \to \langle u^*, u \rangle$, which is (iii). By the assumptions on (u_n), we have $u_n \rightharpoonup u$ in W_0, which implies $u_n \to u$ in $L^p(Q)$ from the compact embedding $W_0 \subset L^p(Q)$. As X_0 is dense in $L^p(Q)$, we know that $u_n^* \in \partial J(u_n)$ (see [68, p. 47]), and thus $u_n^* \in L^q(Q)$ with $u_n^* \rightharpoonup u^*$ in $L^q(Q)$. Because the mapping $\partial J : L^p(Q) \to 2^{L^q(Q)}$ is weak-closed (cf. [68, p. 29] and note $L^q(Q)$ is reflexive), we deduce that $u^* \in \partial J(u)$, and moreover, the following holds:

$$\langle u_n^*, u_n \rangle_{X_0^*, X_0} = \langle u_n^*, u_n \rangle_{L^q(Q), L^p(Q)} \to \langle u^*, u \rangle_{L^q(Q), L^p(Q)} = \langle u^*, u \rangle_{X_0^*, X_0},$$

which completes the proof. $\square$

Corollary 6.23. *Assume hypotheses (A1)–(A3) and (H)(ii). Then the operator $A + \partial(J|_{X_0}) : X_0 \to 2^{X_0^*}$ is pseudomonotone w.r.t. $D(L)$ and bounded.*

Proof: The Leray–Lions conditions (A1)–(A3) imply that the operator A is pseudomonotone w.r.t. $D(L)$, and by Lemma 6.22, the multivalued operator $\partial(J|_{X_0}) : X_0 \to 2^{X_0^*}$ is pseudomonotone w.r.t. $D(L)$. To prove that $A + \partial(J|_{X_0}) : X_0 \to 2^{X_0^*}$ is pseudomonotone w.r.t. $D(L)$, note first that $A + \partial(J|_{X_0}) : X_0 \to 2^{X_0^*}$ is bounded. Thus, we only need to verify property (iii) of Definition 2.154. To this end, assume $(u_n) \subset D(L)$ with $u_n \rightharpoonup u$ in X_0, $Lu_n \rightharpoonup Lu$ in X_0^*, $u_n^* \in (A + \partial(J|_{X_0}))(u_n)$ with $u_n^* \rightharpoonup u^*$ in X_0^*, and

$$\limsup_n \langle u_n^*, u_n - u \rangle \le 0. \tag{6.89}$$

We need to show that $u^* \in (A + \partial(J|_{X_0}))(u)$ and $\langle u_n^*, u_n \rangle \to \langle u^*, u \rangle$. From $u_n^* \in (A + \partial(J|_{X_0}))(u_n)$, we have $u_n^* = Au_n + \eta_n$ with $\eta_n \in \partial(J|_{X_0})(u_n)$, and (6.89) reads

$$\limsup_n \langle Au_n + \eta_n, u_n - u \rangle \le 0. \tag{6.90}$$

Because the sequence $(\eta_n) \subset L^q(Q)$ is bounded and $u_n \to u$ in $L^p(Q)$, we obtain

$$\langle \eta_n, u_n - u \rangle = \int_Q \eta_n (u_n - u) \, dx dt \to 0 \quad \text{as} \quad n \to \infty. \tag{6.91}$$

From (6.90) and (6.91), we deduce

$$\limsup_n \langle Au_n, u_n - u \rangle \le 0. \tag{6.92}$$

The sequence $(Au_n) \subset X_0^*$ is bounded, so that there is some subsequence (Au_k) with $Au_k \rightharpoonup v$. As A is pseudomonotone w.r.t. $D(L)$, it follows that $v = Au$ and $\langle Au_k, u_k \rangle \to \langle Au, u \rangle$. This result shows that each weakly convergent subsequence of (Au_n) has the same limit Au, and thus, the entire sequence (Au_n) satisfies

$$Au_n \rightharpoonup Au \quad \text{and} \quad \langle Au_n, u_n \rangle \to \langle Au, u \rangle. \tag{6.93}$$

From (6.93) and $u_n^* = Au_n + \eta_n \rightharpoonup u^*$, we obtain $\eta_n = u_n^* - Au_n \rightharpoonup u^* - Au$, which in view of (6.91) and the pseudomonotonicity of $\partial(J|_{X_0})$ implies $u^* - Au \in \partial(J|_{X_0})(u)$, and thus $u^* \in (A + \partial(J|_{X_0}))(u)$, and moreover,

$$\langle u_n^* - Au_n, u_n \rangle \to \langle u^* - Au, u \rangle,$$

which yields $\langle u_n^*, u_n \rangle \to \langle u^*, u \rangle$.   □

The main result of this section is the following theorem.

Theorem 6.24. *Let hypotheses (A1)–(A4) and (H) be satisfied. Given subsolutions $\underline{u}_i$ and supersolutions $\bar{u}_i$, $i = 1, 2$, of (6.77) such that $\max\{\underline{u}_1, \underline{u}_2\} =: \underline{u} \le \bar{u} := \min\{\bar{u}_1, \bar{u}_2\}$. Then solutions of (6.77) exist within the order interval $[\underline{u}, \bar{u}]$.*

Proof: The proof will be carried out in three steps.

Step 1: Auxiliary Hemivariational Inequality.

Let us first introduce the cutoff function $b : Q \times \mathbb{R} \to \mathbb{R}$ related with the ordered pair of functions $\underline{u}$, $\bar{u}$, and given by

$$b(x, t, s) = \begin{cases} (s - \bar{u}(x,t))^{p-1} & \text{if } s > \bar{u}(x,t), \\ 0 & \text{if } \underline{u}(x,t) \le s \le \bar{u}(x,t), \\ -(\underline{u}(x,t) - s)^{p-1} & \text{if } s < \underline{u}(x,t). \end{cases} \tag{6.94}$$

As we know, b is a Carathéodory function satisfying the growth condition

$$|b(x, t, s)| \le k_2(x, t) + c_3 |s|^{p-1} \tag{6.95}$$

for a.e. $(x, t) \in Q$, for all $s \in \mathbb{R}$, with some function $k_2 \in L_+^q(Q)$ and a constant $c_3 > 0$. Moreover, we have the following estimate:

$$\int_Q b(x, t, u(x,t)) \, u(x,t) \, dxdt \ge c_4 \|u\|_{L^p(Q)}^p - c_5, \quad \forall \, u \in L^p(Q), \tag{6.96}$$

where c_4 and c_5 are some positive constants. In view of (6.95), the Nemytskij operator $B : L^p(Q) \to L^q(Q)$ defined by

$$Bu(x, t) = b(x, t, u(x, t))$$

is continuous and bounded, and thus, from the compact embedding $W_0 \subset L^p(Q)$ it follows that $B : W_0 \to L^q(Q) \subset X_0^*$ is completely continuous, which implies that $B : X_0 \to X_0^*$ is compact w.r.t. $D(L)$. Let us consider the following auxiliary evolution hemivariational inequality:

$$u \in D(L) : \langle Lu + A(u) + \lambda B(u) - f, v - u \rangle + \int_Q j^o(u; v - u) \, dxdt \ge 0 \tag{6.97}$$

for all $v \in X_0$, where λ is some positive constant to be specified later. The existence of solutions of (6.97) will be proved by using Theorem 2.156 (see Chap. 2, Sect. 2.4.4). To this end, consider the multivalued operator

$$A + \lambda B + \partial(J|_{X_0}) : X_0 \to 2^{X_0^*},$$

where J is the locally Lipschitz functional defined in (6.87) and $\partial(J|_{X_0})$ is the generalized Clarke's gradient of the restriction $J|_{X_0}$. By Corollary 6.23 and the property of B, we readily see that $A + \lambda B + \partial(J|_{X_0}) : X_0 \to 2^{X_0^*}$ is pseudomonotone w.r.t. $D(L)$ and bounded. To apply Theorem 2.156, we need to show the coercivity of $A + \lambda B + \partial(J|_{X_0}) : X_0 \to 2^{X_0^*}$. For any $v \in X_0 \setminus \{0\}$ and any $w \in \partial(J|_{X_0})(v)$, we obtain by applying (A3), (H)(ii) and (6.96) the estimate

$$\frac{1}{\|v\|_{X_0}} \langle Av + \lambda B(v) + w, v \rangle$$

$$= \frac{1}{\|v\|_{X_0}} \left[\int_Q \sum_{i=1}^N a_i(\cdot, \cdot, v, \nabla v) \frac{\partial v}{\partial x_i} \, dx dt + \lambda \langle B(v), v \rangle + \int_Q wv \, dx dt \right]$$

$$\geq \frac{1}{\|v\|_{X_0}} \left[\nu \int_Q |\nabla v|^p \, dx dt - \int_Q k_1 \, dx dt + c_4 \lambda \|v\|_{L^p(Q)}^p - c_5 \lambda \right.$$

$$\left. - c_2 \int_Q (1 + |v|^{p-1}) |v| \, dx dt \right]$$

$$\geq \frac{1}{\|v\|_{X_0}} \left[\nu \|v\|_{X_0}^p - C_0 \right],$$

for some constant $C_0 > 0$, by choosing the constant λ sufficiently large such that $c_4 \lambda > c_2$, which implies the coercivity. Thus, we may apply Theorem 2.156 to ensure that range $(L + A + \lambda B + \partial(J|_{X_0})) = X_0^*$, which yields the existence of $u \in D(L)$ such that $f \in Lu + A(u) + \lambda B(u) + \partial(J|_{X_0})(u)$; i.e., an $\xi \in \partial(J|_{X_0})(u)$ exists such that

$$u \in D(L) : \quad Lu + A(u) - f + \lambda B(u) + \xi = 0 \quad \text{in } X_0^*. \tag{6.98}$$

As X_0 is dense in $L^p(Q)$, we get $\xi \in \partial J(u)$, and thus, by the characterization (6.88) of $\partial J(u)$, it follows that $\xi \in L^q(Q)$ and $\xi(x,t) \in \partial j(u(x,t))$, so that from (6.98), we get

$$\langle Lu + A(u) - f + \lambda B(u), \varphi \rangle + \int_Q \xi(x,t)\varphi(x,t) \, dx dt = 0, \quad \forall \, \varphi \in X_0. \tag{6.99}$$

By definition of Clarke's generalized gradient ∂j, it follows that

$$\int_Q \xi(x,t) \, \varphi(x,t) \, dx dt \leq \int_Q j^0(u(x,t); \varphi(x,t)) \, dx dt, \quad \forall \, \varphi \in X_0. \tag{6.100}$$

In view of (6.99) and (6.100), the auxiliary problem (6.97) has $u \in D(L)$ as solution. Next we shall show that any solution u of the auxiliary evolution hemivariational inequality (6.97) satisfies $\underline{u} \leq u \leq \bar{u}$.

Step 2: Comparison $u \in [\underline{u}, \bar{u}]$.

Let u be any solution of (6.97). We are going to show that $\underline{u}_k \leq u \leq \bar{u}_j$ holds, where $k, j = 1, 2$, which implies the assertion. Let us first prove that $u \leq \bar{u}_j$ is true. By Definition 6.19 $\bar{u}_j$ satisfies $\bar{u}_j(\cdot, 0) \geq 0$ in Ω, $\bar{u}_j \geq 0$ on Γ, and

$$\left\langle \frac{\partial \bar{u}_j}{\partial t} + A\bar{u}_j - f, v - \bar{u}_j \right\rangle + \int_Q j^\circ(\bar{u}_j; v - \bar{u}_j)\, dxdt \geq 0, \quad \forall\, v \in \bar{u}_j \vee X_0,$$

(6.101)

which implies from $v = \bar{u}_j \vee \varphi = \bar{u}_j + (\varphi - \bar{u}_j)^+$ with $\varphi \in X_0$ the following inequality:

$$\left\langle \frac{\partial \bar{u}_j}{\partial t} + A\bar{u}_j - f, (\varphi - \bar{u}_j)^+ \right\rangle + \int_Q j^\circ(\bar{u}_j; (\varphi - \bar{u}_j)^+)\, dxdt \geq 0, \quad \forall\, \varphi \in X_0.$$

(6.102)

If $M := \{(\varphi - \bar{u}_j)^+ : \varphi \in X_0\}$, then by the density result Lemma 5.36 (see Chap. 5), it follows that $\overline{M}^{X_0} = X_0 \cap L^p_+(Q)$. As $s \mapsto j^\circ(r; s)$ is continuous, we get from (6.102) by using Fatou's lemma the inequality

$$\left\langle \frac{\partial \bar{u}_j}{\partial t} + A\bar{u}_j - f, \psi \right\rangle + \int_Q j^\circ(\bar{u}_j; \psi)\, dxdt \geq 0, \quad \forall\, \psi \in X_0 \cap L^p_+(Q). \quad (6.103)$$

Taking in the auxiliary problem (6.97) the special test function $v = u - \psi$ and adding (6.97) and (6.103), we obtain

$$\left\langle \frac{\partial u}{\partial t} - \frac{\partial \bar{u}_j}{\partial t} + A(u) - A(\bar{u}_j) + \lambda B(u), \psi \right\rangle$$
$$\leq \int_Q \left(j^\circ(\bar{u}_j; \psi) + j^\circ(u; -\psi) \right) dxdt \quad (6.104)$$

for all $\psi \in X_0 \cap L^p_+(Q)$. Now we construct a special test function in (6.104). By (A4), for any fixed $\varepsilon > 0$, $\delta(\varepsilon) \in (0, \varepsilon)$ exists such that

$$\int_{\delta(\varepsilon)}^\varepsilon \frac{1}{\omega(r)}\, dr = 1.$$

We use the function $\theta_\varepsilon : \mathbb{R} \to \mathbb{R}_+$ defined by

$$\theta_\varepsilon(s) = \begin{cases} 0 & \text{if } s < \delta(\varepsilon) \\ \displaystyle\int_{\delta(\varepsilon)}^s \frac{1}{\omega(r)}\, dr & \text{if } \delta(\varepsilon) \leq s \leq \varepsilon \\ 1 & \text{if } s > \varepsilon. \end{cases}$$

The function θ_ε has already been used in previous chapters to prove comparison results. We readily verify that, for each $\varepsilon > 0$, the function θ_ε is continuous, piecewise differentiable and the derivative is nonnegative and bounded. Therefore, the function θ_ε is Lipschitz continuous and nondecreasing. In addition, it satisfies

$$\theta_\varepsilon \to \chi_{\{s>0\}} \quad \text{as } \varepsilon \to 0, \tag{6.105}$$

where $\chi_{\{s>0\}}$ is the characteristic function of the set $\{s \in \mathbb{R} : s > 0\}$. Moreover, we have

$$\theta'_\varepsilon(s) = \begin{cases} \dfrac{1}{\omega(s)} & \text{if } \delta(\varepsilon) < s < \varepsilon \\ 0 & \text{if } s \notin [\delta(\varepsilon), \varepsilon]. \end{cases}$$

Taking in (6.104) the test function $\psi = \theta_\varepsilon(u - \bar{u}_j) \in X_0 \cap L^p_+(Q)$, we get

$$\left\langle \frac{\partial(u - \bar{u}_j)}{\partial t}, \theta_\varepsilon(u - \bar{u}_j) \right\rangle + \left\langle A(u) - A(\bar{u}_j), \theta_\varepsilon(u - \bar{u}_j) \right\rangle$$

$$+ \lambda \int_Q B(u)\,\theta_\varepsilon(u - \bar{u}_j)\,dx dt$$

$$\leq \int_Q \Big(j^\circ(\bar{u}_j; \theta_\varepsilon(u - \bar{u}_j)) + j^\circ(u; -\theta_\varepsilon(u - \bar{u}_j)) \Big)\,dx dt. \tag{6.106}$$

Let Θ_ε be the primitive of the function θ_ε defined by

$$\Theta_\varepsilon(s) = \int_0^s \theta_\varepsilon(r)\,dr.$$

We obtain for the first term on the left-hand side of (6.106) (see Lemma 2.146)

$$\left\langle \frac{\partial(u - \bar{u}_j)}{\partial t}, \theta_\varepsilon(u - \bar{u}_j) \right\rangle = \int_\Omega \Theta_\varepsilon(u - \bar{u}_j)(x, \tau)\,dx \geq 0. \tag{6.107}$$

Using (A4) and (A2), the second term on the left-hand side of (6.106) can be estimated as follows:

$$\langle A(u) - A(\bar{u}_j), \theta_\varepsilon(u - \bar{u}_j) \rangle$$

$$= \sum_{i=1}^N \int_Q (a_i(x, t, u, \nabla u) - a_i(x, t, \bar{u}_j, \nabla \bar{u}_j)) \frac{\partial}{\partial x_i} \theta_\varepsilon(u - \bar{u}_j)\,dx\,dt$$

$$\geq \sum_{i=1}^N \int_Q (a_i(x, t, u, \nabla u) - a_i(x, t, u, \nabla \bar{u}_j)) \frac{\partial(u - \bar{u}_j)}{\partial x_i} \theta'_\varepsilon(u - \bar{u}_j)\,dx\,dt$$

$$- N \int_Q (k_2 + |u|^{p-1} + |\bar{u}_j|^{p-1} + |\nabla \bar{u}_j|^{p-1})\,\omega(|u - \bar{u}_j|) \times$$

$$\times\, \theta'_\varepsilon(u - \bar{u}_j)|\nabla(u - \bar{u}_j)|\,dx\,dt$$

$$\geq -N \int_{\{\delta(\varepsilon)<u-\bar{u}_j<\varepsilon\}} \gamma \, |\nabla(u - \bar{u}_j)| \, dx \, dt, \tag{6.108}$$

where $\gamma = k_2 + |u|^{p-1} + |\bar{u}_j|^{p-1} + |\nabla\bar{u}_j|^{p-1} \in L^q(Q)$. The term on the right-hand side of (6.108) tends to zero as $\varepsilon \to 0$. By using (6.105) and applying Lebesgue's dominated convergence theorem, it follows that

$$\lim_{\varepsilon \to 0} \int_Q B(u) \, \theta_\varepsilon(u - \bar{u}_j) \, dx \, dt = \int_Q B(u) \, \chi_{\{u-\bar{u}_j>0\}} \, dx \, dt. \tag{6.109}$$

Again by applying Fatou's lemma and the continuity of $s \mapsto j^\circ(r; s)$, we obtain the following estimate for the right-hand side of (6.106):

$$\limsup_{\varepsilon \to 0} \left(\int_Q \Big(j^\circ(\bar{u}_j; \theta_\varepsilon(u - \bar{u}_j)) + j^\circ(u; -\theta_\varepsilon(u - \bar{u}_j)) \Big) \, dx \, dt \right)$$

$$\leq \int_Q \Big(j^\circ(\bar{u}_j; \chi_{\{u-\bar{u}_j>0\}}) + j^\circ(u; -\chi_{\{u-\bar{u}_j>0\}}) \Big) \, dx \, dt. \tag{6.110}$$

Finally, from (6.106) to (6.110), we get the inequality

$$\lambda \int_Q B(u) \, \chi_{\{u-\bar{u}_j>0\}} \, dx \, dt$$

$$\leq \int_Q \Big(j^\circ(\bar{u}_j; \chi_{\{u-\bar{u}_j>0\}}) + j^\circ(u; -\chi_{\{u-\bar{u}_j>0\}}) \Big) \, dx \, dt. \tag{6.111}$$

Note that $\bar{u} = \min\{\bar{u}_1, \bar{u}_2\}$, which by definition of the operator B yields

$$\lambda \int_Q B(u) \, \chi_{\{u-\bar{u}_j>0\}} \, dx \, dt = \lambda \int_{\{u>\bar{u}_j\}} (u - \bar{u})^{p-1} dx \, dt$$

$$\geq \lambda \int_{\{u>\bar{u}_j\}} (u - \bar{u}_j)^{p-1} dx \, dt. \tag{6.112}$$

The function $r \mapsto j^\circ(s; r)$ is finite and positively homogeneous, $\partial j(s)$ is a nonempty, convex, and compact subset of $\mathbb{R}$, and we have

$$j^\circ(s; r) = \max\{\xi \, r : \xi \in \partial j(s)\}. \tag{6.113}$$

By using (H)(i), (6.113), and the properties of j° and ∂j, we get for certain $\xi(x, t) \in \partial j(u(x, t))$ and $\bar{\xi}_j(x, t) \in \partial j(\bar{u}_j(x, t))$ with $\xi, \bar{\xi}_j \in L^q(Q)$, the following estimate:

$$\int_Q \Big(j^\circ(\bar{u}_j; \chi_{\{u-\bar{u}_j>0\}}) + j^\circ(u; -\chi_{\{u-\bar{u}_j>0\}}) \Big) \, dx \, dt$$

$$= \int_{\{u>\bar{u}_j\}} \Big(j^\circ(\bar{u}_j; 1) + j^\circ(u; -1) \Big) \, dx \, dt$$

$$= \int_{\{u > \bar{u}_j\}} (\bar{\xi}_j(x,t) - \xi(x,t)) \, dx \, dt$$

$$\leq c_1 \int_{\{u > \bar{u}_j\}} (u(x,t) - \bar{u}_j(x,t))^{p-1} \, dx \, dt. \tag{6.114}$$

Thus, (6.111), (6.112), and (6.114) result in

$$(\lambda - c_1) \int_{\{u > \bar{u}_j\}} (u - \bar{u}_j)^{p-1} \, dx \, dt \leq 0. \tag{6.115}$$

Selecting λ large enough such that $\lambda > c_1$, then (6.115) implies that meas $\{u > \bar{u}_j\} = 0$, and thus $u \leq \bar{u}_j$ in Q, where $j = 1, 2$, which shows that $u \leq \bar{u}$. The proof of the inequality $\underline{u} \leq u$ can be done analogously.

Step 3: Completion of the Proof of the Theorem.

From Step 1 and Step 2, it follows that any solution u of the auxiliary evolution hemivariational inequality (6.97) with $\lambda > 0$ sufficiently large satisfies $u \in [\underline{u}, \bar{u}]$, which implies $B(u) = 0$, and hence u is a solution of the original evolution hemivariational inequality (6.77) within the interval $[\underline{u}, \bar{u}]$.   □

The following corollaries are immediate consequences of Theorem 6.24.

Corollary 6.25. *Let $\underline{w}$ and $\bar{w}$ be any subsolution and supersolution, respectively, of (6.77) satisfying $\underline{w} \leq \bar{w}$. Then solutions of (6.77) exist within the order interval $[\underline{w}, \bar{w}]$.*

Proof: Set $\underline{w} = \underline{u}_1 = \underline{u}_2$ and $\bar{w} = \bar{u}_1 = \bar{u}_2$, and apply Theorem 6.24.   □

Let $\mathcal{S}$ denote the set of all solutions of (6.77) within the interval $[\underline{w}, \bar{w}]$ of an ordered pair of sub- and supersolutions; then the following corollary holds.

Corollary 6.26. *The solution set $\mathcal{S}$ of (6.77) is a directed set.*

Proof: Let $u_1, u_2 \in \mathcal{S}$. As any solution of (6.77) is a subsolution and a supersolution as well, by Theorem 6.24, solutions of (6.77) exist within $[\max\{u_1, u_2\}, \bar{w}]$ and within $[\underline{w}, \min\{u_1, u_2\}]$. This process proves the directedness.   □

6.3.3 Compactness and Extremality Results

In this subsection, we are going to show that the solution set $\mathcal{S}$ of (6.77) within the interval of an ordered pair of sub-and supersolutions $[\underline{w}, \bar{w}]$ possesses the smallest and greatest elements with respect to the given natural partial ordering of functions. The smallest and greatest elements of $\mathcal{S}$ are called the extremal solutions of (6.77) within $[\underline{w}, \bar{w}]$. We shall assume hypotheses (A1)–(A4) and (H) throughout this subsection.

Theorem 6.27. *The solution set S is weakly sequentially compact in W_0 and compact in X_0.*

Proof: The solution set $S \subset [\underline{w}, \overline{w}]$ is bounded in $L^p(Q)$. We next show that S is bounded in W_0. Let $u \in S$ be given, and take as a special test function in (6.77) $v = 0$. This process leads to

$$\langle u_t + Au, u \rangle \leq \langle f, u \rangle + \int_Q j^o(u; -u) \, dx dt. \tag{6.116}$$

As $u(\cdot, 0) = 0$, it follows that

$$\langle u_t, u \rangle = \frac{1}{2}\|u(\cdot, \tau)\|^2_{L^2(\Omega)} \geq 0,$$

and (H)(ii) results in

$$\int_Q j^o(u; -u) \, dx dt \leq c_2 \int_Q (1 + |u|^{p-1}) \, |u| \, dx dt.$$

Thus, by means of (A3) and taking the $L^p(Q)$-boundedness of S into account, the following uniform estimate follows from (6.116):

$$\|u\|_{X_0} \leq C, \quad \forall \, u \in S. \tag{6.117}$$

Taking in (6.77) the special test function $v = u - \varphi$, where $\varphi \in B = \{v \in X_0 : \|v\|_{X_0} \leq 1\}$, we obtain

$$|\langle u_t, \varphi \rangle| \leq |\langle f, \varphi \rangle| + |\langle Au, \varphi \rangle| + \left| \int_Q j^o(u; -\varphi) \, dx dt \right|. \tag{6.118}$$

In view of (6.117) from (6.118), we get

$$|\langle u_t, \varphi \rangle| \leq const., \quad \forall \, \varphi \in B, \tag{6.119}$$

where the constant on the right-hand side of (6.119) does not depend on u, which yields $\|u_t\|_{X^*} \leq const$, and thus from (6.117) and (6.119), we get

$$\|u\|_{W_0} \leq const., \quad \forall \, u \in S. \tag{6.120}$$

Now let $(u_n) \subset S$ be any sequence. Then by (6.120), a weakly convergent subsequence (u_k) exists with

$$u_k \rightharpoonup u \quad \text{in } W_0.$$

As u_k are solutions of (6.77), we have

$$\left\langle \frac{\partial u_k}{\partial t} + Au_k - f, v - u_k \right\rangle + \int_Q j^o(u_k; v - u_k) \, dx dt \geq 0, \quad \forall \, v \in X_0. \tag{6.121}$$

Taking as a special test function the weak limit u, we obtain

$$\langle Au_k, u_k - u \rangle \leq \left\langle \frac{\partial u_k}{\partial t} - f, u - u_k \right\rangle + \int_Q j^\circ(u_k; u - u_k)\, dx dt$$

$$\leq \left\langle \frac{\partial u}{\partial t} - f, u - u_k \right\rangle + \int_Q j^\circ(u_k; u - u_k)\, dx dt. \quad (6.122)$$

The weak convergence of (u_k) in W_0 implies $u_k \to u$ in $L^p(Q)$ from the compact embedding $W_0 \subset L^p(Q)$, and thus by applying (H)(ii), the right-hand side of (6.122) tends to zero as $k \to \infty$, which yields

$$\limsup_k \langle Au_k, u_k - u \rangle \leq 0. \quad (6.123)$$

As A is pseudomonotone w.r.t. $D(L)$, from (6.123) we get

$$Au_k \rightharpoonup Au \quad \text{and} \quad \langle Au_k, u_k \rangle \to \langle Au, u \rangle, \quad (6.124)$$

and moreover, because A has the (S_+)-property w.r.t. $D(L)$, the strong convergence $u_k \to u$ in X_0 holds. The convergence properties of the subsequence (u_k) obtained so far and the upper semicontinuity of $j^\circ : \mathbb{R} \times \mathbb{R} \to \mathbb{R}$ finally allow the passage to the limit in (6.121), which completes the proof. $\quad \square$

Theorem 6.28. *The solution set S possesses extremal elements.*

Proof: We prove the existence of the greatest solution of (6.77) within $[\underline{w}, \overline{w}]$; i.e., the greatest element of S. The proof of the smallest element can be done in a similar way. As W_0 is separable, $S \subset W_0$ is separable as well, and a countable, dense subset $Z = \{z_n : n \in \mathbb{N}\}$ of S exists. By Corollary 6.26, S is a directed set. This result allows the construction of an increasing sequence $(u_n) \subset S$ as follows. Let $u_1 = z_1$. Select $u_{n+1} \in S$ such that

$$\max\{z_n, u_n\} \leq u_{n+1} \leq \overline{w}.$$

The existence of u_{n+1} is from Corollary 6.26. As (u_n) is increasing and both bounded and order-bounded, we deduce by applying Lebesgue's dominated convergence theorem that $u_n \to w := \sup_n u_n$ strongly in $L^p(Q)$. By Theorem 6.27, we find a subsequence (u_k) of (u_n), and an element $u \in S$ such that $u_k \rightharpoonup u$ in W_0, and $u_k \to u$ in $L^p(Q)$ and in X_0. Thus, $u = w$ and each weakly convergent subsequence must have the same limit w, which implies that the entire increasing sequence (u_n) satisfies

$$u_n, w \in S: \quad u_n \rightharpoonup w \text{ in } W_0, \quad u_n \to w \text{ in } X_0. \quad (6.125)$$

By construction, we see that

$$\max\{z_1, z_2, \ldots, z_n\} \leq u_{n+1} \leq w, \quad \forall\, n;$$

thus, $Z \subset [\underline{w}, w]$. As the interval $[\underline{w}, w]$ is closed in W_0, we infer

$$\mathcal{S} \subset \overline{Z} \subset \overline{[\underline{w}, w]} = [\underline{w}, w],$$

which in conjunction with $w \in \mathcal{S}$ ensures that w is the greatest element of $\mathcal{S}$.   □

Remark 6.29. It should be noted that the main results of this section remain valid also when the operator A involves quasilinear first-order terms, i.e., operators A in the form

$$Au(x,t) = -\sum_{i=1}^{N} \frac{\partial}{\partial x_i} a_i(x,t,u(x,t),\nabla u(x,t)) + a_0(x,t,u(x,t),\nabla u(x,t)),$$

$$(6.126)$$

where $a_0 : Q \times \mathbb{R} \times \mathbb{R}^N \to \mathbb{R}$ satisfies the same regularity and growth condition as a_i, $i = 1, \ldots, N$.

Next we provide examples to demonstrate the applicability of the theory developed in this section.

Example 6.30. Let c_P denote the best constant in Poincaré's inequality; i.e., the greatest constant $c_P > 0$ satisfying

$$\int_Q |\nabla v|^p \, dx dt \geq c_P \int_Q |v|^p \, dx dt, \quad \forall \, v \in X_0.$$

(c_P is in fact the first eigenvalue of $-\Delta_p$ on X_0.) Assume that (A1)–(A4) and (H) are fulfilled, and suppose in addition:

(a) $a_i(x,t,0,0) = 0$ for a.e. $(x,t) \in Q$, $i = 1, \ldots, N$.
(b) $f \in L^q(Q)$ satisfying

$$f(x,t) \geq \max\{0, \min_{\zeta \in \partial j(0)} \zeta\} \text{ for a.e. } (x,t) \in Q.$$

(c) $k_1 = 0$ in assumption (A3).
(d) $c_P \nu > c_2$, where ν and c_2 are the constants in (A3) and (H)(ii), respectively.

Under these assumptions, problem (6.77) admits extremal nonnegative solutions.

First, we check that $\underline{u} = 0$ is a subsolution of problem (6.77). Indeed, using Definition 6.18, we have to check the inequality

$$\langle A0 - f, v \rangle + \int_Q j^\circ(0; v) \, dx dt \geq 0,$$

for all $v \in 0 \wedge X_0 = \{\min\{0, w\} : w \in X_0\} = \{-w^- : w \in X_0\}$ (where $w^- = \max\{0, -w\}$). Taking into account assumption (a), this reduces to

$$\int_Q (j^o(0; -1) + f)w^- \, dxdt \geq 0, \quad \forall \, w \in X_0.$$

This result is true from assumption (b) because

$$f(x, t) \geq \min_{\zeta \in \partial j(0)} \zeta = - \max_{\zeta \in \partial j(0)} \zeta(-1) = -j^o(0; -1) \text{ for a.e. } (x, t) \in Q.$$

The claim that $\underline{u} = 0$ is a subsolution of (6.77) is verified. Consider now the initial boundary value problem: Find $u \in W_0$ with $u(\cdot, 0) = 0$ in Ω such that

$$\frac{\partial u}{\partial t} - \sum_{i=1}^N \frac{\partial}{\partial x_i} a_i(x, t, u, \nabla u) - c_2(1 + |u|^{p-1}) = f \quad \text{in } Q, \tag{6.127}$$

which may be rewritten as the following abstract problem:

$$u \in D(L) : Lu + A(u) + G(u) = f \quad \text{in } X_0^*, \tag{6.128}$$

where $G : X_0 \to X_0^*$ is defined by

$$\langle G(u), v \rangle = -c_2 \int_Q (1 + |u|^{p-1}) v \, dxdt.$$

We easily verify that $A + G : X_0 \to X_0^*$ is bounded, continuous, and pseudomonotone w.r.t. $D(L)$, and from condition (d) given above, $A + G : X_0 \to X_0^*$ is also coercive. Thus, $L + A + G : D(L) \subset X_0 \to X_0^*$ is surjective, which implies that (6.128) and hence (6.127) possesses solutions.

We are going to show that any solution of (6.127) is nonnegative and a supersolution of (6.77). Let $\bar{u} \in W_0$ be any solution of (6.127). Testing the equation by $-\bar{u}^-$, we find

$$\int_Q \frac{\partial \bar{u}}{\partial t}(-\bar{u}^-) \, dxdt + \sum_{i=1}^N \int_Q a_i(x, t, \bar{u}, \nabla \bar{u}) \frac{\partial}{\partial x_i}(-\bar{u}^-) \, dxdt$$

$$= \int_Q (c_2(1 + |\bar{u}|^{p-1}) + f)(-\bar{u}^-) \, dxdt.$$

As

$$\int_Q \frac{\partial \bar{u}}{\partial t}(-\bar{u}^-) \, dxdt = \frac{1}{2} \int_\Omega (\bar{u}^-)^2(x, \tau) \, dx \geq 0,$$

and using assumption (A3), it follows that

$$\nu \int_{\{\bar{u} \leq 0\}} |\nabla \bar{u}|^p \, dxdt + c_2 \int_{\{\bar{u} \leq 0\}} |\bar{u}|^p \, dxdt$$

$$\leq c_2 \int_{\{\bar{u} \leq 0\}} \bar{u} \, dxdt + \int_{\{\bar{u} \leq 0\}} f\bar{u} \, dxdt \leq 0.$$

Here we used also the assumptions (b) and (c). Taking into account that $\nu > 0$, we conclude that $\bar{u} \geq 0$. To obtain the desired conclusion concerning the existence of extremal nonnegative solutions of (6.77), it is sufficient to show that $\bar{u}$ is a supersolution of problem (6.77). Toward this, we see that every $v \in \bar{u} \vee X_0$ can be written as $v = \bar{u} + (w - \bar{u})^+$ with $w \in X_0$. Then we have

$$\left\langle \frac{\partial \bar{u}}{\partial t} + A\bar{u} - f, (w - \bar{u})^+ \right\rangle + \int_Q j^\circ(\bar{u}; (w - \bar{u})^+) \, dx dt$$

$$\geq \left\langle \frac{\partial \bar{u}}{\partial t} + A\bar{u} - f, (w - \bar{u})^+ \right\rangle - c_2 \int_Q (1 + |\bar{u}|^{p-1})(w - \bar{u})^+ \, dx dt = 0,$$

for all $w \in X_0$, where hypothesis (H)(ii) has been used as well as the fact that $\bar{u}$ solves the initial boundary value problem (6.127). Therefore, $\bar{u} \geq 0$ is a supersolution of problem (6.77). Consequently, Theorem 6.28 yields extremal solutions within the ordered interval $[0, \bar{u}]$.

Remark 6.31. When we have $p = 2$ in Example 6.30, then condition (d) is not needed, because we can always transform the problem into an equivalent coercive one by performing the exponential shift transformation

$$u(x, t) = e^{\lambda t} w(x, t),$$

with $\lambda > 0$ sufficiently large.

Example 6.32. Here we provide sufficient conditions for obtaining constants as sub-supersolutions. Let us assume that $a_i(x, t, u, 0) = 0$ for a.e. $(x, t) \in Q$, all $u \in \mathbb{R}$, $i = 1, \ldots, N$. Then we have the following proposition.

Proposition 6.33. *Let $D \in \mathbb{R}$.*

(a) *If $D \leq 0$ and $f(x, t) \geq -j^\circ(D; -1)$ for a.e. $(x, t) \in Q$, then $\underline{u} = D$ is a subsolution of (6.77).*

(b) *If $D \geq 0$ and $f(x, t) \leq j^\circ(D; 1)$ for a.e. $(x, t) \in Q$, then $\bar{u} = D$ is a supersolution of (6.77).*

Proof: (a) We only need to check (ii) in Definition 6.18. Note that $\underline{u}_t = 0$ and $A\underline{u} = 0$. Let $v \in D \wedge X_0$. As $v - \underline{u} \leq 0$ in Q, we have

$$\langle \underline{u}_t + A\underline{u} - f, v - \underline{u} \rangle + \int_Q j^\circ(\underline{u}; v - \underline{u}) dx dt$$

$$= \int_Q [j^\circ(D; v - \underline{u}) - f(v - \underline{u})] dx dt$$

$$= \int_Q [j^\circ(D; -1) + f] |v - \underline{u}| dx dt \geq 0.$$

(b) Similarly, in the second case, we have $v - D \geq 0$ for $v \in D \vee X_0$ and

$$\langle \bar{u}_t + A\bar{u} - f, v - \bar{u} \rangle + \int_Q j^\circ(\bar{u}; v - \bar{u}) dx dt$$

$$= \int_Q [j^\circ(D; v - \bar{u}) - f(v - \bar{u})] dx dt$$

$$= \int_Q [j^\circ(D; 1) - f](v - \bar{u}) dx dt \geq 0.$$

$\square$

As consequences, for example, if $D > 0$ exists such that

$$-j^\circ(0; -1) \leq f(x, t) \leq j^\circ(D; 1) \quad \text{for a.e. } (x, t) \in Q, \tag{6.129}$$

then (6.77) has a nonnegative bounded solution (in the interval $[0, D]$). Similarly, if there is $D < 0$ such that

$$-j^\circ(D; -1) \leq f(x, t) \leq j^\circ(0; 1) \quad \text{for a.e. } (x, t) \in Q, \tag{6.130}$$

then (6.77) has a nonpositive bounded solution (in $[D, 0]$).

It should be noted that, e.g., condition (6.129) may also be formulated in terms of the generalized gradient as follows:

$$\min_{\zeta \in \partial j(0)} \zeta \leq f(x, t) \leq \max_{\zeta \in \partial j(D)} \zeta \quad \text{for a.e. } (x, t) \in Q. \tag{6.131}$$

Example 6.34. Finally, here we characterize a class of locally Lipschitz functions j satisfying the hypothesis (H).

Let $j_1 : (-\infty, 0) \to \mathbb{R}$ be a convex function, and let $j_2 : [0, +\infty) \to \mathbb{R}$ be a continuously differentiable function such that

(1) $\lim_{s \to 0} j_1(s) = j_2(0)$;
(2) For all $t < 0$ and for all $s \geq 0$ let

$$-c_2(1 + |t|^{p-1}) \leq \min_{\xi \in \partial j_1(t)} \xi \leq \max_{\xi \in \partial j_1(t)} \xi \leq j_2'(s) \leq c_2(1 + |s|^{p-1});$$

(3)

$$\sup_{0 \leq s_1 < s_2} \frac{j_2'(s_1) - j_2'(s_2)}{(s_2 - s_1)^{p-1}} \leq c_1.$$

Here c_1 and c_2 are positive constants. Then $j : \mathbb{R} \to \mathbb{R}$ defined as $j(s) = j_1(s)$ for $s < 0$ and $j(s) = j_2(s)$ for $s \geq 0$ satisfies (H).

6.4 Notes and Comments

Our main goal in this chapter was to extend the comparison principles established for variational equations (see Chap. 3) to variational problems involving nonsmooth and nonconvex integral functionals in the form

$$J(u) = \int_\Omega j(u(x)) \, dx \quad \text{respectively,} \quad J(u) = \int_Q j(u(x,t)) \, dx dt.$$

We note that without difficulties the methods developed here can be modified to deal with hemivariational inequalities in the form (for the elliptic case)

$$u \in V_\Gamma : \langle Au - f, v - u \rangle + \int_\Gamma j^\circ(\gamma u; \gamma v - \gamma u) \, d\Gamma \geq 0 \qquad (6.132)$$

for all $v \in V_\Gamma$, where $\Gamma \subset \partial\Omega$ is some portion of the boundary $\partial\Omega$, and the space V_Γ is defined by

$$V_\Gamma := \{ u \in V = W^{1,p}(\Omega) : \gamma u = 0 \text{ on } \partial\Omega \setminus \Gamma \},$$

with $\gamma : V \to L^p(\partial\Omega)$ denoting the trace operator. Problem (6.132) is closely related to the following boundary inclusion problem:

$$Au = f \quad \text{in } \Omega \qquad (6.133)$$

$$u = 0 \quad \text{on } \partial\Omega \setminus \Gamma, \quad -\frac{\partial u}{\partial\nu} \in \partial j(u) \quad \text{on } \Gamma. \qquad (6.134)$$

Boundary inclusion problems in the form (6.133), (6.134), and their corresponding evolutionary counterparts have been treated, e.g., in [36, 39]. To preserve the characteristic features of the comparison principles of the previous chapters, a one-sided growth condition on Clarke's generalized gradient ∂j [see condition (H)(ii)] plays an important role. In fact, this condition allows us to prove the equivalence of hemivariational inequalities and their corresponding inclusion problems, and it provides a new analytical framework to deal with differential inclusion problems. Moreover, as we will see in Chap. 7, by means of the variational methods developed here in conjunction with that for variational inequalities in Chap. 5, we can consider hemivariational inequalities (respectively, differential inclusion problems) under constraints. In this sense, Chap. 6 may be considered also as a transition to a unified treatment of variational and hemivariational inequalities, which leads to the subject of variational–hemivariational inequalities to be treated in Chap. 7. An example of this kind is the following problem:

$$u \in K : \quad \langle -\Delta_p u - f, v - u \rangle + \int_\Omega j^\circ(u; v - u) \, dx \geq 0, \quad \forall\, v \in K, \quad (6.135)$$

where $j^\circ(s; r)$ denotes the generalized directional derivative of some locally Lipschitz function $j : \mathbb{R} \to \mathbb{R}$ at s in the direction r, and $K \subset V_0$ is some closed and convex subset. The operator $\Delta_p u = \text{div}\,(|\nabla u|^{p-2}\nabla u)$ is the p-Laplacian, $1 < p < \infty$, and $f \in V_0^*$. Problem (6.135) includes various special cases:

(i) For $K = V_0$ and $j : \mathbb{R} \to \mathbb{R}$ smooth, (6.135) is the weak formulation of the Dirichlet problem

$$u \in V_0 : \quad -\Delta_p u + j'(u) = f \quad \text{in } V_0^*,$$

which is a special variational equation considered in Chap. 3.

(ii) If $K = V_0$, and $j : \mathbb{R} \to \mathbb{R}$ is locally Lipschitz (not necessarily smooth) and satisfies condition (H), then (6.135) is a hemivariational inequality of the form

$$u \in V_0 : \quad \langle -\Delta_p u - f, v - u \rangle + \int_\Omega j^\circ(u; v - u)\, dx \geq 0, \quad \forall\, v \in V_0,$$

or equivalently a differential inclusion problem of the form

$$u \in V_0 : \quad -\Delta_p u + \partial j(u) \ni f \quad \text{in } V_0^*,$$

which is the subject of the current chapter.

(iii) If $j = 0$, then (6.135) becomes a variational inequality for which a sub-supersolution method has been developed in Chap. 5.

We introduce the following notion of sub-supersolutions of (6.135).

Definition 6.35. *A function $\underline{u} \in V$ is called a subsolution of (6.135) if the following holds:*

(i) $\underline{u} \leq 0$ on $\partial\Omega$.

(ii) $\langle -\Delta_p \underline{u} - f, v - \underline{u} \rangle + \int_\Omega j^\circ(\underline{u}; v - \underline{u})\, dx \geq 0, \quad \forall\, v \in \underline{u} \wedge K.$

Definition 6.36. *$\bar{u} \in V$ is a supersolution of (6.135) if the following holds:*

(i) $\bar{u} \geq 0$ on $\partial\Omega$.

(ii) $\langle -\Delta_p \bar{u} - f, v - \bar{u} \rangle + \int_\Omega j^\circ(\bar{u}; v - \bar{u})\, dx \geq 0, \quad \forall\, v \in \bar{u} \vee K.$

Under condition (H) of this chapter, the following comparison principle has been proved in [37].

Theorem 6.37. *Let $\underline{u}$ and $\bar{u}$ be sub- and supersolutions of (6.135), respectively, satisfying $\underline{u} \leq \bar{u}$, and assume $\bar{u} \wedge K \subset K$ and $\underline{u} \vee K \subset K$. Then under hypothesis (H), solutions of (6.135) exist within the order interval $[\underline{u}, \bar{u}]$.*

In Chap. 7, we are going to extend this result to a wider class of variational–hemivariational inequalities.

7

Variational–Hemivariational Inequalities

In the previous chapters, the BVPs we considered in the form of hemivariational inequalities were formulated on the whole space. We are now taking into account problems subject to constraints for hemivariational inequalities, which means dealing with variational–hemivariational inequalities. The aim of this chapter is three-fold: (a) to develop the method of sub- and supersolutions for quasilinear elliptic variational–hemivariational inequalities; (b) to treat an evolution variational–hemivariational inequality by the method of sub- and supersolutions; and (c) to study variational–hemivariational inequalities by minimax methods in the nonsmooth critical point theory viewing the (weak) solutions as critical points of the corresponding nonsmooth functionals. The two general methods, namely the sub-supersolutions approach and the nonsmooth critical point theory, are complementary and permit us to investigate various types of problems. Specifically, Sect. 7.1 and Sect. 7.2 deal with the method of sub- and supersolutions for hemivariational inequalities, whereas Sect. 7.3, Sect. 7.4, and Sect. 7.5 present applications of nonsmooth critical point results for this kind of problem emphasizing the treatment for corresponding eigenvalue problems. In both methods, an essential feature consists of the use of comparison arguments. They allow us to provide location information for the solutions.

7.1 Elliptic Variational–Hemivariational Inequalities

In this section, we study general quasilinear elliptic variational–hemivariational inequalities through the method of sub- and supersolutions. The problem under consideration is the following: Find $u \in \mathrm{dom}\,(\psi) \cap V_0$ such that

$$\langle Au - f, v - u \rangle + \psi(v) - \psi(u) + \int_\Omega j^o(u; v - u)\,dx \geq 0, \quad \forall\, v \in V_0. \quad (7.1)$$

Let us explain the meaning of the data entering problem (7.1). Here $\Omega \subset \mathbb{R}^N$ is a bounded domain with Lipschitz boundary $\partial\Omega$. Denote $V = W^{1,p}(\Omega)$

and $V_0 = W_0^{1,p}(\Omega)$, for some $1 < p < \infty$, with their dual spaces V^* and V_0^*, respectively. Let $f \in V_0^*$. Given a locally Lipschitz function $j : \mathbb{R} \to \mathbb{R}$, the notation $j^\circ(s; r)$ represents the generalized directional derivative of j at $s \in \mathbb{R}$ in the direction $r \in \mathbb{R}$ (cf. Definition 2.161). In (7.1), we also have a function $\psi : V \to \mathbb{R} \cup \{+\infty\}$ that is convex, lower semicontinuous, and satisfies $\mathrm{dom}\,(\psi) \cap V_0 \neq \emptyset$, where $\mathrm{dom}\,(\psi)$ stands for the effective domain of ψ; i.e., $\mathrm{dom}\,(\psi) = \{v \in V : \psi(v) < +\infty\}$. The operator $A : V \to V_0^*$ is a second-order quasilinear differential operator in divergence form

$$Au(x) = -\sum_{i=1}^{N} \frac{\partial}{\partial x_i} a_i(x, \nabla u(x))$$

being of Leray–Lions type as described in Definition 2.103 whose coefficient functions a_i, $i = 1, \dots, N$, verify conditions (H1), (H2), (H3) in Sect. 2.3.2, dropping the dependence with respect to s.

7.1.1 Comparison Principle

For the sake of clarity, we write conditions (H1)–(H3) in Sect. 2.3.2 in the particular case of the operator A in (7.1):

(A1) Each $a_i : \Omega \times \mathbb{R}^N \to \mathbb{R}$ is a Carathéodory condition; i.e., $a_i(x, \xi)$ is measurable in $x \in \Omega$ for all $\xi \in \mathbb{R}^N$ and continuous in ξ for almost all $x \in \Omega$. A constant $c_0 > 0$ and a function $k_0 \in L^q(\Omega), 1/p + 1/q = 1$ exist, such that

$$|a_i(x, \xi)| \leq k_0(x) + c_0 |\xi|^{p-1},$$

for a.e. $x \in \Omega$ and for all $\xi \in \mathbb{R}^N$.

(A2) $\sum_{i=1}^{N}(a_i(x, \xi) - a_i(x, \xi'))(\xi_i - \xi_i') > 0$ for a.e. $x \in \Omega$, and for all $\xi, \xi' \in \mathbb{R}^N$ with $\xi \neq \xi'$.

(A3) $\sum_{i=1}^{N} a_i(x, \xi)\xi_i \geq \nu |\xi|^p - k_1(x)$ for a.e. $x \in \Omega$, and for all $\xi \in \mathbb{R}^N$ with some constant $\nu > 0$ and a function $k_1 \in L^1(\Omega)$.

Conditions (A1), (A2) ensure that

$$\langle Au, \varphi \rangle = \int_\Omega \sum_{i=1}^{N} a_i(x, \nabla u) \frac{\partial \varphi}{\partial x_i} \, dx \in \mathbb{R}, \quad \forall \, \varphi \in V_0,$$

for any $u \in V$, and the operator $A : V_0 \to V_0^*$ is continuous, bounded, and strictly monotone (see Theorem 2.109).

Remark 7.1. There are various important special cases of problem (7.1) such as the following:

(i) For $\psi(u) \equiv 0$ and $j : \mathbb{R} \to \mathbb{R}$ smooth with its derivative $j' : \mathbb{R} \to \mathbb{R}$, (7.1) reduces to the weak formulation of the Dirichlet problem: Find $u \in V_0$ such that

$$Au + j'(u) = f \quad \text{in } V_0^*.$$

(ii) For $\psi(u) \equiv 0$ and $j : \mathbb{R} \to \mathbb{R}$ locally Lipschitz (not necessarily smooth), then (7.1) is a hemivariational inequality of the form: Find $u \in V_0$ such that

$$\langle Au - f, v - u \rangle + \int_\Omega j^\circ(u; v - u)\, dx \geq 0, \quad \forall\, v \in V_0.$$

(iii) For $j : \mathbb{R} \to \mathbb{R}$ smooth, then (7.1) becomes the variational inequality: Find $u \in \mathrm{dom}\,(\psi) \cap V_0$ such that

$$\langle Au + j'(u) - f, v - u \rangle + \psi(v) - \psi(u) \geq 0, \quad \forall\, v \in V_0.$$

First we introduce our basic notion of sub-supersolution for problem (7.1). To this end, for functions w, $z : \Omega \to \mathbb{R}$ and sets W and Z of functions defined on Ω, we use the notations already given in the previous chapters: $w \wedge z = \min\{w, z\}$, $w \vee z = \max\{w, z\}$, $W \wedge Z = \{w \wedge z : w \in W,\ z \in Z\}$, $W \vee Z = \{w \vee z : w \in W,\ z \in Z\}$, and $w \wedge Z = \{w\} \wedge Z$, $w \vee Z = \{w\} \vee Z$.

Definition 7.2. *A function $\underline{u} \in V$ is called a subsolution of (7.1) if the following conditions are fulfilled:*

(i) $\underline{u} \leq 0$ *on $\partial\Omega$.*
(ii) $\underline{u} \vee (\mathrm{dom}\,(\psi) \cap V_0) \subset \mathrm{dom}\,(\psi) \cap V_0$.
(iii) *A mapping $\hat{\psi} : V \to \mathbb{R} \cup \{+\infty\}$ and a constant $\hat{c} \geq 0$ exists such that the following holds:*
  (a) $\underline{u} \in \mathrm{dom}\,(\hat{\psi})$.
  (b) $\psi(v \vee \underline{u}) + \hat{\psi}(v \wedge \underline{u}) - \psi(v) - \hat{\psi}(\underline{u}) \leq \hat{c} \int_\Omega [(\underline{u} - v)^+]^p\, dx$ *for all $v \in \mathrm{dom}\,(\psi) \cap V_0$.*
  (c) $\langle A\underline{u} - f, v - \underline{u} \rangle + \hat{\psi}(v) - \hat{\psi}(\underline{u}) + \int_\Omega j^\circ(\underline{u}; v - \underline{u})\, dx \geq 0$ *for all $v \in \underline{u} \wedge (\mathrm{dom}\,(\psi) \cap V_0)$.*

Definition 7.3. *A function $\bar{u} \in V$ is a supersolution of (7.1) if the following conditions are fulfilled:*

(i) $\bar{u} \geq 0$ *on $\partial\Omega$.*
(ii) $\bar{u} \wedge (\mathrm{dom}\,(\psi) \cap V_0) \subset \mathrm{dom}\,(\psi) \cap V_0$.
(iii) *A mapping $\tilde{\psi} : V \to \mathbb{R} \cup \{+\infty\}$ and a constant $\tilde{c} \geq 0$ exists such that*
  (a) $\bar{u} \in \mathrm{dom}\,(\tilde{\psi})$.
  (b) $\psi(v \wedge \bar{u}) + \tilde{\psi}(v \vee \bar{u}) - \psi(v) - \tilde{\psi}(\bar{u}) \leq \tilde{c} \int_\Omega [(v - \bar{u})^+]^p\, dx$ *for all $v \in \mathrm{dom}\,(\psi) \cap V_0$.*
  (c) $\langle A\bar{u} - f, v - \bar{u} \rangle + \tilde{\psi}(v) - \tilde{\psi}(\bar{u}) + \int_\Omega j^\circ(\bar{u}; v - \bar{u})\, dx \geq 0$ *for all $v \in \bar{u} \vee (\mathrm{dom}\,(\psi) \cap V_0)$.*

Remark 7.4. The above definitions of sub-supersolutions requiring the existence of functionals $\hat{\psi}$ and $\tilde{\psi}$ that satisfy conditions (a)–(c) in Definitions 7.2 and 7.3, respectively, extend the ones for variational inequalities and inclusions of hemivariational type as given in the previous chapters (see also the references [142, 58, 59] and [53, 56, 141, 142]).

To see the applicability of the above concepts of sub- and super-solutions, we discuss three relevant examples.

Example 7.5. Assume $\psi(u) \equiv 0$ and $j : \mathbb{R} \to \mathbb{R}$ smooth. Then, as pointed out in Remark 7.1(i), problem (7.1) reduces to the Dirichlet problem: Find $u \in V_0$ such that

$$Au + j'(u) = f \quad \text{in } V_0^*.$$

According to Definition 7.2, a function $\underline{u} \in V$ with $\underline{u} \leq 0$ on $\partial\Omega$ is a subsolution if (ii) and (iii) of Definition 7.2 can be fulfilled. As $\operatorname{dom}(\psi) = V$, we see that by choosing $\hat{\psi} = 0$, the conditions (ii) and (iii), (a)–(b) are trivially satisfied. Thus, $\underline{u}$ is only required to satisfy condition (iii)(c); i.e.,

$$\langle A\underline{u} - f, v - \underline{u} \rangle + \int_\Omega j'(\underline{u})(v - \underline{u})\, dx \geq 0, \quad \forall\, v \in \underline{u} \wedge V_0.$$

Given $\varphi \in V_0$, setting $v = \underline{u} \wedge \varphi = \underline{u} - (\underline{u} - \varphi)^+$ yields

$$\langle A\underline{u} - f, -(\underline{u} - \varphi)^+ \rangle + \int_\Omega j'(\underline{u})(-(\underline{u} - \varphi)^+)\, dx \geq 0.$$

Thus, we obtain

$$\langle A\underline{u} - f, w \rangle + \int_\Omega j'(\underline{u})w\, dx \leq 0, \quad \forall\, w \in W,$$

where $W = \{w = (\underline{u} - \varphi)^+ : \varphi \in V_0\}$. Observing that W is dense in $V_0 \cap L_+^p(\Omega)$ (see Lemma 5.4 in Chap. 5), we get the usual notion of weak subsolution of the Dirichlet problem. Similarly, Definition 7.3 reduces to the ordinary concept for a weak supersolution of the above Dirichlet problem.

Example 7.6. Let $K \subset V_0$ be a nonempty, closed, and convex set, and let $\psi = I_K$, where $I_K : V \to \mathbb{R} \cup \{+\infty\}$ denotes the indicator function related to K; i.e.,

$$I_K(u) = \begin{cases} 0 & \text{if } u \in K, \\ +\infty & \text{if } u \notin K, \end{cases}$$

which is proper, convex, and lower semicontinuous. Problem (7.1) then becomes: Find $u \in K$ such that

$$\langle Au - f, v - u \rangle + I_K(v) - I_K(u) + \int_\Omega j^\circ(u; v - u)\, dx \geq 0, \quad \forall\, v \in V_0. \quad (7.2)$$

In this case, $\underline{u} \in V$ is a subsolution of (7.2) according to Definition 7.2 if the following is satisfied:

(1) $\underline{u} \leq 0$ on $\partial\Omega$.
(2) $\underline{u} \vee K \subset K$.
(3) $\langle A\underline{u} - f, v - \underline{u} \rangle + \int_\Omega j^\circ(\underline{u}; v - \underline{u})\, dx \geq 0, \ \forall\, v \in \underline{u} \wedge K$.

Specifically, taking $\hat{\psi}(v) \equiv 0$ and $\hat{c} = 0$, all conditions of Definition 7.2 are fulfilled. Analogous conditions can be found for a supersolution $\bar{u}$ of (7.2):

(1') $\bar{u} \geq 0$ on $\partial\Omega$

(2') $\bar{u} \wedge K \subset K$.

(3') $\langle A\bar{u} - f, v - \bar{u} \rangle + \int_\Omega j^\circ(\bar{u}; v - \bar{u}) \, dx \geq 0, \ \forall \, v \in \bar{u} \vee K$.

Remark 7.7. Conditions (1)–(3) and (1')–(3'), which were introduced in [141] (see Chap. 5) to define sub-supersolutions turn out to be special cases of Definition 7.2 and Definition 7.3, respectively. Problem (7.2) is the object of the work in [56].

Example 7.8. Given a convex lower semicontinuous function $h : \mathbb{R} \to \mathbb{R}$, we introduce $g : V \to \mathbb{R} \cup \{+\infty\}$ by

$$g(v) = \begin{cases} \int_\Omega h(v(x)) dx & \text{if } h(v) \in L^1(\Omega), \\ +\infty & \text{if } h(v) \notin L^1(\Omega). \end{cases}$$

The function g is known to be proper, convex, and lower semicontinuous. Consider problem (7.1) with $\psi = g$; i.e., find $u \in \text{dom}\,(g) \cap V_0$ such that

$$\langle Au - f, v - u \rangle + g(v) - g(u) + \int_\Omega j^\circ(u; v - u) \, dx \geq 0, \quad \forall \, v \in V_0.$$

The following conditions on a function $\underline{u} \in V$ imply that $\underline{u}$ is a subsolution according to Definition 7.2:

(1) $\underline{u} \leq 0$ on $\partial\Omega$,

(2) $\underline{u} \vee (\text{dom}\,(g) \cap V_0) \subset \text{dom}\,(g) \cap V_0$,

(3) $\underline{u} \in \text{dom}\,(g)$, and

$$\langle A\underline{u} - f, v - \underline{u} \rangle + g(v) - g(\underline{u}) + \int_\Omega j^\circ(\underline{u}; v - \underline{u}) \, dx \geq 0$$

for all $v \in \underline{u} \wedge (\text{dom}\,(g) \cap V_0)$.

Indeed, taking $\hat{\psi} = g$ and $\hat{c}$ any nonnnegative constant, we can see that in view of the above assumptions (1)–(3), all conditions of Definition 7.2 are satisfied. This result is because for all $v \in \text{dom}\,(g) \cap V_0$, we have

$$g(v \vee \underline{u}) + g(v \wedge \underline{u}) - g(v) - g(\underline{u}) = 0. \tag{7.3}$$

Identity (7.3) can easily be proved by splitting up Ω into $\Omega = \Omega_1 \cup \Omega_2$, where

$$\Omega_1 = \{x \in \Omega : v(x) < \underline{u}(x)\}, \quad \Omega_2 = \{x \in \Omega : v(x) \geq \underline{u}(x)\},$$

and by considering the resulting integrals. For example, if $f \in L^{p^{*'}}(\Omega)$ (with p^* the critical Sobolev exponent and $p^{*'}$ being its Hölder conjugate) and $a_i(x, 0) = 0$ for $i = 1, \ldots, N$, then $\underline{u} = 0$ is a subsolution if for some $\xi \in \partial h(0)$ the following inequality holds:

$$f(x) \geq -j^\circ(0; -1) + \xi, \quad \text{for a.e. } x \in \Omega.$$

For a supersolution $\bar{u}$, we can easily find the corresponding sufficient conditions.

We proceed in studying problem (7.1) by assuming the following hypothesis for j, which was used already in previous chapters:

(H) The function $j : \mathbb{R} \to \mathbb{R}$ is locally Lipschitz, and its generalized gradient ∂j satisfies the following growth conditions:

(i) A constant $c_1 \geq 0$ exists such that

$$\xi_1 \leq \xi_2 + c_1(s_2 - s_1)^{p-1}$$

for all $\xi_i \in \partial j(s_i)$, $i = 1, 2$, and for all s_1, s_2 with $s_1 < s_2$;

(ii) A constant $c_2 \geq 0$ exists such that

$$|\xi| \leq c_2 \left(1 + |s|^{p-1}\right)$$

for all $\xi \in \partial j(s)$ and for all $s \in \mathbb{R}$.

Like in the previous chapters, let $L^p(\Omega)$ be equipped with the natural partial ordering of functions defined by $u \leq w$ if and only if $w - u$ belongs to the positive cone $L^p_+(\Omega)$ of all nonnegative elements of $L^p(\Omega)$. This finding induces a corresponding partial ordering also in the subspace V of $L^p(\Omega)$, and if u, $w \in V$ with $u \leq w$, then

$$[u, w] = \{z \in V : u \leq z \leq w\}$$

denotes the order interval formed by u and w.

In the proofs of our main results, we again make use of the cutoff function $b : \Omega \times \mathbb{R} \to \mathbb{R}$ related with an ordered pair of functions $\underline{u} \leq \bar{u}$ and given by

$$b(x, s) = \begin{cases} (s - \bar{u}(x))^{p-1} & \text{if } s > \bar{u}(x), \\ 0 & \text{if } \underline{u}(x) \leq s \leq \bar{u}(x), \\ -(\underline{u}(x) - s)^{p-1} & \text{if } s < \underline{u}(x). \end{cases}$$

Recall that b is a Carathéodory function satisfying the growth condition

$$|b(x, s)| \leq k(x) + c_3 |s|^{p-1} \tag{7.4}$$

for a.e. $x \in \Omega$, for all $s \in \mathbb{R}$, with some function $k \in L^q_+(\Omega)$ and a constant $c_3 \geq 0$, and we have the following estimate:

$$\int_\Omega b(x, u(x)) u(x) \, dx \geq c_4 \|u\|^p_{L^p(\Omega)} - c_5, \quad \forall u \in L^p(\Omega), \tag{7.5}$$

where c_4 and c_5 are positive constants. In view of (7.4), the Nemytskij operator $B : L^p(\Omega) \to L^q(\Omega)$ defined by $Bu(x) = b(x, u(x))$ is continuous and bounded, and thus from the compact embedding $V \subset L^p(\Omega)$, it follows that $B : V_0 \to V_0^*$ is compact.

The main result of the present section is given by the following theorem in [55] (also see [56] for a special case), which provides an existence and comparison result for the elliptic variational–hemivariational inequality (7.1).

Theorem 7.9. *Let $\underline{u}$ and $\bar{u}$ be sub- and supersolutions of (7.1), respectively, satisfying $\underline{u} \le \bar{u}$. Then under hypotheses (A1)–(A3) and (H), solutions of (7.1) exist within the ordered interval $[\underline{u}, \bar{u}]$.*

Proof: As we are looking for solutions of (7.1) within $[\underline{u}, \bar{u}]$, we consider the following auxiliary problem: Find $u \in \mathrm{dom}\,(\psi) \cap V_0$ such that

$$\langle Au - f + \lambda B(u), v - u \rangle + \psi(v) - \psi(u) + \int_\Omega j^\circ(u; v - u)\, dx \ge 0, \quad \forall\, v \in V_0, \tag{7.6}$$

where B is the cutoff operator as above and $\lambda \ge 0$ is some parameter to be specified later.

The proof consists in two steps.

Step 1: Existence for (7.6).

Let us introduce the functional $J : L^p(\Omega) \to \mathbb{R}$ defined by

$$J(v) = \int_\Omega j(v(x))\, dx, \quad \forall\, v \in L^p(\Omega).$$

In view of hypothesis (H), by Theorem 2.181, the functional J is locally Lipschitz, and for each $u \in L^p(\Omega)$, we have

$$\xi \in \partial J(u) \implies \xi \in L^q(\Omega) \text{ with } \xi(x) \in \partial j(u(x)) \text{ for a.e. } x \in \Omega.$$

Consider now the multivalued operator

$$A + \lambda B + \partial(J|_{V_0}) + \partial(\psi|_{V_0}) : V_0 \to 2^{V_0^*},$$

where $J|_{V_0}$ and $\psi|_{V_0}$ denote the restriction of J and ψ, respectively, to V_0. The notation $\partial(J|_{V_0})$ and $\partial(\psi|_{V_0})$ means the generalized gradient of $J|_{V_0}$ and the subdifferential of $\psi|_{V_0}$ in the sense of convex analysis, respectively. It is well known that $\partial(\psi|_{V_0}) : V_0 \to 2^{V_0^*}$ is a maximal monotone operator (cf. [13] or [222]). As $A : V_0 \to V_0^*$ is strictly monotone, bounded, and continuous, and $\lambda B : V_0 \to V_0^*$ is bounded, continuous, and compact, we obtain that $A + \lambda B : V_0 \to V_0^*$ is a (singlevalued) pseudomonotone, continuous, and bounded operator. It has already been shown that $\partial(J|_{V_0}) : V_0 \to 2^{V_0^*}$ is a (multivalued) pseudomonotone operator (see Lemma 4.16, Chap. 4), which, from (H), is bounded. Thus, from Theorem 2.124 (ii), $A_0 = A + \lambda B + \partial(J|_{V_0}) : V_0 \to 2^{V_0^*}$ is a pseudomonotone and bounded operator. Hence, it follows by Theorem

2.127 that range $(A_0 + \partial(\psi|_{V_0})) = V_0^*$ provided A_0 is u_0-coercive for some $u_0 \in D(\partial(\psi|_{V_0}))$. Toward this end, for any $v \in V_0$ and any $w \in \partial(J|_{V_0})(v)$, we obtain by applying (A3), (H)(ii), and (7.5) the estimate

$$\langle Av + \lambda B(v) + w, v - u_0 \rangle$$

$$= \int_\Omega \sum_{i=1}^N a_i(x, \nabla v) \frac{\partial v}{\partial x_i}\, dx + \lambda \langle B(v), v \rangle$$

$$+ \int_\Omega wv\, dx - \langle Av + \lambda B(v) + w, u_0 \rangle$$

$$\geq \nu \int_\Omega |\nabla v|^p\, dx - \|k_1\|_{L^1(\Omega)} + c_4 \lambda \|v\|_{L^p(\Omega)}^p - c_5 \lambda$$

$$-c_2 \int_\Omega (1 + |v|^{p-1})|v|\, dx - |\langle Av + \lambda B(v) + w, u_0 \rangle|$$

$$\geq \nu \|v\|_{V_0}^p - C (1 + \|v\|_{V_0}^{p-1}), \tag{7.7}$$

for some constant $C > 0$. By choosing the constant λ in such a way that $c_4\lambda > c_2$, the coercivity of A_0 follows from (7.7). In view of the surjectivity of the operator $A_0 + \partial(\psi|_{V_0})$ a $u \in D(\partial(\psi|_{V_0})) \subset D(\psi) \cap V_0$, an $\xi \in \partial(J|_{V_0})(u)$ with $\xi \in L^q(\Omega)$ and $\xi(x) \in \partial j(u(x))$ for a.e. $x \in \Omega$, and an $\eta \in \partial(\psi|_{V_0})(u)$ exist such that

$$Au - f + \lambda B(u) + \xi + \eta = 0 \quad \text{in } V_0^*, \tag{7.8}$$

where

$$\langle \xi, \varphi \rangle = \int_\Omega \xi(x)\, \varphi(x)\, dx, \quad \forall\, \varphi \in V_0, \tag{7.9}$$

and

$$\psi(v) \geq \psi(u) + \langle \eta, v - u \rangle, \quad \forall\, v \in V_0. \tag{7.10}$$

By definition of the generalized gradient ∂j (see Definition 2.166) from (7.9), we get

$$\langle \xi, \varphi \rangle \leq \int_\Omega j^\circ(u(x); \varphi(x))\, dx, \quad \forall\, \varphi \in V_0. \tag{7.11}$$

Thus, from (7.8)–(7.11) with φ replaced by $v - u$, we obtain (7.6), which proves the existence of solutions of problem (7.6).

Step 2: $\underline{u} \leq u \leq \bar{u}$ for any solution u of (7.6).

Let us first show $u \leq \bar{u}$. By Definition 7.3, the supersolution $\bar{u}$ satisfies $\bar{u} \in \text{dom}\,(\tilde{\psi})$, $\bar{u} \geq 0$ on $\partial\Omega$, and for all $v \in \bar{u} \vee (\text{dom}\,(\psi) \cap V_0)$ the following inequality:

$$\langle A\bar{u} - f, v - \bar{u}\rangle + \tilde{\psi}(v) - \tilde{\psi}(\bar{u}) + \int_\Omega j^\circ(\bar{u}; v - \bar{u})\, dx \geq 0. \tag{7.12}$$

Let u be any solution of (7.6). We apply the special test function $v = \bar{u} \vee u = \bar{u} + (u - \bar{u})^+ \in \bar{u} \vee (\mathrm{dom}\,(\psi) \cap V_0)$ in (7.12) and $v = \bar{u} \wedge u = u - (u - \bar{u})^+ \in \mathrm{dom}\,(\psi) \cap V_0$ (from requirement (ii) in Definition 7.3) in (7.6). By adding the resulting inequalities, we get

$$\langle A\bar{u} - Au, (u - \bar{u})^+\rangle + \lambda\langle B(u), -(u - \bar{u})^+\rangle$$
$$+\tilde{\psi}(\bar{u} \vee u) - \tilde{\psi}(\bar{u}) + \psi(\bar{u} \wedge u) - \psi(u)$$
$$+ \int_\Omega \Big(j^\circ(\bar{u}; (u - \bar{u})^+) + j^\circ(u; -(u - \bar{u})^+)\Big)\, dx \geq 0.$$

From

$$\langle Au - A\bar{u}, (u - \bar{u})^+\rangle \geq 0,$$

this yields the inequality

$$\lambda\langle B(u), (u - \bar{u})^+\rangle$$
$$\leq \tilde{\psi}(\bar{u} \vee u) - \tilde{\psi}(\bar{u}) + \psi(\bar{u} \wedge u) - \psi(u)$$
$$+ \int_\Omega \Big(j^\circ(\bar{u}; (u - \bar{u})^+) + j^\circ(u; -(u - \bar{u})^+)\Big)\, dx. \tag{7.13}$$

By using Proposition 2.171 and hypothesis (H)(i), we find the following estimate of the second term on the right-hand side of (7.13):

$$\int_\Omega \Big(j^\circ(\bar{u}; (u - \bar{u})^+) + j^\circ(u; -(u - \bar{u})^+)\Big)\, dx$$
$$= \int_{\{u>\bar{u}\}} \Big(j^\circ(\bar{u}; u - \bar{u}) + j^\circ(u; -(u - \bar{u}))\Big)\, dx$$
$$= \int_{\{u>\bar{u}\}} \Big(\bar{\xi}(x)(u(x) - \bar{u}(x)) + \xi(x)(-(u(x) - \bar{u}(x)))\Big)\, dx$$
$$= \int_{\{u>\bar{u}\}} (\bar{\xi}(x) - \xi(x))(u(x) - \bar{u}(x))\, dx$$
$$\leq \int_{\{u>\bar{u}\}} c_1\, (u(x) - \bar{u}(x))^p\, dx, \tag{7.14}$$

for certain $\bar{\xi}(x) \in \partial j(\bar{u}(x))$ and $\xi(x) \in \partial j(u(x))$. As

$$\langle B(u), (u - \bar{u})^+\rangle = \int_{\{u>\bar{u}\}} (u - \bar{u})^p\, dx,$$

we derive from (7.13), (7.14), and thanks to (iii)(b) in Definition 7.3, the estimate

$$(\lambda - c_1 - \tilde{c}) \int_{\{u > \bar{u}\}} (u - \bar{u})^p \, dx \leq 0. \tag{7.15}$$

Selecting the parameter λ, in addition, such that $\lambda - c_1 - \tilde{c} > 0$, then (7.15) yields

$$\int_\Omega \left((u - \bar{u})^+ \right)^p \, dx \leq 0,$$

which implies $u \leq \bar{u}$.

The proof for the inequality $\underline{u} \leq u$ can be carried out in a similar way. Specifically, by Definition 7.2, the subsolution $\underline{u}$ satisfies: $\underline{u} \in \text{dom}\,(\hat{\psi})$, $\underline{u} \leq 0$ on $\partial\Omega$, and for all $v \in \underline{u} \wedge (\text{dom}\,(\psi) \cap V_0)$,

$$\langle A\underline{u} - f, v - \underline{u} \rangle + \hat{\psi}(v) - \hat{\psi}(\underline{u}) + \int_\Omega j^\circ(\underline{u}; v - \underline{u}) \, dx \geq 0. \tag{7.16}$$

Using the test functions $v = \underline{u} \wedge u = \underline{u} - (\underline{u} - u)^+ \in \underline{u} \wedge (\text{dom}\,(\psi) \cap V_0)$ in (7.16) and $v = \underline{u} \vee u = u + (\underline{u} - u)^+ \in \text{dom}\,(\psi) \cap V_0$ in (7.6), respectively, we get by adding the resulting inequalities the following one:

$$\langle Au - A\underline{u}, (\underline{u} - u)^+ \rangle + \lambda \langle B(u), (\underline{u} - u)^+ \rangle$$
$$+ \hat{\psi}(\underline{u} \wedge u) - \hat{\psi}(\underline{u}) + \psi(\underline{u} \vee u) - \psi(u)$$
$$+ \int_\Omega \left(j^\circ(\underline{u}; -(\underline{u} - u)^+) + j^\circ(u; (\underline{u} - u)^+) \right) dx \geq 0.$$

Along the same lines as above, we arrive at

$$(\lambda - c_1 - \hat{c}) \int_{\{\underline{u} > u\}} (\underline{u} - u)^p \, dx \leq 0.$$

Choosing $\lambda - c_1 - \hat{c} > 0$ implies $\underline{u} \leq u$.

Steps 1 and 2 complete the proof of the theorem. Indeed, Step 1 ensures the existence of a solution to the auxiliary problem (7.6). Taking into account Step 2 and the definition of the cutoff operator B, we conclude that any solution of problem (7.6) becomes a solution of problem (7.1), so a solution of problem (7.1) exists. $\square$

7.1.2 Compactness and Extremality

In this subsection, we focus on some qualitative properties (compactness and existence of extremal solutions) for the set of solutions of problem (7.1).

Theorem 7.10. *Let $\underline{u}$ and $\bar{u}$ be a sub- and supersolution of (7.1), respectively, satisfying $\underline{u} \leq \bar{u}$. Under the hypotheses of Theorem 7.9, the set S of all solutions of (7.1) within the interval $[\underline{u}, \bar{u}]$ is compact in V_0.*

Proof: First we prove that $\mathcal{S}$ is bounded in V_0. As $\mathcal{S}$ is contained in the interval $[\underline{u}, \bar{u}]$, it follows that $\mathcal{S}$ is bounded in $L^p(\Omega)$. Moreover, any $u \in \mathcal{S}$ solves (7.1); i.e., u satisfies $u \in \text{dom}(\psi) \cap V_0$ and

$$\langle Au - f, v - u \rangle + \psi(v) - \psi(u) + \int_\Omega j^\circ(u; v - u)\, dx \geq 0, \quad \forall\, v \in V_0.$$

Let u_0 be any (fixed) element of $\text{dom}(\psi) \cap V_0$. By taking $v = u_0$ in the above inequality, we get

$$\langle Au, u \rangle \leq \langle Au, u_0 \rangle + \langle f, u - u_0 \rangle + \psi(u_0) - \psi(u) + \int_\Omega j^\circ(u; u_0 - u)\, dx.$$

As ψ is bounded below by an affine function on V, we get the following estimate with some nonnegative constant d:

$$\psi(u) \geq -d(\|u\|_V + 1).$$

Taking into account Young's inequality and the equivalence of the norms $\|\cdot\|_V$ and $\|\nabla \cdot\|_{L^p(\Omega)}$ on V_0, it turns out that

$$\psi(u) \geq -\frac{\nu}{2}\|\nabla u\|_{L^p(\Omega)}^p - D$$

for some constant $D > 0$ not depending on u. By means of the last inequality and by applying (A3), (H)(ii), and Young's inequality, we obtain the estimate

$$\frac{\nu}{2}\|\nabla u\|_{L^p(\Omega)}^p \leq \|k_1\|_{L^1(\Omega)} + c(\varepsilon)(\|f\|_{V_0^*}^q + 1) + \varepsilon\|u\|_{V_0}^p + \tilde{\alpha}\left(\|u\|_{L^p(\Omega)}^p + 1\right)$$

for any $\varepsilon > 0$ and a constant $\tilde{\alpha} > 0$. Hence, the boundedness of $\mathcal{S}$ in V_0 follows by choosing ε sufficiently small and by using that $\mathcal{S}$ is bounded in $L^p(\Omega)$.

Let $(u_n) \subset \mathcal{S}$. Knowing the boundedness of $\mathcal{S}$ in V_0, we can pick a subsequence (u_k) of (u_n) such that

$$u_k \rightharpoonup u \text{ in } V_0, \quad u_k \to u \text{ in } \; L^p(\Omega), \quad \text{and} \quad u_k(x) \to u(x) \text{ a.e. in } \; \Omega. \quad (7.17)$$

Obviously $u \in [\underline{u}, \bar{u}]$. As each u_k solves (7.1), we can put $v = u \in V_0$ in (7.1) (with u_k instead of u) and get

$$\langle Au_k - f, u - u_k \rangle + \psi(u) - \psi(u_k) + \int_\Omega j^\circ(u_k; u - u_k)\, dx \geq 0,$$

and thus,

$$\langle Au_k, u_k - u \rangle \leq \langle f, u_k - u \rangle + \psi(u) - \psi(u_k) + \int_\Omega j^\circ(u_k; u - u_k)\, dx. \quad (7.18)$$

From (7.17) and because $(s, r) \mapsto j^\circ(s; r)$ is upper semicontinuous [cf. Proposition 2.162(iii)], we infer by applying Fatou's lemma

$$\limsup_{k} \int_{\Omega} j^{\circ}(u_k; u - u_k)\, dx \leq \int_{\Omega} \limsup_{k} j^{\circ}(u_k; u - u_k)\, dx = 0. \qquad (7.19)$$

In view of (7.19), we thus obtain from (7.17), (7.18), and because ψ is weakly lower semicontinuous

$$\limsup_{k} \langle Au_k, u_k - u \rangle \leq 0. \qquad (7.20)$$

As the operator A has the (S_+)-property (see Theorem 2.109(iii)), the weak convergence of (u_k) in V_0 along with (7.20) imply the strong convergence $u_k \to u$ in V_0. Moreover, the limit u belongs to S as can be seen by passing to the lim sup in the inequality

$$\langle Au_k - f, v - u_k \rangle + \psi(v) - \psi(u_k) + \int_{\Omega} j^{\circ}(u_k; v - u_k)\, dx \geq 0,$$

and using Fatou's lemma, the lower semicontinuity of ψ and the strong convergence of $u_k \to u$ in V_0. This process completes the proof.   □

We are now ready to prove our extremality result for problem (7.1).

Theorem 7.11. *Let the hypotheses of Theorem 7.9 be satisfied, and assume, moreover,*

$$\mathrm{dom}(\psi) \wedge \mathrm{dom}(\psi) \subset \mathrm{dom}(\psi) \quad \text{and} \quad \mathrm{dom}(\psi) \vee \mathrm{dom}(\psi) \subset \mathrm{dom}(\psi). \qquad (7.21)$$

If there is a constant $c \geq 0$ such that

$$\psi(w \vee v) - \psi(w) + \psi(w \wedge v) - \psi(v) \leq c \int_{\{v > w\}} (v - w)^p\, dx \qquad (7.22)$$

for all $w, v \in \mathrm{dom}(\psi)$, then the set S of all solutions of (7.1) within the interval $[\underline{u}, \bar{u}]$ possesses extremal elements.

Proof: Step 1: S is a directed set.

Theorem 7.9 ensures that $S \neq \emptyset$. Given $u_1, u_2 \in S$, let us show that there is a $u \in S$ such that $u_k \leq u$, $k = 1, 2$, which means S is upward directed. To this end, we consider the following auxiliary variational–hemivariational inequality. Find $u \in \mathrm{dom}(\psi) \cap V_0$ such that

$$\langle Au - f + \lambda B(u), v - u \rangle + \psi(v) - \psi(u) + \int_{\Omega} j^{\circ}(u; v - u)\, dx \geq 0, \quad \forall\, v \in V_0, \qquad (7.23)$$

where $\lambda \geq 0$ is a free parameter to be chosen later. Unlike in the proof of Theorem 7.9, the operator B is now given by the following cutoff function $b : \Omega \times \mathbb{R} \to \mathbb{R}$:

$$b(x,s) = \begin{cases} (s - \bar{u}(x))^{p-1} & \text{if } s > \bar{u}(x), \\ 0 & \text{if } u_0(x) \le s \le \bar{u}(x), \\ -(u_0(x) - s)^{p-1} & \text{if } s < u_0(x), \end{cases} \qquad (7.24)$$

where $u_0 = \max(u_1, u_2)$. By arguments similar to those in the proof of Theorem 7.9, we deduce the existence of solutions of (7.23) (see Step 1 in the proof of Theorem 7.9). The set $\mathcal{S}$ is shown to be upward directed provided that any solution u of (7.23) satisfies $u_k \le u \le \bar{u}$, $k = 1, 2$, because then $Bu = 0$ [cf. (7.24)] and thus $u \in \mathcal{S}$.

As $u_k \in \mathcal{S}$, we have $u_k \in \operatorname{dom}(\psi) \cap V_0 \cap [\underline{u}, \bar{u}]$ and

$$\langle Au_k - f, v - u_k \rangle + \psi(v) - \psi(u_k) + \int_\Omega j^o(u_k; v - u_k)\, dx \ge 0, \quad \forall\, v \in V_0.$$
$$(7.25)$$

Note that (7.21) implies

$$u + (u_k - u)^+ = u \vee u_k \in \operatorname{dom}(\psi) \cap V_0$$

and

$$u_k - (u_k - u)^+ = u \wedge u_k \in \operatorname{dom}(\psi) \cap V_0.$$

Set $v = u + (u_k - u)^+$ in (7.23) and $v = u_k - (u_k - u)^+$ in (7.25). Adding the resulting inequalities, we obtain

$$\langle Au_k - Au, (u_k - u)^+ \rangle - \lambda \langle B(u), (u_k - u)^+ \rangle$$
$$\le \psi(u \vee u_k) - \psi(u) + \psi(u \wedge u_k) - \psi(u_k)$$
$$+ \int_\Omega \Big(j^o(u; (u_k - u)^+) + j^o(u_k; -(u_k - u)^+) \Big)\, dx. \qquad (7.26)$$

Arguing as in (7.14), we have the estimate

$$\int_\Omega \Big(j^o(u; (u_k - u)^+) + j^o(u_k; -(u_k - u)^+) \Big)\, dx$$
$$\le \int_{\{u_k > u\}} c_1 (u_k(x) - u(x))^p\, dx. \qquad (7.27)$$

By hypothesis (A2), we know

$$\langle Au_k - Au, (u_k - u)^+ \rangle \ge 0, \qquad (7.28)$$

whereas (7.24) yields

$$\langle B(u), (u_k - u)^+ \rangle = -\int_{\{u_k > u\}} (u_0(x) - u(x))^{p-1}(u_k(x) - u(x))\, dx$$
$$\le -\int_{\{u_k > u\}} (u_k(x) - u(x))^p\, dx. \qquad (7.29)$$

Combining (7.26)–(7.29) and assumption (7.22) leads to

$$(\lambda - c_1 - c) \int_{\{u_k > u\}} (u_k(x) - u(x))^p \, dx \leq 0. \qquad (7.30)$$

Choosing some λ with $\lambda > c_1 + c$, from (7.30), we deduce $u_k \leq u$.

The proof for $u \leq \bar{u}$ follows arguments similar to the ones in Step 2 of the proof of Theorem 7.9. Thus, the ordered set $\mathcal{S}$ is upward directed.

The fact that $\mathcal{S}$ is downward directed can be shown analogously arguing on a corresponding auxiliary problem.

Step 2: Existence of extremal solutions.

We show only the existence of the greatest element of $\mathcal{S}$. The existence of the smallest element of $\mathcal{S}$ can be proved in a similar way. As V_0 is separable, we have that $\mathcal{S} \subset V_0$ is separable too. Fix a countable, dense subset $Z = \{z_n : n \in \mathbb{N}\}$ of $\mathcal{S}$. We construct an increasing sequence $(u_n) \subset \mathcal{S}$ as follows. Let $u_1 = z_1$. Assuming that $u_n \in \mathcal{S}$ is constructed, then Step 1 enables us to select $u_{n+1} \in \mathcal{S}$ such that

$$\max\{z_n, u_n\} \leq u_{n+1} \leq \bar{u}.$$

Theorem 7.10 ensures that the set $\mathcal{S}$ is compact in V_0. Consequently, we can fix a subsequence of (u_n), denoted again (u_n), and an element $u \in \mathcal{S}$ such that $u_n \to u$ in V_0, and $u_n(x) \to u(x)$ a.e. in Ω. This last property of (u_n) combined with its increasing monotonicity implies that the entire sequence is convergent to u in V_0, and moreover, $u = \sup_n u_n$. By construction, we see

$$\max\{z_1, z_2, \ldots, z_n\} \leq u_{n+1} \leq u, \quad \forall\, n;$$

thus $Z \subset [\underline{u}, u]$. As the interval $[\underline{u}, u]$ is closed in V_0, we infer

$$\mathcal{S} \subset \overline{Z} \subset \overline{[\underline{u}, u]} = [\underline{u}, u].$$

In conjunction with $u \in \mathcal{S}$, this guarantees that u is the greatest solution of (7.1) within $[\underline{u}, \bar{u}]$. $\qquad \square$

Remark 7.12. We note for the proof of Theorem 7.11 it is enough to assume instead of (7.21) that

$$\mathrm{dom}\,(\psi) \wedge (\mathrm{dom}\,(\psi) \cap [\underline{u}, \bar{u}]) \subset \mathrm{dom}\,(\psi)$$

and

$$\mathrm{dom}\,(\psi) \vee (\mathrm{dom}\,(\psi) \cap [\underline{u}, \bar{u}]) \subset \mathrm{dom}\,(\psi).$$

Remark 7.13. Condition (7.22) cannot be simplified to have the right-hand side equal to zero [as, for example, in (7.3)]. There are functionals $\psi : V_0 \to \mathbb{R}$ for which condition (7.22) is satisfied provided $c > 0$. For instance, let $\psi :$

$V_0 \to \mathbb{R}$ be the function $\psi = \psi_1|_{V_0}$ with $\psi_1 : L^p(\Omega) \to \mathbb{R}$ differentiable and convex. The differential at $u \in V_0$ is denoted $\psi'(u) \in V_0^*$ and is equal to $\psi'(u) = i^*\psi_1'(u)$ in V_0^*, with $\psi_1'(u) \in L^q(\Omega)$ and the inclusion map $i : V_0 \to L^p(\Omega)$. Assume that a constant $c > 0$ exists such that whenever $v, w \in V_0$, we have

$$\psi_1'(v) - \psi_1'(w) \le c(v-w)^{p-1} \quad \text{for a.e. on } \{w < v\}.$$

For all $w, v \in V_0$, we find that

$$
\begin{aligned}
&\psi(w \vee v) - \psi(w) + \psi(w \wedge v) - \psi(v) \\
&\le \int_\Omega \psi_1'(w \vee v)(w \vee v - w)\, dx + \int_\Omega \psi_1'(w \wedge v)(w \wedge v - v)\, dx \\
&= \int_\Omega (\psi_1'(w + (v-w)^+) - \psi_1'(v - (v-w)^+))(v-w)^+\, dx \\
&= \int_{\{w<v\}} (\psi_1'(v) - \psi_1'(w))(v-w)\, dx \le c \int_{\{w<v\}} (v-w)^p\, dx,
\end{aligned}
$$

so (7.22) is valid with $c > 0$.

In the rest of this section, we illustrate the applicability of our results to a variational–hemivariational inequality with constraints described by an obstacle problem.

Let $f \in L^\infty(\Omega) \subset V_0^*$, and let $K \subset V_0$ represent the following obstacle:

$$K = \{v \in V_0 : v(x) \le \phi(x) \text{ for a.e. } x \in \Omega\}, \tag{7.31}$$

with $\phi : \Omega \to \mathbb{R}$ measurable. Let $g : V \to \mathbb{R} \cup \{+\infty\}$ be the integral functional introduced in Example 7.8 (described by a convex lower semicontinuous function $h : \mathbb{R} \to \mathbb{R}$) and $I_K : V \to \mathbb{R} \cup \{+\infty\}$ be the indicator function related with the set K in (7.31) assuming $K \ne \emptyset$. Then the functional $\psi : V \to \mathbb{R} \cup \{+\infty\}$ defined by

$$\psi = I_K + g$$

is proper, convex, and lower semicontinuous with $\mathrm{dom}\,(\psi) = K \cap \mathrm{dom}\,(g)$. With f and ψ as specified above, we consider the variational–hemivariational inequality (7.1); i.e., we are looking for a $u \in K \cap \mathrm{dom}\,(g)$ such that

$$\langle Au - f, v - u \rangle + \psi(v) - \psi(u) + \int_\Omega j^\circ(u; v - u)\, dx \ge 0, \quad \forall\, v \in V_0. \tag{7.32}$$

The following theorem provides conditions that ensure the existence of an ordered pair of constant sub- and supersolutions of (7.32).

Theorem 7.14. *Let $a_i(x, 0) \equiv 0$ for all $1 \le i \le N$, and let the constants $\alpha \le 0$, $\beta \ge 0$ satisfy the conditions:*

(i) $\alpha \le \phi(x)$ *for a.e. $x \in \Omega$.*

(ii) *For some $\xi \in \partial h(\alpha)$, $\eta \in \partial h(\beta)$, the following inequalities are satisfied:*

$$-j^{\circ}(\alpha; -1) + \xi \leq f(x) \leq j^{\circ}(\beta; 1) + \eta \quad \text{for a.e. } x \in \Omega. \tag{7.33}$$

Then the constant functions $\underline{u} = \alpha$ and $\bar{u} = \beta$ form an ordered pair of sub-and supersolutions of (7.32).

Proof: First let us verify that $\underline{u}(x) \equiv \alpha$ is a subsolution according to Definition 7.2. Recall that $\text{dom}(\psi) = K \cap \text{dom}(g)$. As $\alpha \in \text{dom}(g)$, $\alpha \leq 0$ and $\alpha \leq \phi$ [see (i)], we get $\alpha \vee (\text{dom}(\psi) \cap V_0) \subset \text{dom}(\psi) \cap V_0$, and thus (i) and (ii) of Definition 7.2 are satisfied. To verify (iii) of Definition 7.2, we need to construct an appropriate functional $\hat{\psi}$ with the properties (a)–(c) of Definition 7.2. To this end, we set $\hat{\psi} = g$. Then (a) is satisfied, because $\alpha \in \text{dom}(g)$. For $v \in \text{dom}(\psi) \cap V_0 = K \cap \text{dom}(g)$, we obtain

$$\psi(v \vee \underline{u}) + \hat{\psi}(v \wedge \underline{u}) - \psi(v) - \hat{\psi}(\underline{u}) = g(v \vee \alpha) + g(v \wedge \alpha) - g(v) - g(\alpha) = 0. \tag{7.34}$$

The second equality of (7.34) can easily be shown by splitting up the domain Ω into $\Omega = \Omega_1 \cup \Omega_2 = \{x \in \Omega : v(x) \geq \alpha\} \cup \{x \in \Omega : v(x) < \alpha\}$. It follows from (7.34) that (b) of Definition 7.2 is verified with $\hat{c} = 0$. To see that also (c) of Definition 7.2 is valid, let $v \in \alpha \wedge (K \cap \text{dom}(g))$. Then $v - \alpha \leq 0$ in Ω, and by (7.33), we get

$$\langle A\alpha - f, v - \alpha \rangle + g(v) - g(\alpha) + \int_{\Omega} j^{\circ}(\alpha; v(x) - \alpha) \, dx$$

$$\geq \int_{\Omega} \Big(j^{\circ}(\alpha; -1) + f(x) - \xi \Big)(\alpha - v(x)) \, dx \geq 0,$$

which proves that α is a subsolution.

Let us show that β is a supersolution of (7.32). We readily see that $\beta \wedge K \subset K$ and $\beta \wedge \text{dom}(g) \subset \text{dom}(g)$ holds, and thus (i) and (ii) of Definition 7.3 are satisfied. It remains to check (iii) of Definition 7.3. To this end, we show that with $\tilde{\psi} = g$ and applying (7.33), the conditions (a)–(c) of Definition 7.3 can be fulfilled. We have $\beta \in \text{dom}(g)$, and for $v \in K \cap \text{dom}(g)$, the following equalities are satisfied:

$$\psi(v \wedge \bar{u}) + \tilde{\psi}(v \vee \bar{u}) - \psi(v) - \tilde{\psi}(\bar{u}) = g(v \wedge \beta) + g(v \vee \beta) - g(v) - g(\beta) = 0,$$

which shows that (b) of Definition 7.3 holds with $\tilde{c} = 0$. Finally, to verify (c), let $v \in \beta \vee (K \cap \text{dom}(g))$; then $v \geq \beta$, and we obtain by means of (7.33),

$$\langle A\beta - f, v - \beta \rangle + g(v) - g(\beta) + \int_{\Omega} j^{\circ}(\beta; v(x) - \beta) \, dx$$

$$\geq \int_{\Omega} \Big(j^{\circ}(\beta; 1) - f(x) + \eta \Big)(v(x) - \beta) \, dx \geq 0,$$

which proves that the constant $\beta \geq 0$ is a supersolution. $\qquad \square$

Our results in studying problem (7.32) through an ordered pair of constant sub- and supersolutions are summarized in the next statement.

Corollary 7.15. *Let the hypotheses of Theorem 7.14, (A1)–(A3), and (H) be satisfied. Then the variational–hemivariational inequality (7.32) has the property that the set of solutions belonging to the order interval $[\alpha, \beta]$ possesses extremal elements. Moreover, the set S of all solutions of (7.32) within $[\alpha, \beta]$ is compact.*

Proof: By Theorem 7.14 the constants α and β form an ordered pair of sub- and supersolutions, respectively. Then Theorem 7.9 and Theorem 7.10 provide the existence of solutions within $[\alpha, \beta]$ and the compactness of the set S of such solutions. For the existence of extremal solutions, we apply Theorem 7.11. To this end, we only need to verify conditions (7.21) and (7.22) for the specific functional $\psi = I_K + g$ considered here. It can easily be seen that the following is true: $K \vee K \subset K$, $K \wedge K \subset K$, $\mathrm{dom}\,(g) \vee \mathrm{dom}\,(g) \subset \mathrm{dom}\,(g)$, and $\mathrm{dom}\,(g) \wedge \mathrm{dom}\,(g) \subset \mathrm{dom}\,(g)$, and hence condition (7.21) holds [note that $\mathrm{dom}\,(\psi) = K \cap \mathrm{dom}\,(g)$]. For $w, v \in K \cap \mathrm{dom}\,(g)$, we have

$$\psi(w \vee v) - \psi(w) + \psi(w \wedge v) - \psi(v)$$
$$= g(w \vee v) - g(w) + g(w \wedge v) - g(v) = 0,$$

and thus, (7.22) is satisfied (with $c = 0$). This process completes the proof. $\square$

Finally, we discuss an example that provides a sufficient condition for zero to be a subsolution of problem (7.1). In the proof, we demonstrate the flexibility in the choice of the auxiliary functional $\hat{\psi}$ entering Definition 7.2.

Example 7.16. Assume that the operator A satisfies $a_i(x, 0) \equiv 0$ for all $1 \leq i \leq N$. Let $\psi : V_0 \to \mathbb{R}$ be given by

$$\psi(v) = \frac{\lambda}{p} \int_\Omega |v|^p \, dx, \quad \forall \, v \in V_0,$$

for some $\lambda \geq 0$, and let $f \in L^{p^{*'}}(\Omega)$ (p^* being the critical Sobolev exponent and $p^{*'}$ its conjugate) such that $f(x) \geq -j^\circ(0; -1)$ for a.e. $x \in \Omega$, where $j : \mathbb{R} \to \mathbb{R}$ verifies assumption (H). Then $\underline{u} = 0$ is a subsolution of problem (7.1). Toward this end, we need to verify the conditions of Definition 7.2. As $\mathrm{dom}\,(\psi) = V_0$, (i) and (ii) of Definition 7.2 are trivially satisfied. To check condition (iii), let us choose the function $\hat{\psi} : V \to \mathbb{R}$ in the form

$$\hat{\psi}(v) = \frac{m\lambda}{p} \int_\Omega |v|^p \, dx, \quad \forall \, v \in V,$$

where $m \in [0, +\infty)$. Condition (iii)(a) is evident, whereas condition (iii)(b) is verified because we have

$$\psi(v^+) + \hat{\psi}(-v^-) - \psi(v) - \hat{\psi}(0)$$

$$= \frac{\lambda}{p} \left[\int_{\Omega} |v^+|^p \, dx + m \int_{\Omega} |v^-|^p \, dx - \left(\int_{\Omega} |v^+|^p \, dx + \int_{\Omega} |v^-|^p \, dx \right) \right]$$

$$= \frac{(m-1)\lambda}{p} \int_{\Omega} |v^-|^p \, dx = \frac{(m-1)\lambda}{p} \int_{\Omega} |(-v)^+|^p \, dx, \quad \forall \, v \in V_0,$$

and thus condition (iii)(b) holds with $\hat{c} = 0$ for $m \in [0,1]$, and a positive constant $\hat{c} = \frac{(m-1)\lambda}{p}$ for $m > 1$. It remains to justify condition (iii)(c); that is,

$$\langle -f, v \rangle + \frac{m\lambda}{p} \int_{\Omega} |v|^p \, dx + \int_{\Omega} j^\circ(0; v) \, dx \geq 0, \quad \forall \, v \in 0 \wedge V_0.$$

Setting $v = -w^-$ with $w \in V_0$, this reads

$$\int_{\Omega} \left(f + \frac{m\lambda}{p} (w^-)^{p-1} + j^\circ(0; -1) \right) w^- \, dx \geq 0,$$

which in view of our assumption is true for any $m \in [0, +\infty)$.

7.2 Evolution Variational–Hemivariational Inequalities

To formulate our evolution problem, let $\Omega \subset \mathbb{R}^N$ be a bounded domain with Lipschitz boundary $\partial \Omega$, $Q = \Omega \times (0, \tau)$, and $\Gamma = \partial \Omega \times (0, \tau)$, with $\tau > 0$.

Consider the following quasilinear evolutionary variational–hemivariational inequality:

Find $u \in W \cap K$, $u(\cdot, 0) = 0$ in Ω,

$$\left\langle \frac{\partial u}{\partial t} + Au - f, v - u \right\rangle + \int_{\Gamma} j^\circ(\gamma u; \gamma v - \gamma u) d\Gamma \geq 0, \quad \forall \, v \in K, \quad (7.35)$$

where K is a closed and convex subset of $X = L^p(0, \tau; W^{1,p}(\Omega))$, for some $2 \leq p < \infty$. Here $\langle \cdot, \cdot \rangle$ denotes the duality pairing between X and its dual X^*, whereas $W = \{ w \in X : \partial w / \partial t \in X^* \}$. The derivative $u' := \partial u / \partial t$ is understood in the sense of vector-valued distributions (see Definition 2.138). By $j^\circ(s; r)$, we denote the generalized directional derivative of a locally Lipschitz function $j : \mathbb{R} \to \mathbb{R}$ at s in the direction r. The operator $A : X \to X^*$ is assumed to be $A = -\Delta_p$, where $\Delta_p u = \text{div}\,(|\nabla u|^{p-2} \nabla u)$ is the p-Laplacian, $f \in L^q(Q) \subset X^*$, with q being the Hölder conjugate of p, and $\gamma : X \to L^p(\Gamma)$ denotes the trace operator.

The fact that we have only taken into account the p-Laplacian $A = -\Delta_p :$ $X \to X^*$, i.e.,

$$\langle A(u), v \rangle = \int_{Q} |\nabla u|^{p-2} \nabla u \nabla v \, dx \, dt, \quad \forall \, v \in X,$$

is for emphasizing the main ideas. However, it should be noted that our results can be extended to general second-order quasilinear differential operators A of Leray–Lions type in the form

$$Au(x,t) = -\sum_{i=1}^{N} \frac{\partial}{\partial x_i} a_i(x,t,u(x,t),\nabla u(x,t)) + a_0(x,t,u(x,t),\nabla u(x,t))$$

(see also Definition 2.103).

We point out that problem (7.35) includes various important special cases:

(i) If $K = X$ and $j : \mathbb{R} \to \mathbb{R}$ is smooth with its derivative $j' : \mathbb{R} \to \mathbb{R}$, then (7.35) reduces to the weak formulation of the following parabolic initial boundary value problem:

$$u \in W : \quad \frac{\partial u}{\partial t} + Au = f \quad \text{in } X^*,$$

$$u(\cdot,0) = 0 \quad \text{in } \Omega, \quad \text{and} \quad -\frac{\partial u}{\partial \nu} = j'(u) \text{ on } \Gamma,$$

where $\partial/\partial \nu$ denotes the exterior conormal derivative on Γ associated with the operator A. The method of sub-supersolution for quasilinear parabolic initial boundary value problems is well established and was the subject of Chap. 3. Even though in Chap. 3 Dirichlet boundary conditions have been treated only, the method can easily be extended to nonlinear boundary conditions.

(ii) If $K = X$ and the locally Lipschitz function $j : \mathbb{R} \to \mathbb{R}$ is regular (see Definition 2.163), then (7.35) expresses the weak formulation of the following parabolic initial boundary inclusion problem:

$$u \in W : \quad \frac{\partial u}{\partial t} + Au = f \quad \text{in } X^*,$$

$$u(\cdot,0) = 0 \quad \text{in } \Omega, \quad \text{and} \quad -\frac{\partial u}{\partial \nu} \in \partial j(u) \text{ on } \Gamma,$$

where $\partial j : \mathbb{R} \to 2^{\mathbb{R}} \setminus \{\emptyset\}$ denotes the generalized gradient of j (cf. Definition 2.166). Existence and comparison results for parabolic inclusions with Clarke's gradient by using appropriately defined sub- and super-solutions have been obtained in Chap. 4 (see also recent papers by the authors [39, 58]).

(iii) If $K = X$ and $j : \mathbb{R} \to \mathbb{R}$ is a general locally Lipschitz function, then (7.35) reduces to an evolutionary hemivariational inequality of the form:

Find $u \in W$, $u(\cdot,0) = 0$ in Ω,

$$\left\langle \frac{\partial u}{\partial t} + Au - f, v - u \right\rangle + \int_{\Gamma} j^{\circ}(\gamma u; \gamma v - \gamma u)\, d\Gamma \geq 0, \quad \forall\, v \in X.$$

Evolutionary hemivariational inequalities under homogeneous Dirichlet boundary conditions have been studied in Chap. 6 and recently in [54]. Here the problem is different, because the boundary condition is of a hemivariational type.

(iv) For $j : \mathbb{R} \to \mathbb{R}$ smooth and for K being a closed and convex subset of X, (7.35) becomes a parabolic variational inequality with a nonlinear Robin type boundary condition:

$$\text{Find } u \in W \cap K, \ u(\cdot,0) = 0 \ \text{ in } \Omega, \ \text{ and } \ -\frac{\partial u}{\partial \nu} = j'(u) \text{ on } \Gamma$$

$$\left\langle \frac{\partial u}{\partial t} + Au - f, v - u \right\rangle \geq 0, \quad \forall \, v \in K.$$

Existence and comparison results for parabolic variational inequalities under homogeneous Dirichlet boundary condition have been obtained in in Chap. 5 (see also [51]).

7.2.1 Definitions and Hypotheses

Making use of the linear operator $L : D(L) \subset X \to X^*$ defined by $L := \partial/\partial t$ with the domain

$$D(L) = \{u \in X : u' \in X^* \text{ and } u(0) = 0\},$$

we note that the evolutionary variational–hemivariational inequality (7.35) may be rewritten as follows: Find $u \in D(L) \cap K$ such that

$$\langle Lu + A(u) - f, v - u \rangle + \int_\Gamma j^\circ(\gamma u; \gamma v - \gamma u) \, d\Gamma \geq 0, \quad \forall \, v \in K. \qquad (7.36)$$

We proceed by recalling some notational conventions related to the partial ordering in $L^p(Q)$ defined by $u \leq w$ if and only if $w - u$ belongs to the positive cone $L^p_+(Q)$ of all nonnegative elements of $L^p(Q)$. This result induces a corresponding partial ordering in the subspaces W and X of $L^p(Q)$ and implies a corresponding natural partial ordering for the traces; i.e., if $u, w \in X$ and $u \leq w$, then $\gamma u \leq \gamma w$ in $L^p(\Gamma)$. Given $u, w \in W$ with $u \leq w$, we put

$$[u, w] = \{v \in W : u \leq v \leq w\}.$$

Furthermore, for $u, v \in X$, and $U_1, U_2 \subset X$, we use the usual notation $u \wedge v = \min\{u, v\}$, $u \vee v = \max\{u, v\}$, $U_1 * U_2 = \{u * v : u \in U_1, v \in U_2\}$, and $u * U_1 = \{u\} * U_1$ with $* \in \{\wedge, \vee\}$.

Our basic notion of sub-and supersolution of (7.35) is now defined.

Definition 7.17. *A function $\underline{u} \in W$ is called a subsolution of (7.35) if the following holds:*

(i) $\underline{u}(\cdot, 0) \leq 0$ *in Ω.*
(ii) $\langle \underline{u}' + A\underline{u} - f, v - \underline{u} \rangle + \int_\Gamma j^\circ(\gamma \underline{u}; \gamma v - \gamma \underline{u}) \, d\Gamma \geq 0, \quad \forall \, v \in \underline{u} \wedge K.$

Definition 7.18. *$\bar{u} \in W$ is a supersolution of (7.35) if the following holds:*

(i) $\bar{u}(\cdot, 0) \geq 0$ in Ω.

(ii) $\langle \bar{u}' + A\bar{u} - f, v - \bar{u} \rangle + \int_\Gamma j^\circ(\gamma\bar{u}; \gamma v - \gamma\bar{u})\,d\Gamma \geq 0, \quad \forall\, v \in \bar{u} \vee K.$

Remark 7.19. The notions of sub- and supersolutions introduced here extend those of the special cases (i)–(iv) presented above. For example, let us show this in case (i); i.e., when $K = X$ and $j : \mathbb{R} \to \mathbb{R}$ is continuously differentiable with its derivative $j' : \mathbb{R} \to \mathbb{R}$. Let $\underline{u}$ be a subsolution of (7.35) according to Definition 7.17. By Example 2.168 and Proposition 2.171, it is clear that Definition 7.17 reads as

$$\langle \underline{u}' + A\underline{u} - f, v - \underline{u} \rangle + \int_\Gamma j'(\gamma\underline{u})\,(\gamma v - \gamma\underline{u})\,d\Gamma \geq 0, \quad \forall\, v \in \underline{u} \wedge X. \quad (7.37)$$

Given $\varphi \in X$; then $v = \underline{u} \wedge \varphi \in X$ is expressed by $v = \underline{u} - (\underline{u} - \varphi)^+$, where $w^+ := w \vee 0$. Thus, (7.37) yields

$$\langle \underline{u}' + A\underline{u} - f, -(\underline{u} - \varphi)^+ \rangle + \int_\Gamma j'(\gamma\underline{u})\,(-\gamma(\underline{u} - \varphi)^+)\,d\Gamma \geq 0, \quad \forall\, \varphi \in X. \quad (7.38)$$

It is not very hard to see that the set $Y = \{y \in X : y = (\underline{u} - \varphi)^+, \ \varphi \in X\} \subset X \cap L^p_+(Q)$ is dense in $X \cap L^p_+(Q)$. Then (7.38) implies

$$\langle \underline{u}' + A\underline{u} - f, \psi \rangle + \int_\Gamma j'(\gamma\underline{u})\,\gamma\psi\,d\Gamma \leq 0, \quad \forall\, \psi \in X \cap L^p_+(Q),$$

which verifies that $\underline{u}$ is a subsolution of the parabolic initial boundary value problem in the usual sense.

In our approach, we need the following notion.

Definition 7.20. *Let $C \neq \emptyset$ be a closed and convex subset of a reflexive Banach space X. A bounded, hemicontinuous, and monotone operator $P : X \to X^*$ is called a penalty operator associated with C if*

$$P(u) = 0 \iff u \in C. \quad (7.39)$$

We recall that if $p \geq 2$, there is a constant $c(p) > 0$ such that

$$(|\xi|^{p-2}\xi - |\xi'|^{p-2}\xi') \cdot (\xi - \xi') \geq c(p)|\xi - \xi'|^p, \quad \forall\, \xi, \xi' \in \mathbb{R}^N \quad (7.40)$$

(see also Sect. 2.2.4). Evidently, if $p = 2$, we have $c(p) = 1$. In the sequel, we shall use the positive constant $c(p)$ in (7.40).

We assume in this section the following hypotheses:

(H1) The generalized gradient ∂j of the locally Lipschitz function $j : \mathbb{R} \to \mathbb{R}$ verifies the growth conditions:

(i) A constant $c_1 \geq 0$ exists satisfying $c_1 \|\gamma\|^p < c(p)$ such that

$$\xi_1 \leq \xi_2 + c_1(s_2 - s_1)^{p-1}$$

for all $\xi_i \in \partial j(s_i)$, $i = 1, 2$, and for all s_1, s_2 with $s_1 < s_2$.
(ii) A constant $c_2 \geq 0$ exists satisfying $c_2 \|\gamma\|^p < 1$ such that

$$\xi \in \partial j(s): \quad |\xi| \leq c_2 \left(1 + |s|^{p-1}\right), \quad \forall s \in \mathbb{R}.$$

(H2) A penalty operator $P : X \to X^*$ exists associated with K fulfilling the requirement:
For each $u \in D(L)$, $w = w(u) \in X$, $w \neq 0$ if $P(u) \neq 0$, exists satisfying:
(i) A constant $\alpha > 0$ exists independent of u and w such that

$$\langle u' + Au, w \rangle \geq -\alpha(\|w\|_{L^p(Q)} + \|\gamma w\|_{L^p(\Gamma)}). \qquad (7.41)$$

(ii) A constant $D > 0$ exists independent of u and w such that

$$\langle P(u), w \rangle \geq D\|P(u)\|_{X^*}(\|w\|_{L^p(Q)} + \|\gamma w\|_{L^p(\Gamma)}). \qquad (7.42)$$

Remark 7.21. We shall see later that hypothesis (H2) can easily be satisfied for the obstacle problem described by the obstacle function ψ provided we have $\psi' + A\psi \geq 0$ in X^*.

7.2.2 Preliminary Results

Consider the functional $J : L^p(\Gamma) \to \mathbb{R}$ defined by

$$J(v) = \int_\Gamma j(v(x, t))\, d\Gamma, \quad \forall v \in L^p(\Gamma).$$

Using the growth condition (H1)(ii) and Theorem 2.177, we note that the functional J is well defined. Moreover, Theorem 2.181 ensures that it is Lipschitz continuous on bounded sets in $L^p(\Gamma)$, and for each $v \in L^p(\Gamma)$, its generalized gradient $\partial J : L^p(\Gamma) \to 2^{L^q(\Gamma)}$ $(1/p + 1/q = 1)$ satisfies

$$\xi \in \partial J(v) \implies \xi \in L^q(\Gamma) \text{ with } \xi(x, t) \in \partial j(v(x, t)) \text{ for a.e. } (x, t) \in \Gamma.$$

Let us introduce the multivalued mapping $\partial^\gamma J : X \to 2^{X^*}$ defined by

$$\partial^\gamma J(u) = \{u^* \in X^* : J^\circ(\gamma u; \gamma\varphi) \geq \langle u^*, \varphi \rangle, \quad \forall \varphi \in X\}.$$

Lemma 7.22. *The operator $\partial^\gamma J : X \to 2^{X^*}$ is bounded and pseudomonotone w.r.t. $D(L)$, where $L := \partial/\partial t$ and $D(L) = \{u \in X : u' \in X^* \text{ and } u(0) = 0\}$.*

Proof: Let us verify the conditions of Definition 2.154. As the functional $J : L^p(\Gamma) \to \mathbb{R}$ is locally Lipschitz, $\partial J(\gamma u)$ is nonempty for each $u \in X$; i.e., there is a $\xi \in L^q(\Gamma)$ such that

$$J^o(\gamma u; v) \geq \langle \xi, v \rangle, \quad \forall \, v \in L^p(\Gamma).$$

If $\gamma^* : L^q(\Gamma) \to X^*$ stands for the adjoint operator of γ, then we get

$$J^o(\gamma u; \gamma \varphi) \geq \langle \gamma^* \xi, \varphi \rangle, \quad \forall \, \varphi \in X,$$

which shows that $\gamma^* \xi \in \partial^\gamma J(u)$, so $\partial^\gamma J(u) \neq \emptyset$. To prove that $\partial^\gamma J(u)$ is bounded, we observe that for each $u \in X$, there is a constant $C_u \geq 0$ depending on u such that $|J^o(\gamma u; v)| \leq C_u \|v\|_{L^p(\Gamma)}$ for all $v \in L^p(\Gamma)$. Knowing that the trace operator $\gamma : X \to L^p(\Gamma)$ is linear and bounded, there is some positive constant C_γ such that $\|\gamma \varphi\|_{L^p(\Gamma)} \leq C_\gamma \|\varphi\|_X$ and we obtain

$$|J^o(\gamma u; \gamma \varphi)| \leq C \|\varphi\|_X, \tag{7.43}$$

where $C = C_u \, C_\gamma$. In view of the definition of $\partial^\gamma J(u)$ and by applying (7.43), we derive for $u^* \in \partial^\gamma J(u)$ the estimate

$$|\langle u^*, \varphi \rangle| \leq C \|\varphi\|_X, \quad \forall \, \varphi \in X. \tag{7.44}$$

Inequality (7.44) expresses that $\|u^*\| \leq C$, thus, the boundedness of $\partial^\gamma J(u)$. The convexity and closedness of the set $\partial^\gamma J(u)$ are obvious.

Let $M \subset X$ be bounded. We are going to show that the set

$$M^* = \bigcup_{u \in M} \partial^\gamma J(u)$$

is bounded in X^*. For a constant $C > 0$, we have $\|\gamma u\|_{L^p(\Gamma)} \leq C, \forall \, u \in M$. In view of the Lipschitz continuity of $J : L^p(\Gamma) \to \mathbb{R}$ on bounded sets, this implies the existence of some positive constants C_M and $\tilde{C}_M$ such that the following inequalities hold:

$$|J^o(\gamma u; \gamma v)| \leq C_M \|\gamma v\|_{L^p(\Gamma)} \leq \tilde{C}_M \|v\|_X, \quad \forall \, u \in M \text{ and } \forall \, v \in X. \tag{7.45}$$

By definition of $\partial^\gamma J$ and applying (7.45), we obtain for any $u^* \in M^*$ that

$$|\langle u^*, v \rangle| \leq |J^o(\gamma u; \pm \gamma v)| \leq \tilde{C}_M \|v\|_X, \quad \forall \, v \in X,$$

which leads to $\|u^*\| \leq \tilde{C}_M$; that is, M^* is bounded.

Next we show that $\partial^\gamma J : X \to 2^{X^*}$ satisfies (ii) of Definition 2.154. We prove an even stronger result, namely that $\partial^\gamma J$ is strongly-weakly upper semicontinuous at $u \in X$. Assume $\partial^\gamma J$ fails to have this property. Then there is a sequence $(u_k) \subset X$ with $u_k \to u$ in X and a sequence $(u_k^*) \in X^*$ with $u_k^* \rightharpoonup u^*$ (weakly) in X^* such that $u_k^* \in \partial^\gamma J(u_k)$ for each k, but u^* does not belong to $\partial^\gamma J(u)$. Thus, $\gamma u_k \to \gamma u$ in $L^p(\Gamma)$, which together with the weak convergence of (u_k^*) and the upper semicontinuity of $J^o : L^p(\Gamma) \times L^p(\Gamma) \to \mathbb{R}$ [see Proposition 2.162(iii)] results in

$$J^o(\gamma u; \gamma v) \geq \limsup_{k \to \infty} J^o(\gamma u_k; \gamma v) \geq \lim_{k \to \infty} \langle u_k^*, v \rangle = \langle u^*, v \rangle, \quad \forall \, v \in X.$$

The last inequality implies $u^* \in \partial^\gamma J(u)$ contradicting the assumption.

It remains to verify condition (iii) of Definition 2.154. To this end, let $(u_n) \subset D(L)$ with $u_n \rightharpoonup u$ in X, $Lu_n \rightharpoonup Lu$ in X^*, $u_n^* \in \partial^\gamma J(u_n)$ with $u_n^* \rightharpoonup u^*$ in X^*, and $\limsup \langle u_n^*, u_n - u \rangle \leq 0$. As $u_n^* \in \partial^\gamma J(u_n)$, we have

$$J^o(\gamma u_n; \gamma v) \geq \langle u_n^*, v \rangle, \quad \forall \, v \in X. \tag{7.46}$$

From the weak convergence of (u_n) and (Lu_n), it follows that $u_n \rightharpoonup u$ in W. By Proposition 2.143, we know that the trace operator $\gamma : W \to L^p(\Gamma)$ is compact, and thus we get $\gamma u_n \to \gamma u$ in $L^p(\Gamma)$. Taking into account inequality (7.46) and the fact that $J : L^p(\Gamma) \to \mathbb{R}$ is locally Lipschitz, we deduce

$$|\langle u_n^*, u_n - u \rangle| \leq C_u \|\gamma u_n - \gamma u\|_{L^p(\Gamma)}, \tag{7.47}$$

with a constant $C_u > 0$. From (7.47), we directly obtain $\langle u_n^*, u_n \rangle \to \langle u^*, u \rangle$ as $n \to \infty$, and by passing to the $\limsup$ in (7.46), we get $u^* \in \partial^\gamma J(u)$. $\quad\square$

Lemma 2.149 guarantees that the operator $L = \partial/\partial t : D(L) \subset X \to X^*$ is closed, densely defined, and maximal monotone. Related to this we have the following result.

Corollary 7.23. *The operator* $-\Delta_p + \partial^\gamma J : X \to 2^{X^*}$ *is pseudomonotone w.r.t.* $D(L)$ *and bounded.*

Proof: Let $A = -\Delta_p$. The operator $A : X \to X^*$ is continuous, bounded, and monotone. In particular, this result implies that $A : D(L) \subset X \to X^*$ is pseudomonotone w.r.t. $D(L)$ (see [43, Theorem E.3.2]). On the other hand, Lemma 7.22 establishes that the operator $\partial^\gamma J : X \to 2^{X^*}$ is bounded and pseudomonotone w.r.t. $D(L)$. As both operators A and $\partial^\gamma J$ are bounded and pseudomonotone w.r.t. $D(L)$, we only need to verify property (iii) of Definition 2.154 for the sum $A + \partial^\gamma J$. To this end, assume $(u_n) \subset D(L)$ with $u_n \rightharpoonup u$ in X, $Lu_n \rightharpoonup Lu$ in X^*, $u_n^* \in (A + \partial^\gamma J)(u_n)$ with $u_n^* \rightharpoonup u^*$ in X^*, and

$$\limsup_n \langle u_n^*, u_n - u \rangle \leq 0. \tag{7.48}$$

We must show that $u^* \in (A + \partial^\gamma J)(u)$ and $\langle u_n^*, u_n \rangle \to \langle u^*, u \rangle$. From $u_n^* \in (A + \partial^\gamma J)(u_n)$, we have $u_n^* = Au_n + \eta_n^*$ with $\eta_n^* \in \partial^\gamma J(u_n)$, and (7.48) reads as

$$\limsup_n \langle Au_n + \eta_n^*, u_n - u \rangle \leq 0. \tag{7.49}$$

It is known from Proposition 2.143 that the trace operator $\gamma : W \to L^p(\Gamma)$ is compact, so the weak convergence $u_n \rightharpoonup u$ in W allows us to get $\gamma u_n \to \gamma u$ in $L^p(\Gamma)$, and as for obtaining (7.47),

$$|\langle \eta_n^*, u_n - u \rangle| \leq C_u \|\gamma u_n - \gamma u\|_{L^p(\Gamma)} \to 0 \quad \text{as} \quad n \to \infty. \tag{7.50}$$

Combining (7.49) and (7.50) implies

$$\limsup_{n}\langle Au_n, u_n - u\rangle \leq 0.$$

The sequence $(Au_n) \subset X^*$ is bounded, so that there is some subsequence (Au_k) with $Au_k \rightharpoonup v$. By means of Theorem 2.153(i), we see that A is pseudomonotone w.r.t. $D(L)$. It follows that $v = Au$ and $\langle Au_k, u_k\rangle \to \langle Au, u\rangle$. This finding shows that each weakly convergent subsequence of (Au_n) has the same limit Au, and thus, the entire sequence (Au_n) satisfies

$$Au_n \rightharpoonup Au \quad \text{and} \quad \langle Au_n, u_n\rangle \to \langle Au, u\rangle. \tag{7.51}$$

From (7.51) and $u_n^* = Au_n + \eta_n^* \rightharpoonup u^*$, we obtain $\eta_n^* = u_n^* - Au_n \rightharpoonup u^* - Au$. As from (7.49) and (7.51) it follows that

$$\limsup_{n}\langle \eta_n^*, u_n - u\rangle \leq 0,$$

the pseudomonotonicity of $\partial^\gamma J$ (cf. Lemma 7.22) ensures $u^* - Au \in \partial^\gamma J(u)$; thus $u^* \in (A + \partial^\gamma J)(u)$, and

$$\langle u_n^* - Au_n, u_n\rangle \to \langle u^* - Au, u\rangle,$$

which according to (7.51) yields $\langle u_n^*, u_n\rangle \to \langle u^*, u\rangle$.   □

7.2.3 Existence and Comparison Result

Let $\underline{u}$, $\bar{u}$ be an ordered pair of sub- and supersolutions for problem (7.35). We introduce the usual cutoff function $b : Q \times \mathbb{R} \to \mathbb{R}$ related with this pair as follows:

$$b(x, t, s) = \begin{cases} (s - \bar{u}(x,t))^{p-1} & \text{if } s > \bar{u}(x,t), \\ 0 & \text{if } \underline{u}(x,t) \leq s \leq \bar{u}(x,t), \\ -(\underline{u}(x,t) - s)^{p-1} & \text{if } s < \underline{u}(x,t). \end{cases}$$

It is straightforward to verify that b is a Carathéodory function satisfying the growth condition

$$|b(x,t,s)| \leq k(x,t) + c_3 |s|^{p-1} \tag{7.52}$$

for a.e. $(x,t) \in Q$, for all $s \in \mathbb{R}$, with some function $k \in L_+^q(Q)$ and a constant $c_3 > 0$. Moreover, we have the estimate

$$\int_Q b(x,t,u(x,t))\, u(x,t)\, dx\, dt \geq c_4 \|u\|_{L^p(Q)}^p - c_5, \quad \forall\, u \in L^p(Q), \tag{7.53}$$

where c_4 and c_5 are some positive constants. Corresponding to the function b, we introduce the Nemytskij operator $B : L^p(Q) \to L^q(Q)$ defined by

$$Bu(x,t) = b(x,t,u(x,t)), \quad \forall\, u \in L^p(Q).$$

Remark 7.24. The role of the operator B introduced above is twofold. It is used later in some auxiliary problem as a coercivity generating term, and it will allow us to provide some comparison result.

Lemma 7.25. *Let $P : X \to X^*$ be a penalty operator related with the given closed, convex subset K of X in the sense of Definition 7.20. Then the (single-valued) operators $B, P : X \to X^*$ are bounded and pseudomonotone w.r.t. $D(L)$.*

Proof: In view of (7.52), the Nemytskij operator $B : L^p(Q) \to L^q(Q)$ is continuous and bounded. Thus, from the compact embedding $W \subset L^p(Q)$, it follows that $B : W \to L^q(Q) \subset X^*$ is completely continuous, so in particular, pseudomonotone w.r.t. $D(L)$. By definition, the penalty operator $P : X \to X^*$ is bounded, hemicontinuous, and monotone. This result implies that $P : X \to X^*$ is pseudomonotone in the usual sense [cf. Lemma 2.98 (i)] and thus pseudomonotone w.r.t. $D(L)$. $\square$

Using the above operator B, let us consider now the following auxiliary variational–hemivariational inequality: Find $u \in D(L) \cap K$ such that

$$\langle Lu + A(u) + \lambda B(u) - f, v - u \rangle + \int_\Gamma j^o(\gamma u; \gamma v - \gamma u) \, d\Gamma \geq 0, \quad \forall\, v \in K,$$
$$(7.54)$$

with a number $\lambda > 0$ that will be chosen later.

Lemma 7.26. *Let $\underline{u}$ and $\bar{u}$ be sub- and supersolutions of (7.35) satisfying $\underline{u} \leq \bar{u}$. Suppose furthermore that $D(L) \cap K \neq \emptyset$ and the hypotheses (H1) and (H2). Then problem (7.54) has solutions.*

Proof: We state a penalty problem related to (7.54): Find $u \in D(L)$ such that

$$\langle Lu + A(u) + \lambda B(u) + \frac{1}{\varepsilon} P(u) - f, v - u \rangle$$
$$+ \int_\Gamma j^o(\gamma u; \gamma v - \gamma u) \, d\Gamma \geq 0, \quad \forall\, v \in X, \qquad (7.55)$$

where $\varepsilon > 0$ is arbitrarily small, and P is the penalty operator associated with K whose existence is assumed by hypothesis (H2).

(a) Existence of solutions of (7.55).

Denote $\mathcal{A} = A + \lambda B + \frac{1}{\varepsilon} P + \partial^\gamma J : X \to 2^{X^*}$. We infer from Lemma 7.22 and Lemma 7.25 that the operator $\mathcal{A} : X \to 2^{X^*}$ is bounded and pseudomonotone w.r.t. $D(L)$. We claim that $\mathcal{A}$ is coercive in the sense of Definition 2.155. Toward this end, let $v^* \in \partial^\gamma J(v) \neq \emptyset$ (it was shown in the proof of Lemma 7.22 that $\partial^\gamma J(v) \neq \emptyset$). By means of (H1)(ii) and applying Theorem 2.181, we get

$$J^o(\gamma v; \gamma \varphi) \le \int_\Gamma j^o(\gamma v; \gamma \varphi) \, d\Gamma, \quad \forall \, v, \varphi \in X. \tag{7.56}$$

As $v^* \in \partial^\gamma J(v)$, relation (7.56) and assumption (H1)(ii) result in the estimate

$$|\langle v^*, v \rangle| \le c_2 \int_\Gamma (1 + |\gamma v|^{p-1}) |\gamma v| \, d\Gamma. \tag{7.57}$$

Taking into account that $\gamma : X \to L^p(\Gamma)$ is linear and bounded, from (7.57), we get

$$|\langle v^*, v \rangle| \le \tilde{c}_2 \|v\|_X + c_2 \|\gamma\|^p \|v\|_X^p \tag{7.58}$$

with a constant $\tilde{c}_2 > 0$. By Definition 7.20 of the penalty operator, we have

$$\langle P(v), v \rangle \ge \langle P(0), v \rangle \ge -\|P(0)\|_{X^*} \|v\|_X. \tag{7.59}$$

Thus, (7.53), (7.58), and (7.59) yield the estimate

$$\langle A(v) + \lambda B(v) + \frac{1}{\varepsilon} P(v) + v^*, v \rangle$$
$$\ge (1 - c_2 \|\gamma\|^p) \|\nabla v\|_{L^p(Q)}^p + (\lambda c_4 - c_2 \|\gamma\|^p) \|v\|_{L^p(Q)}^p - \tilde{c}_2 \|v\|_X$$
$$- \frac{1}{\varepsilon} \|P(0)\|_{X^*} \|v\|_X - c_5.$$

Selecting $\lambda > 0$ such that

$$\lambda > \frac{c_2 \|\gamma\|^p}{c_4},$$

and using the assumption $c_2 \|\gamma\|^p < 1$ [see (H1)(ii)], this proves the coercivity of $\mathcal{A}$. We are thus in a position to apply Theorem 2.156. It follows that range$(L + \mathcal{A}) = X^*$; i.e., there is a $u \in D(L)$ and an $\eta^* \in \partial^\gamma J(u)$ such that

$$Lu + A(u) + \lambda B(u) + \frac{1}{\varepsilon} P(u) + \eta^* = f \quad \text{in } X^*. \tag{7.60}$$

By definition of $\partial^\gamma J(u)$ and in view of (7.56), we conclude

$$\langle \eta^*, \varphi \rangle \le \int_\Gamma j^o(\gamma u; \gamma \varphi) \, d\Gamma, \quad \forall \, \varphi \in X. \tag{7.61}$$

Finally from (7.60) and (7.61), we derive that for any $\varepsilon > 0$, the penalty problem (7.55) has a solution.

(b) Boundedness of the penalty solutions in W.

According to part (a) for any $\varepsilon > 0$, a solution u_ε of (7.55) exists that satisfies equation (7.60). We show that the family $\{u_\varepsilon : \varepsilon > 0, \text{ small}\}$ is bounded with respect to the graph norm of $D(L)$. To this end, let u_0 be a (fixed) element of $D(L) \cap K$. Multiplying (7.60) (with u replaced by u_ε) by $v = u_\varepsilon - u_0$, we get

$$\langle Lu_\varepsilon + A(u_\varepsilon) + \lambda B(u_\varepsilon) + \frac{1}{\varepsilon} P(u_\varepsilon) + \eta_\varepsilon^*, u_\varepsilon - u_0 \rangle = \langle f, u_\varepsilon - u_0 \rangle,$$

where $\eta_\varepsilon^* \in \partial^\gamma J(u_\varepsilon)$. Using the monotonicity of L and that $Pu_0 = 0$ [cf. (7.39)], we get

$$\langle f - u_0', u_\varepsilon - u_0 \rangle$$
$$= \langle u_\varepsilon' - u_0', u_\varepsilon - u_0 \rangle + \langle (A + \lambda B)(u_\varepsilon), u_\varepsilon - u_0 \rangle + \frac{1}{\varepsilon} \langle Pu_\varepsilon - Pu_0, u_\varepsilon - u_0 \rangle$$
$$+ \langle \eta_\varepsilon^*, u_\varepsilon - u_0 \rangle$$
$$\geq \langle (A + \lambda B)(u_\varepsilon) + \eta_\varepsilon^*, u_\varepsilon - u_0 \rangle.$$

Thus,
$$\frac{\langle (A + \lambda B)(u_\varepsilon) + \eta_\varepsilon^*, u_\varepsilon - u_0 \rangle}{\|u_\varepsilon - u_0\|_X} \leq \|f - u_0'\|_{X^*},$$

for all $\varepsilon > 0$. As the operator $A + \lambda B + \partial^\gamma J : X \to X^*$ is coercive, we deduce that $\|u_\varepsilon\|_X$ is bounded. As a consequence, we see that the sets $(A(u_\varepsilon))$, $(B(u_\varepsilon))$, and (η_ε^*) are bounded in X^*. Moreover, from the growth conditions of b, we readily see that $(B(u_\varepsilon))$ is bounded in $L^q(Q)$. Recall that the penalty solutions u_ε satisfy (7.60); i.e.,

$$\left\langle Lu_\varepsilon + A(u_\varepsilon) + \lambda B(u_\varepsilon) + \frac{1}{\varepsilon} P(u_\varepsilon) + \eta_\varepsilon^*, \varphi \right\rangle = \langle f, \varphi \rangle, \quad \forall \, \varphi \in X. \quad (7.62)$$

We immediately see from (7.62) that (u_ε') is bounded if and only if $(\frac{1}{\varepsilon} P(u_\varepsilon))$ is bounded. Next, we check that the sequence $(\frac{1}{\varepsilon} P(u_\varepsilon))$ is bounded in X^*. To see this, for each ε, we choose $w = w_\varepsilon$ to be an element satisfying (7.41) and (7.42) with $u = u_\varepsilon$. From (7.62), we have

$$\langle u_\varepsilon', w_\varepsilon \rangle + \langle (A + \lambda B)(u_\varepsilon) + \eta_\varepsilon^*, w_\varepsilon \rangle + \frac{1}{\varepsilon} \langle Pu_\varepsilon, w_\varepsilon \rangle = \langle f, w_\varepsilon \rangle.$$

By using (7.41), we get

$$\frac{1}{\varepsilon} \langle P(u_\varepsilon), w_\varepsilon \rangle \leq \langle f - \lambda B(u_\varepsilon), w_\varepsilon \rangle - \langle \eta_\varepsilon^*, w_\varepsilon \rangle + \alpha(\|w_\varepsilon\|_{L^p(Q)} + \|\gamma w_\varepsilon\|_{L^p(\Gamma)}).$$
$$(7.63)$$

Let $c > 0$ be some generic constant. As $(\|B(u_\varepsilon)\|_{L^q(Q)})$ is bounded, we obtain

$$|\langle f - \lambda B(u_\varepsilon), w_\varepsilon \rangle| \leq c\|w_\varepsilon\|_{L^p(Q)}, \, \forall \, \varepsilon,$$

because $\lambda > 0$ is fixed. On the other hand, from (7.44) [see also (7.45)] and the boundedness of $\|u_\varepsilon\|_X$, we find that there is a constant $c > 0$ such that

$$|\langle \eta_\varepsilon^*, w_\varepsilon \rangle| \leq c\|\gamma w_\varepsilon\|_{L^p(\Gamma)}, \quad \forall \, \varepsilon.$$

Hence, we get for some $c > 0$,

$$|\langle f - \lambda B(u_\varepsilon) - \eta_\varepsilon^*, w_\varepsilon\rangle| \le c(\|w_\varepsilon\|_{L^p(Q)} + \|\gamma w_\varepsilon\|_{L^p(\Gamma)}), \quad \forall \varepsilon.$$

This result, (7.63), and (7.42) imply that

$$\frac{1}{\varepsilon}\|Pu_\varepsilon\|_{X^*} \le \frac{\alpha + c}{D}, \quad \forall \varepsilon.$$

Consequently, the set (u_ε) is bounded in W, and thus there is some weakly convergent subsequence (u_n) with $u_n = u_{\varepsilon_n}$ and $\varepsilon_n \to 0$ as $n \to \infty$; i.e.,

$$u_n \rightharpoonup u \quad \text{in } X, \quad u_n' \rightharpoonup u' \quad \text{in } X^*.$$

As $D(L)$ is closed in W and convex, it is weakly closed in W, and so $u \in D(L)$.

(c) The limit u solves (7.54).

We prove that u obtained as the weak limit in W as shown in part (b) above is a solution of inequality (7.54). We have already obtained in part (b) that $Pu_n \to 0$ in X^*. Then, on the basis of the monotonicity of P, it follows that

$$\langle Pv, v - u\rangle \ge 0, \quad \forall\, v \in X.$$

As in the proof of Minty's lemma (cf. [124]), using the hemicontinuity of P, we obtain from this inequality that

$$\langle Pu, v\rangle \ge 0, \quad \forall\, v \in X.$$

Hence, $Pu = 0$ in X^*; that is, $u \in K$ (see (7.39) in Definition 7.20).

Setting $v = u$ in the inequality (7.55) satisfied by the penalty solutions u_n enables us to write

$$\left\langle u_n' + A(u_n) + \lambda B(u_n) + \frac{1}{\varepsilon_n}P(u_n) - f, u - u_n \right\rangle + \int_\Gamma j^\circ(\gamma u_n; \gamma u - \gamma u_n)\, d\Gamma$$

$$\ge 0. \tag{7.64}$$

As

$$\langle u' - u_n', u - u_n\rangle \ge 0 \quad \text{and} \quad -\frac{1}{\varepsilon_n}\langle P(u_n), u - u_n\rangle \ge 0,$$

we derive from (7.64) the inequality

$$\langle A(u_n), u_n - u\rangle \le \langle u' + \lambda B(u_n) - f, u - u_n\rangle + \int_\Gamma j^\circ(\gamma u_n; \gamma u - \gamma u_n)\, d\Gamma.$$

$$\tag{7.65}$$

Theorem 2.141 ensures that the embedding $W \subset L^p(Q)$ is compact, whereas Proposition 2.143 guarantees the compactness of the trace operator $\gamma : W \to L^p(\Gamma)$. Then, from the upper semicontinuity of j° (see Proposition 2.162 (iii)), we get from (7.65)

$$\limsup_{n\to\infty}\langle A(u_n), u_n - u\rangle \le 0.$$

Because A is of class (S_+) with respect to $D(L)$ [cf. Lemma 2.111(i) and Theorem 2.153(ii)], we infer

$$u_n \to u \text{ in } X. \tag{7.66}$$

Inequality (7.55) with $v \in K$ entails that the penalty solutions u_n satisfy

$$\langle u_n' + A(u_n) + \lambda B(u_n) - f, v - u_n \rangle + \int_\Gamma j^o(\gamma u_n; \gamma u - \gamma u_n) \, d\Gamma$$

$$\geq \langle -\frac{1}{\varepsilon_n} P(u_n), v - u_n \rangle \geq 0.$$

The weak convergence of (u_n) in W and (7.66) allow us to pass to the limit as $n \to \infty$, proving that u is a solution of (7.54).   □

The main existence and comparison result of this section is now formulated.

Theorem 7.27. *Let the hypotheses of Lemma 7.26 be satisfied. Suppose furthermore that*

$$\underline{u} \vee K \subset K, \quad \bar{u} \wedge K \subset K.$$

Then the variational–hemivariational inequality (7.35) has solutions within the ordered interval $[\underline{u}, \bar{u}]$ formed by the pair of sub- and supersolutions $\underline{u}$ and $\bar{u}$ with $\underline{u} \leq \bar{u}$.

Proof: In view of Lemma 7.26, the auxiliary variational–hemivariational inequality (7.54) possesses solutions. To justify the assertion of Theorem 7.27, we only need to show that there are solutions of (7.54) lying within the interval $[\underline{u}, \bar{u}]$ of the given sub- and supersolutions, because in this case, $B(u) = 0$ and any solution of (7.54) must be also a solution of (7.35) or equivalently of (7.36). Let us check that $u \leq \bar{u}$, where u is a solution of (7.54) and $\bar{u}$ is the given supersolution of (7.35). We have $u \in D(L) \cap K$,

$$\langle Lu + A(u) + \lambda B(u) - f, v - u \rangle + \int_\Gamma j^o(\gamma u; \gamma v - \gamma u) \, d\Gamma \geq 0, \quad \forall v \in K, \tag{7.67}$$

and $\bar{u} \in W$ with $\bar{u}(\cdot, 0) \geq 0$ in Ω and

$$\langle \bar{u}' + A(\bar{u}) - f, v - \bar{u} \rangle + \int_\Gamma j^o(\gamma \bar{u}; \gamma v - \gamma \bar{u}) \, d\Gamma \geq 0, \quad \forall v \in \bar{u} \vee K. \tag{7.68}$$

Setting $v = \bar{u} \wedge u \in K$ in (7.67) and $v = \bar{u} \vee u$ in (7.68), we obtain

$$\langle u' - \bar{u}', (u - \bar{u})^+ \rangle + \langle A(u) - A(\bar{u}) + \lambda B(u), (u - \bar{u})^+ \rangle$$

$$\leq \int_\Gamma \left(j^o(\gamma \bar{u}; \gamma(u - \bar{u})^+) + j^o(\gamma u; -\gamma(u - \bar{u})^+) \right) d\Gamma. \tag{7.69}$$

For the terms on the left-hand side of (7.69), we have

$$\langle u' - \bar{u}', (u - \bar{u})^+ \rangle \geq 0 \tag{7.70}$$

and

$$\langle A(u) - A(\bar{u}) + \lambda B(u), (u - \bar{u})^+ \rangle \geq c(p) \|(u - \bar{u})^+\|_X^p \tag{7.71}$$

for λ large enough, where relation (7.40) has been used. By means of Proposition 2.171(ii) and on the basis of assumption (H1)(i), we can estimate the right-hand side of (7.69) as follows:

$$\int_\Gamma \left(j^\circ(\gamma\bar{u}; \gamma(u - \bar{u})^+) + j^\circ(\gamma u; -\gamma(u - \bar{u})^+) \right) d\Gamma$$

$$= \int_{\{\gamma u > \gamma\bar{u}\}} \left(j^\circ(\gamma\bar{u}; \gamma(u - \bar{u})) + j^\circ(\gamma u; -\gamma(u - \bar{u})) \right) d\Gamma$$

$$= \int_{\{\gamma u > \gamma\bar{u}\}} \left(\bar{\xi}\gamma(u - \bar{u}) + \xi(-\gamma(u - \bar{u})) \right) d\Gamma$$

$$= \int_{\{\gamma u > \gamma\bar{u}\}} (\bar{\xi} - \xi)(\gamma u - \gamma\bar{u}) \, d\Gamma$$

$$\leq c_1 \int_{\{\gamma u > \gamma\bar{u}\}} (\gamma u - \gamma\bar{u})^p \, d\Gamma = c_1 \|\gamma(u - \bar{u})^+\|_{L^p(\Gamma)}^p$$

$$\leq c_1 \|\gamma\|^p \|(u - \bar{u})^+\|_X^p, \tag{7.72}$$

where $\bar{\xi} \in \partial j(\gamma\bar{u})$ and $\xi \in \partial j(\gamma u)$. Thus, from (7.69)–(7.72), we get

$$(c(p) - c_1 \|\gamma\|^p)\|(u - \bar{u})^+\|_X^p \leq 0,$$

which implies in view of the relation $c(p) - c_1 \|\gamma\|^p > 0$ in Hypothesis (H1)(i) that $(u - \bar{u})^+ = 0$; i.e., $u \leq \bar{u}$. The proof of the inequality $\underline{u} \leq u$ follows the same arguments, and we omit it. This process completes the proof. $\square$

Remark 7.28. We remark that the restrictions imposed on the constants c_1 and c_2 of hypothesis (H1) have been made only for technical reasons and can be avoided. To this end, the proof presented here that is based on the auxiliary problem (7.54) has to be appropriately modified in that instead of (7.54), a more involved auxiliary variational–hemivariational inequality has to be considered that includes an additional cutoff operator acting on the traces γu of $u \in X$.

In the following example, we demonstrate the applicability of Theorem 7.27 for a convex, closed set K representing an obstacle problem constructing a penalty function associated with K to satisfy assumption (H2).

Example 7.29. We consider an obstacle problem where the set of constraints is given by

$$K = \{u \in X : u \leq \psi \text{ a.e. on } Q\},$$

with the obstacle function ψ required to verify:

(i) $\psi \in W$ and $\psi(\cdot, 0) \geq 0$ on Ω.

(ii) $\psi' + A\psi \geq 0$ in X^*; i.e., $\langle \psi' + A\psi, v \rangle \geq 0$, $\forall\, v \in X \cap L_+^p(Q)$.

The penalty operator $P : X \to X^*$ associated with the convex set K can be chosen as

$$\langle P(u), v \rangle = \int_Q [(u - \psi)^+]^{p-1}\, v\, dx\, dt + \int_\Gamma [(\gamma u - \gamma\psi)^+]^{p-1}\, \gamma v\, d\Gamma, \qquad (7.73)$$

for all $u, v \in X$. It is easy to see that P is bounded, continuous, and monotone from X to X^*. Let us check that it also satisfies (7.39). If $P(u) = 0$, then

$$\int_Q [(u - \psi)^+]^{p-1}\, v\, dx\, dt = \int_\Gamma [(\gamma u - \gamma\psi)^+]^{p-1}\, \gamma v\, d\Gamma = 0, \; \forall\, v \in X.$$

In particular, this result implies that

$$(u - \psi)^+ = 0 \quad \text{a.e. in } Q; \qquad (7.74)$$

i.e.,

$$u \leq \psi \quad \text{a.e. in } Q; \qquad (7.75)$$

that is, $u \in K$. Conversely, assume that $u \in K$; i.e., u satisfies (7.75) or (7.74). Then, by applying Fubini's theorem, for a.a. $t \in (0, \tau)$, we have $u(\cdot, t) \leq \psi(\cdot, t)$ a.e. in Ω, which ensures that

$$\gamma_{\partial\Omega} u(\cdot, t) \leq \gamma_{\partial\Omega} \psi(\cdot, t) \quad \text{a.e. on } \partial\Omega$$

($\gamma_{\partial\Omega}$ is the trace operator on $\partial\Omega$). It means that $\gamma u \leq \gamma\psi$ a.e. on Γ, and thus, $(\gamma u - \gamma\psi)^+ = 0$ a.e. on Γ. Together with (7.74), this shows via (7.73) that $P(u) = 0$ in X^*, so the equivalence in (7.39) holds. We have to check (7.41) and (7.42). In this respect, for each $u \in D(L)$, we choose $w = (u - \psi)^+$. Then, $w \in X$ and $w \neq 0$ whenever $P(u) \neq 0$. As, by (i), $(u - \psi)^+(\cdot, 0) = 0$, we have

$$\langle u' - \psi', (u - \psi)^+ \rangle = \frac{1}{2} \| (u - \psi)^+(\cdot, \tau) \|_{L^2(\Omega)}^2 \geq 0.$$

Combining with $\langle Au - A\psi, (u - \psi)^+ \rangle \geq 0$ yields

$$\langle u' + Au, (u - \psi)^+ \rangle \geq \langle \psi_t + A\psi, (u - \psi)^+ \rangle \geq 0,$$

because $(u - \psi)^+ \in X \cap L_+^p(Q)$ and from (ii). Thus, (7.41) is satisfied for any $\alpha > 0$. To verify (7.42), we note from (7.73) that

$$\begin{aligned}
\langle P(u), w \rangle &= \int_Q [(u - \psi)^+]^p\, dx\, dt + \int_\Gamma [(\gamma u - \gamma\psi)^+]^p\, d\Gamma \\
&= \| (u - \psi)^+ \|_{L^p(Q)}^p + \| (\gamma u - \gamma\psi)^+ \|_{L^p(\Gamma)}^p. \qquad (7.76)
\end{aligned}$$

Using Hölder's inequality and (7.73), we find some constant $c > 0$ such that

$$|\langle P(u), v \rangle| \le \|(u - \psi)^+\|_{L^p(Q)}^{p-1}\|v\|_{L^p(Q)} + \|(\gamma u - \gamma \psi)^+\|_{L^p(\Gamma)}^{p-1}\|\gamma v\|_{L^p(\Gamma)}$$

$$\le c(\|(u - \psi)^+\|_{L^p(Q)}^{p-1} + \|(\gamma u - \gamma \psi)^+\|_{L^p(\Gamma)}^{p-1})\|v\|_X,$$

for all $v \in X$. Hence,

$$\|P(u)\|_{X^*} \le c(\|(u - \psi)^+\|_{L^p(Q)}^{p-1} + \|(\gamma u - \gamma \psi)^+\|_{L^p(\Gamma)}^{p-1}), \ \forall \, u \in X.$$

This result, together with (7.76) and Young's inequality, implies (7.42). Thus, P in (7.73) is a penalty operator for K. For K, in our example, we have $\bar{u} \wedge K \subset K$ whenever $\bar{u} \in W$ and $\underline{u} \vee K \subset K$ if $\underline{u} \le \psi$ on Q. Moreover, the conditions $K \wedge K \subset K$ and $K \vee K \subset K$ are also satisfied, so Theorem 7.27 can be applied for any locally Lipschitz potential j verifying condition (H1).

Finally, we focus on the special case of problem (7.35) when K is the whole space X; i.e., we deal with a hemivariational inequality (see Chap. 6): Find $u \in D(L)$ such that

$$\langle Lu + A(u) - f, v - u \rangle + \int_\Gamma j^\circ(\gamma u; \gamma v - \gamma u) \, d\Gamma \ge 0, \quad \forall \, v \in X. \quad (7.77)$$

As now $K = X$, Hypotheses (H2) is fulfilled with $P = 0$. As an immediate consequence of Theorem 7.27, we have the following existence and comparison result.

Theorem 7.30. *Let $\underline{u}$ and $\bar{u}$ be sub- and supersolutions of (7.77) satisfying $\underline{u} \le \bar{u}$. Then under Hypothesis (H1), problem (7.77) admits at least one solution within the ordered interval $[\underline{u}, \bar{u}]$.*

7.2.4 Compactness and Extremality

Let S denote the set of all solutions u of (7.77) enclosed by given sub- and supersolution $\underline{u}, \bar{u}$; i.e., $\underline{u} \le u \le \bar{u}$. We know from Theorem 7.30 that $S \ne \emptyset$. We point out some compactness properties of the solution set S.

Theorem 7.31. *The solution set S is weakly sequentially compact in W and compact in X.*

Proof: The solution set $S \subset [\underline{u}, \bar{u}]$ is bounded in $L^p(Q)$, so we have the $L^p(\Gamma)$–boundedness of the traces of S. Next we show that S is bounded in W. Let $u \in S$, and take as a test function in (7.77) $v = 0$. This process leads to

$$\langle u' + Au, u \rangle \le \langle f, u \rangle + \int_\Gamma j^\circ(\gamma u; -\gamma u) \, d\Gamma. \quad (7.78)$$

Notice that

$$\langle u', u \rangle = \frac{1}{2} \|u(\cdot, \tau)\|^2_{L^2(\Omega)} \geq 0,$$

and by (H1)(i),

$$\int_\Gamma j^0(\gamma u; -\gamma u)\, d\Gamma \leq c_2 \int_\Gamma (1 + |\gamma u|^{p-1})\, |\gamma u|\, d\Gamma.$$

Then we get from (7.78) the following uniform estimate:

$$\|\nabla u\|^p_{L^p(Q)} \leq \|f\|_{X^*}\|u\|_X + C, \quad \forall\, u \in \mathcal{S},$$

with a constant $C > 0$. This result determines the boundedness of $\mathcal{S}$ in X. Setting in (7.77) the special test function $v = u - \varphi$, with $u \in \mathcal{S}$ and with $\varphi \in B = \{v \in X : \|v\|_X \leq 1\}$, we obtain

$$|\langle u', \varphi \rangle| \leq |\langle f, \varphi \rangle| + |\langle Au, \varphi \rangle| + \left| \int_\Gamma j^0(\gamma u; -\gamma \varphi)\, d\Gamma \right|.$$

In view of the boundedness of $\mathcal{S}$ in X, we derive

$$|\langle u', \varphi \rangle| \leq c, \quad \forall\, \varphi \in B, \tag{7.79}$$

where c on the right-hand side of (7.79) is a constant that does not depend on u. Thus, we conclude

$$\|u\|_W \leq C, \quad \forall\, u \in \mathcal{S}, \tag{7.80}$$

for a constant $C > 0$.

Now let $(u_n) \subset \mathcal{S}$ be any sequence. Then by (7.80), a subsequence (u_k) of (u_n) exists with

$$u_k \rightharpoonup u \quad \text{in} \quad W,$$

for some $u \in W$. As u_k are solutions of (7.77), we have

$$\left\langle \frac{\partial u_k}{\partial t} + Au_k - f, v - u_k \right\rangle + \int_\Gamma j^0(\gamma u_k; \gamma v - \gamma u_k)\, d\Gamma \geq 0, \quad \forall\, v \in X. \tag{7.81}$$

Taking the weak limit u as test function v in (7.81), we get

$$\langle Au_k, u_k - u \rangle \leq \left\langle \frac{\partial u_k}{\partial t} - f, u - u_k \right\rangle + \int_\Gamma j^0(\gamma u_k; \gamma u - \gamma u_k)\, d\Gamma$$

$$\leq \left\langle \frac{\partial u}{\partial t} - f, u - u_k \right\rangle + \int_\Gamma j^0(\gamma u_k; \gamma u - \gamma u_k)\, d\Gamma. \tag{7.82}$$

The weak convergence of (u_k) in W implies $\gamma u_k \to \gamma u$ in $L^p(\Gamma)$ because of the compactness of the trace operator (cf. Proposition 2.143), and thus by applying (H1)(ii) the right-hand side of the last inequality in (7.82) tends to zero as $k \to \infty$. This result yields

$$\limsup_{k} \langle Au_k, u_k - u \rangle \leq 0. \tag{7.83}$$

As A is pseudomonotone w.r.t. $D(L)$, from (7.83), we get

$$Au_k \rightharpoonup Au \quad \text{and} \quad \langle Au_k, u_k \rangle \to \langle Au, u \rangle.$$

Moreover, because A has the (S_+)−property w.r.t. $D(L)$, the strong convergence $u_k \to u$ in X holds. The convergence properties of the subsequence (u_k) obtained so far and the upper semicontinuity of $j^o : \mathbb{R} \times \mathbb{R} \to \mathbb{R}$ [cf. Proposition 2.162(iii)] allow us the passage to the limit in (7.81) leading to $u \in \mathcal{S}$, which completes the proof. □

Next we shall prove some properties of the solution set $\mathcal{S}$ related to the partial order $\leq$ on X.

Lemma 7.32. *Under hypothesis (H1), the solution set $\mathcal{S}$ of (7.77) within $[\underline{u}, \bar{u}]$, for an ordered pair of sub-supersolutions, is directed.*

Proof: For the proof, we only show that $\mathcal{S}$ is upward directed, because the downward directedness can be proved similarly.

Let $u_1, u_2 \in \mathcal{S}$, and denote $u_0 = \max\{u_1, u_2\}$. We introduce a cutoff function $b_0 : Q \times \mathbb{R} \to \mathbb{R}$ as follows:

$$b_0(x, t, s) = \begin{cases} (s - \bar{u}(x,t))^{p-1} & \text{if } s > \bar{u}(x,t), \\ 0 & \text{if } u_0(x,t) \leq s \leq \bar{u}(x,t), \\ -(u_0(x,t) - s)^{p-1} & \text{if } s < u_0(x,t). \end{cases}$$

Corresponding to the function b_0, we introduce the Nemytskij operator $B_0 : L^p(Q) \to L^q(Q)$ defined by

$$B_0 u(x,t) = b_0(x, t, u(x,t)), \quad \forall u \in L^p(Q).$$

as the function b_0 satisfies (7.52) and (7.53) with b replaced by b_0, the Nemytskij operator B_0 has the same properties as the previously used operator B. Consider the following auxiliary problem: find $u \in D(L)$ such that

$$\langle Lu + A(u) + \lambda B_0(u) - f, v - u \rangle + \int_\Gamma j^o(\gamma u; \gamma v - \gamma u) \, d\Gamma \geq 0, \quad \forall v \in X, \tag{7.84}$$

with $\lambda > 0$ that will be later chosen sufficiently large. The existence proof for solutions of (7.84) follows the same idea as for the existence of the penalty solutions of problem (7.55), making a suitable choice for $\lambda > 0$. Let u be a solution of (7.84). We are going to show next that u verifies the inequality: $u_0 \leq u \leq \bar{u}$.

Recalling that u_k are solutions of (7.77), they satisfy

$$u_k \in D(L): \quad \langle Lu_k + A(u_k) - f, v - u_k \rangle + \int_\Gamma j^\circ(\gamma u_k; \gamma v - \gamma u_k)\, d\Gamma$$

$$\geq 0, \quad \forall\, v \in X. \tag{7.85}$$

If we take the special test function $v = u + (u_k - u)^+$ in (7.84) and $v = u_k - (u_k - u)^+$ in (7.85), we obtain by adding the resulting inequalities the following:

$$\langle u_k' - u', (u_k - u)^+ \rangle + \langle A(u_k) - A(u) - \lambda B_0(u), (u_k - u)^+ \rangle$$

$$\leq \int_\Gamma \Big(j^\circ(\gamma u; \gamma(u_k - u)^+) + j^\circ(\gamma u_k; -\gamma(u_k - u)^+) \Big)\, d\Gamma.$$

As in the proof of Theorem 7.27, the terms on the left-hand side can be estimated below by

$$\langle u_k' - u', (u_k - u)^+ \rangle \geq 0$$

and

$$\langle A(u_k) - A(u) - \lambda B_0(u), (u_k - u)^+ \rangle \geq c(p)\, \|(u - \bar{u})^+\|_X^p, \tag{7.86}$$

and the right-hand side can be estimated above by

$$\int_\Gamma \Big(j^\circ(\gamma u; \gamma(u_k - u)^+) + j^\circ(\gamma u_k; -\gamma(u_k - u)^+) \Big)\, d\Gamma$$

$$\leq c_1 \|\gamma\|^p \|(u_k - u)^+\|_X^p. \tag{7.87}$$

Thus, from (7.86)–(7.87), we get

$$(c(p) - c_1 \|\gamma\|^p) \|(u_k - u)^+\|_X^p \leq 0.$$

In view of Hypothesis (H1)(i), this implies that $(u_k - u)^+ = 0$; i.e., $u_k \leq u$. The proof of $u \leq \bar{u}$ is carried over similarly. $\qquad\square$

On the basis of Lemma 7.32, we establish our extremality result.

Theorem 7.33. *Assume the hypotheses of Lemma 7.32. The solution set S has extremal solutions; i.e., a greatest solution u^* and a smallest solution u_* of S exist.*

Proof: We only prove the existence of the greatest solution of (7.77) within $[\underline{u}, \bar{u}]$, i.e., the greatest element of S. The proof of the smallest element can be done in a similar way. As W is separable, the subset $S \subset W$ is separable. Therefore, a countable, dense subset $Z = \{z_n : n \in \mathbb{N}\}$ of S exists. It is known from Lemma 7.32 that S is a directed set, which allows the construction of an increasing sequence $(u_n) \subset S$ as follows. We pose $u_1 = z_1$. Select $u_{n+1} \in S$ such that

$$\max\{z_n, u_n\} \leq u_{n+1} \leq \bar{u}.$$

The existence of u_{n+1} is from Lemma 7.32. As (u_n) is increasing and both bounded in $L^p(Q)$ and order-bounded, we deduce by applying Lebesgue's dominated convergence theorem that $u_n \to u^* := \sup_n u_n$ strongly in $L^p(Q)$. By Theorem 7.31, we find a subsequence (u_k) of (u_n), and an element $u \in \mathcal{S}$ such that $u_k \rightharpoonup u$ in W, and $u_k \to u$ in $L^p(Q)$ and in X. Thus, $u = u^*$ and each weakly convergent subsequence must have the same limit u^*, which implies that the entire increasing sequence (u_n) satisfies

$$u_n, u^* \in \mathcal{S}: \quad u_n \rightharpoonup u^* \text{ in } W, \ u_n \to u^* \text{ in } X.$$

By construction, we see that

$$\max\{z_1, z_2, \ldots, z_n\} \leq u_{n+1} \leq u^*, \quad \forall\, n;$$

thus $Z \subset [\underline{u}, u^*]$. As the interval $[\underline{u}, u^*]$ is closed in W, we infer

$$\mathcal{S} \subset \overline{Z} \subset \overline{[\underline{u}, u^*]} = [\underline{u}, u^*],$$

which together with $u^* \in \mathcal{S}$ ensures that u^* is the greatest element of $\mathcal{S}$. $\qquad\square$

7.3 Nonsmooth Critical Point Theory

The goal of this section is to show how the hemivariational inequalities can be investigated, alternatively to the sub-supersolution method, by using the nonsmooth critical point theory. The study of hemivariational inequalities has been initiated and developed by P. D. Panagiotopoulos (see [101, 103, 104, 108, 171, 177, 179, 180]) to treat phenomena arising in mechanics and engineering problems where unilateral nonmonotone boundary value conditions are present. For a recent use of the variational approach in the frame of hemivariational inequalities, we refer to [78, 80, 97, 103, 104, 109, 154, 155, 169, 170, 171, 172, 173].

We are concerned with the following vector-valued hemivariational inequality: Find $u \in V$ such that

$$(P) \qquad a(u, v) + \int_\Omega j^o(x, u(x); v(x))dx \geq 0, \ \forall\, v \in V.$$

Here V stands for a reflexive Banach space, endowed with the norm $\|\cdot\|_V$, which is densely and compactly embedded in $L^p(\Omega; \mathbb{R}^m)$, $2 < p < +\infty$:

$$V \subset L^p(\Omega; \mathbb{R}^m), \tag{7.88}$$

for a bounded domain $\Omega \subset \mathbb{R}^N$ with a Lipschitz boundary $\partial\Omega$. The mapping $a: V \times V \to \mathbb{R}$ is a continuous, bilinear, symmetric form that is coercive:

$$a(v, v) \geq \alpha\|v\|_V^2, \ \forall\, v \in V, \tag{7.89}$$

with a constant $\alpha > 0$. The function $j: \Omega \times \mathbb{R}^m \to \mathbb{R}$ is supposed to satisfy the following conditions:

(a) $j(\cdot, y) : \Omega \to \mathbb{R}$ is measurable, $\forall\, y \in \mathbb{R}^m$.
(b) $j(x, \cdot) : \mathbb{R}^m \to \mathbb{R}$ is locally Lipschitz, a.e. $x \in \Omega$.

As usual, the notation $j^\circ(x, \cdot; \cdot)$ in the formulation of (P) means the generalized directional derivative of $j(x, \cdot)$ (cf. Definition 2.161). In view of condition (b), the integrand $j^\circ(x, u(x); v(x))$ in (P) is finite a.e. in Ω. For a later use, we recall that the symbol $\partial j(x, \cdot) \subset \mathbb{R}^m$ designates the generalized gradient of $j(x, \cdot)$ (cf. Definition 2.166). In addition, the function $j : \Omega \times \mathbb{R}^m \to \mathbb{R}$ is assumed to fulfill the following hypotheses:

(H1) A constant $c > 0$ exists such that

$$|w| \le c(1 + |y|^{p-1}), \quad \forall\, w \in \partial j(x, y), \quad \text{a.e } x \in \Omega, \; \forall\, y \in \mathbb{R}^m.$$

(H2) Constants $\mu > 2$, $a_1 \ge 0$, $a_2 \ge 0$ and $0 \le \sigma < 2$ exist such that

$$\mu j(x, y) - j^\circ(x, y; y) \ge -a_1 |y|^\sigma - a_2, \quad \text{a.e. } x \in \Omega, \; \forall\, y \in \mathbb{R}^m.$$

(H3) $\displaystyle\liminf_{y \to 0} \frac{j(x, y)}{|y|^2} \ge 0$ uniformly with respect to $x \in \Omega$, and $j(\cdot, 0) = 0$.
(H4) $v_0 \in V \setminus \{0\}$ exists such that

$$\liminf_{s \to +\infty} s^{-\sigma} \int_\Omega j(x, sv_0(x))dx < \frac{a_1}{\sigma - \mu} \int_\Omega |v_0(x)|^\sigma dx.$$

Notice first that (H1) ensures that the integral in Problem (P) exists. Indeed, by Proposition 2.171(ii) and (H1), we can write

$$|j^\circ(x, u(x); v(x))| = |\max\{w \cdot v(x) : \; w \in \partial j(x, u(x))\}|$$
$$\le c(1 + |u(x)|^{p-1})|v(x)|$$

for a.e. $x \in \Omega$ and for all $u, v \in V$. According to (7.88), we have $|u|^{p-1} \in L^q(\Omega)$, with $1/p + 1/q = 1$, and $v \in L^p(\Omega)$; thus, the integral in (P) is finite. We remark also that the integrals in (H4) make sense. This is easily seen from relation $j(\cdot, 0) = 0$ [cf. (H3)], Theorem 2.177, and (H1).

Our result for the existence of solutions to problem (P) is the following.

Theorem 7.34. *Assume that conditions (H1)–(H4) are satisfied. Then problem (P) possesses at least one (nontrivial) solution $u \in V \setminus \{0\}$.*

Proof: In view of the application of a variational approach, we consider the functional $I : V \to \mathbb{R}$ given by

$$I(v) = \frac{1}{2}a(v, v) + \int_\Omega j(x, v(x))dx, \quad \forall\, v \in V \tag{7.90}$$

and the functional $J : L^p(\Omega; \mathbb{R}^m) \to \mathbb{R}$ defined as follows:

$$J(v) = \int_\Omega j(x, v(x))dx, \quad \forall\, v \in L^p(\Omega; \mathbb{R}^m). \tag{7.91}$$

Assumption (H1) guarantees that J is Lipschitz continuous on the bounded subsets of $L^p(\Omega; \mathbb{R}^m)$ and its generalized gradient $\partial J(v) \subset L^q(\Omega; \mathbb{R}^m)$ has the property

$$\partial J(v) \subseteq \{w \in L^q(\Omega; \mathbb{R}^m) : w(x) \in \partial j(x, v(x)) \text{ for a.e. } x \in \Omega\} \qquad (7.92)$$

(see Theorem 2.181). The inclusion in (7.92) ensures that every element $z \in \partial J(v)$ verifies

$$\langle z, v \rangle = \int_\Omega z(x) \cdot v(x) dx, \ \forall \, v \in L^p(\Omega; \mathbb{R}^m) \qquad (7.93)$$

and

$$z(x) \in \partial j(x, v(x)) \text{ for a.a. } x \in \Omega. \qquad (7.94)$$

Taking into account (7.88), (7.90), and (7.91), it is clear that the locally Lipschitz functional $I : V \to \mathbb{R}$ is expressed by

$$I(v) = \frac{1}{2} a(v, v) + (J|_V)(v), \ \forall \, v \in V, \qquad (7.95)$$

and its generalized gradient $\partial I(v) \subset V^*$ satisfies

$$\partial I(v) = Av + i^* \partial J(v), \ \forall \, v \in V, \qquad (7.96)$$

where $A : V \to V^*$ is the continuous linear operator corresponding to the bilinear form $a : V \times V \to \mathbb{R}$; i.e. $\langle Av, w \rangle_{V^*, V} = a(v, w), \ \forall \, v, w \in V$, and $i : V \to L^p(\Omega; \mathbb{R}^m)$ is the embedding in (7.88).

Our goal is to show that the functional $I : V \to \mathbb{R}$ has a critical point $u \in V$ in the sense of Definition 2.182 with $\Psi = 0$ (which in fact is the sense of Chang [64] as noticed in Example 2.188); that is,

$$0 \in \partial I(u). \qquad (7.97)$$

To this end, we apply Theorem 2.197 with $\Phi = I$ and $\Psi = 0$. We first show that the functional $I : V \to \mathbb{R}$ in (7.95) satisfies the (PS) condition in the sense of Definition 2.190 with $\Psi = 0$ (which now takes the form in Proposition 2.193). For checking condition (PS) for the functional I, let $(v_n) \subset V$ be a sequence such that

$$|I(v_n)| \leq M, \quad \forall \, n \geq 1, \qquad (7.98)$$

with a constant $M > 0$, and let $(w_n) \subset V^*$ be a sequence satisfying

$$w_n \in \partial I(v_n), \quad \forall \, n \geq 1, \qquad (7.99)$$

and

$$w_n \to 0 \text{ in } V^* \text{ as } n \to \infty. \tag{7.100}$$

By (7.96) and (7.99), we find that

$$z_n \in \partial J(v_n) \subset L^q(\Omega; \mathbb{R}^m), \quad \forall\, n \geq 1, \tag{7.101}$$

such that

$$w_n = Av_n + i^* z_n, \quad \forall\, n \geq 1. \tag{7.102}$$

For n sufficiently large, by (7.98), (7.100), (7.95), and (7.102), we have

$$
\begin{aligned}
M + \|v_n\|_V &\geq I(v_n) - \frac{1}{\mu}\langle w_n, v_n\rangle_{V^*,V} \\
&= \left(\frac{1}{2} - \frac{1}{\mu}\right) a(v_n, v_n) + \frac{1}{\mu}\int_\Omega \big[\mu j(x, v_n(x)) - z_n(x)\cdot v_n(x)\big]\,dx,
\end{aligned}
\tag{7.103}
$$

where $\mu > 2$ is given in (H2). Using (7.89), (7.94), $(H2)$ and the continuity of embedding (7.88), we obtain from (7.103) that constants $b_1 \geq 0$ and $b_2 \geq 0$ exist such that

$$M + \|v_n\|_V \geq \alpha\left(\frac{1}{2} - \frac{1}{\mu}\right)\|v_n\|_V^2 - b_1\|v_n\|_V^\sigma - b_2. \tag{7.104}$$

As $\mu > 2$ and $\sigma < 2$, the estimate in (7.104) enables us to deduce that the sequence (v_n) is bounded in V. The reflexivity of V and the compactness of the embedding (7.88) ensure the existence of a subsequence of (v_n) denoted again by (v_n) and of an element $v \in V$ with $v_n \rightharpoonup v$ in V and

$$v_n \to v \text{ in } L^p(\Omega; \mathbb{R}^m) \text{ as } n \to \infty. \tag{7.105}$$

As $J : L^p(\Omega; \mathbb{R}^m) \to \mathbb{R}$ is locally Lipschitz, we see from (7.101) and (7.105) that

$$(z_n) \text{ is bounded in } L^q(\Omega; \mathbb{R}^m). \tag{7.106}$$

Furthermore, the compactness of embedding (7.88) and (7.106) implies that along a relabeled subsequence

$$(i^* z_n) \text{ converges strongly in } V^*. \tag{7.107}$$

From (7.100), (7.102), and (7.107), we see that for a subsequence of (v_n), (Av_n) converges strongly in V^*. Because of (7.89), it follows that a strongly convergent subsequence of (v_n) exists, which yields the (PS) condition for the locally Lipschitz functional I.

In the next step of the proof, we show that

$$\lim_{t \to +\infty} I(tv_0) = -\infty, \tag{7.108}$$

where $v_0 \in V$ is the element from assumption (H4). To prove (7.108), we remark that, given $x \in \Omega$ and $y \in \mathbb{R}^m$, the following differentiation formula holds:

$$\frac{d}{d\tau}(\tau^{-\mu}j(x,\tau y))$$

$$= \mu\tau^{-\mu-1}\left[-j(x,\tau y) + \frac{1}{\mu}j_y'(x,\tau y)(\tau y)\right], \quad \text{for a.a. } \tau \in \mathbb{R}, \quad (7.109)$$

where j_y' denotes the differential with respect to y (it exists a.e. because $j(x,\cdot)$ is locally Lipschitz). Integrating (7.109) over $[1,t]$, with $t > 1$, and taking into account that the differential always belongs to the generalized gradient (see [68, p. 32]), we see that

$$t^{-\mu}j(x,ty) - j(x,y) \le -\int_1^t \tau^{-\mu-1}\left[\mu j(x,\tau y) - j^\circ(x,\tau y;\tau y)\right]d\tau \quad (7.110)$$

holds for all $t > 1$, for a.a. $x \in \Omega$, and for all $y \in \mathbb{R}^m$. Thus, (7.110) and (H2) yield

$$t^{-\mu}j(x,ty) - j(x,y) \le \int_1^t \tau^{-\mu-1}(a_1\tau^\sigma|y|^\sigma + a_2)d\tau$$

$$= a_1|y|^\sigma\frac{1}{\sigma-\mu}(t^{\sigma-\mu}-1) - \frac{a_2}{\mu}(t^{-\mu}-1)$$

$$\le \frac{a_1}{\mu-\sigma}|y|^\sigma + \frac{a_2}{\mu} \quad (7.111)$$

for all $t > 1$, for a.a. $x \in \Omega$, and for all $y \in \mathbb{R}^m$. Setting $y = sv_0(x)$ in (7.111), for $x \in \Omega$ and $s > 0$, it turns out that

$$j(x,tsv_0(x)) \le t^\mu\left[j(x,sv_0(x)) + \frac{a_1}{\mu-\sigma}s^\sigma|v_0(x)|^\sigma + \frac{a_2}{\mu}\right] \quad (7.112)$$

for all $t > 1$, $s > 0$, and for a.a. $x \in \Omega$. On the basis of (7.95) and (7.112), we get

$$I(tsv_0) \le \frac{1}{2}t^2s^2a(v_0,v_0) + t^\mu s^\sigma\left[s^{-\sigma}\int_\Omega j(x,sv_0(x))dx\right.$$

$$\left.+\frac{a_1}{\mu-\sigma}\int_\Omega|v_0(x)|^\sigma dx + \frac{a_2}{\mu}|\Omega|s^{-\sigma}\right] \quad (7.113)$$

for all $t > 1$, $s > 0$, and for a.a. $x \in \Omega$. By assumption (H4), there is a number $s > 0$ such that

$$s^{-\sigma}\int_\Omega j(x,sv_0(x))dx + \frac{a_1}{\mu-\sigma}\int_\Omega|v_0(x)|^\sigma dx + \frac{a_2}{\mu}|\Omega|s^{-\sigma} < 0. \quad (7.114)$$

Fixing $s > 0$ in (7.114) and passing to the limit in (7.113) as $t \to +\infty$, we arrive at (7.108), because $\mu > 2$.

The proof continues by now checking that constants $b > 0$ and $\rho > 0$ exist for which we have

$$I(v) \geq b, \quad \forall\, v \in V \text{ with } \|v\|_V = \rho. \tag{7.115}$$

To this end, we make use of $(H3)$. Given $\varepsilon > 0$, by $(H3)$, there is $\delta = \delta(\varepsilon) > 0$ such that

$$j(x, y) \geq -\varepsilon|y|^2, \text{ a.a. } x \in \Omega, \ \forall\, y \in \mathbb{R}^m, \ |y| \leq \delta. \tag{7.116}$$

The function $j : \Omega \times \mathbb{R}^m \to \mathbb{R}$ can be estimated from (H1) as follows:

$$|j(x, y)| \leq c_1|y|^p + c_2, \text{ a.a. } x \in \Omega, \ \forall\, y \in \mathbb{R}^m,$$

with constants $c_1 > 0$ and $c_2 > 0$, implying that

$$|j(x, y)| \leq \left(c_1 + \frac{c_2}{\delta^p}\right)|y|^p, \text{ a.a. } x \in \Omega, \ \forall\, y \in \mathbb{R}^m, \ |y| \geq \delta. \tag{7.117}$$

Relations (7.116) and (7.117) lead to

$$j(x, y) \geq -\varepsilon|y|^2 - \left(c_1 + \frac{c_2}{\delta^p}\right)|y|^p, \text{ a.a. } x \in \Omega, \ \forall\, y \in \mathbb{R}^m. \tag{7.118}$$

From the continuity of the embedding in (7.88) and using (7.118), we see that constants $c_0 > 0$ exist and $\bar{c} > 0$ such that

$$I(v) \geq \left[\frac{1}{2}\alpha - c_0\varepsilon - \bar{c}\left(c_1 + \frac{c_2}{\delta^p}\right)\|v\|_V^{p-2}\right]\|v\|_V^2, \ \forall\, v \in V. \tag{7.119}$$

Choosing $\varepsilon > 0$ sufficiently small and using that $p > 2$, estimate (7.119) allows us to get constants $b > 0$ and $\rho > 0$ establishing assertion (7.115).

On the other hand, by relation $j(\cdot, 0) = 0$ in assumption (H3), we know that $I(0) = 0$, whereas from (7.108), we can find a number $t_0 > 0$ satisfying

$$t_0\|v_0\|_V > \rho \text{ and } I(t_0v_0) < 0, \tag{7.120}$$

where $\rho > 0$ is as in (7.115). We are now in a position to apply Theorem 2.197 for $S = \{v \in V : \|v\|_V = \rho\}$ and $Q = [0, t_0v_0] = \{tt_0v_0 \in V : 0 \leq t \leq 1\}$, with $\partial Q = \{0, t_0v_0\}$. According to the first relation in (7.120), the sets S and Q link in the sense of Definition 2.195 (making in fact the linking situation in the mountain pass theorem). We have from the second relation in (7.120) and from (7.115) that

$$\max_{\partial Q} I \leq 0 < b \leq \inf_S I.$$

Taking into account that the (PS) condition is satisfied, we may apply Theorem 2.197. So $u \in V$ exists such that (7.97) holds. From the above inequalities and the last assertion of Theorem 2.197, we have $u \neq 0$.

We now show that u in (7.97) solves problem (P). In view of (7.96), relation (7.97) becomes

$$Au + i^*z = 0 \text{ for some } z \in \partial J(u). \tag{7.121}$$

Using (7.92), (7.93), (7.94), and Proposition 2.171(ii), equality (7.121) yields the desired result.   □

Remark 7.35. Theorem 7.34 extends Theorem 2.15 in [64] (and a fortiori Theorem 3.10 in [3]). The most important aspect in this direction is that assumption (H2) significantly weakens the requirement $j^o(x, y; y) \leq \mu j(x, y) < 0$ for large y used extensively in many works (see Ambrosetti and Rabinowitz [3], Chang [64], Rabinowitz [191]) by dropping the sign condition $j(x, y) < 0$ and allowing a more general growth for $\mu j(x, y) - j^o(x, y; y)$ as well as vector-values for y [see hypothesis (H2)]. Notice that in the works [3, 64, 191], the potential $-j$ is used in place of our j. In comparison with the cited works, the relaxed conditions that we assume permit to treat problems involving both superlinear and sublinear terms under the integral in (P). It is also worth to point out that problem (P) deals with vector-valued generalized gradients ∂j that enables us to cover systems of hemivariational inequalities (see [177] for a different approach but where the superlinear case cannot be treated).

The following examples provide nonsmooth function j satisfying (H1)–(H4). For the sake of simplicity, we drop the dependence of j with respect to $x \in \Omega$.

Example 7.36. Let the function $j : \mathbb{R} \to \mathbb{R}$ be defined by

$$j(y) = \max\left\{ -\frac{1}{h}|y|^h, -\frac{1}{r}|y|^r \right\}, \ \forall \, y \in \mathbb{R},$$

with numbers h and r satisfying $h, r \in (2, p]$, where p is as in (7.88). It is clear that the function j is locally Lipschitz and verifies (H3). Using the differentiation formula for the generalized gradient of the maximum of finitely many functions (cf. [68, p. 47]), we see that

$$j^o(y; z) = \max\{-|y|^{h-2}yz, -|y|^{r-2}yz\}, \ \forall \, y, z \in \mathbb{R}.$$

Thus, the generalized gradient of j verifies the growth condition (H1). A direct computation shows that (H2) is satisfied by choosing any number μ with $2 < \mu \leq \min\{h, r\}$ and any $\sigma \in [0, 2)$. We have for all $v_0 \in V \setminus \{0\}$ and $\sigma < 2$ that

$$\liminf_{s \to +\infty} s^{-\sigma} \int_\Omega j(sv_0(x)) dx = -\infty.$$

Therefore, assumption (H4) is verified and Theorem 7.34 can be applied to the corresponding problem (P).

Example 7.37. Consider the function $j : \mathbb{R}^2 \to \mathbb{R}$ defined by

$$j(y) = -\frac{1}{p}|y_1|^p + \int_0^{y_2} \beta(t) dt, \ \forall \, y = (y_1, y_2) \in \mathbb{R}^2,$$

with $p > 2$ as in (7.88) and a function $\beta : \mathbb{R} \to \mathbb{R}$ satisfying $\beta \in L^\infty_{loc}(\mathbb{R})$, $t\beta(t) \geq 0$ for $t \in \mathbb{R}$ near 0 and $|\beta(t)| \leq c(1 + |t|^\gamma)$, for all $t \in \mathbb{R}$, with $c > 0$ and $0 \leq \gamma < 1$. It is readily seen that the function j is locally Lipschitz and satisfies the growth condition (H1). Assumption (H2) is verified for any $2 < \mu \leq p$ and $\sigma = \gamma + 1$. The hypothesis $t\beta(t) \geq 0$ for $t \in \mathbb{R}$ near 0 implies that (H3) is valid. Taking $v_0 = (v_1, 0) \in V \setminus \{0\}$, we note that

$$\lim_{s \to +\infty} s^{-(\gamma+1)} \int_\Omega j(sv_0(x))dx = -\infty,$$

so condition (H4) is verified. Theorem 7.34 can be applied to the corresponding problem (P).

7.4 A Constraint Hemivariational Inequality

In contrast to the first two sections of this chapter, we now treat the variational-hemivariational inequalities by variational methods in place of the sub-supersolution method. Specifically, we take advantage here of the nonsmooth critical point theory developed in the sense of Definition 2.182.

We introduce the functional setting of our problem. Let Ω be a bounded domain in $\mathbb{R}^N$ with a Lipschitz boundary $\partial \Omega$. Let the Sobolev space $H^1_0(\Omega)$ be endowed with the scalar product

$$(u, v)_{H^1_0(\Omega)} = \int_\Omega \nabla u \cdot \nabla v dx, \quad \forall\, u, v \in H^1_0(\Omega),$$

which makes it a Hilbert space. The associated norm is denoted $\|\cdot\|$. The cone of nonnegative functions in $H^1_0(\Omega)$, denoted

$$K = \{u \in H^1_0(\Omega) : u(x) \geq 0 \text{ for a.e. } x \in \Omega\}, \tag{7.122}$$

is a convex and closed set. Corresponding to K in (7.122), we consider its indicator function $\Psi : H^1_0(\Omega) \to \mathbb{R} \cup \{+\infty\}$; that is,

$$\Psi(u) = \begin{cases} 0 & \text{if } u \in K \\ +\infty & \text{if } u \notin K. \end{cases} \tag{7.123}$$

It follows that Ψ is a proper, convex, lower semicontinuous function.

In our problem, there are also given $g \in L^2(\Omega)$ such that

$$g \leq 0, \quad \text{a.e. in } \Omega \tag{7.124}$$

and a (Carathéodory) function $j : \Omega \times \mathbb{R} \to \mathbb{R}$ satisfying

(a) $j(\cdot, y) : \Omega \to \mathbb{R}$ is measurable, for all $y \in \mathbb{R}$.
(b) $j(x, \cdot) : \mathbb{R} \to \mathbb{R}$ is locally Lipschitz, for a.a. $x \in \Omega$.

In view of (b), the generalized directional derivative denoted $j^{\circ}(x, y; \cdot)$ of $j(x, y)$ with respect to the second variable $y \in \mathbb{R}$ and the corresponding generalized gradient $\partial j(x, y)$ of j with respect to $y \in \mathbb{R}$ are well defined (cf. Definition 2.161 and Definition 2.166).

We impose the following hypotheses on the function $j : \Omega \times \mathbb{R} \to \mathbb{R}$ to be fulfilled:

(j_1) $|\xi| \leq c(1 + |y|^{p-1})$ for a.a. $x \in \Omega$, for all $y \in \mathbb{R}$, and for all $\xi \in \partial j(x, y)$, with constants $c > 0$ and $1 \leq p < 2N/(N-2)$ if $N \geq 3$ and an arbitrary $p \geq 1$ if $N = 1$ or $N = 2$.

(j_2) $\liminf_{y \to 0+} y^{-2} j(x, y) \geq 0$ uniformly for a.e. $x \in \Omega$, and $j(\cdot, 0) = 0$.

(j_3) $\mu^{-1} j^{\circ}(x, y; y) \leq j(x, y)$ for a.a. $x \in \Omega$ and for all $y \in \mathbb{R}$, $y \geq 0$, where μ is a constant with $\mu > 2$.

(j_4) there is an element $u_0 \in K$ such that

$$\int_{\Omega} j(x, u_0(x)) dx < 0.$$

Denote by λ_1 the first eigenvalue of $-\Delta$ on $H_0^1(\Omega)$. We state the following result.

Theorem 7.38. *Let the subset K of $H_0^1(\Omega)$ be the one in (7.122). Under the above assumptions for $g \in L^2(\Omega)$ and $j : \Omega \times \mathbb{R} \to \mathbb{R}$, whenever $\lambda < \lambda_1$ the variational–hemivariational inequality: Find $u \in K$ such that*

$$\int_{\Omega} \nabla u \cdot (\nabla v - \nabla u) \, dx + \int_{\Omega} j^{\circ}(x, u(x); v(x) - u(x)) \, dx$$

$$\geq \lambda \int_{\Omega} u(v - u) \, dx + \int_{\Omega} g(v - u) \, dx, \quad \forall \, v \in K, \tag{7.125}$$

has a nontrivial solution.

Proof: The conclusion will be achieved through Theorem 2.197. First, we remark that without loss of generality, we may assume $p > 2$ in (j_1). With the number p in (j_1), we introduce the functional $J : L^p(\Omega) \to \mathbb{R}$ by

$$J(u) = \int_{\Omega} j(x, u(x)) dx. \tag{7.126}$$

By means of (j_1) and applying Theorem 2.181, we see that the functional J in (7.126) is Lipschitz continuous on the bounded subsets of $L^p(\Omega)$; thus, it is locally Lipschitz on $L^p(\Omega)$. Moreover, Theorem 2.181 ensures that the generalized gradient $\partial J(u)$ of J at any $u \in L^p(\Omega)$ satisfies

$$\partial J(u) \subseteq \{w \in L^q(\Omega; \mathbb{R}) : w(x) \in \partial j(x, u(x)) \text{ for a.e. } x \in \Omega\}, \tag{7.127}$$

with $1/p + 1/q = 1$. Let us now define for any fixed number $\lambda \in \mathbb{R}$ the functional $\Phi : H_0^1(\Omega) \to \mathbb{R}$ by

$$\Phi(u) = \frac{1}{2}\|u\|^2 + \int_\Omega j(x, u(x))\, dx - \frac{\lambda}{2}\|u\|_{L^2(\Omega)}^2 - \int_\Omega gu\, dx$$

$$= \frac{1}{2}\|u\|^2 + J(u) - \frac{\lambda}{2}\|u\|_{L^2(\Omega)}^2 - \int_\Omega gu\, dx. \tag{7.128}$$

In writing (7.128), we have used (7.126) and Theorem 2.74 (i) according to the assumption for p in (j$_1$). Furthermore, we consider the functional I : $H_0^1(\Omega) \to \mathbb{R} \cup \{+\infty\}$ given by

$$I = \Phi + \Psi, \tag{7.129}$$

with $\Phi : H_0^1(\Omega) \to \mathbb{R}$ and $\Psi : H_0^1(\Omega) \to \mathbb{R} \cup \{+\infty\}$ expressed in (7.128) and (7.123), respectively. As $\Phi : H_0^1(\Omega) \to \mathbb{R}$ is locally Lipschitz and $\Psi : H_0^1(\Omega) \to \mathbb{R} \cup \{+\infty\}$ is proper, convex, and lower semicontinuous, it turns out that the functional $I : H_0^1(\Omega) \to \mathbb{R} \cup \{+\infty\}$ in (7.129) satisfies the structural hypothesis (H) in Sect. 2.5.3.

We show that the functional $I : H_0^1(\Omega) \to \mathbb{R} \cup \{+\infty\}$ in (7.129) fulfills the (PS) condition in the sense of Definition 2.190. To see this, let a sequence $(u_n) \subset H_0^1(\Omega)$ satisfy $I(u_n) \to c$, with a $c \in \mathbb{R}$, and

$$\Phi^\circ(u_n; v - u_n) + \Psi(v) - \Psi(u_n) \geq -\varepsilon_n\|v - u_n\|, \quad \forall\, v \in H_0^1(\Omega), \tag{7.130}$$

for a sequence $(\varepsilon_n) \subset (0, +\infty)$ with $\varepsilon_n \to 0$. It is clear that $(u_n) \subset K$. Assume first that $\lambda \in [0, \lambda_1)$. Using the expressions of Φ and Ψ given in (7.128) and (7.123), respectively, inequality (7.130) becomes

$$\int_\Omega \nabla u_n \cdot \nabla(v - u_n) dx + J^\circ(u_n; v - u_n) - \lambda \int_\Omega u_n(v - u_n) dx - \int_\Omega g(v - u_n) dx$$

$$\geq -\varepsilon_n\|v - u_n\|, \quad \forall\, v \in K. \tag{7.131}$$

Notice that by (7.122) it is permitted to put $v = 2u_n$ in (7.131). Thus, we derive

$$\|u_n\|^2 + J^\circ(u_n; u_n) - \lambda \int_\Omega u_n^2\, dx - \int_\Omega gu_n\, dx - \varepsilon_n\|u_n\|, \quad \forall\, n \in \mathbb{N}. \tag{7.132}$$

Then, for any sufficiently large n, in view of (7.128), (7.132), that $I(u_n) \to c$, and making use of the constant $\mu > 2$ in (j$_3$), we obtain

$$c + 1 + \frac{1}{\mu}\|u_n\| \geq \Phi(u_n) + \frac{1}{\mu}\varepsilon_n\|u_n\|$$

$$\geq \left(\frac{1}{2} - \frac{1}{\mu}\right)\|u_n\|^2 + \lambda\left(\frac{1}{\mu} - \frac{1}{2}\right)\|u_n\|_{L^2(\Omega)}^2 + \left(\frac{1}{\mu} - 1\right)\int_\Omega gu_n\, dx$$

$$+ \int_\Omega j(x, u_n)\, dx - \frac{1}{\mu}J^\circ(u_n; u_n). \tag{7.133}$$

On the basis of (7.127), Proposition 2.171(ii), and because $\mu > 2$ and $\lambda \geq 0$, it follows that inequality (7.133) leads to

$$c + 1 + \frac{1}{\mu}\|u_n\| \geq \left(\frac{1}{2} - \frac{1}{\mu}\right)(1 - \lambda_1^{-1}\lambda)\|u_n\|^2$$

$$+ \int_\Omega \left(j(x, u_n) - \frac{1}{\mu}j^\circ(x, u_n; u_n)\right) dx$$

$$+ \left(\frac{1}{\mu} - 1\right)\lambda_1^{-\frac{1}{2}}\|g\|_{L^2(\Omega)}\|u_n\|. \tag{7.134}$$

In writing (7.134), we have also used the Rayleigh–Ritz variational characterization of λ_1

$$\lambda_1 = \min_{v \in H_0^1(\Omega),\ v \neq 0} \frac{\|u_n\|^2}{\|u_n\|_{L^2(\Omega)}^2},$$

as well as Theorem 2.181. Then (7.134) and (j_3) imply

$$c + 1 + \frac{1}{\mu}\|u_n\| \geq \left(\frac{1}{2} - \frac{1}{\mu}\right)(1 - \lambda_1^{-1}\lambda)\|u_n\|^2 + \left(\frac{1}{\mu} - 1\right)\lambda_1^{-\frac{1}{2}}\|g\|_{L^2(\Omega)}\|u_n\|. \tag{7.135}$$

As $\mu > 2$ and $\lambda < \lambda_1$, from estimate (7.135), it follows that the sequence (u_n) is bounded in $H_0^1(\Omega)$. If $\lambda < 0$, it is seen from (7.133) (valid for every λ) and (j_3) that the same conclusion holds. Then, by Theorem 2.74(i), $u \in K$ such that along a relabelled subsequence, we have

$$u_n \rightharpoonup u \text{ in } H_0^1(\Omega), \text{ and } u_n \to u \text{ in } L^2(\Omega) \text{ and in } L^p(\Omega). \tag{7.136}$$

Consequently, from (7.136) and (7.131) with $v = u$, in conjunction with Proposition 2.162(iii), we get

$$\limsup_{n \to \infty} \|u_n\|^2 \leq \|u\|^2.$$

This result yields that $u_n \to u$ strongly in $H_0^1(\Omega)$, which enables us to conclude that the functional I in (7.129) satisfies the (PS) condition in the sense of Definition 2.190.

Toward the application of Theorem 2.197 to the functional I in (7.129), we claim that we can find constants $\alpha > 0$ and $\rho > 0$ such that

$$I(v) \geq \alpha \quad \text{whenever} \quad \|v\| = \rho. \tag{7.137}$$

The argument is carried out as follows. Fix an $\varepsilon > 0$. Assumption (j_2) ensures the existence of some $\delta > 0$ such that

$$|y|^{-2}j(x, y) \geq -\varepsilon \text{ for a.a. } x \in \Omega \text{ and for all } y \in \mathbb{R} \text{ with } |y| \leq \delta. \tag{7.138}$$

Theorem 2.177 (j_1) and (j_2) imply the estimate

$$|j(x, y)| = |j(x, y) - j(x, 0)| \leq c_1(1 + |y|^p)$$

for all $(x, y) \in \Omega \times \mathbb{R}$, with a constant $c_1 > 0$. Combining with (7.138), we find

$$-j(x,y) \leq \varepsilon |y|^2 + c_1(\delta^{-p} + 1)|y|^p, \quad \text{for a.a. } x \in \Omega, \text{ for all } y \in \mathbb{R}.$$

The following estimate for the functional J in (7.126) is then available:

$$-J(u) \leq \varepsilon \|u\|_{L^2(\Omega)}^2 + c_1(\delta^{-p} + 1)\|u\|_{L^p(\Omega)}^p, \quad \forall u \in L^p(\Omega). \tag{7.139}$$

By Theorem 2.74(i) and because $p > 2$, we infer from (7.139) that a constant $a > 0$ exists independent of ε such that the inequality

$$-J(u) \leq \varepsilon a \|u\|^2$$

holds for all $u \in H_0^1(\Omega)$ with $\|u\|$ sufficiently small. Suppose $\lambda \geq 0$. Then (7.128), (7.124), and (7.139) yield

$$I(u) = \Phi(u) \geq \frac{1}{2}(1 - \lambda \lambda_1^{-1} - \varepsilon a)\|u\|^2$$

for all $u \in K$ provided $\|u\|$ is small enough. As $\lambda < \lambda_1$ and $\varepsilon > 0$ can be chosen arbitrarily small, the claim in (7.137) is valid. If $\lambda < 0$, from (7.128) and (7.124), we obtain

$$I(u) \geq \frac{1}{2}(1 - \varepsilon a)\|u\|^2,$$

whenever $u \in K$ with sufficiently small $\|u\|$. Choosing $\varepsilon < a^{-1}$ leads to (7.137).

The next step in the proof is to show that for the element $u_0 \in H_0^1(\Omega)$ given in assumption (j$_4$) we have

$$\lim_{t \to +\infty} I(tu_0) = -\infty. \tag{7.140}$$

For proving this result, we need the formula below involving the generalized gradient ∂_t with respect to $t \in \mathbb{R}$ of the locally Lipschitz function $t \in (0, +\infty) \mapsto t^{-\mu} j(x, ty) \in \mathbb{R}$, where $y \in \mathbb{R}$ is fixed:

$$\partial_t(t^{-\mu} j(x, ty)) = \mu t^{-1-\mu}(\mu^{-1} ty \partial j(x, ty) - j(x, ty))$$

for a.a. $x \in \Omega$, all $y \in \mathbb{R}$ and $t > 0$. By Theorem 2.177 and Proposition 2.171(ii), the previous relation implies

$$t^{-\mu} j(x, ty) - j(x, y) \leq \mu \tau^{-1-\mu}(\mu^{-1} j^\circ(x, \tau y; \tau y) - j(x, \tau y))(t - 1) \tag{7.141}$$

for a.a. $x \in \Omega$ and all $y \in \mathbb{R}$, $t > 1$, with some $\tau \in (1, t)$. Here μ designates the constant $\mu > 2$ entering condition (j$_3$). We note that assumptions (j$_1$), (j$_3$), and relation (7.141) ensure

$$j(x, ty) - t^\mu j(x, y) \leq 0 \tag{7.142}$$

for a.a. $x \in \Omega$, for all $y \in \mathbb{R}$, $y \geq 0$, and $t > 1$. Using the element $u_0 \in K$ given in assumption (j$_4$), from (7.142), we deduce

$$I(tu_0) = \Phi(tu_0)$$

$$\leq \frac{t^2}{2}\left(\|u_0\|^2 - \lambda\|u_0\|^2_{L^2(\Omega)}\right) + t^\mu \int_\Omega j(x, u_0(x))\,dx - t\int_\Omega g(x)u_0(x)dx.$$

$$(7.143)$$

Letting $t \to +\infty$ in (7.143), and taking into account hypothesis (j$_4$) as well as $\mu > 2$, we arrive at the claim in (7.140).

On the basis of (7.140), we fix $t_0 > 0$ such that

$$I(t_0 u_0) \leq 0 \quad \text{and} \quad t_0\|u_0\| > \rho, \qquad (7.144)$$

for $\rho > 0$ entering (7.137). Our goal is to apply Theorem 2.197 choosing

$$S = \{v \in H^1_0(\Omega) : \|v\| = \rho\}, \qquad (7.145)$$

with $\rho > 0$ as in (7.137), and

$$Q = [0, t_0 u_0] = \{tt_0 u_0 \in H^1_0(\Omega) : 0 \leq t \leq 1\}, \qquad (7.146)$$

where $\partial Q = \{0, t_0 u_0\}$. It is clear from the second relation in (7.144) that S and Q link in the sense of Definition 2.195. In fact, this is the linking situation in the mountain pass theorem.

Let us show that the functional $I : H^1_0(\Omega) \to \mathbb{R}\cup\{+\infty\}$ in (7.129) satisfies the hypotheses of Theorem 2.197 with the choices for S and Q in (7.145) and (7.146), respectively. We have already proved that I complies with the structural hypothesis (H) in Sect. 2.5.3, and it satisfies condition (PS) in the sense of Definition 2.190. As $Q = [0, t_0 u_0] \subset K$, it follows from the convexity of Ψ that $\sup_Q I \in \mathbb{R}$ and $\inf_S I \in \mathbb{R}$. Notice, from (7.137), the first relation in (7.144) and in view of $I(0) = 0$ [cf. the second part of (j$_2$)], it follows that

$$\max_{\partial Q} I \leq 0 < \alpha \leq \inf_S I.$$

Consequently, all assumptions of Theorem 2.197 are fulfilled. Applying Theorem 2.197, we obtain a nontrivial critical point $u \in K$ of $I = \Phi + \Psi$ in the sense of Definition 2.182, which reads

$$\int_\Omega \nabla u \cdot (\nabla v - \nabla u)dx + J^\circ(u; v - u) \geq \lambda \int_\Omega u(v - u)dx + \int_\Omega g(v - u)dx$$

$$(7.147)$$

for all $v \in K$, where $\lambda < \lambda_1$. On the other hand, (7.127) ensures

$$J^\circ(u; w) \leq \int_\Omega j^\circ(x, u(x); w(x))dx, \quad \forall\, w \in L^p(\Omega). \qquad (7.148)$$

Combining (7.147) and (7.148) results in (7.125), which completes the proof. $\qquad\square$

Example 7.39. Let the locally Lipschitz function $j : \mathbb{R} \to \mathbb{R}$ be as follows:

$$j(t) = -\frac{1}{p}t^p + \frac{1}{r}t^r, \quad \forall\, t \geq 0,$$

with $2 < r < p < 2^*$ and $j|_{(-\infty,0)}$ satisfies condition (j_1). For the function j, assumptions (j_1)–(j_4) are satisfied. Indeed, assumptions (j_1) and (j_2) are clearly fulfilled. Condition (j_3) is true for any $r \leq \mu \leq p$. Condition (j_4) holds with any $u_0 = k_0 v_0$ with $v_0 \in H_0^1(\Omega) \setminus \{0\}$, $v_0 \geq 0$ a.e. and a constant $k_0 > 0$ sufficiently large. Notice that $j(t)$ contains both convex and concave terms. Another locally Lipschitz function j verifying (j_1)–(j_4), which is not continuously differentiable is given in Example 7.36. For such functions j, Theorem 7.38 can be applied to the corresponding problems (7.125).

Remark 7.40. Theorem 7.38 can be formulated for more general second-order uniformly elliptic operators in place of $-\Delta$. For the situation $\lambda < \lambda_1$, it extends Theorem 5.1 of Szulkin [211], which considers the potential $j(t) = (1/p)|t|^p$ and strict inequality $g < 0$ a.e. It also extends results in [3, 64, 171, 191].

7.5 Eigenvalue Problem for a Variational–Hemivariational Inequality

The aim of this section is to study nonlinear eigenvalue problems for general variational–hemivariational inequalities that depend on a parameter. The motivation for such a study comes, for instance, from the investigation of perturbations, usually determined in terms of parameters. The variational–hemivariational inequalities considered here are expressed in an abstract form compatible with the nonsmooth critical point theory developed in Sect. 2.5.3 and Sect. 2.5.4.

We proceed by formulating two problems for variational–hemivariational inequalities, one being stationary (in the sense that it is independent of a parameter) and the other one stated with eigenvalues and containing a parameter. For this purpose, we first describe the functional setting. Let H be a real Hilbert space endowed with the inner product $(\cdot,\cdot)_H$ and the associated norm $\|\cdot\|$. Let $J : H \times \mathbb{R} \to \mathbb{R}$ be a locally Lipschitz functional, and let $\psi : H \to \mathbb{R} \cup \{+\infty\}$ be a convex, proper, and lower semicontinuous function. Fix real numbers $a > 0$, $a_1 > 0$, $\alpha > 0$, $p \geq 0$, $\rho > 0$, and $r > 0$ with $\rho < r$.

The first problem is the following variational–hemivariational inequality on H with constraints on the solutions: Find $u \in H$ such that

(P_0) $\begin{cases} (-J)^0(u,0; v - u, 0) + \psi(v) - \psi(u) + a(u, v - u)_H \geq 0, \quad \forall\, v \in H; \\ \alpha \leq -J(u,0) + \psi(u) + \dfrac{a}{2}\|u\|^2 \leq \alpha + a_1. \end{cases}$

Our second problem is an eigenvalue problem for a variational–hemivariational inequality with constraints on the eigensolutions $(u, \lambda) \in H \times \mathbb{R}$ and depending on a real parameter denoted s: Find the eigenfunction $u \in H$, the eigenvalue $\lambda \in \mathbb{R}$, and the parameter $s \in \mathbb{R}$ such that

$$(P) \quad \begin{cases} (-J)^{\circ}(u, s; v - u, 0) + \psi(v) - \psi(u) + \lambda(u, v - u)_H \geq 0, \quad \forall\, v \in H; \\[2mm] \rho \leq s \leq r; \\[2mm] \rho^2 \|u\|^p \leq \lambda - a \leq r^2 \|u\|^p; \\[2mm] -a_1 \leq \dfrac{s^2}{p+2} \|u\|^{p+2} - J(u, s) + \dfrac{a}{2} \|u\|^2 + \psi(u) \leq \alpha + a_1. \end{cases}$$

Here $(-J)^{\circ}$ stands for the generalized directional derivative in Definition 2.161 of the locally Lipschitz function $-J$. We use $-J$ instead of J to achieve compatibility with the framework of nonlinear boundary value problems for semilinear elliptic equations as discussed in the example below.

In the following example, we outline the applicability of our approach to nonlinear elliptic eigenvalue problems, with Dirichlet boundary conditions, possessing a parameter in the equation and that are subject to constraints.

Example 7.41. Let $\Omega \subset \mathbb{R}^N$ be a bounded domain with a Lipschitz boundary $\partial\Omega$. Consider the Sobolev space $H_0^1(\Omega)$ endowed with the scalar product $(\cdot, \cdot)_{H_0^1(\Omega)}$ as in Sect. 7.4. Let a nonempty, closed, convex subset $K \subset H_0^1(\Omega)$, an element $f \in H^{-1}(\Omega)$, a locally Lipschitz function $G : \mathbb{R} \to \mathbb{R}$, and a continuous function $q : \mathbb{R} \to \mathbb{R}$ be given. Assume that the generalized gradient ∂G of G and the function q verify subcritical growth conditions as in (H1) of Sect. 7.3 or (j$_1$) of Sect. 7.4. With these data, we state the eigenvalue problem: Find $u \in K$, $\mu \in \mathbb{R}$ and $\nu \in \mathbb{R}$ such that

$$(P_{\mu,\nu}) \quad \begin{aligned} &\int_{\Omega} \nabla u(x) \cdot \nabla(v - u)(x)\, dx + \mu \int_{\Omega} (-G)^{\circ}(u(x); v(x) - u(x))\, dx \\ &\quad - \nu \left[\int_{\Omega} q(u(x))(v(x) - u(x))\, dx + \langle f, v - u \rangle \right] \geq 0, \quad \forall\, v \in K. \end{aligned}$$

Assuming $\mu > 0$, dividing by μ, and denoting $\lambda = \frac{1}{\mu}$ and $s = \frac{\nu}{\mu}$, problem $(P_{\mu,\nu})$ can be put in the form of a variational–hemivariational inequality on $H = H_0^1(\Omega)$ [which will represent the first relation in problem (P)]:

$$(P_1) \quad \begin{aligned} &\int_{\Omega} \left[(-G)^{\circ}(u(x); v(x) - u(x)) - s q(u(x))(v(x) - u(x)) \right] dx \\ &\quad - s \langle f, v - u \rangle + \lambda(u, v - u)_{H_0^1(\Omega)} \geq 0, \quad \forall\, v \in K. \end{aligned}$$

Let us introduce the locally Lipschitz function $J : H_0^1(\Omega) \times \mathbb{R} \to \mathbb{R}$ by

$$J(v, t) = \int_{\Omega} (G(v(x)) + tQ(v(x)))\, dx + t\langle f, v \rangle,$$

where

$$Q(t) = \int_0^t q(\tau)\,d\tau$$

for all $t \in \mathbb{R}$, and let $\psi : H_0^1(\Omega) \to \mathbb{R}$ be the indicator function of the subset K of $H_0^1(\Omega)$ [see (7.123)], which is convex, proper, and lower semicontinuous. Using Theorem 2.181, we see that (P_1) fits the formulation of the first inequality in (P). Important situations are included in problems of type $(P_{\mu,\nu})$. For example, choosing $K = H_0^1(\Omega)$, $f \in L^{2^{*'}}(\Omega)$, with $1/2^* + 1/2^{*'} = 1$, and

$$G(t) = \int_0^t g(\tau)\,d\tau$$

for all $t \in \mathbb{R}$, with $g : \mathbb{R} \to \mathbb{R}$ continuous, problem $(P_{\mu,\nu})$ reduces to an eigenvalue problem for a semilinear elliptic equation with Dirichlet boundary condition:

$$\begin{cases} -\Delta u = \mu g(u) + \nu\,[\,q(u) + f\,] & \text{on } \Omega \\ \quad u = 0 & \text{in } \partial\Omega. \end{cases}$$

As an illustration, we indicate the choice $g(u) = u^+$ and $q(u) = -u^-$, where $u^+ = \max\{u, 0\}$ and $u^- = \max\{-u, 0\}$, which involves the Fučik spectrum.

We return to our abstract setting in problems (P_0) and (P). Suppose that the following conditions hold:

(C_1) There is a constant $a_2 > 0$ such that

$$-J(v,t) + \psi(v) \geq -a_1 - a_2\|v\|^{p+2}, \quad \forall\,(v,t) \in H \times \mathbb{R}.$$

(C_2) $\rho \geq \sqrt{a_2(p+2)}$ and $\displaystyle\sup_{t \in [0,1]} (-J(0,tr)) + \psi(0) \leq 0.$

(C_3) $J^\circ(v,t;0,t) \leq 0, \quad \forall\,(v,t) \in H \times \mathbb{R}.$

(C_4) Every sequence $(v_n) \subset H$ for which there is a sequence $(t_n) \subset \mathbb{R}$ such that $t_n \to 0$, $-J(v_n, t_n) + \psi(v_n) + \frac{a}{2}\,\|v_n\|^2$ is bounded in $\mathbb{R}$ and which satisfies

$$(-J)^\circ(v_n, t_n; v - v_n, 0) + \psi(v) - \psi(v_n) + a(v_n, v - v_n)_H \geq -\varepsilon_n\|v - v_n\|$$

for all $v \in H$, and for some $(\varepsilon_n) \subset \mathbb{R}$ with $\varepsilon_n > 0$, $\varepsilon_n \to 0$, is bounded in H.

(C_5) If $u_n \rightharpoonup u$ in H and $t_n \to t$ in $\mathbb{R}$, then a subsequence of (u_n, t_n) exists, which is still denoted by (u_n, t_n), such that

$$\limsup_{n \to \infty} J^\circ(u_n, t_n; u_n - u, t_n - t) \leq 0.$$

(C_6) For every $t \in [\rho, r]$, there is $v_t \in H$ such that

$$(-J)^\circ(0, t; v_t, 0) + \psi(v_t) - \psi(0) < 0.$$

Our result on the problems (P_0) and (P) is expressed as an alternative.

Theorem 7.42. *With given constants $a > 0$, $a_1 > 0$, $p \geq 0$, and $0 < \rho < r$, assume that the conditions (C_1)–(C_6) are fulfilled. Then, for any number $\alpha > 0$, either problem (P_0) has a solution $u \in H \setminus \{0\}$ or problem (P) admits a solution $(u, \lambda, s) \in H \times \mathbb{R} \times \mathbb{R}$ in the sense that (P) is satisfied with the eigensolution $(u, \lambda) \in (H \setminus \{0\}) \times \mathbb{R}$ corresponding to the parameter $s \in \mathbb{R}$.*

Proof: Fix a constant $\alpha > 0$ and then choose a function $\beta \in C^1(\mathbb{R})$ with the properties:

(β_1) $\beta(0) = \beta(r) = 0$, $\frac{2}{p+2}\beta(\rho) = a_1 + \alpha$.

(β_2) $\displaystyle\lim_{|t| \to +\infty} \beta(t) = +\infty$.

(β_3) $\beta'(t) < 0 \Longleftrightarrow t < 0$ or $\rho < t < r$.

(β_4) $\beta'(t) = 0 \Longrightarrow t \in \{0, \rho, r\}$.

We apply Corollary 2.198, with $E = H$, to the functional $F : H \times \mathbb{R} \to \mathbb{R} \cup \{+\infty\}$ defined as follows:

$$F(v, t) = \frac{t^2}{p+2}\|v\|^{p+2} + \frac{2}{p+2}\beta(t) - J(v, t) + \frac{a}{2}\|v\|^2 + \psi(v). \quad (7.149)$$

Note that the functional F in (7.149) complies with the structural hypothesis (H) in Sect. 2.5.3, where the locally Lipschitz functional $\Phi : H \times \mathbb{R} \to \mathbb{R}$ is given by

$$\Phi(v, t) = \frac{t^2}{p+2}\|v\|^{p+2} + \frac{2}{p+2}\beta(t) - J(v, t) + \frac{a}{2}\|v\|^2, \quad (7.150)$$

and the functional $\Psi : H \times \mathbb{R} \to \mathbb{R} \cup \{+\infty\}$ is defined by

$$\Psi(v, t) = \psi(v), \quad (7.151)$$

which is convex, proper, and lower semicontinuous. By the second part of (C_2) and (β_1), we see from (7.149) that the assumptions $F(0, 0) \leq 0$ and $F(0, r) \leq 0$ in Corollary 2.198 are verified. Furthermore, from (C_1), (β_1), and the condition $\rho \geq \sqrt{a_2(p+2)}$ in (C_2), we obtain the estimate

$$F(v, \rho) \geq \left(\frac{\rho^2}{p+2} - a_2\right)\|v\|^{p+2} + \frac{2}{p+2}\beta(\rho) - a_1$$

$$\geq \frac{2}{p+2}\beta(\rho) - a_1 = \alpha, \quad (7.152)$$

for all $v \in H$, and thus, $\inf_{v \in H} F(v, \rho) > 0$. We check that the functional $F : H \times \mathbb{R} \to \mathbb{R} \cup \{+\infty\}$ satisfies the (PS) condition in the sense of Definition 2.190. Let (u_n, t_n) be a sequence in $H \times \mathbb{R}$ such that there is a constant $M > 0$ with $|F(u_n, t_n)| \leq M$, for all $n \in \mathbb{N}$, and there is a sequence $(\varepsilon_n) \subset \mathbb{R}$, $\varepsilon_n > 0$, $\varepsilon_n \to 0$, for which we have

$$\Phi^o(u_n, t_n; v - u_n, t - t_n) + \Psi(v, t) - \Psi(u_n, t_n)$$

$$\geq -\varepsilon_n(\|v - u_n\| + |t - t_n|)$$

for all $(v,t) \in H \times \mathbb{R}$, with Φ and Ψ given in (7.150) and (7.151), respectively. This finding reads as follows:

$$\left| \frac{t_n^2}{p+2}\|u_n\|^{p+2} + \frac{2}{p+2}\beta(t_n) - J(u_n, t_n) + \frac{a}{2}\|u_n\|^2 + \psi(u_n) \right| \leq M \quad (7.153)$$

and

$$\frac{2}{p+2}\left[t_n\|u_n\|^{p+2} + \beta'(t_n) \right](t - t_n) + (-J)^\circ(u_n, t_n; v - u_n, t - t_n)$$
$$+\psi(v) - \psi(u_n) + (a + t_n^2\|u_n\|^p)(u_n, v - u_n)_H$$
$$\geq -\varepsilon_n(\|v - u_n\| + |t - t_n|) \quad (7.154)$$

for all $(v,t) \in H \times \mathbb{R}$. Setting $v = u_n$ in (7.154) leads to

$$\frac{2}{p+2}\left[t_n\|u_n\|^{p+2} + \beta'(t_n) \right](t - t_n) + (-J)^\circ(u_n, t_n; 0, t - t_n)$$
$$\geq -\varepsilon_n|t - t_n| \quad (7.155)$$

whenever $t \in \mathbb{R}$. For $t = 0$, inequality (7.155) becomes

$$\frac{2}{p+2}\left[t_n^2\|u_n\|^{p+2} + t_n\beta'(t_n) \right] - J^\circ(u_n, t_n; 0, t_n) \leq \varepsilon_n|t_n|. \quad (7.156)$$

We obtain from (7.149) and (C_1) that

$$M \geq \left(\frac{t_n^2}{p+2} - a_2 \right)\|u_n\|^{p+2} + \frac{2}{p+2}\beta(t_n) - a_1, \quad \forall\, n \in \mathbb{N}.$$

On the basis of condition (β_2), we derive from the above inequality that

$$(t_n) \text{ is bounded in } \mathbb{R}. \quad (7.157)$$

First, we consider in (7.157) the situation

$$t_n \to 0 \text{ in } \mathbb{R} \text{ as } n \to \infty. \quad (7.158)$$

In view of (7.158), we may suppose $|t_n| < \rho$ for n sufficiently large. Then from (β_3), it follows that $t_n\beta'(t_n) \geq 0$. The obtained inequality enables us to get from (7.156) and (C_3) that $t_n^2\|u_n\|^{p+2} \to 0$ as $n \to \infty$. This result can be used in conjunction with (7.153) to derive

$$-J(u_n, t_n) + \psi(u_n) + \frac{a}{2}\|u_n\|^2 \quad \text{is a bounded sequence in } \mathbb{R}. \quad (7.159)$$

Setting $t = t_n$ in (7.154) yields

$$(-J)^\circ(u_n, t_n; v - u_n, 0) + \psi(v) - \psi(u_n) + a(u_n, v - u_n)_H$$
$$\geq -(\varepsilon_n + t_n^2 \|u_n\|^{p+1}) \|v - u_n\| \tag{7.160}$$

for all $v \in H$. Writing

$$t_n^2 \|u_n\|^{p+1} = |t_n|^{\frac{2}{p+2}} (t_n^2 \|u_n\|^{p+2})^{\frac{p+1}{p+2}},$$

we see that $t_n^2 \|u_n\|^{p+1} \to 0$ as $n \to \infty$. Making use of (7.158)–(7.160), we may invoke assumption (C_4), with $\varepsilon_n + t_n^2 \|u_n\|^{p+1}$ in place of ε_n, to deduce that (u_n) is bounded in H. So $u \in H$ exists such that, along a relabelled subsequence,

$$u_n \rightharpoonup u \text{ in } H \text{ as } n \to \infty. \tag{7.161}$$

Without loss of generality, we may replace the sequence (u_n, t_n) by the re-labelled subsequence given in assumption (C_5). Setting $v = u$ and $t = 0$ in (7.154), we obtain

$$\frac{2}{p+2} \left[t_n \|u_n\|^{p+2} + \beta'(t_n) \right] (-t_n) + (-J)^\circ(u_n, t_n; u - u_n, -t_n)$$
$$\psi(u) - \psi(u_n) + (a + t_n^2 \|u_n\|^p)(u_n, u - u_n)_H$$
$$\geq \varepsilon_n(\|u - u_n\| + |t_n|).$$

Letting $n \to \infty$, from (7.158) and (7.161), we find

$$0 \leq \limsup_{n \to \infty} \left[(-J)^\circ(u_n, t_n; u - u_n, -t_n) + \psi(u) - \psi(u_n) - a\|u_n - u\|^2 \right].$$

Assumption (C_5) and the lower semicontinuity of ψ ensure that

$$\liminf_{n \to \infty} \|u_n - u\|^2 \leq 0,$$

which guarantees that the sequence (u_n) possesses a strongly convergent subsequence (with limit u). Therefore, the (PS) condition is verified for the functional F in (7.149) in the case where (7.158) holds true.

It remains to check the (PS) condition for F in (7.149) when there is a constant $\delta > 0$ such that, up to a subsequence of (t_n) in (7.157), we have

$$|t_n| \geq \delta, \quad \forall \, n \in \mathbb{N}. \tag{7.162}$$

Through relation (7.156) and assumption (C_3), it is seen that

$$t_n^2 \|u_n\|^{p+2} + t_n \beta'(t_n) \leq \frac{p+2}{2} \varepsilon_n |t_n|,$$

which, taking into account (7.157), shows that $(t_n^2 \|u_n\|^{p+2})$ is a bounded sequence. It turns out from (7.162) that

$$\|u_n\|^{p+2} = (t_n^2\|u_n\|^{p+2})\frac{1}{t_n^2} \le \frac{1}{\delta^2}(t_n^2\|u_n\|^{p+2}),$$

so we have the boundedness of (u_n) in H. Therefore, passing to a relabelled subsequence, we may assume that (7.161) holds and $t_n \to \tau$ in $\mathbb{R}$ as $n \to \infty$ for some $\tau \in \mathbb{R}$. Let us put $v = u$ and $t = \tau$ in (7.154). Then letting $n \to \infty$ in (7.154), from assumption (C_5) and the lower semicontinuity of ψ, we arrive at the conclusion

$$0 \le \limsup_{n\to\infty}\left[(a + t_n^2\|u_n\|^p)(u_n, u - u_n)_H\right] \le -a\liminf_{n\to\infty}\|u - u_n\|^2.$$

Thus, along a subsequence still denoted (u_n), we have $u_n \to u$ strongly in H, which proves that the (PS) condition for F is verified. As all assumptions are fulfilled, we may apply Corollary 2.198 to the functional F in (7.149), and therefore, we find a point $(u, s) \in H \times \mathbb{R}$ such that

$$\inf_{v\in H} F(v, \rho) \le F(u, s) \le \sup_{t\in[0,1]} F(0, tr) \tag{7.163}$$

and, with Φ introduced in (7.150),

$$\Phi^\circ(u, s; v - u, t - s) + \psi(v) - \psi(u) \ge 0 \tag{7.164}$$

for all $(v, t) \in H \times \mathbb{R}$. Using estimate (7.152), the second part of assumption (C_2), and properties (β_1), (β_3), we deduce from (7.163) that

$$\alpha \le \frac{s^2}{p+2}\|u\|^{p+2} + \frac{2}{p+2}\beta(s) - J(u, s) + \frac{a}{2}\|u\|^2 + \psi(u) \le \alpha + a_1. \tag{7.165}$$

Explicitly, inequality (7.164) means

$$\frac{2}{p+2}\left[s\|u\|^{p+2} + \beta'(s)\right](t - s) + (-J)^\circ(u, s; v - u, t - s)$$
$$+ (a + s^2\|u\|^p)(u, v - u)_H + \psi(v) - \psi(u) \ge 0 \tag{7.166}$$

for all $(v, t) \in H \times \mathbb{R}$. Setting in (7.166) $t = s$ and $v = u$ yields, respectively,

$$(-J)^\circ(u, s; v - u, 0) + \psi(v) - \psi(u) + (a + s^2\|u\|^p)(u, v - u)_H \ge 0 \tag{7.167}$$

for all $v \in H$, and

$$\frac{2}{p+2}\left[s\|u\|^{p+2} + \beta'(s)\right](t - s) + (-J)^\circ(u, s; 0, t - s) \ge 0 \tag{7.168}$$

for all $t \in \mathbb{R}$. Putting $t = 0$ in (7.168) gives

$$\frac{2}{p+2}\left[s^2\|u\|^{p+2} + s\beta'(s)\right] - J^\circ(u, s; 0, s) \le 0. \tag{7.169}$$

Combining (7.169) and hypothesis (C_3) ensures that

$$s\beta'(s) \leq 0. \qquad (7.170)$$

First, we regard the case $s = 0$ in (7.170). Writing (7.165) and (7.167), for $s = 0$, expresses that $u \in H$ solves problem (P_0). We point out that $u \neq 0$. This result is true because of the inequalities

$$0 < \alpha \leq -J(u, 0) + \psi(u) + \frac{\alpha}{2}\|u\|^2$$

[cf. (P_0)] and $-J(0, 0) + \psi(0) \leq 0$ derived from assumption (C_2). It remains to analyze the situation $s \neq 0$ in (7.170). The monotonicity properties of the function β required in (β_3) and (β_4) enable us to conclude from (7.170) that $\rho \leq s \leq r$. Denoting $\lambda = a + s^2\|u\|^p$, we see then from (7.165) and (7.167) that the triple $(u, \lambda, s) \in H \times \mathbb{R} \times \mathbb{R}$ is a solution of problem (P). As $\rho \leq s \leq r$ and in view of hypothesis (C_6) applied for $t = s$, it is clear that $u \neq 0$. The proof is thus complete. $\qquad \square$

Remark 7.43. Hypothesis (C_6) is needed only to guarantee that $u \neq 0$ for every solution (u, λ, s) of eigenvalue problem (P).

Remark 7.44. We briefly comment about the meaning of assumptions (C_1)–(C_6) and how they can be practically verified. Condition (C_1) requires a unilateral polynomial growth for the functional $-J + \psi$. Condition (C_2) demands that the constant $\rho > 0$ be sufficiently large and an upper estimate for the functional $-J(0, \cdot)$ on the interval $[0, r]$. Condition (C_3) expresses a monotonicity property for the function $J(v, \cdot)$ on $\mathbb{R}$ for every $v \in H$. Condition (C_4) asserts a weak form of Palais–Smale condition for the perturbation of our functional $-J + \psi$ with the fixed quadratic term $\frac{a}{2}\|v\|^2$. Condition (C_5) claims to have the generalized directional derivative J° sequentially weakly upper semicontinuous. For instance, this result can be achieved if the space H is compactly embedded in a Banach space X and $J = \tilde{J}|_{H \times \mathbb{R}}$ with $\tilde{J} : X \times \mathbb{R} \to \mathbb{R}$ being locally Lipschitz. Condition (C_6) presents a kind of nonvanishing requirement with respect to the function ψ for the generalized directional derivative of $-J(\cdot, t)$ at $0 \in H$ when t runs on the prescribed interval $[\rho, r]$ in $\mathbb{R}$.

7.6 Notes and Comments

In this chapter, we discussed two general methods to treat nonlinear boundary value problems: sub-supersolution method and minimax method in the critical point theory. We illustrated them in the general framework of nonsmooth problems expressed as variational–hemivariational inequalities. The two methods are complementary and as they are used in our approach allow obtaining location information for the solutions. A significant combined application of the two methods has been given in Sect. 3.4.

The exposition of the sub-supersolution method for quasilinear elliptic variational–hemivariational inequalities of the form (7.1) as presented in Sect. 7.1 relies on [55]. The important particular case of constraint hemivariational inequality, which can be specified in our problem by taking ψ to be the indicator function, was the object of [56]. Other important classes of problems incorporated in our setting are those of variational inequalities and of differential inclusions with Clarke's generalized gradient. The study of sub-supersolution method for variational inequalities was initiated in [141, 142] and was developed in Chap. 5. For the approach based on the sub-supersolution method in the case of differential inclusions of hemivariational type, we refer to [53, 58, 59]. This was the object of Chap. 4.

Section 7.2 constructs the basic frame to handle the sub-supersolution method for evolutionary variational–hemivariational inequalities of type (7.35). The development follows [57]. Some special cases can be found in [39, 43, 51, 54, 58]. The applicability of our abstract results is illustrated to an obstacle problem under hemivariational boundary conditions.

Section 7.3 deals with the nonsmooth critical point theory in the setting of hemivariational inequalities where it is allowed to contain both superlinear and sublinear terms. This result is mainly achieved by relaxing the celebrated condition of Ambrosetti–Rabinowitz type (see [3] for the smooth case and [64] for the nonsmooth case) in assuming our weaker conditions $(H2)$ and $(H4)$. The existence result given here for nontrivial solutions of hemivariational inequalities is taken from [109]. Concerning the applicability of our result, we remark that two adhesively connected von Kármán plates subjected to elastoplastic boundary conditions, to unilateral contact boundary conditions, or to friction boundary conditions lead to hemivariational inequalities in the form studied here (cf. [181]). Various models are analyzed in [177, 180].

Section 7.4, as well as Sect. 7.5, makes full use of the nonsmooth critical point theory developed in Sect. 2.5.3. The geometric situation encountered in Sect. 7.4 is the one of a mountain pass theorem for variational–hemivariational inequalities. In the case of smooth functions, the mountain pass theorem was given in [3] (see also [117, 160, 191]), whereas for the locally Lipschitz function, this result is found in [64] (and with Cerami condition in [128]). The extension as applied in Sect. 7.4 was obtained in [171] (and in a slightly less general version in [102]). The limiting case in the minimax principle has been treated in [156]. For the situation $\lambda < \lambda_1$, Theorem 7.38 extends Theorem 5.1 of Szulkin [211] as well as other results in [3, 64, 171, 191].

In Sect. 7.5, we treat one-parameter families of nonlinear eigenvalue boundary value problems with nonsmooth potentials and including constraints on the solutions. The pairs $(\lambda, s) \in \mathbb{R}^2$ considered in the variational–hemivariational inequality can be interpreted like a generalized version of the Fučik spectrum as outlined in Example 7.41. Our main result, stated as Theorem 7.42, is expressed as an alternative between having a given number as an eigenvalue [see problem (P_0)] or solving an eigenvalue problem involving parameters [see problem (P)]. Our exposition follows [168]. In the case where the function J does not depend on parameter, this theorem extends a result in [2] in the nonsmooth case (and the corresponding property in [166] for the smooth case). In [167], it is shown for the smooth potentials that this approach allows us to deduce various qualitative properties for the eigensolutions including location information. It is worth pointing out that the location of eigensolutions is achieved by using the graph of a function enjoying the properties (β_1)–(β_4) in the proof of Theorem 7.42. For applications to study further properties of eigensolutions for eigenvalue problems involving semilinear Dirichlet problems, we refer to [167]. A recent new method in treating nonlinear eigenvalue problems was initiated by Ricceri [193, 194, 195]. This method has been applied to different nonlinear elliptic problems with smooth or nonsmooth potentials (see [21, 154, 155, 130]).

Correction to: Nonsmooth Variational Problems and Their Inequalities

Siegfried Carl, Vy Khoi Le and Dumitru Motreanu

Correction to:

S. Carl, V.K. Le and D. Montreanu, *Nonsmooth Variational Problems and Their Inequalities*, https://doi.org/10.1007/978-0-387-46252-3

The first name and last name of author Vy Khoi Le was unfortunately published with an error. The First Name should be Vy Khoi and Le should be the Last Name. The initially published version has now been corrected.

The updated online version of this book can be found at
https://doi.org/10.1007/978-0-387-46252-3

List of Symbols

$\mathbb{N}$	natural numbers
$\mathbb{N}_0$	$\mathbb{N} \cup \{0\}$
$\mathbb{R}$	real numbers
$\mathbb{R}_+$	nonnegative real numbers
$\mathbb{R}^N$	N-dimensional Euclidean space
Ω	open domain in $\mathbb{R}^N$
$\partial\Omega$	boundary of Ω
$|E|$	Lebesgue-measure of a subset $E \subset \mathbb{R}^N$
X	real normed linear space
X^*	dual space of X
X^{**}	bidual space of X
X_+	positive (or order) cone of X
X_+^*	dual-order cone of X
$x \wedge y$	$\min\{x, y\}$
$x \vee y$	$\max\{x, y\}$
x^+	$\max\{x, 0\}$
x^-	$\max\{-x, 0\}$
"iff"	stands for "if and only if"
$X \subset Y$	X is a subset of Y including $X = Y$
2^X	power set of the set X, i.e., the set of all subsets of X
$\mathrm{cl}(K)$ or $\overline{K}$	closure of a subset K of X
$\mathrm{int}(K)$ or $\overset{\circ}{K}$	interior of K
$L(X, Y)$	space of bounded linear mappings from X to Y
$D(A)$	domain of the operator A
$\mathrm{dom}(A)$	effective domain of the mapping A
I_K	indicator function, i.e., $I_K(x) = 0$ if $x \in K$, $+\infty$ otherwise
χ_E	characteristic function of the set E
$\mathrm{Gr}(A)$	graph of the mapping A
A^*	adjoint or dual operator to A

$\rightharpoonup$	weak convergence
$\rightharpoonup^*$	weak* convergence
$\delta f(u;h)$ or $f'(u;h)$	directional derivative
$D_G f$	Gâteaux derivative
$D_F f$ or f'	Fréchet derivative
$f^\circ(u;h)$	generalized directional derivative
∂f	subdifferential of f or Clarke's generalized gradient
∇f	$(\partial f/\partial x_1, \partial f/\partial x_2, \ldots, \partial f/\partial x_N)$, the gradient of f
Δf	$\partial^2 f/\partial x_1^2 + \partial^2 f/\partial x_2^2 + \cdots + \partial^2 f/\partial x_N^2$, the Laplacian of f
$\Delta_p f$	the p-Laplacian of f
$C_0^\infty(\Omega)$	space of infinitely differentiable functions with compact support in Ω
$\|f\|_{L^p(\Omega)}$	$\left(\int_\Omega |f|^p dx\right)^{1/p}$, the L^p norm
$L^p(\Omega)$	space of p integrable functions (whose L^p norm is bounded)
$L^p_{\mathrm{loc}}(\Omega)$	space of locally p integrable functions
$\|f\|_{W^{m,p}(\Omega)}$	$\left(\sum_{|\beta|\le m} \int_\Omega |D^\beta f|^p dx\right)^{1/p}$, the Sobolev norm
$W^{m,p}(\Omega)$	space of functions with bounded $W^{m,p}(\Omega)$ Sobolev norm
$W_0^{m,p}(\Omega)$	$W^{m,p}(\Omega)$-functions with generalized homogeneous boundary values
$\gamma(u)$ or γu	trace of u or generalized boundary values of u
$L^p(0,\tau;B)$	space of p integrable vector-valued functions $u:(0,\tau)\to B$
$C([0,\tau];B)$	space of continuous vector-valued functions $u:[0,\tau]\to B$
$C^1([0,\tau];B)$	space of continuously differentiable vector-valued functions $u:[0,\tau]\to B$

References

1. Addou, A., Mermri, B.: Topological degree and application to a parabolic variational inequality problem. IJMMS, **25**, 273–287 (2001)
2. Adly, S., Motreanu, D.: Location of eigensolutions to variational-hemivariational inequalities. J. Nonlinear Convex Anal., **1**, 255-270 (2000)
3. Ambrosetti, A., Rabinowitz, P.H.: Dual variational methods in critical point theory and applications. J. Func. Anal., **14**, 349-381 (1973)
4. Akô, K.: On the Dirichlet problem for quasilinear elliptic differential equations of second order. J. Math. Soc. Japan, **13**, 45–62 (1961)
5. Alt, H.W.: Lineare Funktionalanalysis. Springer-Verlag, Berlin (2002)
6. Anane, A.: Simplicity and isolation of the first eigenvalue of the p-Laplacian with weight. C. R. Acad. Sci. Paris Sér. I Math., **305**(16), 725–728 (1987)
7. Ang, D.D., Schmitt, K., Le, V.K.: Noncoercive variational inequalities: Some applications. Nonlinear Analysis, TMA, **15**, 497–512 (1990)
8. Attouch, H., Azé, D.: Approximation and regularization of arbitrary functions in Hilbert spaces by the Lasry–Lions method. Ann. Inst. H. Poincaré Anal. Non Linéaire, **10**, 289–312 (1993)
9. Aubin, J. P., Clarke, F.H.: Shadow prices and duality for a class of optimal control problems. SIAM J. Control Optim., **17**, 567–586 (1979)
10. Baiocchi, C., Buttazzo, G., Gastaldi, F., Tomarelli, F.: General existence theorems for unilateral problems in continuum mechanics. Arch. Rational Mech. Anal., **100**, 149–189 (1988)
11. Baiocchi, C., Capelo, A.: Variational and Quasivariational Inequalities: Applications to Free Boundary Problems. Wiley, New York (1984)
12. Baiocchi, C., Gastaldi, F., Tomarelli, F.: Some existence results on noncoercive variational inequalities. Ann. Scuola Norm. Sup. Pisa Cl. Sci., **13**, 617–659 (1986)
13. Barbu, V.: Nonlinear Semigroups and Differential Equations in Banach Spaces. Noordhoff International Publishing, Leiden (1976)
14. Barbu, V., Korman, P.: Approximating optimal controls for elliptic obstacle problems by monotone iteration schemes. Numer. Funct. Anal. And Optimiz., **12**, 429–442 (1991)
15. Bebernes, J.W., Schmitt, K.: On the existence of maximal and minimal solutions for parabolic partial differential equations. Proc. Amer. Math. Soc., **73**, 211–218 (1979)

382   References

16. Berestycki, H., Lions, P.L.: Some applications of the method of super and sub-solutions. Bifurcation and nonlinear eigenvalueproblems. Proc., Session, Univ. Paris XIII, Villetaneuse, 1978, Springer, 16–41 (1980)
17. Berkovits, J., Mustonen, V.: On the topological degree for mappings of monotone type. Nonlinear Anal., 10, 1373–1383 (1986)
18. Berkovits, J., Mustonen, V.: Nonlinear mappings of monotone type. Report, University of Oulu, Oulu (1988)
19. Berkovits, J., Mustonen, V.: Topological degree for perturbations of linear maximal monotone mappings and applications to a class of parabolic problems. Rend. Mat. Appl., Serie VII 12, 597–621 (1992)
20. Berkovits, J., Mustonen, V.: Monotonicity methods for nonlinear evolution equations. Nonlinear Anal., 27, 1397–1405 (1996)
21. Bonanno, G.: Some remarks on a three critical points theorem. Nonlinear Anal., 54, 651–665 (2003)
22. Borwein, J.M., Treiman, J.S., Zhu, Q.J.: Necessary conditions for constrained optimization problems with semicontinuous and continuous data. Trans. Amer. Math. Soc., 350, 2409–2429 (1998)
23. Brezis, H.: Équations et inéquations non linéaires dans les espaces vectoriels en dualité. Ann. Inst. Fourier, 18, 115–175 (1968)
24. Brézis, H.: Analyse Fonctionnelle - Théorie et Applications. Masson, Paris (1983)
25. Brézis, H., Nirenberg, L.: Characterizations of the ranges of some nonlinear operators and applications to boundary value problems. Ann. Scuola Norm. Sup. Pisa, 5, 225–326 (1978)
26. Browder, F.: Nonlinear operators and nonlinear equations of evolution in Banach spaces. Proc. Sympos. Pure Math., Vol. 18, Part 2, Am. Math. Soc., Providence (1976)
27. Browder, F.E.: Fixed point theory and nonlinear problems. Bull. Amer. Math. Soc., 9, 1–39 (1983)
28. Buttazzo, G., Tomarelli, F.: Compatibility conditions for nonlinear Neumann problems. Adv. Math., 89, 127–143 (1991)
29. Carl, S.: On the existence of extremal weak solutions for a class of quasilinear parabolic problems. Diff. Integral Equations, 6, 1493–1505 (1993)
30. Carl, S.: Leray–Lions operators perturbed by state-dependent subdifferentials. Nonlinear World, 3, 505–518 (1996)
31. Carl, S.: Existence of extremal periodic solutions for quasilinear parabolic equations. Abstr. Appl. Anal., 2, 257–270 (1997)
32. Carl, S.: Quasilinear elliptic equations with discontinuous nonlinearities in $\mathbb{R}^N$. Nonlinear Anal., 30, 1743–1751 (1997)
33. Carl, S.: An extremality result for quasilinear elliptic equations in $\mathbb{R}^N$. Comm. Appl. Nonlinear Anal. 4, 43–51 (1997)
34. Carl, S.: Extremal solutions for quasilinear elliptic inclusions in all of $\mathbb{R}^N$ with state-dependent subdifferentials. J. Optim. Theory Appl., 104, 323–342 (2000)
35. Carl, S.: Extremal solutions of hemivariational inequalities with d.c.-superpotentials. In: "Differential Equations and Nonlinear Mechanics", 11–25, Math. Appl., 528, Kluver Acad. Publ., Dordrecht (2001)
36. Carl, S.: Existence of extremal solutions of boundary hemivariational inequalities. J. Differential Equations, 171, 370–396 (2001)
37. Carl, S.: Existence and comparison results for variational-hemivariational inequalities. J. Inequal. Appl., 2005:1, 33–44 (2005)

38. Carl, S., Diedrich, H.: The weak upper and lower solution method for quasilinear elliptic equations with generalized subdifferentiable perturbations. Appl. Anal., **56**, 263–278 (1995)
39. Carl, S., Gilbert, R.P.: Extremal solutions of a class of dynamic boundary hemivariational inequalities. J. Inequal. Appl., **7**, 479–502 (2002)
40. Carl, S., Grossmann, C.: Monotone enclosure for elliptic and parabolic systems with nonmonotone nonlinearities. J. Math. Anal. Appl., **151**, 190–202 (1990)
41. Carl, S., Heikkilä, S.: On extremal solutions of an elliptic boundary value problem involving discontinuous nonlinearities. Diff. Integral Equations, **5**, 581–589 (1992)
42. Carl, S., Heikkilä, S.: A free boundary problem for quasilinear elliptic equations in exterior domains. Diff. Integral Equations, **11**, 409–423 (1998)
43. Carl, S., Heikkilä, S.: Nonlinear Differential Equations in Ordered Spaces. Chapman & Hall/CRC, Boca Raton (2000)
44. Carl, S., Heikkilä, S.: Discontinuous reaction-diffusion equations under discontinuous and nonlocal flux condition. Math. Comput. Modelling, **32**, 1333–1344 (2000)
45. Carl, S., Heikkilä, S., Jerome, J.W.: Trapping regions for discontinuously coupled systems of evolution variational inequalities and applications. J. Math. Anal. Appl., **282**, 421–435 (2003)
46. Carl, S., Heikkilä, S., Lakshmikantham, V.: Nonlinear elliptic differential inclusions governed bystate-dependent subdifferentials. Nonlinear Anal., **25**, 729–745 (1995)
47. Carl, S., Jerome, J.W.: Trapping region for discontinuous quasilinear elliptic systems of mixed monotone type. Nonlinear Analysis, **51**, 843–863 (2002)
48. Carl, S., Lakshmikantham, V.: Generalized quasilinearization for quasilinear parabolic equations with nonlinearities of dc type. J. Optim. Theory Appl., **109**, 27–50 (2001)
49. Carl, S., Lakshmikantham, V.: Generalized quasilinearization method for reaction-diffusion equations under nonlinear and nonlocal flux conditions. J. Math. Anal. Appl., **271**, 182–205 (2002)
50. Carl, S., Le, V. K.: Enclosure results for quasilinear systems of variational inequalities. J. Differential Equations, **199**, 77–95 (2004)
51. Carl, S., Le, V. K.: Sub-supersolution method for quasilinear parabolic variational inequalities. J. Math. Anal. Appl., **293**, 269–284 (2004)
52. Carl, S., Le, V.K.: Monotone penalty approximation of extremal solutions for quasilinear noncoercive variational inequalities. Nonlinear Anal., **57**, 311–322 (2004)
53. Carl, S., Le, V.K., Motreanu, D.: The sub-supersolution method and extremal solutions for quasilinear hemivariational inequalities. Diff. Integral Equations, **17**, 165–178 (2004)
54. Carl, S., Le, V. K., Motreanu, D.: Existence and comparison results for quasilinear evolution hemivariational inequalities. Electron. J. Differential Equations, **2004**, No. 57, 1–17 (2004)
55. Carl, S., Le, V.K., Motreanu, D.: Existence and comparison principles for general quasilinear variational-hemivariational inequalities. J. Math. Anal. Appl., **302**, 65–83 (2005)
56. Carl, S., Le, V.K., Motreanu, D.: Existence, comparison and compactness results for quasilinear variational-hemivariational inequalities. Int. J. Math. Mathematical Sci., **3**, 401–417 (2005).

57. Carl, S., Le, V.K., Motreanu, D.: Evolutionary variational-hemivariational in-
    equalities: Existence and comparison results, submitted
58. Carl, S., Motreanu, D.: Extremal solutions of quasilinear parabolic inclusions
    with generalized Clarke's gradient. J. Differential Equations, **191**, 206–233
    (2003)
59. Carl, S., Motreanu, D.: Quasilinear elliptic inclusions of
    hemivariational type: Extremality and compactness of the solution set. J.
    Math. Anal. Appl., **286**, 147–159 (2003)
60. Carl, S., Motreanu, D.: Extremality in solving general quasilinear parabolic
    inclusions. J. Optim. Theory Appl., **123**, 463–477 (2004)
61. Carl, S., Perera, K.: Sign-changing and multiple solutions for the p-Laplacian.
    Abstract Applied Analysis, **7**(12), 613-625 (2002)
62. Casado-Diaz, J., Porretta, A.: Existence and comparison of maximal and min-
    imal solutions for pseudomonotone elliptic problems in L^1. Nonlinear Anal.,
    53, 351–373 (2003)
63. Chang, K.-C.: Infinite-Dimensional Morse Theory and Multiple Solution Prob-
    lems. Progress in Nonlinear Differential Equations and their Applications, **Vol.
    6**, Birkhäuser Boston Inc., Boston (1993)
64. Chang, K.-C.: Variational methods for non-differentiable functionals and their
    applications to partial differential equations. J. Math. Anal. Appl., **80**, 102–129
    (1981)
65. Charrier, P., Troianiello, G.M.: On strong solutions to parabolic unilateral
    problems with obstacle dependent on time. J. Math. Anal. Appl., **65**, 110–125
    (1978)
66. Chipot, M.: Nonlinear Analysis. Birkhäuser Verlag, Basel (2000)
67. Chipot, M., Rodrigues, J. F.: Comparison and stability of solutions
    to a class of quasilinear parabolic problems. Proc. Roy. Soc. Edinburgh, **A
    110**, 275–285 (1988)
68. Clarke, F.H.: Optimization and Nonsmooth Analysis. Society for Industrial
    and Applied Mathematics (SIAM), Philadelphia, PA (1990)
69. Costa, D.G., Goncalves, J.V.A.: On the existence of positive solutions for a
    class of non-seladjoint elliptic boundary value problems. Appl. Anal., **31**, 309–
    320 (1989)
70. Costa, D.G., Miyagaki, O.H.: Nontrivial solutions for perturbations of the
    p-Laplacian on unbounded domains. J. Math. Anal. Appl., **193**(3), 737–755
    (1995)
71. Cuesta, M., de Figueiredo, D., Gossez, J.-P.: The beginning of the Fučík spec-
    trum for the p-Laplacian. J. Differential Equations, **159**(1), 212–238 (1999)
72. Cuesta, M., Gossez, J.-P.: A variational approach to nonresonance with respect
    to the Fučik spectrum. Nonlinear Anal., **19**(5), 487–500 (1992)
73. Dancer, E.N.: On the Dirichlet problem for weakly non-linear elliptic par-
    tial differential equations. Proc. Roy. Soc. Edinburgh, **Sect. A76**(4), 283–300
    (1976/77)
74. Dancer, E.N., Perera, K.: Some remarks on the Fučík spectrum of the p-
    Laplacian and critical groups. J. Math. Anal. Appl., **254**(1), 164–177 (2001)
75. Dancer, E. N., Sweers, G.: On the existence of a maximal weak solution for a
    semilinear elliptic equation. Differential Integral Equations, **2**, 533–540 (1989)
76. del Pino, M.A., Manásevich, R.F.: Global bifurcation from the eigenvalues of
    the p-Laplacian. J. Differential Equations, **92**, 226–251 (1991)

77. de Figueiredo, D., Gossez, J.-P.: On the first curve of the Fučík spectrum of an elliptic operator. Differential Integral Equations, **7**, 1285–1302 (1994)

78. Degiovanni, M., Marzocchi, M., Rădulescu, V.: Multiple solutions of hemivariational inequalities with area-type term. Calc. Var. Partial Differential Equations, **10**, 355–387 (2000)

79. Denkowski, Z., Migórski, S., Papageorgiou, N.S.: An Introduction to Nonlinear Analysis: Theory. Kluwer Academic Publishers, Boston, Dordrecht, London (2003)

80. Denkowski, Z., Migórski, S., Papageorgiou, N.S.: An Introduction to Nonlinear Analysis: Applications. Kluwer Academic Publishers, Boston, Dordrecht, London (2003)

81. Deuel, J.: Nichtlineare parabolische Randwertprobleme mit Unter- und Oberlösungen. ETH Diss. Nr. 5750, Zürich (1976)

82. Deuel, J., Hess, P.: Inéquations variationnelles elliptiques non coercives. C. R. Acad. Sci. Paris Sér. A, **279**, 719–722 (1974)

83. Deuel, J., Hess, P.: A criterion for the existence of solutions of non-linear elliptic boundary value problems. Proc. Roy. Soc. Edinburgh Sect. A, **74**, 49–54 (1974/1975)

84. Deuel, J., Hess, P.: Nonlinear parabolic boundary value problems with upper and lower solution. Israel J. Math., **29**, 92–104 (1978)

85. di Benedetto, E.: $C^{1+\alpha}$ local regularity of weak solutions of degenerate elliptic equations. Nonlinear Anal., **7**(8), 827–850 (1983)

86. Drábek, P.: Solvability and Bifurcations of Nonlinear Equations. Pitman Research Notes in Mathematics Series, **Vol. 264**, Longman Scientific & Technical, Harlow (1992)

87. Drábek, P., Robinson, S.: Resonance problems for the p-Laplacian. J. Funct. Anal., **169**(1), 189–200 (1999)

88. Du, Y.: A deformation lemma and some critical point theorems. Bull. Austral. Math. Soc., **43**, 161–168 (1991)

89. Duvaut, G., Lions, J.L.: Les Inéquations en Mécanique et en Physique. Dunod, Paris (1972)

90. Edmunds, D.E., Triebel, H.: Function Spaces, Entropy Numbers, Differential Operators. Cambridge Univ. Press, Cambridge (1996)

91. Ekeland, I.: Nonconvex minimization problems. Bull. (New Series) Amer. Math., **1**, 443–474 (1979)

92. El Hachimi, A., De Thelin, F.: Supersolutions and stabilization of the solutions of the equation $\partial u/\partial t - \operatorname{div}(|\nabla u|^{p-2}\nabla u) = h(x, u)$. Publicacions Matematiques, **35**, 347–362 (1991)

93. Evans, L.C.: Partial Differential Equations. **19** AMS, Providence (1998)

94. Evans, L.C., Gariepy, R.: Measure Theory and Fine Properties of Functions. CRC Press, Boca Raton (1992)

95. Friedman, A.: Variational Principles and Free Boundary Value Problems. Wiley-Interscience, New York (1983)

96. Fučik, S.: Boundary value problems with jumping nonlinearities. Časopis Pešt. Mat., **101**(1), 69–87 (1976)

97. Gasiński, L., Papageorgiou, N.S.: Nonsmooth Critical Point Theory and Nonlinear Boundary Value Problems. Chapman & Hall/CRC, Boca Raton (2005)

98. Giaquinta, M., Giusti, E.: Global $C^{1,\alpha}$-regularity for second order quasilinear elliptic equations in divergence form. J. Reine Angew. Math., **351**, 55–65 (1984)

386 References

99. Gilbarg, D., Trudinger, N.S.: Elliptic Partial Differential Equations of Second Order. Springer-Verlag, Berlin (1983)
100. Giuffré, S.: Global Hölder regularity for discontinuous elliptic equations in the plane. Proc. Amer. Math. Soc., **132**(5), 1333–1344 (2003)
101. Goeleven, D., Miettinen, M., Panagiotopoulos, P.D.: Dynamic hemivariational inequalities and their applications. J. Optim. Theory Appl., **103**, 567–601 (1999)
102. Goeleven D., Motreanu D.: Minimax methods of Szulkin's type in unilateral problems. In: P.K. Jain (ed.), Functional Analysis - Selected Topics. Narosa Publishing House. New Delhi, India, pp. 169-183 (1998)
103. Goeleven, D., Motreanu, D., Dumont, Y., Rochdi, M.: Variational and Hemivariational Inequalities, Theory, Methods and Applications, Volume I: Unilateral Analysis and Unilateral Mechanics. Kluwer Academic Publishers, Boston, Dordrecht, London (2003)
104. Goeleven, D., Motreanu, D.: Variational and Hemivariational Inequalities, Theory, Methods and Applications, Volume II: Unilateral Problems. Kluwer Academic Publishers, Boston, Dordrecht, London (2003)
105. Gossez, J.-P., Mustonen, V.: Pseudomonotonicity and the Leray-Lions condition. Differential Integral Equations, **6**, 37–45 (1993)
106. Grenon, N.: Asymptotic behaviour for some quasilinear parabolic equations. Nonlinear Anal., **20**, 755–766 (1993)
107. Habets, P., Schmitt, K.: Nonlinear boundary value problems for systems of differential equations. Arch. Math., **40**, 441–446 (1983)
108. Haslinger, J., Miettinen, M., Panagiotopoulos, P.D.: Finite Element Method for Hemivariational Inequalities. Theory, Methods and Applications, Kluwer Academic Publishers, Boston, Dordrecht, London, 1999
109. Haslinger, J.; Motreanu, D.: Hemivariational inequalities with a general growth condition: existence and approximation. Appl. Anal., **82**, 629–643 (2003)
110. Heikkilä S., Lakshmikantham, V.: Extension of the method of upper and lower solutions for discontinuous differential equations. Differential Equations Dynam. Systems, **1**, 73–85 (1993)
111. Heikkilä, S., Lakshmikantham, V.: Monotone Iterative Techniques for Discontinuous Nonlinear Differential Equations. Marcel Dekker Inc., New York (1994)
112. Heinonen, J., Kilpeläinen, T., Martio, O.: Nonlinear Potential Theory of Degenerate Elliptic Equations. Clarendon Press, Oxford (1993)
113. Hess, P.: On the solvability of nonlinear elliptic boundary value problems. Indiana Univ. Math. J., **25**, 461–466 (1976)
114. Hewitt, E., Sromberg, K.: Real and Absract Analysis. Springer-Verlag, Berlin (1965)
115. Hörmander, L.: Linear Partial Differential Operators. Springer, Berlin-New York (1976)
116. Huang, Y.: Existence of positive solutions for a class of the p-Laplaceequations. J. Austral. Math. Soc. Ser. B, **36**, 249–264 (1994)
117. Jabri, Y.: The Mountain Pass Theorem. Variants, Generalizations and some Applications. Cambridge University Press, Cambridge, 2003
118. Jiang, M-Y.: Critical groups and multiple solutions of the p-Laplacian. Nonlinear Anal., **59**, 1221–1241 (2004)
119. Jiu, Q., Su, J.: Existence and multiplicity results for Dirichlet problems with p-Laplacian. J. Math. Anal. Appl., **281**, 587–601 (2003)

120. Jost, J.: Postmodern Analysis. Springer-Verlag, Berlin (1998)
121. Kacur, J.: Method of Rothe and Evolution Equations. Teubner-Texte zur Mathematik, Leipzig (1985)
122. Kazdan J., Kramer, R.J.: Invariance criteria for existence of solutions to second order quasilinear equations. Comm. Pure Appl. Anal., **31**, 619–645 (1978)
123. Kinderlehrer, D.: Variational inequalities with lower dimensional obstacles. Israel J. Math., **10**, 339–348 (1971)
124. Kinderlehrer, D., Stampaccia, G.: An Introduction to Variational Inequalities and Their Application. Academic Press, New York (1980)
125. Knobloch, H.W.: Eine neue Methode zur Approximation periodischer Lösungen nicht linearer Differentialgleichungen zweiter Ordnung. Mat. Z., **82**, 177–197 (1963)
126. Knobloch, H.W., Schmitt, K.: Non-linear boundary value problems for systems of differential equations. Proc. Roy. Soc. Edinburgh, **78A**, 139–159 (1977)
127. Korman, P., Leung, A.W., Stojanovic, S.: Monotone iterations for nonlinear obstacle problem. J. Austral. Math. Soc. Ser. B, **31**, 259–276 (1990)
128. Kourogenis, N.C., Papageorgiou, N.S.: Nonsmooth critical point theory and nonlinear elliptic equations at resonance. J. Aust. Math. Soc., Ser. A, **69**, 245–271 (2000)
129. Kreyszig, E.: Introductory Functional Analysis with Applications. John Wiley, New York (1978)
130. Kristály, A.: Infinitely many solutions for a differential inclusion problem in $\mathbb{R}^N$. J. Differential Equations, to appear
131. Kučera, M.: A global continuation theorem for obtaining eigenvalues and bifurcation points. Czech. Math. J., **38**, 120–137 (1988)
132. Kufner, A., John, O., Fučik, S.: Function Spaces. ACADEMIA, Praque (1977)
133. Kura, T.: The weak supersolution-subsolution method for second order quasilinear elliptic equations. Hiroshima Math. J., **19**, 1–36 (1989)
134. Ladde, G.S., Lakshmikantham, V., Vatsala, A.S.: Existence of coupled quasi-solutions of systems of noninear linear reaction-diffusion equations. J. Math. Anal. Appl., **108**, 249–266 (1985)
135. Lakshmikantham, V., Köksal, S.: Monotone Flows and Rapid Convergence For Nonlinear Partial Differential Equations. Taylor & Francis, London (2003)
136. Landes, R., Mustonen, V.: On pseudo-monotone operators and nonlinear non-coercive variational problems on unbounded domains. Math. Ann., **248**, 241–246 (1980)
137. Le, V.K.: On global bifurcation of variational inequalities and applications. J. Diff. Eq., **141**, 254–294 (1997)
138. Le, V.K.: On some equivalent properties of sub-supersolutions in second order quasilinear elliptic equations. Hiroshima Math. J., **28**(2), 373–380 (1998)
139. Le, V.K.: Some existence results for noncoercive nonconvex minimization problems with fast or slow perturbing terms. Num. Func. Anal. Optim., **20**, 37–58 (1999)
140. Le, V.K.: Existence of positive solutions of variational inequalities by a subsolution-supersolution approach. J. Math. Anal. Appl., **252**, 65–90 (2000)
141. Le, V. K.: Subsolution-supersolution method in variational inequalities. Nonlinear Anal., **45**, 775–800 (2001)
142. Le, V. K.: Subsolution-supersolutions and the existence of extremal solutions in noncoercive variational inequalities. JIPAM J. Inequal. Pure Appl. Math. **2**, No 2, article 20, 16p, electronic (2001)

143. Le, V.K., Schmitt, K.: Minimization problems for noncoercive functionals subject to constraints. Trans. Amer. Math. Soc., **347**, 4485–4513 (1995)

144. Le, V.K., Schmitt, K.: Minimization problems for noncoercive functionals subject to constraints, Part II. Adv. Differential Equations, **1**, 453–498 (1996)

145. Le, V.K., Schmitt, K.: Global Bifurcation in Variational Inequalities: Applications to Obstacle and Unilateral Problems. Springer, Appl. Math. Sci., Vol. 123, New York (1997)

146. Le, V. K., Schmitt, K.: On boundary value problems for degenerate quasilinear elliptic equations and inequalities. J. Differential Equations, **144**, 170–218 (1998)

147. Le, V.K., Schmitt, K.: Sub-supersolution theorems for quasilinear elliptic problems: A variational approach. Electron. J. Differential Equations, **2004**(118), 1–7 (2004)

148. Lebourg, G.: Valeur moyenne pour gradient généralisé. C. R. Acad. Sci. Paris, **281**, 795–797 (1975)

149. Leray, J., Lions, J.L.: Quelques résultats de Višik sur des problè- mes elliptiques non linéaires par les méthodes de Minty-Browder. Bull. Soc. Math. France, **93**, 97–107 (1965)

150. Lieberman, L.: Boundary regularity for solutions of degenerate elliptic equations. Nonlinear Anal., **12**(11), 1203–1219 (1988)

151. Lindqvist, P.: On the equation div $|\nabla u|^{p-2}\nabla u) + \lambda |u|^{p-2}u = 0$. Proc. Amer. Math. Soc., **109**, 157–164 (1990), Addendum: Proc. Amer. Math. Soc., **116**, 583–584 (1992)

152. Lions, J.L.: Quelques Méthodes de Résolutions des Problèmes aux Limites Nonlinéaires. Dunod, Paris (1969)

153. Liu, W., Barrett, J.: A remark on the regularity of the solutions of the p-Laplacian and its application to their finite element approximation. J. Math. Anal. Appl., **178**(2), 470–487 (1993)

154. Marano, S.A., Motreanu, D.: Infinitely many critical points of non-differentiable functions and applications to a Neumann-type problem involving the p-Laplacian. J. Differential Equations, **182**, 108–120 (2002)

155. Marano, S.A., Motreanu, D.: On a three critical points theorem for non-differentiable functions and applications to nonlinear boundary value problems. Nonlinear Anal., **48**, 37–52 (2002)

156. Marano, S., Motreanu, D.: A deformation theorem and some critical point results for non-differentiable functions. Topol. Methods Nonlinear Anal., **22**, 139–158 (2003)

157. Margulies, C.A., Margulies, W.: An example of the Fučik spectrum. Nonlinear Analysis, **29**, 1373–1378 (1997)

158. Mawhin, J.: Nonlinear functional analysis and periodic solutions of ordinary differential equations. Tatras Summer School on ODE's, Difford 74, 37–60 (1974)

159. Mawhin, J., Schmitt, K.: Upper and lower solutions and semilinear second order elliptic equations with nonlinear boundary conditions. Proc. Roy. Soc. Edinburgh, **A 97**, 199–207 (1984)

160. Mawhin, J., Willem, M.: Critical Point Theory and Hamiltonian Systems. Springer-Verlag, New York (1989)

161. Micheletti, A., Pistoia, A.: On the Fučík spectrum for the p-Laplacian. Differential Integral Equations, **14**(7), 867–882 (2001)

162. Miersemann, E.: Eigenvalue problems for variational inequalities. Contemporary Mathematics, **4**, 25–43 (1984)

163. Mitidieri, E., Sweers, G.: Existence of a maximal solution for quasimonotone elliptic systems. Differential Integral Equations, **7**, 1495–1510 (1994)

164. Mordukhovich, B.S.: Maximum principle in the problem of time optimal response with nonsmooth constraints. J. Appl. Math. Mech., **40**, 960–969 (1976)

165. Mordukhovich, B.S., Shao, Y.H.: Nonsmooth sequential analysis in Asplund spaces. Trans. Amer. Math. Soc., **348**, 1235–1280 (1996)

166. Motreanu, D.: A saddle-point approach to nonlinear eigenvalue problems. Math. Slovaca, **47**, 463-477 (1997)

167. Motreanu, D.: A new approach in studying one parameter nonlinear eigenvalue problems with constraints. Nonlinear Anal., **60**, 443-463 (2005)

168. Motreanu, D.: Parametric eigenvalue problems with constraints for variational-hemivariational inequalities. Nonlinear Anal., **63**, 966–976 (2005).

169. Motreanu, D., Naniewicz, Z.: Discontinuous semilinear problems in vector-valued function spaces. Differential Integral Equations, **9**, 581–598 (1996)

170. Motreanu, D., Panagiotopoulos, P.D.: Double eigenvalue problems for hemivariational inequalities. Arch. Rational Mech. Anal., **140**, 225–251 (1997)

171. Motreanu, D., Panagiotopoulos, P.D.: Minimax Theorems and Qualitative Properties of the Solutions of Hemivariational Inequalities and Applications. Kluwer Academic Publishers, Boston, Dordrecht, London (1999)

172. Motreanu, D., Pavel, N.H.: Tangency, Flow-Invariance for Differential Equations and Optimization Problems. Marcel Dekker, Inc., New York, Basel (1999)

173. Motreanu, D., Radulescu, V.: Variational and Non-variational Methods in Nonlinear Analysis and Boundary Value Problems. Kluwer Academic Publishers, Boston, Dordrecht, London (2003)

174. Mustonen, V.: On pseudomonotone operators and nonlinear parabolic initial-boundary value problems on unbounded domains. Ann. Acad. Sci. Fenn. Ser. AI, **6**, 225–232 (1981)

175. Nagase, H.: On an application of Rothe's method to nonlinear parabolic variational inequalities. Funk. Ekvac., **32**, 273–299 (1989)

176. Nagumo, M.: On principally linear elliptic differential equations of the second order. Osaka Math. J., **6**, 207–229 (1954)

177. Naniewicz, Z., Panagiotopoulos, P.D.: Mathematical Theory of Hemivariational Inequalities and Applications. Marcel Dekker, New York (1995)

178. Ôtani, M., Teshima, T.: On the first eigenvalue of some quasilinear elliptic equations. Proc. Japan Acad., **64**, 8–10 (1988)

179. Panagiotopoulos, P.D.: Inequality Problems in Mechanics and Applications. Convex and Nonconvex Energy Functions. Birkhäuser Verlag, Boston (1985)

180. Panagiotopoulos, P.D.: Hemivariational Inequalities and Applications in Mechanics and Engineering. Springer-Verlag, New York (1993)

181. Panagiotopoulos, P.D., Stavroulakis, G.: A variational-hemivariational inequality approach to the laminated plate theory under subdifferential boundary conditions. Quart. Appl. Math. **46**, 409-430 (1988)

182. Pao, C. V.: Nonlinear Parabolic and Elliptic Equations. Plenum Press, New York (1992)

183. Papageorgiou, N.: On the existence of solutions for nonlinear parabolic problems with nonmonotone discontinuities. J. Math. Anal. Appl., **205**, 434–453 (1997)

184. Papageorgiou, N., Papalini, F., Vercillo, S.: Minimal solutions of nonlinear parabolic problems with unilateral constraints. Houston J. Math., **23**, 189–201 (1997)
185. Perera, K.: On the Fučík spectrum of the p-Laplacian. NoDEA Nonlinear Differential Equations Appl., to appear
186. Perera, K.: Multiple positive solutions for a class of quasilinear elliptic boundary-value problems. Electronic J. Differential Equations, **2003**(7), 1–5, (2003)
187. Protter, M.H., Weinberger, H.F.: Maximum Principles in Differential Equations. Prentice-Hall, Englewood Cliffs, N.J. (1967)
188. Puel, J.-P.: Existence, comportement a l'infini et stabilité dans certains problemes quasilinéaires elliptiques et paraboliques d'ordre 2. Ann. Scuola Norm. Sup. Pisa Cl. Sci., **3**, 89–119 (1976)
189. Puel, J.-P.: Some results on quasi-linear elliptic equations. Pitman Res. Notes Math. Ser., **208**, 306–318 (1989)
190. Quittner, P.: Solvability and multiplicity results for variational inequalities. Comment. Math. Univ. Carolin., **30**, 281–302 (1989)
191. Rabinowitz, P. H.: Minimax methods in critical point theory with applications to differential equations. CBMS Regional Conference Series in Mathematics, **65**, Amer. Math. Soc., Providence, RI (1986)
192. Reddy, B., Tomarelli, F.: The obstacle problem for an elastoplastic body. Appl. Math. Optim., **21**, 89–110 (1990)
193. Ricceri, B.: On a three critical points theorem. Arch. Math., **75**, 220–226 (2000)
194. Ricceri, B.: Three solutions for a Neumann problem. Topol. Methods Nonlinear Anal., **20** 275–281 (2002)
195. Ricceri, B.: Sublevel sets and global minima of coercive functionals and local minima of their perturbations. J. Nonlinear Convex Anal., **5**, 157–168 (2004)
196. Rockafellar, T.R.: Convex Analysis. Princeton University Press, Princeton (1970)
197. Rodrigues, J.F.: Obstacle Problems in Mathematical Physics. North-Holland, Amsterdam (1987)
198. Rudd, M., Schmitt, K.: Variational inequalities of elliptic and parabolic type. Taiwanese J. Math., **6**, 287–322 (2002)
199. Rudin, W.: Real and Complex Analysis. Mc Graw-Hill, New York (1966)
200. Rudin, W.: Functional Analysis. McGraw-Hill, New York (1973)
201. Rudin, W.: Real and Complex Analysis. McGraw-Hill, New York (1987)
202. Sattinger, D.H.: Monotone methods in nonlinear elliiptic and parabolic boundary value problem. Indiana Univ. Math. J., **21**, 979–1000 (1972)
203. Saint Raymond, J.: personal communication
204. Schechter, M.: The Fučik spectrum. Indiana Univ. Math. J., **43**, 1139–1157 (1994)
205. Schmitt, K.: Periodic solutions of nonlinear second order differential equations. Math. Z., **98**, 200–207 (1967)
206. Schmitt, K.: Periodic solutions of second order equations - a variational approach. In: Delgado, M., Suárez, A., López-Gómez, J., Ortega, R. (eds.), The First 60 Years of Nonlinear Analysis of Jean Mawhin. World Scientific, Singapore, pp. 213–220, (2004)
207. Schuricht, F.: Minimax principle for eigenvalue problems of variational inequalities in convex sets. Math. Nachr., **163**, 117–132 (1993)

208. Showalter, R.E.: Monotone Operators in Banach Space and Nonlinear Partial Differential Equations. **49** AMS, Providence (1997)
209. Szulkin, A.: On a class of variational inequalities involving gradient operators. J. Math. Anal. Appl., **100**, 486–499 (1982)
210. Szulkin, A.: Positive solutions of variational inequalities: A degree theoretic approach. J. Differential Equations, **57**, 90–111 (1985)
211. Szulkin, A.: Minimax principles for lower semicontinuous functions and applications to nonlinear boundary value problems. Ann. Inst. Henri Poincaré. Anal. Non Linéaire, **3**, 77-109 (1986)
212. Triebel, H.: Interpolation Theory, Function Spaces, Differential Operators. North-Holland, Amsterdam (1978)
213. Triebel, H.: The Structure of Functions. Birkhäuser Verlag, Basel (2001)
214. Troianiello, G.M.: Bilateral constraints and invariant sets for semilinear parabolic systems. Indiana Univ. Math. J., **32**, 563-577 (1983)
215. Troianiello, G.M.: Elliptic Differential Equations and Obstacle Problems. Plenum Press, New York (1987)
216. Trudinger, N.: On Harnack type inequalities and their application to quasilinear elliptic equations. Comm. Pure Appl. Math., **20**, 721–747 (1967)
217. Vázquez, J.L.: A strong maximum principle for some quasilinear elliptic equations. Appl. Math. Optim., **12**(3), 191–202 (1984)
218. Vivaldi, M.G.: Existence of strong solutions for nonlinear parabolic variational inequalities. Nonlinear Anal., **11**. 285–295 (1987)
219. Wloka, J.: Partielle Differentialgleichungen. B.G. Teubner, Stuttgart (1982)
220. Zhang, Z., Chen, J., Li, S.: Construction of pseudo-gradient vector field and sign-changing multiple solutions involving p-Laplacian. J. Differential Equations, **201**(2), 287–303 (2003)
221. Zhang, Z., Li, S.: On sign-changing and multiple solutions of the p-Laplacian. J. Functional Anal., **197**(2), 447–468 (2003)
222. Zeidler, E.: Nonlinear Functional Analysis and Its Applications, Vols. II A/B. Springer-Verlag, Berlin (1990)
223. Zeidler, E.: Nonlinear Functional Analysis and Its Applications, Vol. III. Springer-Verlag, New York (1985)

Index

adjoint operator, 17
Aubin–Clarke's Theorem, 71

Banach's Closed Graph Theorem, 13
Banach's Continuous Inverse Theorem, 14
Banach's Open Mapping Theorem, 14
Banach–Steinhaus Theorem, 13
bidual space, 16
biobstacle problem, 218
Brouwer's Fixed-Point Theorem, 12

canonical embedding, 16
Carathéodory function, 42
chain, 27
chain rule, 34, 69
Clarkson's inequalities, 38
closed, 13
coercive, 40, 62, 216
compact, 12
comparison principle, 2, 229
complementarity problem, 236
constraint, 236
continuous, 11
convex cone, 269
convex functional, 213, 236
convex set, 213
critical point, 73, 131
critical value, 78, 132
cutoff function, 95, 251

deformation lemma, 133
demicontinuous, 39
differential inclusion, 7

directed, 27, 87
directional derivative, 22
domain, 28
downward directed, 27
dual operator, 17

Eberlein–Smulian Theorem, 18
effective domain, 24
Egorov's Theorem, 29
eigenvalue problem, 247, 370
elasto-plastic torsion problem, 218
Embedding Operator, 14
epigraph, 25
equilibrium problem, 214
Euler–Lagrange equation, 213
evolution equation, 49
evolution triple, 52, 55
evolutionary variational inequality, 248
existence and comparison principles, 215
extremal solutions, 91, 103, 122, 213, 216, 263

Fatou's Lemma, 29
free boundary, 236
Fréchet derivative, 22
Fučik spectrum, 124

generalized chain rule, 34
generalized derivative, 55
generalized directional derivative, 64
generalized gradient, 65
generalized pseudomonotone operator, 48

394   Index

graph, 13, 46
graph norm, 13
graph-closed, 13
greatest element, 27
greatest lower bound, 27
greatest solution, 225, 243, 256
Gâteaux derivative, 22

Hahn–Banach Theorem, 15
hemicontinuous, 39
hemivariational inequality, 7
Hölder's inequality, 37

indicator function, 7
integral of vector-valued functions, 53

lattice, 27
lattice structure, 35, 218
Lax–Milgram's Theorem, 41
least upper bound, 27
Lebesgue's Dominated Convergence
    Theorem, 28
Lebourg's Theorem, 69
Leray–Lions operator, 41
linking, 77
Lipschitz boundary, 31
local minimizer, 131
locally Lipschitz, 63
locally uniformly convex, 21
lower semicontinuous, 23, 46

Main Theorem on Monotone Operators,
    40
Main Theorem on Pseudomonotone
    Operators, 40
maximal element, 27
maximal monotone, 47, 249
maximum principle, 1
Milman–Pettis Theorem, 21
minimal element, 27
minimization problem, 213
Minkowski's inequality, 37
modulus of continuity, 99
monotone, 40
monotone approximation, 213
monotone multivalued operator, 47
monotone scheme, 264
multivalued evolution equation, 62
multivalued pseudomonotone operator
    w.r.t. $D(L)$, 62

Nemytskij operator, 42, 248
noncoercive, 213
noncoercive variational inequality, 236
nonlocal flux boundary condition, 141
nonsmooth variational problems, 4

obstacle, 214, 236
obstacle problem, 218, 257
one-sided growth condition, 243
operator of Leray–Lions type, 248

p-Laplacian, 104, 110
Palais–Smale condition, 75
parabolic variational inequality, 213,
    248
partially ordered set, 27
penalized equation, 254
penalty operator, 254, 264, 341
penalty problem, 269
periodic boundary conditions, 134
Poincaré–Friedrichs inequality, 38
positive cone, 3
preimage, 46
proper, 24
pseudomonotone, 40
pseudomonotone multivalued operator,
    47
pseudomonotone w.r.t. $D(L)$, 61

quasilinear evolutionary variational-
    hemivariational inequality,
    338
quasilinearization method, 142

reaction-diffusion equation, 141
reflexive, 16
regular, 65
Riesz' Lemma, 18

S_+-condition, 40
Schauder's Fixed-Point Theorem, 12
semicontinuous multifunctions, 45
Separation Theorem, 20
sequentially lower semicontinuous, 23
sequentially upper semicontinuous, 24
sign-changing solution, 124, 130
smallest element, 27
smallest solution, 228, 243, 256

Sobolev embedding theorems, 31
step function, 53
strictly convex, 21, 24
strictly differentiable, 65
strictly monotone, 40
strong convergence, 18
strongly continuous, 20
strongly measurable, 53
strongly monotone, 40
sub-supersolution, 82, 94, 136, 213
subdifferential, 25
subgradient, 25
subsolution, 217, 323, 340
sum rule, 26
superpositionally measurable, 43
supersolution, 217, 323, 340
system of variational inequalities, 269

trace, 32
Trace Theorem, 32
truncated vector function, 272
truncation method, 220
truncation operator, 83, 85, 251

unbounded domain, 141
Uniform Boundedness Theorem, 13
uniformly convex, 21
uniformly monotone, 40
unilateral condition, 236
unilateral constraint, 280

upper semicontinuous, 24, 45
upward directed, 27

variational equation, 2
variational inequality, 7, 213, 216
variational–hemivariational inequalities, 8, 321
vector-valued distribution, 248
vector-valued function, 53

weak convergence, 18
weak derivative, 30
weak sequentially lower semicontinuous, 23
weak sequentially upper semicontinuous, 24
weak subsolution, 3
weak supersolution, 3
weak* convergence, 19
weakly coupled quasimonotone system, 270
weakly measurable, 53
weakly sequentially closed, 21
weakly sequentially continuous, 20
Weierstrass' Theorem, 25

Young's inequality, 36

Zorn's Lemma, 27, 91, 225

Printed in the United States
by Baker & Taylor Publisher Services